Biology

for schools and colleges

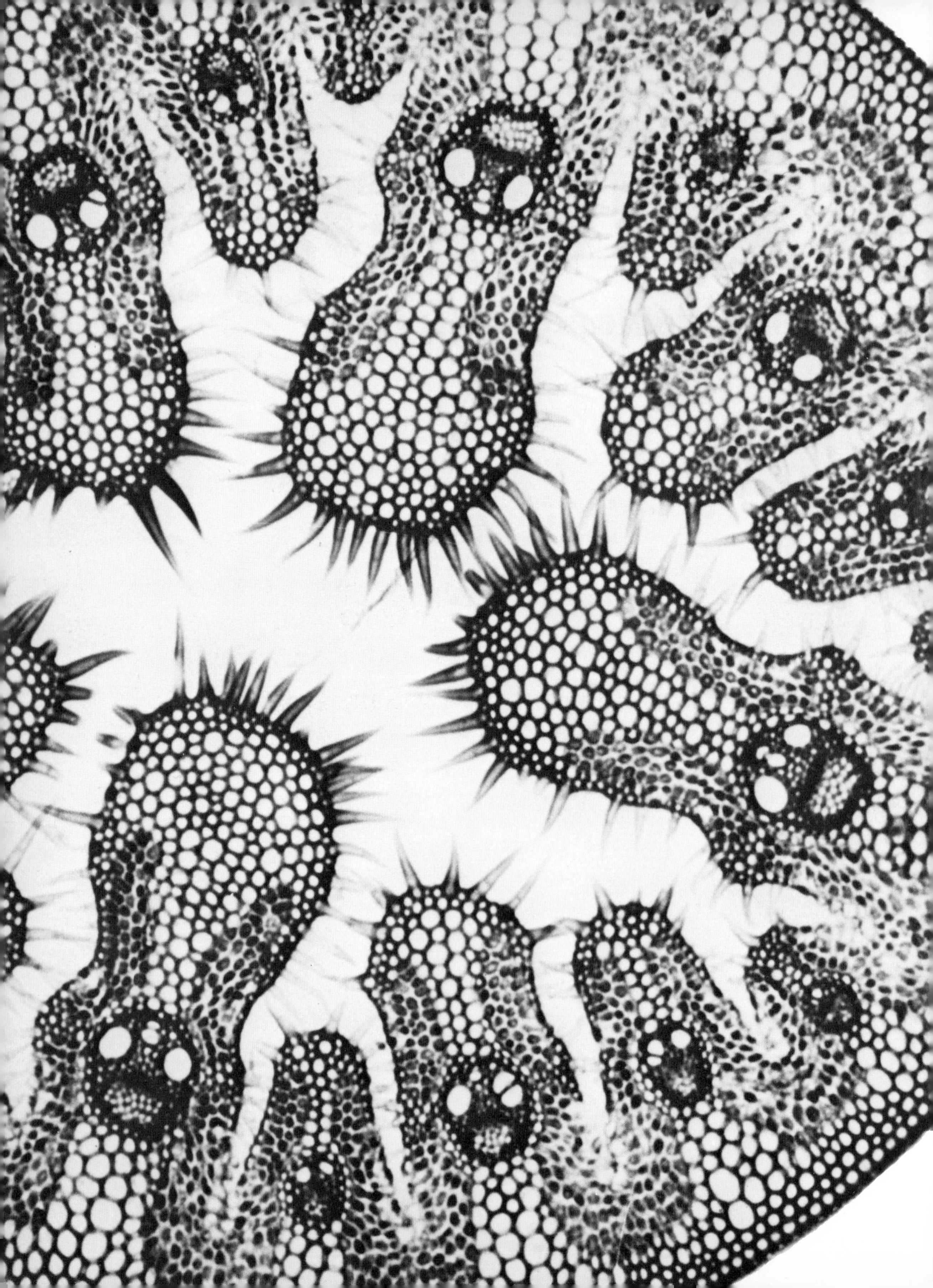

Biology

for schools and colleges

Colin Clegg

with illustrations by Sam Denley

Heinemann Educational Books

Heinemann Educational Books Ltd
22 Bedford Square, London WC1B 3HH

LONDON EDINBURGH MELBOURNE AUCKLAND
HONG KONG SINGAPORE KUALA LUMPUR NEW DELHI
IBADAN NAIROBI JOHANNESBURG
EXETER (NH) KINGSTON PORT OF SPAIN

ISBN 0 435 60171 7

First published 1980

British Library Cataloguing in Publication Data

Clegg, Colin
Biology.
1. Biology
I. Title
574 QH308.7
ISBN 0-435-60171-7

Printed in Great Britain by
Fletcher & Son, Ltd, Norwich

Preface

The aim of this book is to provide concise but comprehensive coverage of the material studied in post 'O' Level courses, especially those leading to the GCE 'A' Level examination. The more traditional aspects of biology are dealt with thoroughly; the applied and environmental aspects, which are of increasing importance both in examination syllabuses and for a general appreciation of the world in which we live, are also investigated.

To facilitate the learning process the material is presented in a variety of ways. *Experimental and applied aspects*, included at the end of each chapter, provide a background and some additional material. It is hoped that these sections will prove interesting whilst not complicating the main text. *Guided examples* are designed to provide information outside the main body of the text in a manner that will stimulate thought. *Questions* of examination standard are chosen to reinforce the learning process, and many of those selected present further information, especially experimental data, from which scientific deductions must be made.

Bold type is used to highlight terms of particular importance and will aid the student in developing the necessary vocabulary of the subject. Cross-references in the text are indicated by superior numerals thus ([107]), giving the relevant page number.

To help achieve the aim of conciseness with clarity, much information is set out in tables, flow-charts, and diagrams. The preparation of the artwork involved the closest co-operation between the author and artist over a number of years in an attempt to include the maximum amount of information, and to be both stimulating and accurate. Indeed the creativity and conscientious attention to detail by the artist Sam Denley in the preparation and design of the illustrations has been of central importance to the book as a whole.

I would also like to express my gratitude to my wife Ann for typing the manuscript under far from ideal conditions, and for making many constructive suggestions relating to the subject matter.

I am indebted to Mrs Dorothy Band, Mr David Parkin, Dr R. G. Smith, and Mr C. J. Deane for their help with various parts of the text. I would also like to thank Mr Graham Taylor and his associates at Heinemann Educational Books who have helped to make the preparation and production of this book so enjoyable.

C. A. Clegg
1980

Contents

Acknowledgements

The author wishes to thank the following people for permission to reproduce their photographs:

Acrodermatitis Enteropathica, **54**, 384 (p. 135)
Heather Angel (p. 33)
British Museum (Natural History) (p. 1)
Professor Lucien Caro (p. 289)
C. Clegg (p. 285)
Gene Cox (pp. 46, 47, 50, 51, 185 lower, 206, 210)
Crown copyright: Dr. D. A. Griffiths and Miss S. V. Cowper, Ministry of Agriculture, Fisheries and Food (pp. 109, 309)
S. Denley (pp. 120, 125, 144, 221, 235, 243, 261, 282, 318 left)
Department of Scientific and Industrial Research, New Zealand (pp. 48, 165)
Courtesy of the Department of Scientific and Industrial Research, from *Probing Plant Structure* by Troughton and Donaldson (London: Chapman & Hall, 1972) (p. 185 upper)
Courtesy of W. T. Hall from *Plant Symbiosis* by G. D. Scott (London: Edward Arnold, 1969) (p. 108)
S. Jenkinson (p. 255)
Robert E. Kuntz (p. 310)
Dr. S. M. Lewis (p. 94)
Natural History Photographic Agency (p. 29)
K. Rogers (p. 306)
A. W. Robards, Philip Harris Biological Ltd (p. 12)
Courtesy of South Air Quality Management District, Los Angeles (p. 317)
R. Turner, Rothamsted Experimental Station (p. 128)
U.S. Department of Agriculture (Soil Conservation Division) (p. 277)
Dr. R. J. H. Williams, U.W.I.S.T. (p. 318 right)
Douglas Wilson (p. 112)

The following examination boards kindly gave permission to reproduce questions from past examination papers:

Associated Examination Board (AEB)
University of Cambridge Local Examinations Syndicate (C)
Joint Matriculation Board (JMB)
The Senate of the University of London (L)
Oxford and Cambridge Schools Examination Board (O & C)
Oxford Delegacy of Local Examinations (O)
Southern Universities Joint Board (SUJB)

PART ONE

Origins and interactions

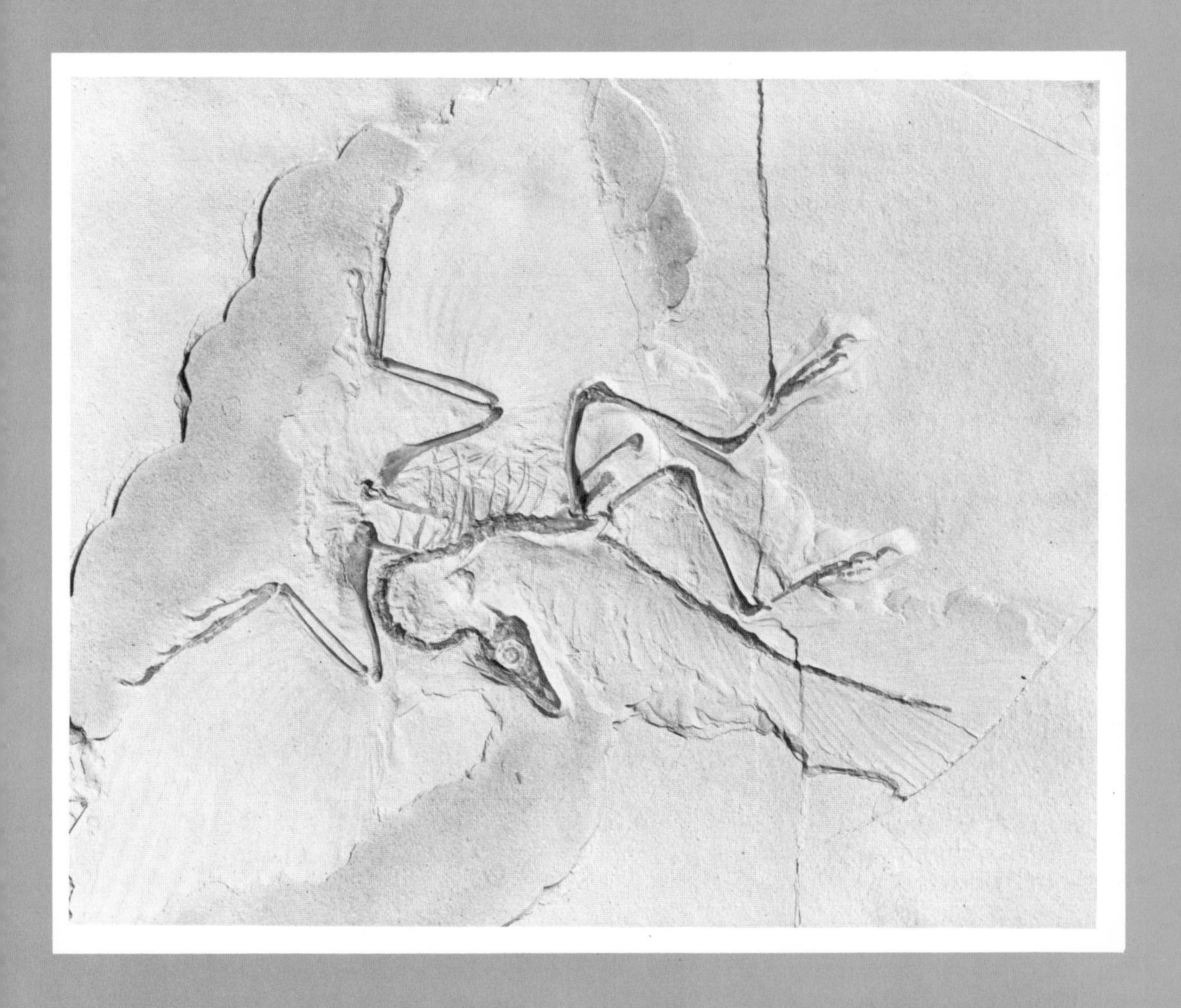

1 Cells and tissues

CELLS

The cell is the basic protoplasmic unit. Its boundaries are limited by the **plasmalemma** or cell membrane. This basic unit constitutes the entire organism in unicellular organisms, such as the protozoa, simple algae, and bacteria. In multicellular organisms the body is composed of many such cells, the functions of which are specialized to a greater or lesser degree to provide a division of labour within the body. This specialization prevents the description of a 'typical' cell, but throughout the living world there is a remarkable uniformity of structure which enables a 'generalized' cell to be described.

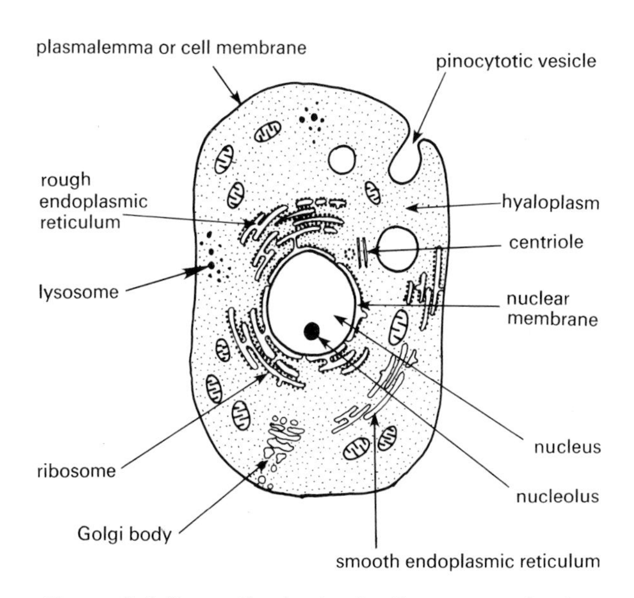

Figure 1.1 *Generalized animal cell as seen under the electron microscope*
Only samples of the different organelles are shown, and these are not drawn to scale either in relation to each other or to the overall size of the cell.

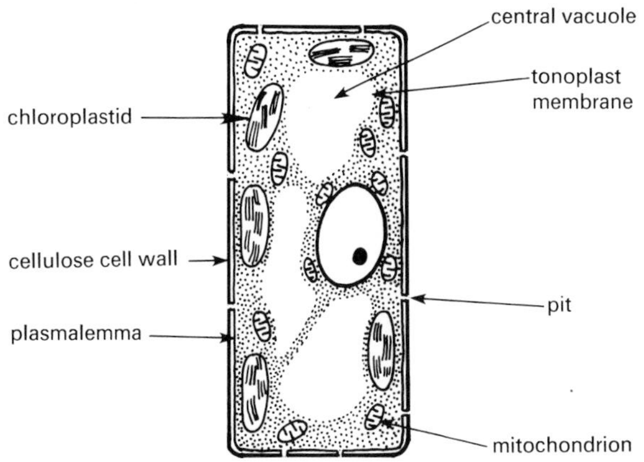

Figure 1.2 *Generalized plant cell as seen under the electron microscope*

The static picture of cell structure suggested by dead cells viewed under the light and electron microscopes is very misleading; it suggests that the particulate bodies known as **organelles** have a distinct fixed state, and a clear and separate identity from the **hyaloplasm**, or clear cytoplasm, in which they are suspended. In fact, the organelles are in a continual state of dynamic change, and although some organelles can be 'separated' by certain techniques, this 'separation' really only isolates fractions of the cytoplasm which contain a predominance of a particular organelle, rather than an absolutely discrete and distinct organellar structure.

The organelles and hyaloplasm are all part of the cytoplasm, in which there is a continual interchange of materials as the organelles are formed, broken down, and reformed.

Cell organelles

Almost all the organelles are membranous structures, and these membranes provide a large internal surface area for the ordered spatial arrangement of the enzymes[8] and co-enzymes[8] which control complex chemical reactions in a sequential order. Each organelle has a specific function, and these specialized areas of cytoplasm are an example of the division of labour within the cell.

Endoplasmic reticulum

The endoplasmic reticulum is a network of internal membranes found mainly in the deeper-lying cytoplasm or endoplasm, but which may be continuous with the plasmalemma in places. These membranes are in a state of flux, being continually broken and remade by cytoplasmic activity. They provide a large internal surface area for surface-catalysed reactions and allow for rapid intra-cellular transport of materials. There are two forms of endoplasmic reticulum; the granular or rough endoplasmic reticulum, and the agranular or smooth endoplasmic reticulum. In any particular cell one type usually dominates, although in mammalian liver cells they are present in roughly equal proportions. The two are not normally connected.

The **granular endoplasmic reticulum** (GER) is found in most types of cell and consists of layered collections of flattened cavities, or **cisternae**, interconnected by tubular canals. The granular appearance is due to the presence of protein-synthesizing ribosomes attached on the cytoplasmic side of the membrane. The cisternae are usually closely associated with mitochondria[5] which supply the GER's synthetic reactions with energy-rich ATP.[367] The GER transports and stores the products of the ribosomes and other intra-cellular substances. It is best seen in glandular cells secreting protein-rich products, e.g. enzymes, antibodies,[291] and some hormones.[212]

The cisternae of this system are also modified to form the nuclear membrane, composed of **perinuclear cisternae**, which enclose the nuclear region.

The **agranular endoplasmic reticulum** (AER) is a tubular system of membranes lacking associated ribosomes. No single clearly defined function has yet been demonstrated.

Ribosomes

Ribosomes are dense, slightly angular structures, about 15 nm in diameter, found in every type of cell. They occur singly or in clusters known as **polyribosomes**, either freely in the hyaloplasm or attached to the GER; those attached to the GER may dissociate themselves and become free. Similar particles are also found in the nucleus and the nucleolus, and it has been suggested that ribosomes may be synthesized in the nucleolus. They are rich in ribonucleic acid (RNA)[74] and are involved in protein synthesis, the details of which are discussed under genetics.

The Golgi complex

The Golgi complex is a concentrated system of smooth membranous cisternae and vesicles found in all animal cells, and is particularly well developed in secretory cells. In plant cells it is less distinct. In secretory cells it continually produces secretory vesicles which move to the plasmalemma. Simultaneously, it is added to by the coalescence of vesicles pinched off from the GER. Its functions are diverse and not absolutely clear. It is well developed in secretory cells; for example in nerve cells, which are very active secretory cells, the Golgi complex surrounds the nucleus completely. It is also well developed in goblet cells and digestive gland cells and, since these cells secrete enzymes (which are proteins), the function of the concentration of protein secretory products seems clear in these cases. Some maintain that the Golgi complex produces lysosomes,[5] since a positive reaction for an enzyme found in lysosomes is obtained in vacuoles apparently arising from the Golgi complex. Moreover, it has an independent synthetic function of its own. A polysaccharide component is manufactured by the Golgi complex and joined or conjugated with protein synthesized by the ribosomes of the GER to form glycoproteins prior to their secretion. It also seems to be involved in lipid metabolism, since fat tends to accumulate in the cisternae of the Golgi complex, and in animal cells the Golgi complex changes significantly during fasting or changes in diet.

The Golgi complex has a very interesting role in the transformation of spermatids[22] into spermatozoa. Membrane-bound granules, known as proacrosomal granules, in the interior of the Golgi complex, similar to the secretory granules in gland cells, run together to form an **acrosome**. This is used in the breakdown, or lysis, of the cell membrane and cytoplasm of the egg cell as the spermatozoa penetrates towards nuclear fertilization.[24] Thus the acrosome acts, in a way, as a giant lysosome, and this function of the Golgi complex is interesting to note in the light of the association with lysosomes mentioned above.

Although there is no single clear function of the Golgi complex which is common to all cells, it generally seems to be involved in the concentration and packaging of GER secretions, in their transport to the exterior of the cell, and perhaps in the growth and repair of the plasmalemma.

Lysosomes

Lysosomes are membrane-bound vesicles containing enzymes, and are sometimes referred to as 'suicide bags' since the powerful enzymes they contain are capable of digesting the cell constituents. As yet they have not been identified in plant cells.

Cell organelles and engulfed particles which need to be broken down are surrounded by a membrane and digested by the lysosomal enzymes. Rupture of the lysosome membranes and the liberation of their contents into the general cytoplasm leads to the destruction of the cell. Indeed, their numbers increase in cells destined to be broken down, e.g. old or damaged cells that need removing or replacing. This is by no means an inconsiderable task as the cells of any body are continually being broken down in this way. In man, for example, the daily cell turnover rate has been estimated as 10^8–10^{10} cells per day.

Lysosomes are not a distinct group of organelles with a common structure, indeed no structure seen in an electron micrograph can be positively identified as a lysosome on the basis of its appearance alone. Their identification depends mainly on biochemical tests determining the presence of certain hydrolytic enzymes.

Mitochondria

Mitochondria are rod-shaped organelles which appear as spherical or elliptical bodies from 0.2 μ to 7.0 μ in diameter when sectioned and viewed under the electron microscope. They are found in all cells except bacteria, blue-green algae, and mature red blood corpuscles.

They are generally seen to be long slender rods, which, in the living cell, appear to be in constant motion—continuously executing slow sinuous movement and sometimes undergoing marked changes in shape as they branch to form interconnections with other organelles. In certain sections of muscle tissue under the electron microscope there appears to be a mitochondrial network in which the mitochondria form an interconnecting system of membranous tubules.

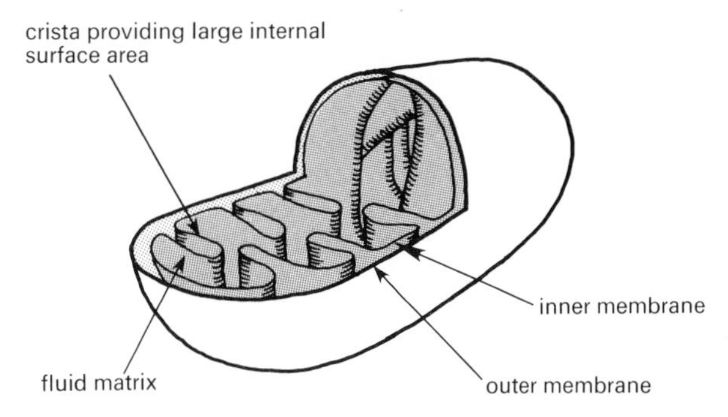

Figure 1.3 *Mitochondrion structure*

The number of mitochondria per cell varies tremendously; thus the unicellular green alga *Microsterias* has a single mitochondrion per cell; whereas the giant *Amoeba chaos* has up to 500 000.

Mitochondria are the centres of cellular respiration,[158] and contain all the enzymes and other compounds involved in the aerobic part of respiration. About seventy enzymes are known to be present. Some are inseparable from the membrane, whereas others are liberated in solution when mitochondria are ruptured, and must therefore be in the fluid matrix. When mitochondria are disrupted, structural sub-units are obtained. These contain complete electron transport chain enzyme assemblies,[160] sometimes called **electron transport particles**.

The number of mitochondria and the complexity of their internal structure increases with the energy requirements of the cell. It is interesting to note that the hormone thyroxine,[214] which has such a profound effect in increasing metabolic rate in vertebrates, has a swelling effect on mitochondria, possibly associated with their increased activity. Generally, plant cells have fewer mitochondria than animal cells, since they not only have a slower metabolic rate and therefore a lower energy requirement than animals, but also because a surplus of energy-rich ATP is produced by the chloroplastids[6] during photosynthesis.[127] They are often randomly distributed, but in some cells are found close to organelles that require energy-rich ATP, for example GER, myofibrils,[226] and microvilli.[10]

During cell division, the mitochondria are roughly equally distributed between the daughter cells, and as the new cell enlarges the mitochondria divide until the normal level is reached; thus they seem to arise only from pre-existing mitochondria. Moreover, they have been shown to contain some of the genetic material DNA[73] (different in molecular nature from nuclear DNA, however) which indicates that they may govern their own heritable qualities to some extent, although estimates of the total amount of genetic information in mitochondrial DNA show that there is not enough to code for all the necessary information.

Microtubules

Microtubules are widespread in plant and animal cells. They are straight, variable in length, with a wall 5 nm thick, and a centre of low density. They are cytoskeletal elements which are involved in the internal movement of cytoplasm and in the maintenance and alteration of cell shape. They also form the fibres of the nuclear spindle, which controls the movement of the chromosomes.[59] In fact, their walls consist of about twelve longitudinal filamentous sub-units similar in structure to those found in cilia,[224] flagella,[224] and centrioles.[60] Their support function is seen in a wide variety of tissues; for example, nucleated red blood cells (in fish, amphibia, and birds) have a marginal band of microtubules which are cytoskeletal in function and maintain the cells' flattened shape. The podocytes[191] in the epithelium of the renal glomerulus have many tubules and filaments which are cytoskeletal in maintaining the position and shape of the inter-digitating processes, and in nerve cells[203] the microtubules are arranged longitudinally to support the long, tenuous neurone fibres. In plant cells they extend along strands of cytoplasm along which cellulose fibres are laid down during cell wall formation.[7]

Filaments

Filaments are protein in nature, consisting of solid rods 4–5 nm in diameter. There could be several distinct fibrous proteins of similar size, but it seems likely that they are a common structural protein found in most cells. In stratified squamous epithelium they are associated into interlacing bundles known as tonofibrils that end in the desmosomes[10] on the cell surface.

Plant cells

Plant cells contain many of the organelles described above, although they are sometimes less distinct and not so well developed; a reflection perhaps of their reduced metabolic rate when compared to animal cells. In addition, photosynthesizing cells contain chloroplastids, and the cytoplasm of plant cells secretes a cellulose cell wall outside the plasmalemma.

Chloroplastids

These have a wide variety of shapes, but in higher plants they are disc-shaped and 4–6 μ in diameter. They are basically internal membrane systems providing surfaces for the ordered spatial arrangement of the enzymes and co-enzymes involved in photosynthesis, the process by which the energy from sunlight is absorbed by the chlorophyll and utilized in synthesizing carbon dioxide and water into glucose. Thus the chloroplastids are ultimately the source of all food material and organic substances on Earth.

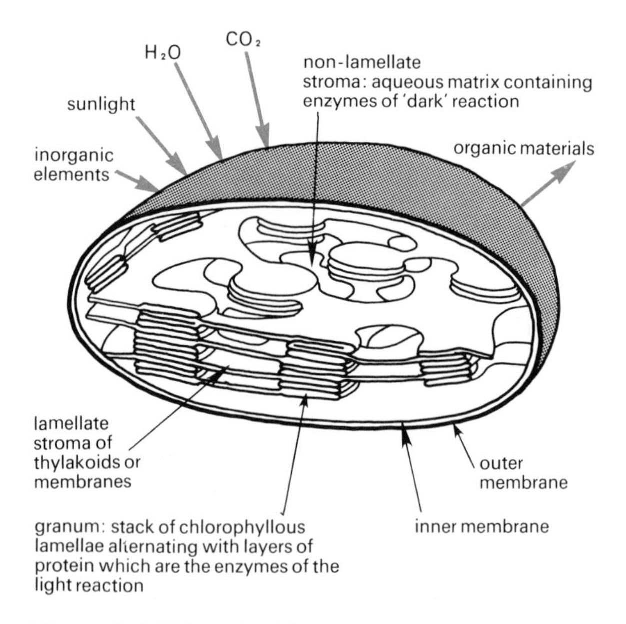

Figure 1.4 *Chloroplastid structure*

Most chloroplastids arise from pre-existing **proplastids**, which are small self-replicating crystalline structures within the cytoplasm. Their division is not correlated with cell division, and there is no mechanism to ensure equal division of proplastids between daughter cells.

The proplastids contain a **prolamellar body**, which in the light develops grana, intergrana, and a precursor molecule of chlorophyll, which is in turn converted to chlorophyll by a light-dependent reduction process requiring iron. This is a good example of interaction between cell and environment to effect control of cell function, since in this case the genes[75] responsible for the development of the chloroplastids are only active in light (it is interesting to note that this process is not light-dependent in Gymnosperms).

The cell wall

The cell wall is an extracellular secretion of the cytoplasm, and its formation and structure are intimately bound up with the growth and development of the plant cell. The first layer of the cell wall consists of amorphous pectic substances, and in adjacent cells these layers fuse to form the **middle lamella** between the two cell walls. It is soft and jelly-like and allows cells to slide against each other during elongation. Calcium and magnesium pectates are formed later to cement the cells together. The next layers, consisting mainly of cellulose, are laid down on this middle lamella. However, certain small areas remain unthickened to form **pits**. These usually coincide in adjacent walls to form pit-pairs in which the two cells are separated only by the middle lamella, and through which most of the **plasmodesmata**, or cytoplasmic strands, pass.

The primary wall is composed mainly of cellulose which may become impregnated with different materials during differentiation. Cellulose is a complex, long-chain polysaccharide[365] carbohydrate, composed of up to 3000 glucose sub-units. Cellulose molecules are arranged in bundles of about 2000 which are known as **microfibrils** and can only be seen under the electron microscope. Within a microfibril the cellulose molecules are arranged irregularly, except in regions called **micelles** where they have a very regular crystalline arrangement.

The first layers of cellulose are laid down whilst the cell is still enlarging and the microfibrils are laid down in a loose meshwork which allows extension. As the cell elongates the microfibrils are drawn out in the direction of elongation. During this process the thickness of the wall is maintained by the laying down of more microfibrils.

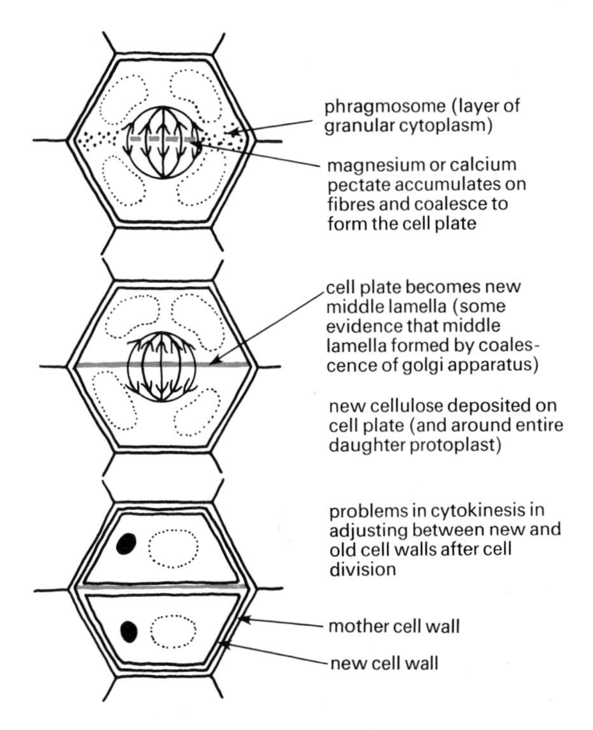

Figure 1.5 *Growth of the cell wall in plants*
As the cell expands, mother cell wall material is stretched and possibly broken down, whilst new cell wall material is still being laid down.

During cell extension there is an increase in cell wall plasticity which decreases the wall pressure,[184] so that water enters by osmosis[310] and causes the cell to swell. At the same time the many small vacuoles originally present in the dense cytoplasm of the young cell run together to form a single, large, central vacuole surrounded by the semi-permeable **tonoplast** membrane. When the cell stops expanding, the microfibrils are laid down in bundles known as **macrofibrils**. These lie parallel to each other in layers with the macrofibrils in each successive layer at a different angle. Macrofibrils are visible under the light microscope.

Hemicelluloses are often found in association with the cellulose of cells walls. They consist of a variety of polymers, or long chain molecules, of different monosaccharides. One type, known as xylan, is a polymer of xylose and is typically found associated with cellulose in the walls of secondary xylem elements. Polymerized mannose (mannan) and polymerized galactose (galactan) are laid down in cell walls but act as food reserves rather than structural elements.

Lignin is amorphous and is deposited in the inter-fibrillar spaces between the cellulose fibrils of the cell walls of certain tissues. The combination of tensile cellulose fibre and amorphous lignin confers great strength and toughness on lignified cell walls. Lignification occurs in the walls of sclerenchyma fibres and sclereids, and xylem vessels and tracheids.[14]

The hyaloplasm

As mentioned at the beginning of this section, the organelles are the particulate fraction of the cytoplasm and are specialized membrane systems within the general mass of the cytoplasm which is known as the hyaloplasm and is a **colloidal system** of great complexity and extreme variability. A colloid is composed of fine particles (the disperse phase) which are suspended in a solution (the continuous phase). The **continuous phase** consists of a dilute solution of various organic and inorganic substances, while the **disperse phase** consists of protein molecules associated in groups. The enormous interfacial surface area between the disperse and continuous phases is of great importance for surface-limited reactions within the cytoplasm.

Although the hyaloplasm has characteristics shown by non-living colloids, it also has unique characteristics by which it differs greatly from them, and which preclude its description in simple terms of colloid behaviour.

(a) It is **thixotropic**, that is, it shows sol/gel phase reversal at constant temperature, utilizing energy from respiration.
(b) Linked with this ability is the phenomenon of **streaming** (shown in all living cytoplasm) which carries organelles around the cell.
(c) Its **permeability** is highly variable and, physiologically, this is one of its most important properties; indeed, it can act as a semi-permeable membrane with regard to osmosis. It can also profoundly affect the entry of solutes into the cell, thus further confusing membrane permeability studies.
(d) It combines **elastic** and **fluid** properties to an unusual degree and its viscosity decreases with increased temperature, as expected, until 60 °C, when it **coagulates** due to the disruption of the internal molecular configuration.
(e) Much of the protein of the hyaloplasm and organelles is enzymatic.

Enzymes

Structure

Enzymes are complex, organic, colloidal catalysts produced by living cells for use in or near their site of production. All enzymes are protein in nature, and each has a particular 3-dimensional tertiary[366] structure, upon which its activity depends. Within this tertiary structure is a special area which combines with the particular substance or substrate with which it reacts. This reactive centre is termed the **active site**.

Some enzymes consist solely of protein, for example pepsin, trypsin, and amylase. Others also have a non-protein part which is essential to their structure. If this non-protein part is an integral part of the enzyme structure and cannot be removed it is termed a **prosthetic group**. If it is removable it is termed a **co-enzyme**, and the protein part is known as an **apoenzyme** (many B-group vitamins[139] are important as precursors of co-enzymes). The detachable non-protein part is called an **activator** if it is a trace element.

Characteristics

Enzymes act as catalysts. That is, they speed up the rate of spontaneous reactions and can start up non-spontaneous reactions (if they are thermodynamically possible).

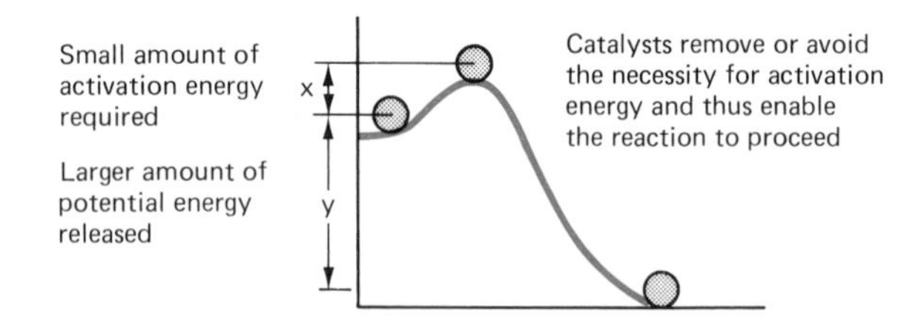

Enzymes enter into combination with their substrate to form an **enzyme–substrate (ES) complex**. As a result, the reaction products are formed and the enzyme is liberated unchanged to catalyse further reactions.

$$\text{Enzyme} + \text{Substrate} \rightleftharpoons \text{ES} \rightleftharpoons \text{Enzyme} + \text{Products}$$

In this way relatively small amounts of enzyme can catalyse the reaction of relatively large amounts of substrate; for example, 1 molecule of catalase can catalyse the breakdown of 100 000 molecules of hydrogen peroxide per second.

However, unlike inorganic catalysts their effectiveness slowly decreases with time. Under certain conditions the formation of an ES complex in enzyme-controlled reactions demonstrates what are known as saturation kinetics.

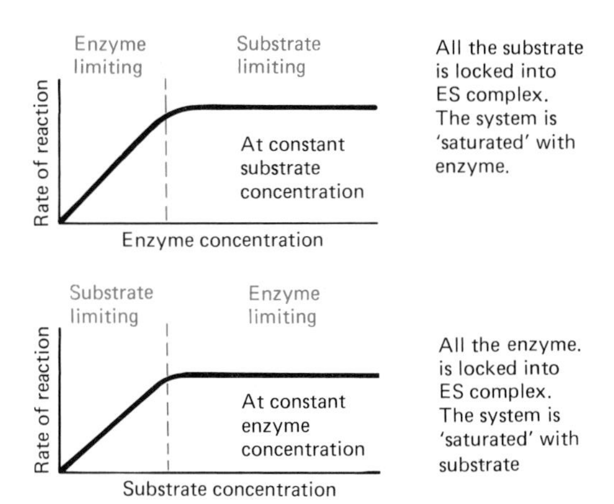

Enzymes are usually very *specific* catalysts. In other words, each enzyme only catalyses a particular reaction. This specificity is explained by assuming that the active centre has a specific shape which will only 'fit' a correspondingly shaped substrate to form the enzyme–substrate complex. This is known as the **'Lock and Key'** theory.

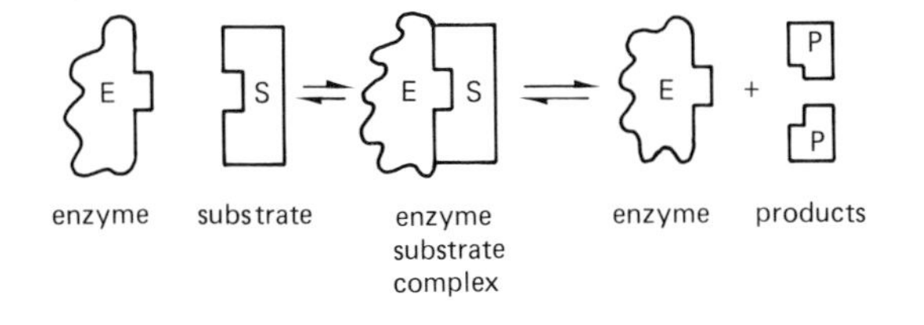

Denaturation

Any disturbance of the 3-dimensional shape (secondary and tertiary structure) of enzymes decreases their efficiency, and, if the disturbance is sufficient, the enzyme may be denatured and rendered inactive, as for example by heat and changes in acidity (pH).[363] Such denaturing is irreversible.

At temperatures above 60–70 °C the secondary linkages that hold the complex protein in shape are broken and the molecule loses its compact 'globular' form and precipitates out of solution.

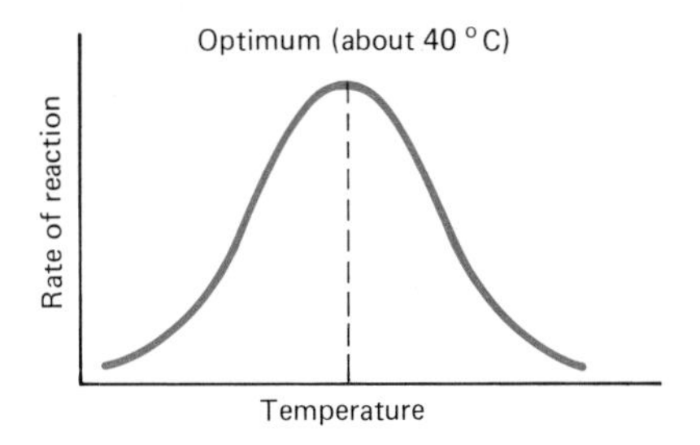

Changes in pH alter the distribution of charged groups on the molecule, and, since these charges exert an internal force in the molecule which helps to maintain its shape, any change in pH will result in a change of enzyme shape and a subsequent denaturation.

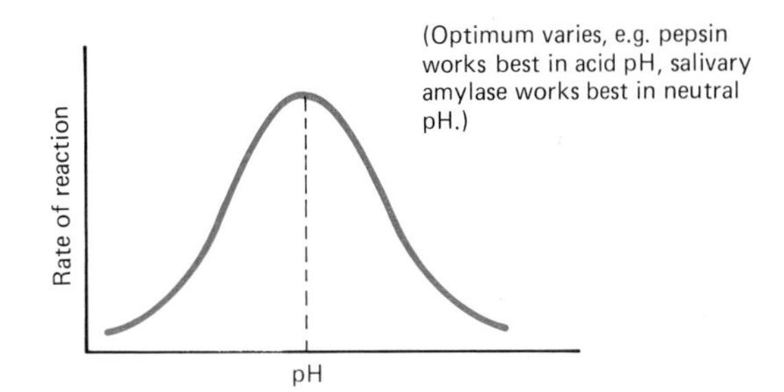

Inhibition

Enzymes with metal ion activators are inhibited if the necessary metal ion is replaced by another. Inhibition may be reversible or irreversible, in which case the enzyme is said to be poisoned. Furthermore, enzymes which depend on sulphydryl groups for their activity are inactivated by heavy metals such as silver, lead, and mercury.

In some cases the end-products of an enzyme-catalysed reaction inhibit the reaction. This end-product inhibition is a form of negative feedback, controlling the rate of reaction and hence the amount of end-product formed. When there is insufficient product, the inhibition is removed and the process proceeds to generate more. The combination of the end-product with the enzyme alters its shape and prevents it from forming the ES complex. Such proteins, whose shape is altered by the binding of small molecules to them, are known as **allosteric** proteins.

P E S P E S

product — enzyme substrate complex — enzyme product complex — substrate displaced

Competitive inhibition occurs when an altered substance or substrate, with a shape very similar to the original substrate, combines with the enzyme to form an ES complex, thus preventing the true substrate from combining. In this case there is competition for combination with the active centre. The sulphonamide drugs exert their controlling effect on bacteria by mimicking an essential substrate for a key bacterial enzyme. The bacteria require this substrate, which is found in traces in the blood and the tissues, but the host cells do not; thus the competitive inhibition by the drug disrupts the metabolism of the bacteria but not that of the host cells, and can effect a cure of certain bacterial diseases.

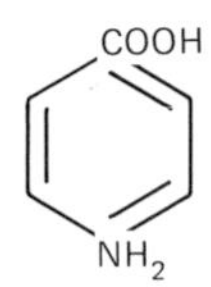

para-amino benzoic acid

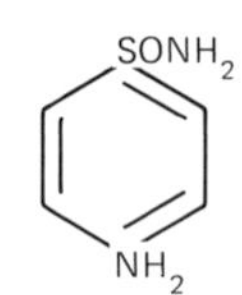

sulphanilamide
(sulphonamide drug)

Classification

There are many ways of classifying enzymes, but none of the systems includes all the different types of enzyme entirely satisfactorily. One simple classification divides them into five groups based on the type of reaction that they catalyse.

(a) Hydrolysing enzymes (hydrolases)

$$AB + H_2O \rightleftharpoons AOH + BH$$

(b) Oxidizing enzymes

(i) Dehydrogenases

$$AH_2 + B \rightleftharpoons A + BH_2$$

(ii) Oxidases

$$BH_2 + \tfrac{1}{2}O_2 \rightleftharpoons B + H_2O$$

(c) Transferring enzymes (transferases)

$$AB + C \rightleftharpoons A + BC$$

(d) Isomerizing enzymes (isomerases)

$$AB \rightleftharpoons BA$$

(e) 'Adding' enzymes

$$A + CO_2 \rightleftharpoons ACO_2$$

Enzymes are often named by adding the suffix **-ase** after the name of the substrate on which they act, for example, cellulase acts on cellulose, urease acts on urea, and maltase acts on maltose.

Summary

By their specific effect on specific reactions enzymes can regulate and control complex metabolic pathways. This control of metabolism is further organized spatially, so that certain enzymes are located in and on specialized membranous organelles within the cell, for example in the mitochondria and chloroplastids. Enzymes are sometimes produced in the form of an **inactive precursor** which is only subsequently activated, either in the presence of the correct substrate or by those conditions in which the substrate is required to be acted upon. About 200 enzymes are common to all living cells, a fact which emphasizes their central role in all the vital life processes.

Plasmalemma

Structure

The plasmalemma, or cell membrane, is similar in appearance to the internal membranes, but has a different structure and different permeabilities. Studies of the plasmalemma suggest that its basic structure consists of a protein/phospholipid/protein 'sandwich', about 7.0 nm thick, perforated at intervals by pores, and with some proteins spanning the width of the membrane. In cells requiring a large surface area for the absorption or exchange of materials it is thrown into folds known as **microvilli**. Where animal cells need to be tightly joined together, specialized regions of the membrane known as **tight junctions** and fibrillar **desmosomes** are seen.

Functions

The functions of the plasmalemma are associated with its position as a boundary between the cytoplasm and the immediate environment of the cell. It **maintains the shape** of the cell, it **separates cells** from their neighbours, thus allowing specialization, it helps to **join cells** together to form tissues, and it is involved in **surface reactions** since many enzymes are bound into the membrane structure. It is also involved in **movement**, particularly of mobile cells like protozoa and mammalian white blood cells, and acts as a surface for **identification** of the cell, for example it provides receptor sites for hormones which generally only act on specific target cells.

The plasmalemma also acts as a **differentially permeable selective barrier** in controlling the passage of substances into and out of the cell. Some substances do pass through it by simple diffusion and osmosis, but the passage of many others is carefully regulated by the membrane via various energy-dependent active processes. One of the best examples is the 'sodium pump'[205] of nerve cells which actively removes sodium (Na^+) from the cells.

Cell membranes as regulators of solute and water movement

It is important to remember that the passage of materials into and out of cells is dependent on the metabolism of the whole cell and cannot therefore be entirely explained by a consideration of membrane properties alone.

UPTAKE OF SOLUTES

As mentioned above, some substances appear to pass across the membrane by simple passive diffusion,[369] while others show an 'uphill' movement against their concentration gradients which involves the utilization of metabolic energy from cellular respiration. However, even the so-called 'passive' processes are indirectly dependent on energy since the membrane must be maintained, and this requires synthetic reactions which are energy-dependent. Furthermore, a passive mechanism for one solute may cause another to enter by a process of 'drag', by which the second is drawn in, against its prevailing diffusion gradient, as a result of its association with the 'target' material. For example, in the mucosal cells of the small intestine it is thought that sodium ions diffuse down a passive gradient by means of a carrier which also complexes with glucose, which is therefore transported against its prevailing gradient, and appears to be taken up actively.

In the transfer of other ions, however, the case for carrier-linked active uptake is clearer. Generally, cells take up K^+ and expel Na^+, and it can be seen that nearly all cells have an **ATP-ase** enzyme in the membrane which is activated by the presence of Na^+ and K^+. Thus the uptake of K^+ is linked, presumably via a carrier, to the expulsion of Na^+, and the process is dependent upon energy released by the action of the ATP-ase on ATP. The nature of the carrier molecule is far from clear, but it is tempting to think that the proteins that span the membrane somehow act in this capacity. Indeed, some postulate a membrane structure in which small sub-units of the membrane bind to materials to carry them into the cytoplasm; they are immediately replaced by new membrane sub-units.

Where an electrical potential exists across a membrane, it is very difficult to determine whether an electrically charged ion is passing passively across as a result of this electrical gradient, or if it is being actively transferred and thus generating the potential. Indeed, potentials can develop passively as a result of the differing mobilities of ions across the membrane; for example, Na^+ ions have a slower mobility than Cl^- ions and this can lead to the establishment of membrane potentials.

The concept of active transport across membranes is obviously far from clear: the variety of uptake mechanisms points to a complex membrane structure not yet fully described. A further complication is provided by the **hormonal control** of cell permeability. Examples are insulin[215] facilitating the passage of glucose across muscle cell membranes, and ADH[213] altering the permeability to water of the kidney collecting duct cell membranes. It is possible that these primary hormones attach themselves to the membrane and cause the release of secondary intracellular messengers that change membrane permeability, and alter metabolic pathways by activating or depressing enzymes. Again, this behaviour certainly indicates that membranes are structurally more complicated than simple protein–lipid permeability barriers.

UPTAKE OF WATER

All water uptake by living cells is achieved by **osmosis**, which is the passage or diffusion of water from a dilute solution to a more concentrated solution through a semi-permeable membrane (one which allows the passage of water but prevents the passage of solute molecules). However, the cell membrane is not perfectly semi-permeable and the simultaneous passage of solutes and water complicates the picture.

There is no evidence in the plant or animal kingdom for the direct active transport of water across membranes. There can sometimes appear to be active uptake of water, but this can always be shown to be the active uptake of salts being followed by the passive osmotic uptake of water. The membrane must have pores which are significant in water uptake since the fatty layers of the membrane are a barrier to the passage of water. However, the size, shape, and number of these pores are unknown; and their existence is deduced from observations of the passage of solute molecules, and of the rate of the passage of water itself. Recent work suggests that the pores may be temporary in nature, continually forming and re-forming as the lipid layers of the membrane alter their configuration to allow the passage of materials. As a result of its lipid nature the membrane is remarkably 'waterproof': for instance, it has been calculated in *Amoeba* that, with an osmotic pressure difference of 1 atmosphere between the inside and the outside of the cell, it takes six weeks for 1 cm^3 of H_2O to pass through 1 cm^2 of membrane.

Water itself presents an additional barrier to the passage of solutes by its physical association with the membrane. The components of the cell membrane cause a structuring of the water in contact with it, and these stable, **unstirred layers** are thought to be of great importance to overall membrane permeability. These layers, which may be up to 500 μ thick, present a further barrier to the free diffusion of solute molecules. Osmotic flow, however, has a sufficient velocity to sweep these layers away.

The plasmalemma is also involved in the macro-transfer of larger molecular masses via **phagocytosis** and **pinocytosis**, where these are defined as the visible transfer of solids and fluids respectively. It may be either into the cell (endocytosis) as in absorptive cells or out of the cell (exocytosis) as in secretory cells. Both involve membrane fusion in which the continuous plasmalemma is broken and then repaired with new materials in a process known as 'membrane flow'. This can be a major cellular event; for example it has been established that an actively pinocytosing *Amoeba* may replace all of its original plasmalemma in under 5 minutes.

The nucleus

All cells, except bacteria, have at least one nucleus at some stage in their development; and even bacteria have a fairly clearly defined 'nuclear area'. The nucleus is usually spherical and occupies about one-tenth of the cell volume. It is limited by the nuclear membrane which is continuous with the granular endoplasmic reticulum. The nucleus is the 'control centre' of the cell, controlling the development and functioning of the cytoplasm via the DNA/RNA mechanism.[74] In the non-dividing cell there is little apparent internal structure besides the small spherical **nucleolus**, the centre of nuclear RNA. When fixed and stained, fibrils of 'chromatin', which consist of DNA and nucleoproteins, can be seen dispersed throughout the nuclear sap. At the onset of nuclear division this scattered chromatin material is condensed into discrete chromosomes. Unfortunately, the electron microscope does not reveal much extra detailed structure within the nucleus.

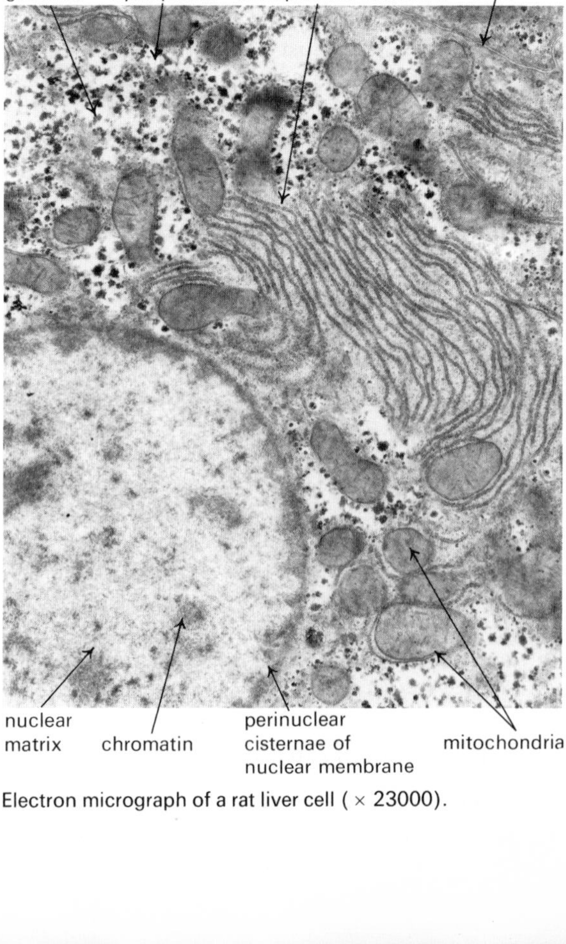

glycogen granules — hyaloplasm — granular or rough endoplasmic reticulum — cell membranes of 2 adjacent cells

nuclear matrix — chromatin — perinuclear cisternae of nuclear membrane — mitochondria

Electron micrograph of a rat liver cell (× 23000).

TISSUES

A tissue can be simply defined as a collection of similar cells, associated together in large numbers to perform a particular function (although many tissues consist of several different types of cells); an organ can be defined as a collection of different tissues co-ordinated towards a common function.

The different types of tissue will only be surveyed briefly here as more detailed descriptions will be found related to their functions in the appropriate chapters.

Plant tissues

Meristematic tissues

Meristematic tissues consist of undifferentiated cells which are capable of repeated mitotic cell division.

PRIMARY MERISTEMS

A primary meristem is one that is derived directly from the original embryonic meristems without ever having lost the power of continual cell division. There are three types of meristem:

(a) Apical meristems
These occur at the tips of roots and shoots. They are made up of small, many-sided cells, with dense granular cytoplasm, small scattered vacuoles, and no inter-cellular spaces. They are usually responsible for longitudinal growth of the stem and root.

(b) Lateral meristems or cambium tissues
These occur along the length of the stems and roots. They appear to consist of small rectangular cells in cross-section, but are in fact long thin cells, many times longer than they are wide. They are responsible for growth in girth[47] of the stems and roots.

(c) Intercalary meristems
These occur isolated between regions of permanent tissues, for example at the base of leaves in many monocotyledons.[351]

SECONDARY MERISTEMS

A secondary meristem is one that is derived from a permanent tissue, which, under some influence as yet not understood, de-differentiates and regains the power to divide continually. This can occur in many places in plants. One example is the **inter-fascicular cambium**, which arises from parenchyma tissue between vascular bundles in dicotyledons undergoing secondary thickening.[48] Another is the **cork cambium**[48] or phellogen which occurs either in the epidermis or cortex of Dicotyledons[351] undergoing secondary thickening and gives rise to the bark.

Permanent tissues

These consist of fully differentiated cells which are usually incapable of any further cell division.

GROUND TISSUE

This is a relatively simple tissue that makes up the bulk of the primary structure of a plant. There is but one main type, known as parenchyma tissue.

Parenchyma
Parenchyma consists of relatively large cells with thin cellulose cell walls and many inter-cellular spaces. They can develop chloroplastids and become photosynthetic (for example, in green stems); they can store starch (for example, in storage organs such as swollen roots and tubers); and they can de-differentiate and become secondarily meristematic (as in the inter-fascicular cambium).

SUPPORTING TISSUE

All tissues have some supporting function in the plant, either due to their turgidity or by means of especially thickened cell walls. However, so-called 'supporting tissues' have support as their main, if not only, function.

Collenchyma
This consists of elongated cells whose cellulose cell walls are especially thickened at the angles of the cells. It is the only mechanical tissue present in actively growing regions, where it is often found just under the epidermis.

Sclerenchyma
There are two elements of sclerenchyma; the fibres and the sclereids. The **fibres** are elongated, with lignified walls and no living contents.

On average they are 1–2 mm in length, but in some species can reach up to 50 cm in length. They have great mechanical strength, and are usually found in association with vascular bundles. The **sclereids**, or stone cells, have a variety of shapes but are not elongated. They can occur scattered, as in pear fruits, but more typically they form the hard shells of nuts and stones of other fruits.

VASCULAR TISSUE

Vascular tissue consists of elongated tubular elements specialized for the internal transport of materials up and down the plant. There are two types, xylem and phloem.

Xylem

This is the tissue in which water and dissolved substances travel from the root to the stem and leaves in the transpiration stream.[168] It consists of elongated tubular elements with waterproof lignified cell walls and no living contents. The lignin is laid down in a variety of patterns, and, when the wall is completely lignified, a variety of pits are found which allow for some lateral transport between the elements. The lignified walls make the xylem important for the support of the plant. The tubular elements are of two types: tracheids and vessels.

(a) **Tracheids** are derived from single cambium cells, and therefore have cross walls at each end to separate them from those above and below.

(b) **Vessels** are derived from a column of cambium cells; the end cross walls break down so that they all combine to form a continuous tube which may be several feet long. They are only found in the flowering plants.

Phloem

This is the living tissue in which organic materials synthesized in the leaves are transported or translocated[171] to various points of usage, or **sinks**, which are mainly the roots. It consists of two types of cell: the sieve-tube elements and their companion cells.

(a) **Sieve tubes** consist of elongated tubular elements separated from each other by perforated end walls known as sieve plates. When first formed there is active cytoplasmic streaming, and dense cytoplasmic connections exist between elements through the sieve plates. As the cell ages the nucleus is lost, the cytoplasmic streaming stops, and the sieve plates become blocked with callose, a carbohydrate deposited on the sieve plates. At this stage the tube is considered non-functional.

(b) **Companion cells** are active cells, with a nucleus and many mitochondria, found in close association with the sieve-tube elements in flowering plants. Up to six may lie alongside one sieve-tube element. They are thought to contribute to the maintenance and function of the sieve-tube elements.

DERMAL TISSUES

These are protective tissues usually found in continuous 'cylinders' which are penetrated at intervals by specialized openings. There are three types, epidermis, endodermis, and periderm.

The epidermis

The epidermis forms a continuous layer over the surface of unthickened plants, and has no intercellular spaces. On the aerial parts of the plant the cells are covered with a cuticle of waterproof cutin of variable thickness, and there are pores or stomata[169] at intervals. Specialized cells known as **guard cells** control these openings; guard cells are usually the only epidermal cells to contain chloroplastids. Epidermal cells are often modified to form hairs and secretory glands.

The endodermis

Endodermis cells form a continuous protective cylinder around the vascular tissues of the root. The mature cell walls are thickened with suberin, except in the thin-walled passage cells.[167]

The periderm

This consists of the cork cambium or phellogen and the tissues derived from it, namely the cork or phellem and the secondary cortex or phelloderm. The cork cambium is formed when the outer epidermis and cortical layers are ruptured by an increase in girth of the stem or root (caused by secondary thickening).

PHOTOSYNTHETIC TISSUE

Many tissues, including parenchyma and collenchyma, can be photosynthetic, but the mesophyll tissues of the leaf[129] are particularly adapted for this function.

SECRETORY TISSUES

Secretory tissues are particularly adapted for the production and secretion of various substances. Examples are found in nectaries, oil glands, resin ducts, latex tubes, glandular epidermal hairs, and digestive glands in insectivorous plants.

Animal tissues (vertebrates)

The four types of vertebrate tissue are described briefly below and summarized, with examples, in Table 1.1.

EPITHELIAL TISSUES

Epithelial tissues cover both external and internal surfaces, and typically form sheets of large surface area but of thin cross-section. The cells are held together by an inter-cellular cement rich in calcium ions, which are important in cell adhesion; and by numerous desmosomes and tight junctions between adjacent membranes. At their base the cells are fixed to a basement membrane. They have one free surface which is often highly specialized with, for example, microvilli or cilia. Glands are derived from epithelial tissue, but their specialization often obscures this origin.

CONNECTIVE TISSUES

Connective tissues connect, bind, support, and protect. They are characterized by a predominance of non-living inter-cellular matrix which is secreted by the connective tissue cells themselves (except in blood: the blood cells do not secrete the plasma).

NERVOUS TISSUES

Apart from the non-conducting satellite cells, nervous tissue consists of cells known as neurones, in which the property of all cells to develop an electrical potential difference across their cell membranes has been developed into the ability to initiate and transmit a nervous impulse.[204] Neurones are also active secretory cells producing neurosecretions.

MUSCLE TISSUES

In these tissues the contractile properties of the cytoplasm are developed to an extreme.

Table 1.1 *Summary of animal tissues*

Tissue	*Form*	*Type*	*Example*
Epithelial tissue	Simple (single layer of cells)	Squamous (pavement)	Bowman's capsule
		Cuboidal	Kidney collecting duct
		Columnar	Stomach (non-ciliated)
			Trachea (ciliated)
	Stratified (more than one layer of cells)	Squamous	Skin (keratinized)
			Oesophagus (non-keratinized)
		Columnar	Male urethra
		Transitional	Urinary bladder
	Glandular	Exocrine	Sweat gland
		Endocrine	Thyroid
Connective tissue	General	Mesenchyme	Embryo
		Areolar	Mesenteries
		White fibrous	Tendons
		Yellow elastic	Ligaments
		Adipose	Fat deposits
	Special	Cartilage	Trachea (hyaline)
			Intervertebral discs (fibrous)
			External ear (elastic)
		Bone	
		Blood	
Nervous tissue	Conducting	Neurones	Nervous system
	Non-conducting	Schwann cells	Around neurones
	Satellite or glial cells	Oligodendrocytes	Central nervous system
		Astrocytes	Brain
Muscle tissue		Smooth	Gut wall
		Cardiac	Heart
		Striated	Skeletal muscles

Experimental and applied aspects

Cytological techniques

Our view of the structure of the cell is determined by the techniques used in investigating that structure. With any cytological technique there is a danger of producing **artefacts**, that is artificial 'structures' that appear as a result of the technique used. Thus some knowledge of the instruments and procedures used in this study is an aid to our interpretation of cell structure.

THE OPTICAL MICROSCOPE

Although living cells can be observed by means of the phase contrast microscope, in most cases the cells are fixed (killed) and stained before observation so that structures which stain differently can be distinguished from each other. Both processes usually involve chemical treatments that distort the cells in some way so that one is no longer seeing cells in their natural state.

In addition, the light microscope can only magnify objects with reasonable clarity up to a maximum of 1500 times.

THE ELECTRON MICROSCOPE

The development of the electron microscope in the 1940s was as big a breakthrough in cytological techniques as was the development of the light microscope in the 1670s; a whole new world of 'sub-microscopic' or 'ultra' structure was revealed. However, the cells must always be fixed and stained by even more drastic methods than with the optical microscope, so that the danger of artefact formation is correspondingly greater.

The electron microscope utilizes a stream of electrons[361] focused by electromagnets on to a fluorescent screen and photographic plate; in this way **electron micrographs** are obtained. The **scanning electron microscope** produces three-dimensional electron micrographs. The use of electrons instead of light rays allows a magnification of up to 500 000 times to be obtained, without loss of clarity.

Membrane structure

Neither the light microscope nor the electron microscope reveals much of the structure of the cell membrane. Models of membrane structure are constructed on the basis of the membrane's behaviour under certain conditions; in other words, it is impossible to separate structure and function since our knowledge of the structure derives from our knowledge of function.

Early investigations were based on studies of the rate of penetration of various materials into the cell, e.g. urea, sugars, alcohols, etc. It was found that fat-soluble substances penetrated more easily than others, which suggested that the membrane was fatty in nature. This hypothesis was tested by extracting lipids from red blood corpuscles, with the underlying assumption that all the lipids resided in the membrane; and then measuring the surface area that this lipid would have if it formed a layer one molecule thick. By comparing this area with the surface area of the red blood corpuscle it was possible to estimate the number of lipid layers there were in the membrane. The results indicated that there were two layers, and so the **bimolecular lipid leaflet** model was constructed.

Further work indicated that 'smaller' molecules penetrated the membrane faster than would have been expected on the basis of their fat solubility alone. Thus the membrane appeared to act both as a selective fat solvent and as a molecular sieve, and the cell membrane was further described as a mosaic of lipid area and aqueous pores. Danielli and Davson showed that the surface tension of the membrane was lower than expected of an exposed lipid layer, they therefore postulated that protein was adsorbed on to the membrane surface, and constructed the **protein/lipid/protein sandwich** model with stable, protein-lined 'pores'. Electron microscopy appeared to confirm this structure, and, since it seemed to be of universal occurrence, led Robertson to the **'unit'** membrane concept—that all membranes showed this common structure. However, the wide variety of functions shown by various membranes seemed to argue against this, and Green refuted the unit membrane concept when he identified the **nesting repeating** structure of the inner mitochondrial membrane. In this model the membrane is seen as consisting of a line of block-like units, which allows an easier explanation of the 'pore' concept since the permeability of such a membrane can be varied according to the tightness with which the repeating units are packed.

More recent work on the red blood corpuscle membrane indicates that many protein molecules span the membrane from its internal to its external surface, and that the structure is far more complicated than the unit model would suggest. As techniques of investigation proliferate, so the apparent structure of the membrane becomes more complex; an excellent example of the way in which our understanding is determined by available techniques.

Guided example

Functions of cell organelles are investigated using a variety of techniques. One such technique is biochemical 'interference', whereby cells are exposed to some substance and the subsequent reactions of a particular organelle are observed.

1 When liver cells are administered fat-soluble drugs, it is found that the amount of agranular endoplasmic reticulum (AER) increases, along with the amount of drug-metabolizing enzymes. How would you interpret these observations in relation to the function of the agranular endoplasmic reticulum?
One interpretation is that the AER is involved in the production of drug-metabolizing enzymes, since both the amount of AER and the enzymes increase. However, the enzymes could well be synthesized elsewhere, and the proliferation of the AER could be a drug-induced artefact.

2 It is seen that fat absorbed by the microvilli of the intestinal epithelium accumulates in the AER, near to the surface, and is transported to the Golgi apparatus. Indeed, isolated 'fractions' of the AER are capable of synthesizing fats if supplied with the required materials. How do these observations further complicate the interpretation of the proliferation of the AER when fat-soluble drugs are administered?
A further complication could be that the fat-soluble drugs were carried to the AER along with any fat in the cell, and that consequently fat-soluble drug-metabolizing enzymes would also be found here in conjunction with their substrate. The attribution of a specific drug-metabolizing function to the AER would therefore be misleading.

3 The AER is often found in close association with glycogen granules in the liver cells. Does this observation enable any conclusions about the function of the AER to be drawn?
One might be tempted to conclude that the AER plays some role in glycogen metabolism, either in its synthesis or in its hydrolytic breakdown. However, the necessary enzymes are not found in the AER, hence the role of the AER in glycogen metabolism is still not clear.

4 In the cells of parts of the ovary, testes, and adrenal glands, all of which produce hormones, the AER practically fills the hyaloplasm as a mass of tangled tubes. What might you deduce from this observation as to the function of the AER?
This could indicate a role in hormone production. Indeed, AER fractions of the testes do contain enzymes involved in the synthesis of androgenic steroids, and also AER fractions of the adrenal cortex contain enzymes involved in the synthesis of cortical hormones.

5 A wide variety of functions of the AER has been identified in addition to those suggested above, for example in striated muscle it has a role in the transmission of the excitatory impulse throughout the fibres. It forms an intimate plexus around each myofibril, and is known here as the **sarcoplasmic reticulum.** In the teleost gill the chloride-secreting cells are rich in AER, as are the oxyntic cells in the vertebrate stomach, indicating that the AER is here involved in the active transport of chloride ions. What does this wide variety of possible functions suggest to you about the structure of the AER?
This wide variety of functions suggests that, in spite of a similarity in appearance of the AER at the different cell sites, there could be differences in basic structure not visible under the conditions of electron microscopy.

Questions

1 Survey the different types of cytoplasmic organelle visible in an electron micrograph of a typical animal cell. Show how the structure of the various types of organelle described relates to the functions which each performs. (L)

2 Give an illustrated account of the structure of the cytoplasm of a living, photosynthetic parenchyma cell as revealed by electron microscopic studies. (L)

3 Discuss the significance of organelles within cells and of cells within organs and of organs within organisms. (C)

4 A solution of amylase was tested against a solution of starch by means of mixing the two together under different conditions and testing the mixture periodically against drops of iodine in potassium iodide solution. When the test no longer produced a colour change with iodine in potassium iodide solution it was assumed that all the starch had been hydrolysed. The following is a set of class results showing the reciprocals of the time in minutes for complete hydrolysis to occur under varying conditions of temperature and pH.

Table 1 *Experiment carried out at pH 6.7*

Temperature (°C)	0	10	20	30	40	50	60	80
Reciprocal of time for hydrolysis	0.0	0.11	0.17	0.31	0.39	0.25	0.08	0.0

Table 2 *Experiment carried out at 40 °C*

pH	4.7	5.3	5.9	6.5	6.8	7.2	7.5	8.2
Reciprocal of time for hydrolysis	0.14	0.18	0.21	0.32	0.35	0.21	0.18	0.11

(a) Plot a graph for the performance of the enzyme at varying temperatures and a relatively steady pH, using the data in Table 1.
(b) Plot a graph for the performance of the enzyme at varying pH levels and a steady temperature, using the data in Table 2.
(c) What does graph (a) allow you to deduce about the optimum temperature for the activity of this particular enzyme at pH 6.7?
(d) Account for the shape of the curve between 0 °C and 30 °C.
(e) Account for the shape of the curve between 50 °C and 80 °C.
(f) What would you expect to happen if the experiment were repeated using enzyme which had been heated previously to 90 °C for 15 minutes? Justify your answer.
(g) What does graph (b) allow you to deduce about the optimum pH for the activity of this particular enzyme at 40 °C?
(h) Explain how you would refine the experiment to increase the accuracy of the answer which can be given to (g). (AEB)

Further reading

Dodge J. D., *An Atlas of Biological Ultrastructure* (London: Arnold, 1968).
Freeman W. H. and Bracegirdle B., *An Atlas of Histology* (London: Heinemann, 1974).
Grimstone A. V., *The Electron Microscope in Biology*, The Institute of Biology Studies in Biology, No. 9 (London: Arnold, 1976).
Haggis G. H., *The Electron Microscope in Molecular Biology* (London: Longmans, 1970).
Shaw A. C., Lazell S. K., and Foster G. N., *Photomicrographs of the Flowering Plant* (London: Longmans, 1975).
Swanson C. P. and Webster P. L., *The Cell* (New Jersey: Prentice-Hall, 1977).

2 Reproduction, growth, and development

Introduction

Sexual reproduction (amphimixis)

Sexual reproduction involves the fusion of the nuclei of the sex cells, or **gametes**, from the male and female in a process of **fertilization**. Individual organisms can be either single sexed (dioecious), or hermaphrodite (monoecious) bearing both male and female organs. Hermaphrodite organisms may be self-fertilizing, or they may have outbreeding mechanisms which favour or compel cross fertilization with another hermaphrodite individual.

During gamete production, or gametogenesis,[21] in animals and in spore formation which precedes gamete production in plants, the number of chromosomes[59] in the nucleus is halved in a process known as **meiosis,[60]** so that normal gametes have half the normal number of chromosomes, that is they are haploid. A diploid zygote, which has twice the haploid number of chromosomes, is formed by fertilization. Certain genetic 'mixing' events,[61] which occur during meiosis, and the random fusion of the gametes result in **genetic variation** in the offspring which may be of adaptive advantage.

When the gametes from the two different sexes are identical, as in some algae, fungi, and protozoa, they are said to be **isogamous**. However, in most plants and animals the two types of gametes are **anisogamous** or **heterogamous**, that is they have different structures. Usually the male gamete is small and motile and the female gamete is large and non-motile. This type of fertilization is also known as **oogamy**. Where an animal species has separate males and females **sexual dimorphism** nearly always occurs, that is, there is a clear distinction in appearance between the sexes.

POLYEMBRYONY

In some cases more than one embryo can develop from a single zygote, usually by fission of the developing zygote into two or more independently developing embryos at some early stage. This is known as polyembryony. The simplest example is seen in monozygotic twins, when the cells produced by the cleavage of the zygote separate and develop into two identical individuals. The armadillo (*Dasypodus*) characteristically produces monozygotic quads, and some parasitic Hymenoptera can produce up to 2000 embryos from a single zygote.

Asexual reproduction

Asexual reproduction does not involve gametes and requires only one parental organism. It involves the **mitotic[60]** division of cells, during which the complete adult number of chromosomes is exactly replicated and passed on, so that the offspring are genetically identical to the parent. It is more common in plants than in animals. Examples include binary fission and spore formation, budding, and vegetative reproduction. In an asexually reproducing line of descendants, or **clone,[76]** a good combination of genes can only arise if all the genes concerned undergo the necessary changes or mutations in the line of descent. Such mutations can occur in different sexually reproducing individuals at different times, and still be brought together by fertilization.

BINARY FISSION

Binary fission is seen in bacteria and many protozoa, including *Amoeba*. After the parent cell has grown to a certain optimum size the nucleus undergoes mitotic division and the cytoplasm undergoes cleavage into roughly equal halves. The stimulus which causes division is not fully understood, but is thought to involve the nucleus to cytoplasm ratio and the surface area to volume ratio[370] of the parent cell. These ratios are important because the nucleus can only 'control' a certain volume of cytoplasm, and exchanges taking place over a certain surface area can only supply a certain volume of cytoplasm. When these ratios fall below a certain critical value, and other conditions are suitable, binary fission occurs. These basic principles are also thought to be involved in the control of some cell divisions in multicellular organisms.

SPORE FORMATION

Spore formation is more common in plants than in animals and is characteristic of saprophytic and parasitic micro-organisms, especially the bacteria and fungi; it is also seen in the sporophyte generation[29] of higher plants. For parasites and saprophytes the production of vast numbers of easily dispersible spores ensures successful transference to a new food source.

BUDDING

Budding is seen in the lower animals such as the Coelenterates. *Hydra* forms a bud of cells by mitotic cell division, which then forms a body wall, a body cavity or enteron continuous with that of the parent, tentacles, and eventually a mouth. The point of connection with the parent is sealed and the new individual drops off to live an individual existence.

VEGETATIVE REPRODUCTION

During vegetative reproduction, which occurs in plants, some part of the vegetative plant body separates from the parent plant and then grows into another complete plant. Such reproduction is seen in the fragmentation of filamentous algae and of the gametophyte plant body of mosses and liverworts;[30] it also occurs in a wide variety of specialized structures such as runners, stolons, rhizomes, bulbs, corms, and tubers in higher plants.

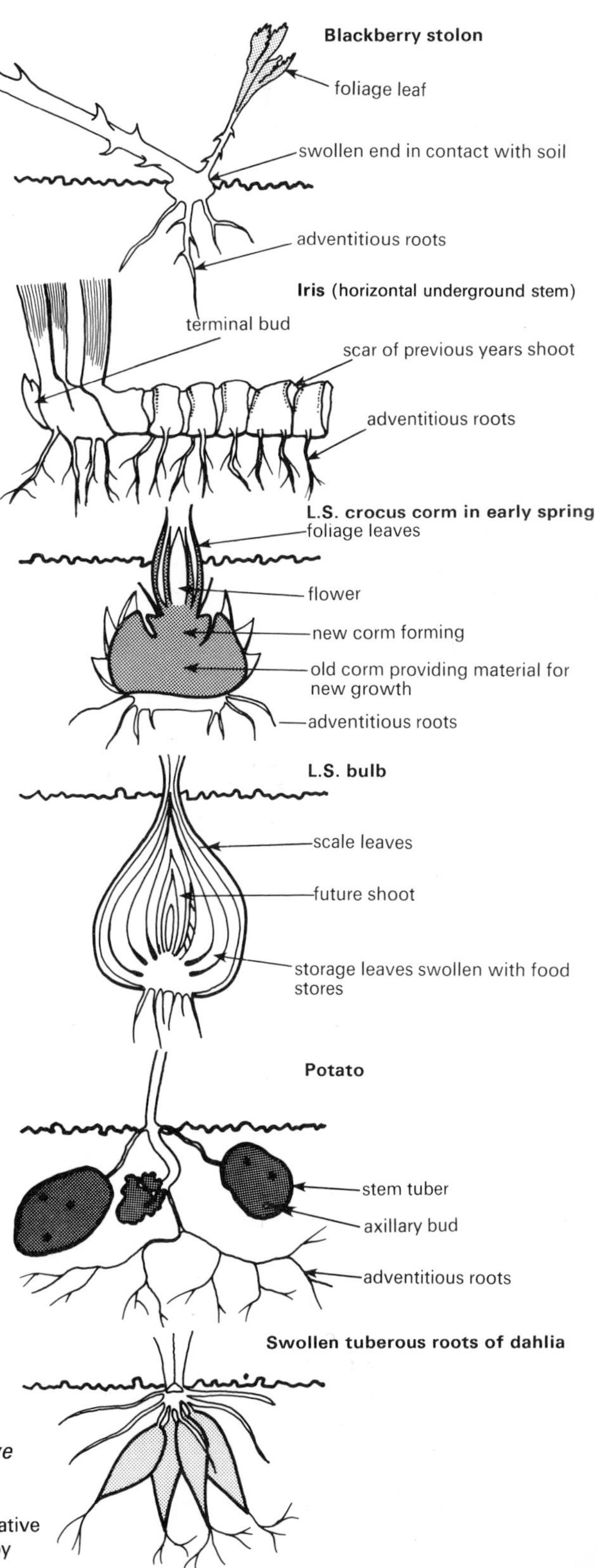

Figure 2.1 *Organs of perennation and vegetative reproduction in Angiosperms*
The ability of plants to survive from year to year on vegetative food stores is known as perennation. Vegetative reproduction only truly occurs when a part separates by fragmentation.

APOMIXIS

Apomixis is a form of reproduction which, although having a superficial similarity to normal sexual reproduction, actually occurs without fertilization. It includes the phenomenon of **parthenogenesis**, in which the female gametes or ova develop without fertilization. There are two kinds of parthenogenesis: diploid and haploid.

In **diploid parthenogenesis** the eggs are produced as a result of mitosis and not meiosis so that, like the parent, they are diploid. Aphids produce parthenogenetic diploid wingless females in the summer months, and thus multiply rapidly. These offspring are genetically identical to the parent. Usually normal sexual reproduction also takes place periodically in such animals, thus introducing variation into the offspring.

In **haploid parthenogenesis** the eggs are produced as normal gametes by meiosis, but something triggers their development into haploid individuals. In the honeybee the fertile male drones develop in this way from unfertilized haploid eggs laid by the queen.

In addition to aphids and bees, parthenogenesis is a normal occurrence for some wasps, flatworms, and dandelions. It can sometimes occur in vertebrates, and cases have been recorded in some lizards, and in turkeys.

Sexual reproduction in mammals

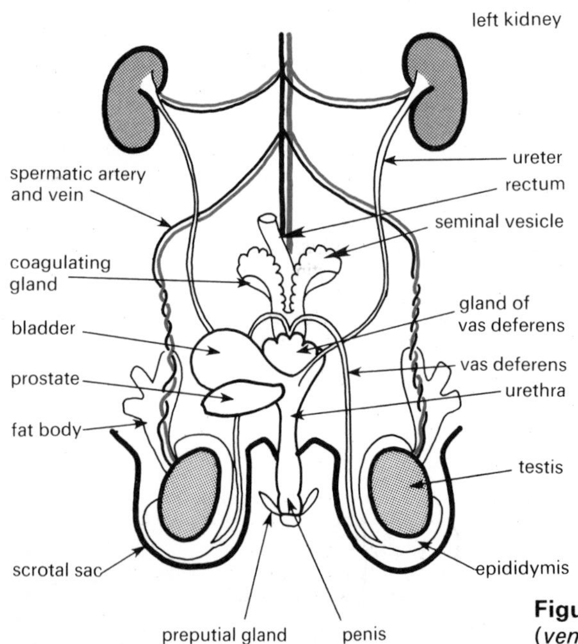

Figure 2.2 *Mammalian male urinogenital system (ventral view)*

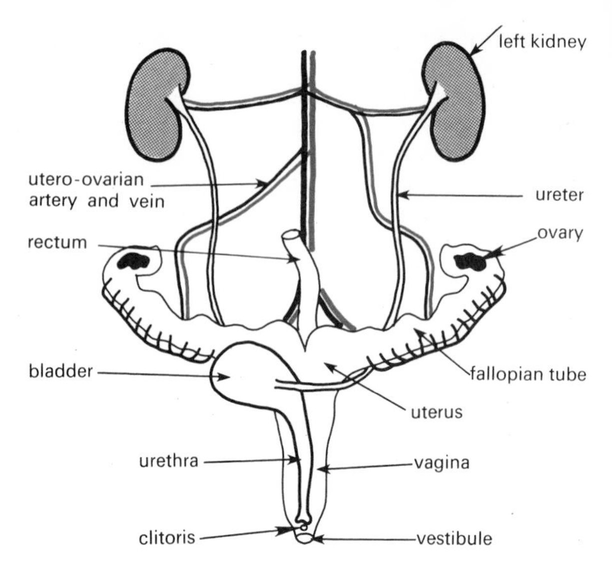

Figure 2.3 *Mammalian female urogenital system (ventral view)*

Mammalian sex hormones

As well as producing the gametes, tissues in the gonads (the ovaries and testes) also secrete sex hormones. The production and secretion of these hormones are themselves under the control of **gonadotrophins**, or gonad controlling hormones, secreted by the adenohypophysis of the pituitary gland.[212] The pituitary gland in turn is controlled by neurosecretions from the hypothalamus[213] of the brain.

The adrenal cortex is a secondary source of sex hormones and produces male androgens, female oestrogens, and progesterone in both sexes. Also, when present, the placenta[25] secretes sex hormones into the maternal blood stream which serve to maintain the pregnancy.

OVARIAN HORMONES

Oestrogens

The term oestrogen refers to a group of substances, of which the main one in mammals is probably oestradiol, but 'oestrogen' is often used as if it referred to a single female hormone. Oestrogens are secreted by the Graafian follicles and corpus luteum of the ovaries, and by the adrenal cortex, and they control the development of the primary and secondary sexual characteristics which appear in the female with the onset of sexual maturity.

The **primary sexual characteristics** are shown by the actual reproductive organs, for example the growth and development of the vagina, uterus, and oviducts. The **secondary sexual characteristics** are not directly involved with the reproductive organs but are characteristic of mature members of that sex, for example the changes in the body shape and appearance that occur.

Oestrogen also has a role in the differentiation of the ovarian follicles, the proliferation of the endometrium, or lining of the uterus, during the **proliferative phase** of the oestrous cycle, and the development of the ducts in the mammary glands. In those mammals where the female is smaller than the male, oestrogen may antagonize the growth hormone, and in birds it stimulates nest-building behaviour.

The ovary is also capable of secreting androgens, or male hormones, but not testosterone. The testes of the male also secrete small amounts of oestrogens, but their role is not clear.

Progesterone

Progesterone is secreted by the corpus luteum, by the placenta, and by the adrenal cortex. It contributes to the control of the ovarian or oestrous cycle by regulating the **secretory phase** of the cycle which follows the oestrogen-evoked proliferative phase. The main feature of the secretory phase is the enlargement of the mucus glands of the uterus. Progesterone also controls the changes involved in pregnancy (especially the embedding of the ovum and the development of the placenta) and suppresses further ovulation.

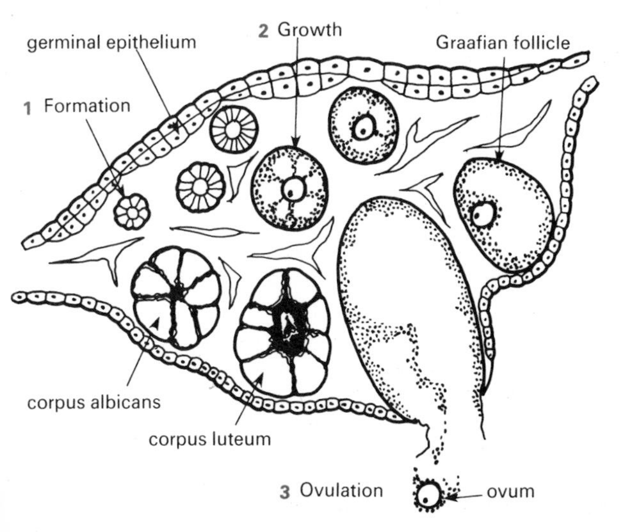

The corpus luteum secretes progesterone which controls changes in the second half of the cycle.

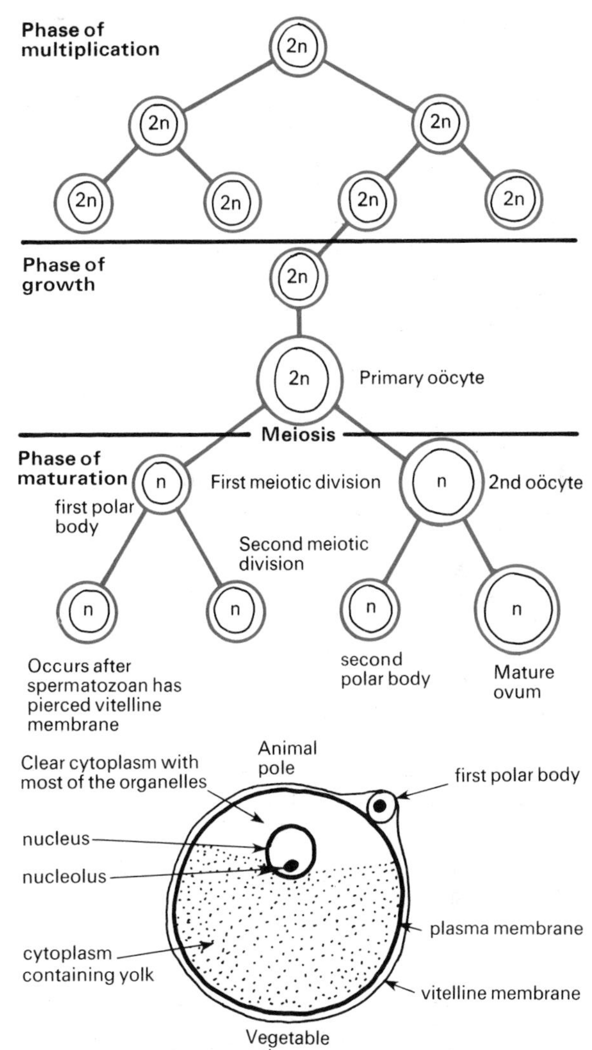

Polar bodies take no part in reproduction and disappear, therefore only one functional cell is produced from the primary oöcyte

Figure 2.5 *Gametogenesis: oogenesis*
In the human female it is estimated that by the fifth month of foetal development the ovary of the female foetus contains about 7 million oogonia, of which only about 400 have the chance of being released.

Figure 2.4 *Structure of the mammalian ovary*
(1) Formation: the germinal epithelium produces follicles.
(2) Growth: the follicle enlarges to form a mature Graafian follicle, the wall of which secretes the oestrogen which controls the changes in the female system in the first half of the menstrual cycle.
(3) Ovulation in the human female occurs on the 14th day of the 28-day cycle.

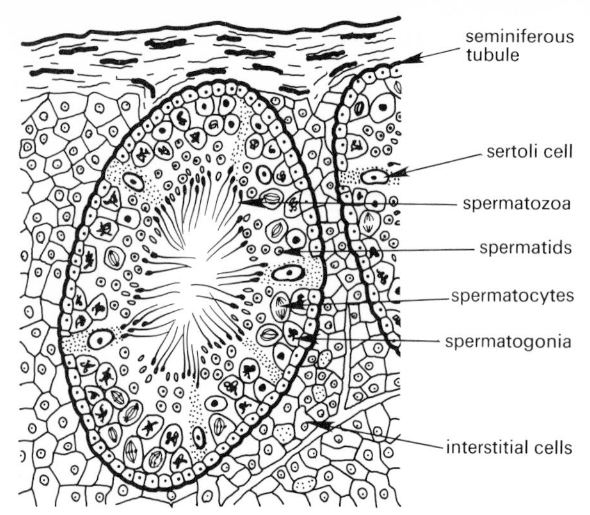

Figure 2.6 *Structure of a seminiferous tubule in the mammalian testis*
(T.S. seen under the light microscope)

Figure 2.7 *Gametogenesis: spermatogenesis*

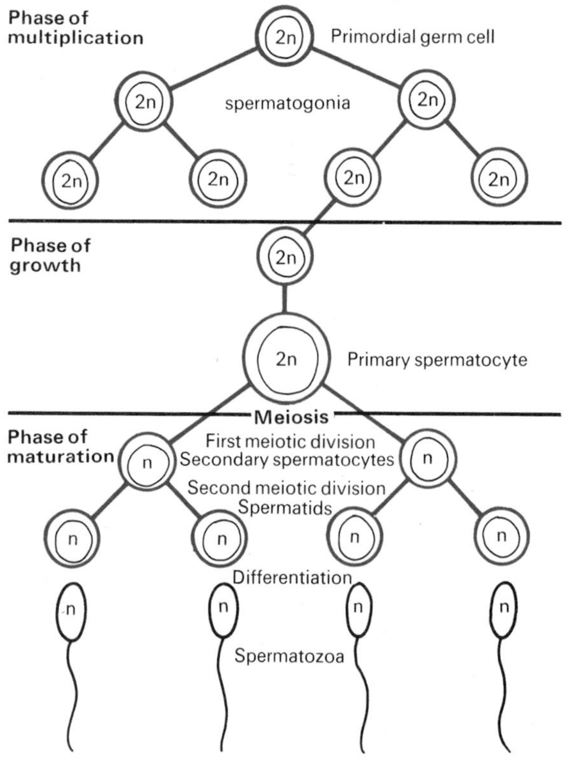

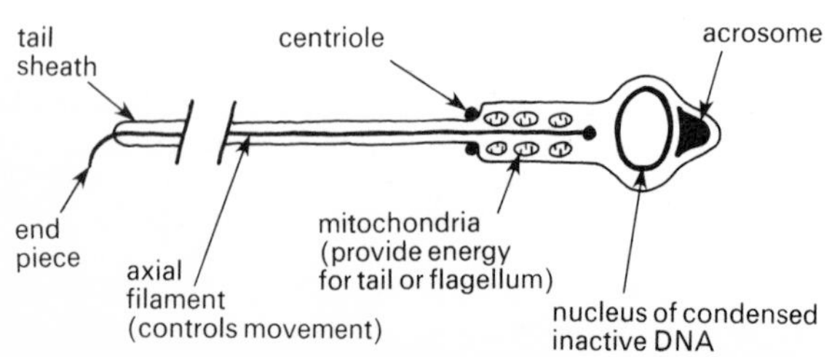

TESTES HORMONES

Androgens

The term androgens refers to a group of substances, of which the main one in the mammal is **testosterone**. They are secreted by the interstitial cells of the testes. Testosterone controls the development of the primary and secondary sexual characteristics that occur in the male with the onset of sexual maturity. By its action on the epididymis of the testes it maintains the fertilizing power of the spermatozoa which are stored there.

Pituitary gonadotrophins

FOLLICLE STIMULATING HORMONE (FSH)

This hormone initiates the development and growth of the Graafian follicles in the ovary. It also stimulates the ovary to secrete oestrogen. In the male it initiates sperm formation in the testes.

INTERSTITIAL CELL STIMULATING HORMONE (ICSH) OR LUTEINIZING HORMONE (LH)

ICSH promotes ovulation, that is the release of the ovum from the ovary, and leads to the development of the corpus luteum from the empty follicle. This in turn secretes oestrogen and progesterone. In the male it stimulates the secretion of testosterone by the testes.

In the human female FSH and ICSH (LH) are the only two hormones required, but in the rat a third gonadotrophin, **prolactin** (**lactogenic hormone**) is identified. Prolactin is also involved in the maintenance of the corpus luteum of the ovary, and has a role in milk production.

LUTEOTROPHIN (LUTEOTROPHIC HORMONE, LTH)

LTH may be a separate hormone. It is thought to stimulate the corpus luteum to secrete progesterone.

SUMMARY

The gonadotrophins, under the control of the hypothalamus, are also involved in the seasonal reproductive activity of many mammals, and their secretion is sequentially arranged so that the young are born at the proper time.

Cyclic changes in the female system

The cyclic changes in the female system are known as the **oestrous cycle** and may occur with varying frequency in different mammals. In some, for example the silver fox, it occurs only once a year, and such types are described as being **monoestrous**. In others, for example rats and mice, it can occur many times a year, and these are described as being **polyoestrous**. In the human female, anthropoid apes, and Old World monkeys it occurs once a month, and in this case the oestrous cycle is also referred to as the **menstrual cycle**. The menstrual cycle has special features, such as the sudden degeneration and elimination of the endometrial lining of the uterus; and the absence of any obvious period of 'heat' or oestrous which, in other mammals, is the only period when they are sexually receptive.

PHASES OF THE MENSTRUAL CYCLE

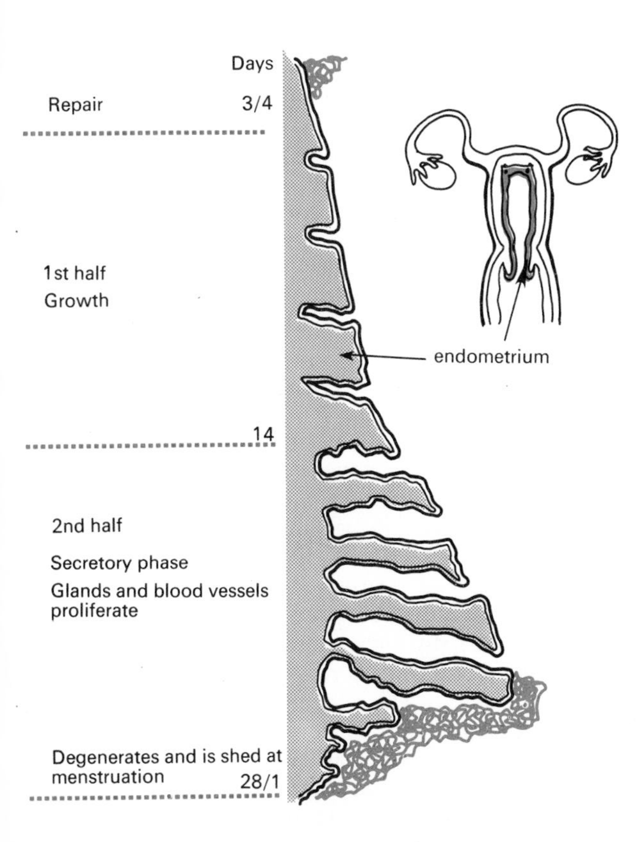

Figure 2.8 *Human menstrual cycle*
The endometrium shows changes during each cycle.

Proliferative phase
After the damage of the preceding menstrual period has been repaired, which takes about six days, the endometrial wall of the uterus thickens, becomes more vascular, and the glands elongate. These changes occur under the control of oestrogen, the secretion of which is, in turn, under the control of FSH.

Ovulation (day 14)
The oocyte in the largest of the Graafian follicles in the ovary, stimulated by a surge of gonadotrophins (especially ICSH (LH)) from the adenohypophysis of the pituitary, starts the process of pre-ovulatory maturation. There is a final growth spurt of the follicle, and ovulation occurs with the secondary oocyte at metaphase of the second meiotic division.[61] The oocyte is released into the abdominal cavity, from where it is swept into the opening of the oviduct (the oviducal funnel) by the ciliary action of the ciliated cells lining the oviduct wall.

Some mammals, for example the rabbit, ferret, and cat, are **induced ovulators**; that is they usually require coitus, or intercourse, in order to ovulate. Others, for example humans, are **spontaneous ovulators** and ovulate without external stimulus.

For most of the oestrous cycle the cervical mucus is very viscous, containing a dense irregular mesh of glycoprotein molecules which present an almost impenetrable barrier to the passage of any spermatozoa. At about the time of ovulation, however, the mucus becomes clear, less viscous, and no longer prevents the passage of spermatozoa.

Secretory phase (days 15–28)
The endometrium continues to thicken, the glands enlarge and become fully secretory, and the lining becomes more vascular. These changes occur under the control of progesterone, the secretion of which is, in turn, under the control of ICSH (LH).

Menstruation (day 1)
As the levels of both oestrogen and progesterone decrease due to the regression of the corpus luteum, the endometrium begins to breakdown gradually, and is eliminated over a period of days as the **menstrual flow**. The cycle itself depends on the cyclic production of gonadotrophic hormones by the adenohypophysis of the pituitary.

Fertilization

The active swimming movements of the ejaculated spermatozoa play only a small part in their ascent of the female reproductive tract. Their passage depends mainly on the peristaltic contractions of the female tract, but it is not understood how ova can be moved down the oviduct at the same time as spermatozoa are moved up.

Before fertilization can occur the spermatozoa must undergo **capacitation**, a process in which the acrosome at the tip of the spermatozoon breaks up and releases enzymes which assist penetration of the egg. This is stimulated in some way by secretions of the oviducts. After the penetration of the vitelline membrane of the secondary oocyte by one spermatozoan, some type of fertilization membrane is formed that prevents the entry of any more.

The entry of a spermatozoon triggers the completion of meiosis, the extrusion of the second polar body and the maturation of the oocyte into an ovum. The diploid nucleus of the fertilized ovum, or zygote, is usually formed on a spindle apparatus and immediately undergoes the first mitotic division prior to the first cleavage of embryological development.

ANTIGENIC ACTIVITY OF SPERMATOZOA

The ejaculation of spermatozoa from the male into the female reproductive tract poses some interesting problems of tissue compatibility. The spermatozoa represent an invasion of foreign protein which can trigger the production of antibodies.[291] However, although females can produce anti-sperm antibodies, generally this does not occur. Spermatozoa also carry blood group antigens[291] which may meet incompatible blood group antibodies in the uterine secretions of the female.

FUNCTION OF ACCESSORY GLANDS

The seminal vesicles secrete a number of substances, known as **prostaglandins**, which are thought somehow to stimulate those contractions of the female reproductive tract that are responsible for the ascent of the spermatozoa. A low level of these substances can be a factor in infertility in males. Fructose is also added to the seminal fluid as a respiratory substrate for the spermatozoa.

After ejaculation the semen coagulates in the female tract, possibly to prevent its subsequent loss from the female. Human semen spontaneously liquifies again within about 20 minutes. The accessory glands of the male reproductive tract, that is the seminal vesicles, prostate, and Cowper's glands, add the substances involved in the phenomenon. For example, in humans, fibrinogen is added by the seminal vesicles and the prostate secretes an enzyme which converts soluble fibrinogen into the insoluble fibrin protein fibres during coagulation. The fibrinolytic enzyme which is responsible for the liquefaction is also secreted by the prostate.

Pregnancy

THE OVARY IN PREGNANCY

When pregnancy occurs, the normal ovarian cycle is suspended and the corpus luteum continues to grow to a size which may occupy about 30–50 per cent of the volume of the ovary. The corpus luteum secretes large amounts of the hormone progesterone which maintains the pregnancy in its early stages and which is essential for the development of the placenta. This secretion reaches its peak about six weeks after conception, begins to decrease at about the second month of pregnancy, and finally ceases at about the fourth month. By this time the allanto-chorionic[25] placenta has begun to secrete placental progesterone which maintains the pregnancy and prepares the mammary glands for lactation.

IMPLANTATION

When the dividing zygote, which now forms a hollow ball of cells known as the blastocyst, enters the uterus it becomes implanted in the uterine wall. **Central implantation** occurs most typically in carnivores and in hoofed mammals. In this case the blastocyst remains unattached for 1–5 weeks, during which time it depends for its nutrition on uterine secretions known as 'uterine milk'. It grows to a large size and the foetus reaches an advanced stage of development before the placental attachments are formed, when nutrients are obtained from the maternal blood.

Excentric implantation occurs in most rodents, in monkeys, and in humans. In this type of implantation the blastocyst remains small and starts to implant within a week of entering the uterus. At first it is nourished by the 'uterine milk', but it soon obtains its nutrients from the maternal blood. In humans the implanting blastocyst becomes completely surrounded by uterine tissues in a special type of implantation known as **interstitial**. The trophoblast of the human placenta secretes a gonadotrophic hormone that can be detected in the urine three or four days after implantation of the blastocyst. This forms the basis of one method of pregnancy testing. The necessary feedback signal to the female's physiology as to whether the changes necessary for maintaining a pregnancy should be sustained is not fully understood.

The development of the foetus in the uterus poses further problems of tissue compatability. The trophoblast appears to have a non-antigenic or 'neutral' layer which does not induce the formation of anti-foetal antibodies by the mother. It thus serves as a barrier between the incompatible tissues of the foetus and the mother.

THE PLACENTA

The placenta consists of a close association of foetal and maternal tissues for the purpose of attachment and physiological exchange. The developing embryo absorbs nutrients, oxygen, water, and some antibodies into its circulation from the maternal circulation; it eliminates waste products such as urea and carbon dioxide in the opposite direction. The placenta produces hormones which maintain the pregnancy, for example gonadotrophins, oestrogen, and progesterone. Chorionic gonadotrophin is secreted by the placenta into the maternal blood stream and helps to maintain the corpus luteum. This in turn continues to secrete the oestrogen and progesterone which prevent ovulation and menstruation. Up to about the third month the ovarian hormones maintain the pregnancy, but after this period the placenta takes over the secretion of oestrogen and progesterone.

Ideally the foetal and maternal circulation should not mix. If the maternal blood circulated through the foetus, the maternal antibodies and white blood cells would recognize the foetus as 'foreign' protein and attack it. Also, since blood groups are determined genetically, the foetus could have a different blood group from the mother, therefore any mixing of the blood could result in problems of Rhesus blood group mismatch (the ABO blood group system does not express itself strongly until after birth). If the two circulations were to mix there would be further problems involving blood pressure, circulating maternal hormones, and waste products.

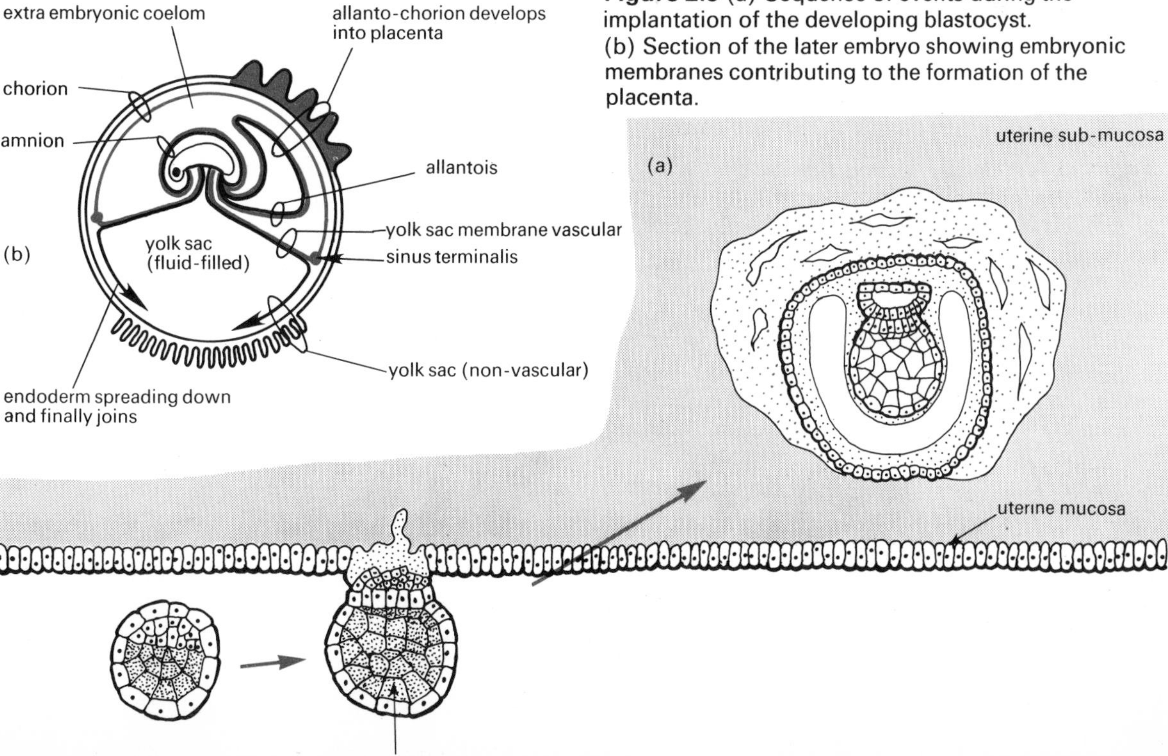

Figure 2.9 (a) Sequence of events during the implantation of the developing blastocyst.
(b) Section of the later embryo showing embryonic membranes contributing to the formation of the placenta.

In actual fact the placenta does allow the passage of some maternal antibodies which serve to provide the new-born with a degree of immunity to certain diseases before its own immune system is fully developed. In addition, placentae are often 'leaky' to the extent that there can be a degree of mixing of the circulations with variable results.

There is a wide variation in the way in which the embryonic membranes of the mammals make contact with the uterine wall for the purpose of physiological exchanges.

However, the true placenta of the Eutheria mammals develops from the fusion and proliferation of the allantois and the chorion, and is known as the **allanto-chorionic placenta**. Basically the allanto-chorionic placenta consists of six layers; namely the allantoic endoderm, the allantoic mesoderm, the chorionic mesoderm, the chorionic ectoderm or trophoblast, the maternal uterine mucosa, and the submucosa.

Although these layers remain intact in those areas which function mainly for attachment, there is a wide variety of placentae distinguished by the degree of tissue breakdown in those areas which function mainly for exchange. Such tissue breakdown decreases the diffusion distance between the maternal and foetal circulations and thus increases the efficiencies of exchanges. For example, in the **haemo-chorial** type of placenta seen in humans, all the maternal layers are eroded, including the endothelial wall of the maternal blood capillaries, so that maternal blood flows into blood spaces. In the **haemo-endothelial** type of placenta the maternal tissues are eroded, as are most of the foetal tissues, leaving only the endothelium of the foetal capillaries separating the two circulations.

After about the fourteenth week of embryonic development in humans, only that part of the chorion and trophoblast with a blood supply from the umbilical cord remains and the rest breaks down and disappears. The mature human placenta covers an area about the size of a dinner plate and has a volume of about 500 cm^3. The foetal blood volume of about 400 cm^3 circulates through the placenta about once every minute, and the maternal blood circulates through the placenta at about 500 cm^3 every minute.

Animals other than mammals can also have their young born alive, but in these cases the developing young are always separated from the parental tissues by the egg membrane and a true placenta is not developed. Such a condition, which occurs in many insects, snails, fish, lizards, and snakes, is known as **ovoviviparity** to distinguish it from the situation seen in the mammals, which is known as **viviparity**.

Embryology

The development of the fertilized egg or zygote is a continuous process, but for convenience it is considered to involve various stages, including cleavage, gastrulation, and organology.

Cleavage involves the mitotic cell division of the zygote usually into a hollow fluid-filled ball of cells known as the blastula or blastocyst.

Gastrulation involves cell migration in morpho-genetic movements, by which the basis of the complex structure of the fully developed animal is established. Morphogenetic movements are characteristic of animal embryological development; in plants the cells are not free to move in the same way, due to their rigid cellulose cell walls. In plants, therefore, morphogenesis is always linked to growth, and plants generally maintain regions of growth, differentiation, and morphogenesis throughout their life.

Organology involves the further development of the organs and systems which culminates in the formation of the fully-developed animal. In animals with larval forms this process may be considered to be spread over a relatively long period of their life, in which the still-developing animal lives an independent existence.

The term **foetus** is used to describe the developing organism after the formation of the main features of the fully-developed animal. Before this time, which in human beings is about two months after conception, the term **embryo** is used. In humans the anatomical differences between male and female embryos are not apparent until seven weeks after conception. The male embryo develops in an environment with high levels of female sex hormones, and to overcome their influence he must produce his own supplies of male sex hormones as soon as possible. The development of the testes is under the control of the Y chromosome. As soon as the testes develop they secrete male sex hormones which control the further development of the male features. In the female embryo the future ovaries do not produce any female hormones; in the absence of the testes, female characteristics develop automatically.

Metamorphosis

Metamorphosis is the change in form, from juvenile stage to adult, seen in the life cycles of organisms. More usually the term is used in relation to animals rather than to plants. The degree of change necessary to qualify for the term metamorphosis is a matter of opinion.

Metamorphosis is, to a large extent, anticipatory. In other words, the changes are often designed to fit the animal to an environment other than that in which it is living at the onset of metamorphosis. The stimulus for the onset of metamorphosis is usually some change in the environment, for example change in day length or temperature. Such changes are detected by the sense organs and relayed to the brain where they stimulate the production of certain neurohormones. These subsequently affect the hormone system which controls metamorphosis.

Metamorphosis in insects

All insects show some metamorphosis of juvenile stages into the adult. If these juvenile stages are similar in appearance to the adults they are referred to as **nymphs**; if they are not, they are referred to as **larvae**. The larval or nymphal stage is one in which the animal feeds almost continuously and in which most, or all, the growth occurs. Indeed some adult insects do not feed at all, and some even lack mouthparts. The adult stage is mainly locomotory and reproductive.

The presence of a hard limiting cuticle, or exoskeleton, makes it necessary for the larva or nymph to undergo a moult or **ecdysis** before it can increase in size. After a moult there is a rapid increase in size whilst the new cuticle is soft and extensible, and before it is hardened by the deposition of tannins. Each larval or nymphal stage between such moults is known as an **instar**. Adults do not usually undergo ecdyses. When considering growth as increase in size, insect growth occurs in a series of steps immediately following each ecdysis, with periods of no increase in size after the cuticle has hardened. However, if growth is measured as increase in dry or wet weight, or total nitrogen, no such steps are apparent. Growth according to these criteria continues in each instar between ecdyses so that the growth curve is continuous.

Insects in which the juvenile stage or nymph does resemble the adult, and into which it changes gradually through a number of instars, emerging as the adult after the final moult, are said to show **incomplete** or **Hemimetabolous metamorphosis**. This group of insects is also known as the **Exopterygota** since wings, when they are present, develop from external buds and are visible throughout most of the instars. Examples of insects showing Hemimetabolous metamorphosis include aphids, locusts, and cockroaches.

Insects in which the juvenile stage or larva does not resemble the adult into which it changes during a pupal stage are said to show **complete** or **Holometabolous metamorphosis**. This group of insects is also known as the **Endopterygota** since wings, when present, develop from internal buds. In these insects there is a great difference in the mode of life of the larva and the adult, and each usually exploits a different environment. This reduces competition between them and increases the chances of survival should there be any disturbance of one of the habitats. In the holometabolous insects the larvae feed and grow, undergoing a number of ecdyses or moults, before **pupating**. In the pupa the larva metamorphoses into the adult. In some insects the structural changes are by differential growth of the larval structures already present. In others the larval structures are broken down and the adult structures are formed by regrowth from imaginal buds.

Figure 2.10 *Growth curves of a Holometabolous insect*

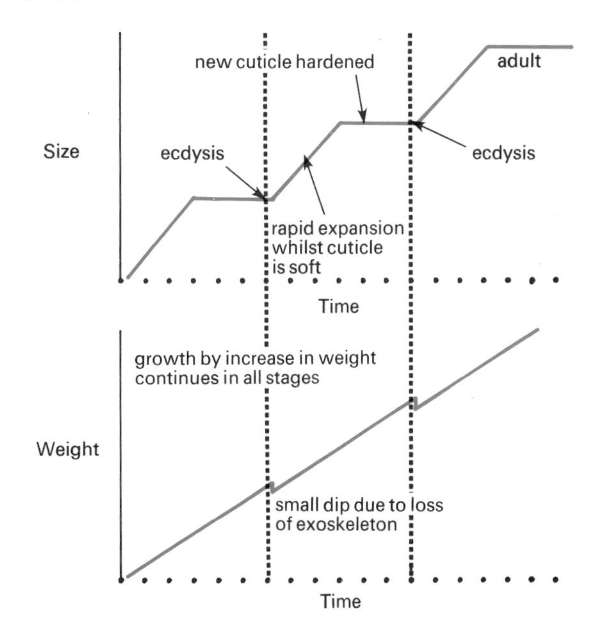

Metamorphosis in the insect is under the control of hormones. In the brain there are groups of neurosecretory cells which pass their secretions along nerve axons to two pairs of bodies, the **corpora cardiaca** and the **corpora allata**, which lie directly behind the brain. These and the **prothoracic** or **ecdysial glands** in the first segment of the thorax make up the insect's endocrine system. The corpora allata store and secrete the hormone **neotenin**, or **juvenile hormone**. This suppresses any tendency to develop pupal or adult structures and thus favours the growth and differentiation of larval structures. Neotenin is therefore essential for the larval moults of both hemimetabolous and holometabolous insects.

Neurosecretory cells in the brain also produce a trophic hormone (one which stimulates another endocrine gland) which is stored in the corpora cardiaca. From here it passes to the thoracic ecdysial glands where it stimulates the release of the hormone **ecdysone** or **moulting hormone**. This initiates moulting and favours the growth and differentiation of adult structures. Both neotenin and ecdysone are involved in the control of growth in the juvenile stages and in the metamorphosis into the adult.

The development of juvenile stages requires a relatively high level of both hormones. Before each moult there is a short-term reduction in the level of neotenin, which enables the moult to occur under the control of ecdysone. The level of neotenin then rises to maintain the next juvenile stage. The neotenin level is decreased permanently for pupation in Holometabolous insects and for the emergence of the adult in both Holometabolous and Hemimetabolous insects.

Metamorphosis in amphibia

Metamorphosis does not occur to the same extent in all amphibia, indeed in neotenous amphibia, such as *Necturus* (the mud puppy), it does not occur at all. It occurs to some extent in the Urodeles, e.g. newts, axolotls, and sala-

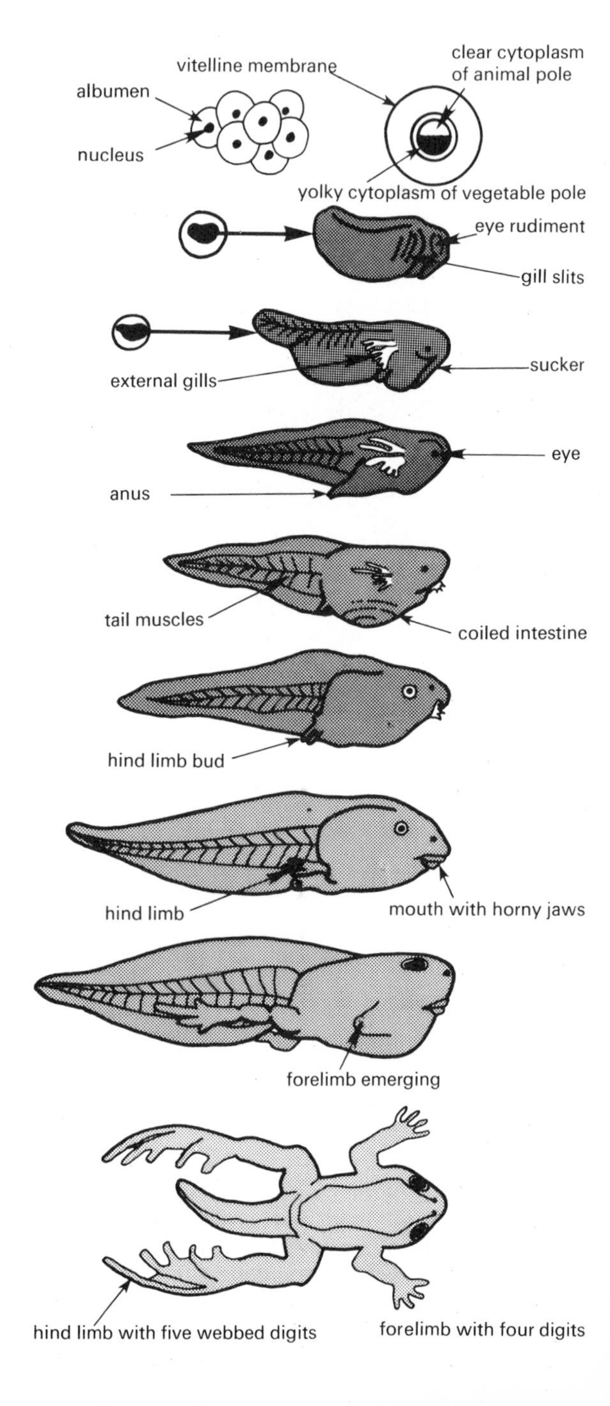

Figure 2.11 *The life cycle of the frog (Rana temporaria)*
Eggs are laid in about March. Fertilization is external, and the albumen swells with water to form the typical frogspawn. About two weeks after fertilization the tadpole emerges and breathes by means of external gills. The tadpole remains fastened to a surface by a ventral sucker. Later, the mouth forms horny jaws for scraping vegetation. At this stage it is purely vegetarian and has a long coiled intestine. Internal gills now develop, covered by an operculum and opening to the exterior by a spiracle. When about 10 weeks old the tadpole becomes carnivorous, the gut shortens, and the hind limbs develop. The forelimbs, developing inside the operculum, now emerge, the left forelimb through the spiracle and the right bursting through the operculum wall. About three months after fertilization the tadpole begins to breathe air. The end of metamorphosis is marked by the complete absorbtion of the tail.

manders, and is a major event in the life cycles of Anurans, e.g. frogs and toads. In amphibia the pituitary secretes the hormone **prolactin** which promotes the growth of tadpoles but retards the onset of metamorphosis into the adult. Under the necessary conditions the hypothalamus of the brain secretes a neurosecretion which stimulates the pituitary to secrete thyroid stimulating hormone (TSH). This in turn stimulates the thyroid gland to secrete **thyroxine**. The change in thyroxine secretion subsequently controls the metamorphosis of the larval tadpole into the adult.

Plant reproduction

Alternation of generations

In all sexually reproducing plants there is an alternation of **haploid** and **diploid** phases in the life history. The cells of the haploid phase contain one set of chromosomes and the cells of the diploid phase contain two sets of chromosomes. This alternation is most clearly seen in the Bryophyta,[349] Pteridophyta,[350] and Spermatophyta,[351] in which a haploid, sexually-reproducing **gametophyte generation** alternates with a diploid asexually-reproducing **sporophyte generation**. The haploid gametophyte produces gametes which fuse in fertilization to produce a diploid zygote. The zygote gives rise to a diploid sporophyte which produces haploid spores by meiosis.[60] The haploid spores germinate to give rise to the haploid gametic generation once again. This basic pattern can be traced right through to the development of the seed habit of the Spermatophyta.

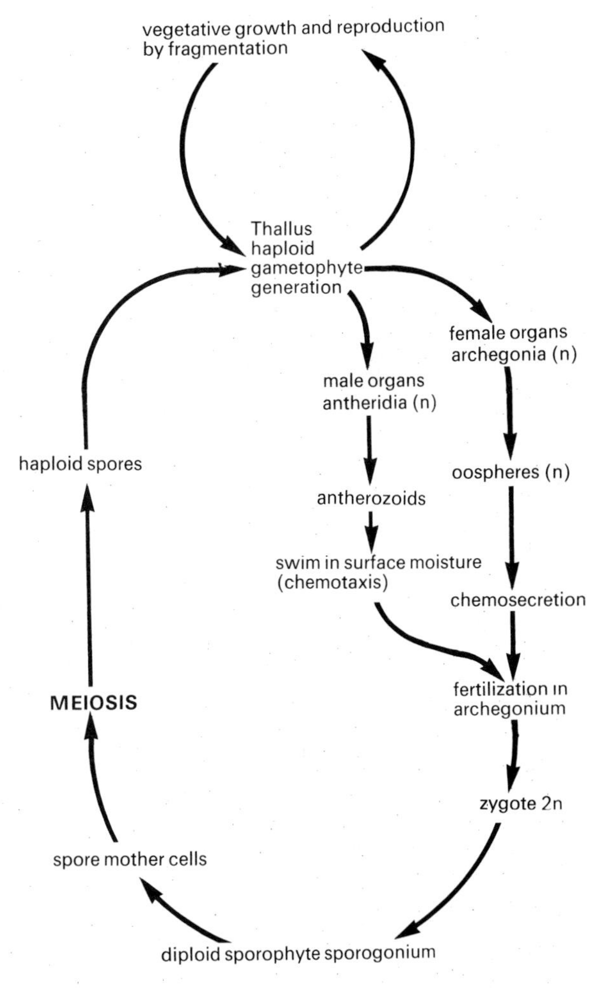

Figure 2.12 *Bryophyte life cycle (alternation of generations series)*

Figure 2.12 (continued)

(b) *Pellia* gametophyte (liverwort).

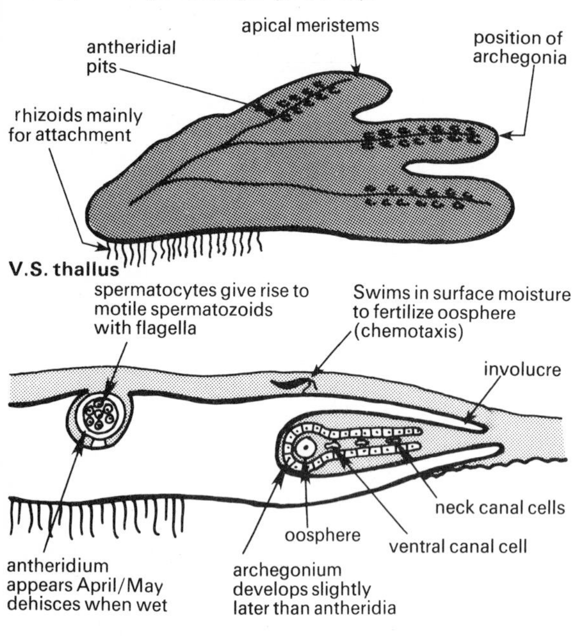

(c) *Pellia* sporophyte.

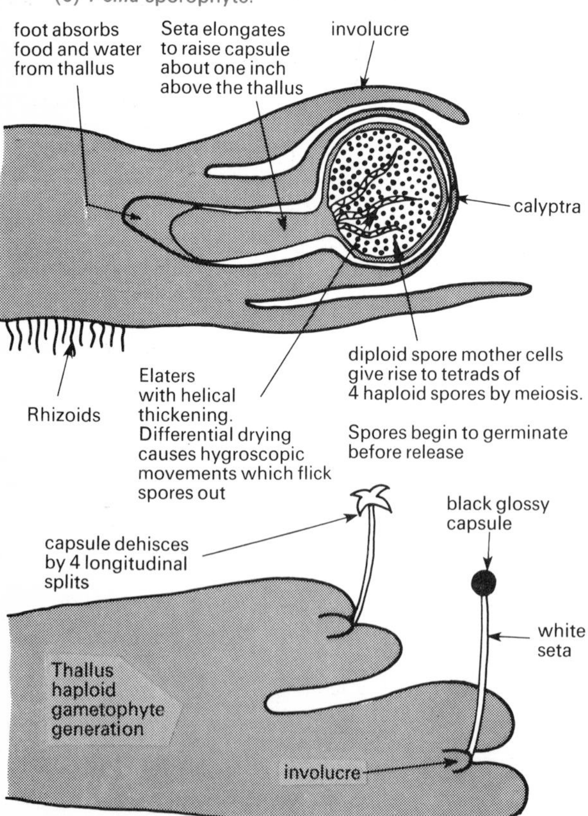

(d) *Funaria* gametophyte and sporophyte (moss).

The leaves of the gametophyte are not homologous with leaves of higher plants as they are of a different generation.

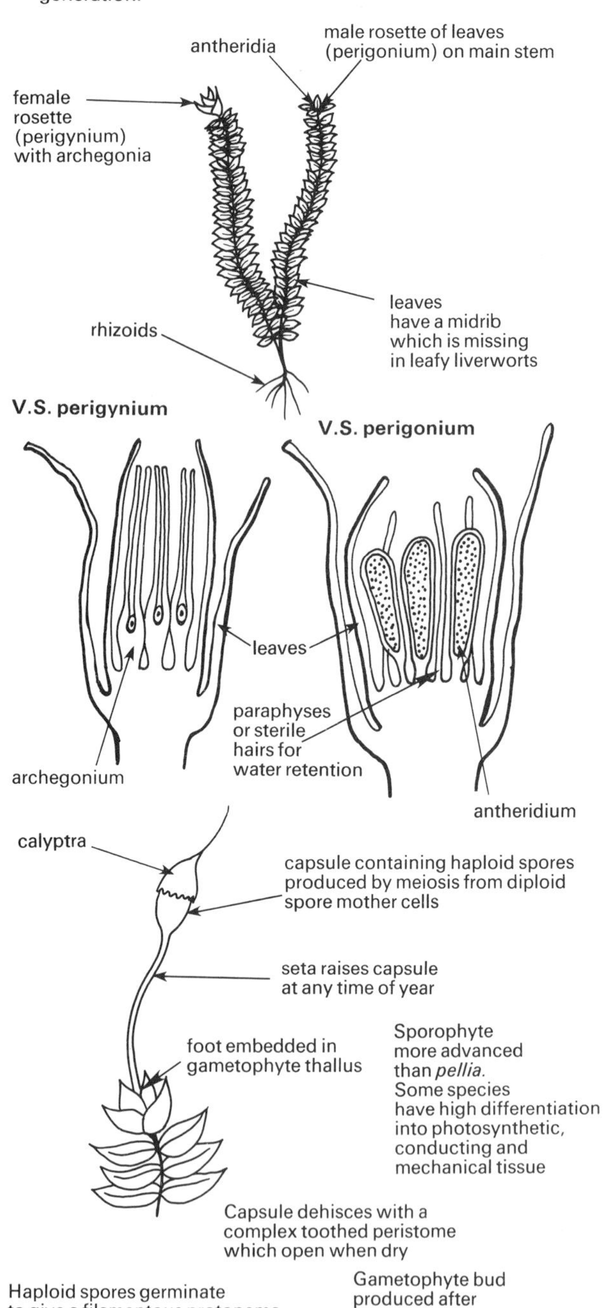

Figure 2.13 *Pteridophyte life cycle (alternation of generations series)*

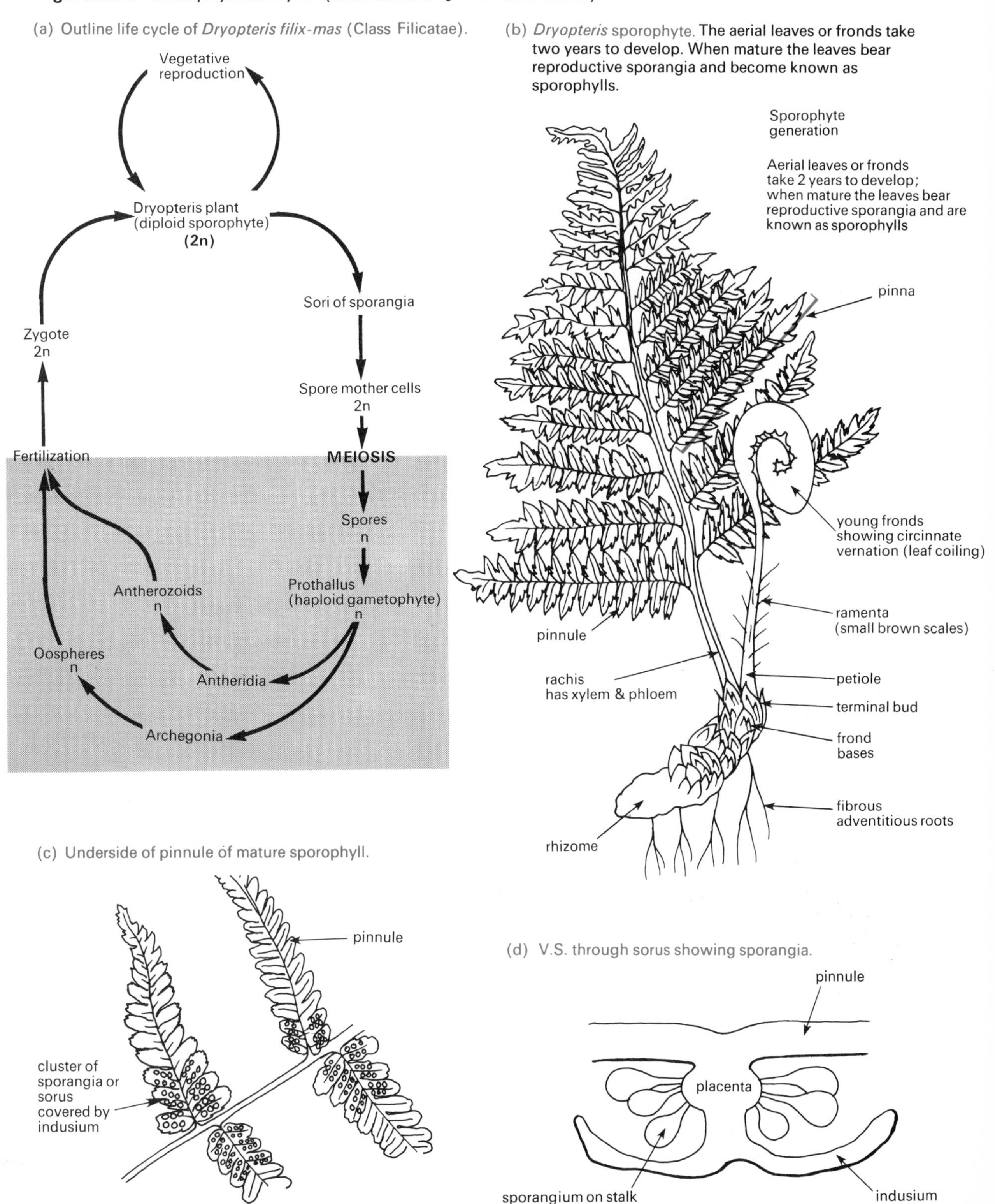

Figure 2.13 (continued)

(e) Sporangium structure and dehiscence. When mature the indusium withers and exposes sporangia to the drying effects of air currents. As the water evaporates the annulus cells are pulled in, due to the cohesive forces of water, and the thin walled stomium is split. Eventually the cohesion breaks and the recoil of the annulus throws spores out into the air currents.

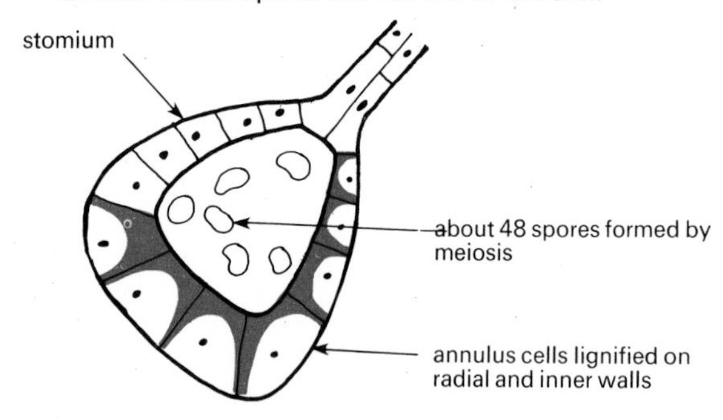

Side view sporangium

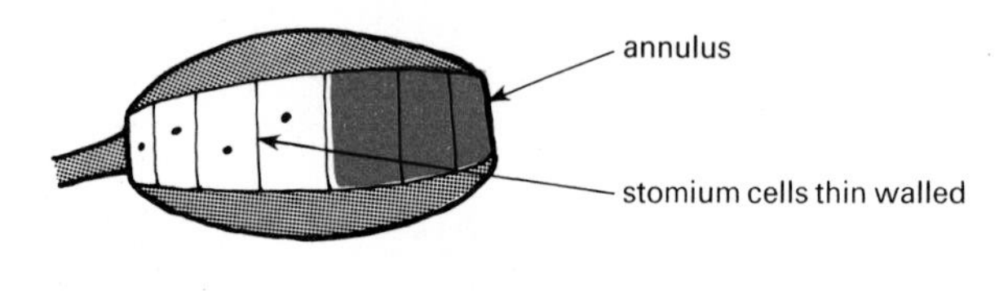

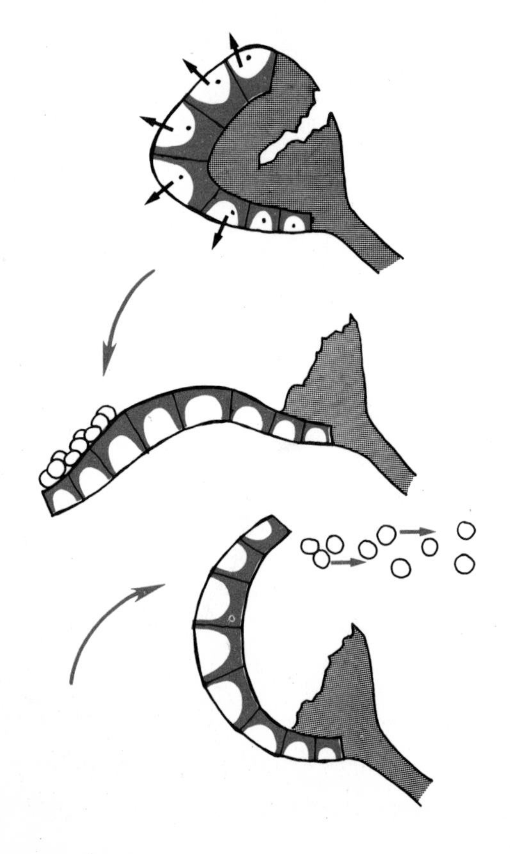

(f) Gametophyte prothallus ventral view (6–13 mm). This is less well adapted to life on land than that of many Bryophytes. The reproductive organs are on the lower surface.

Prothallus ventral view

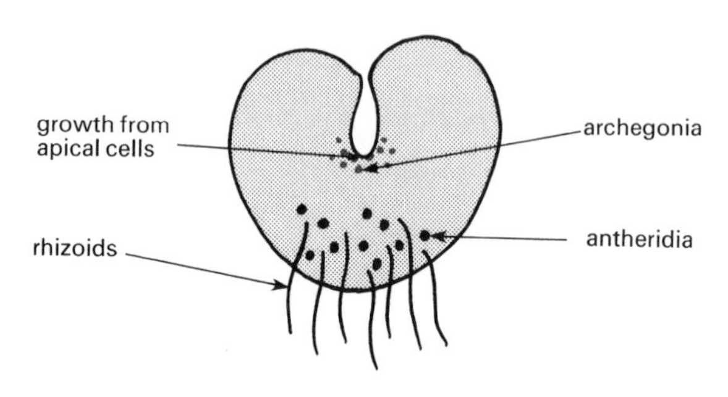

(g) Antheridia, archegonia, and fertilization.

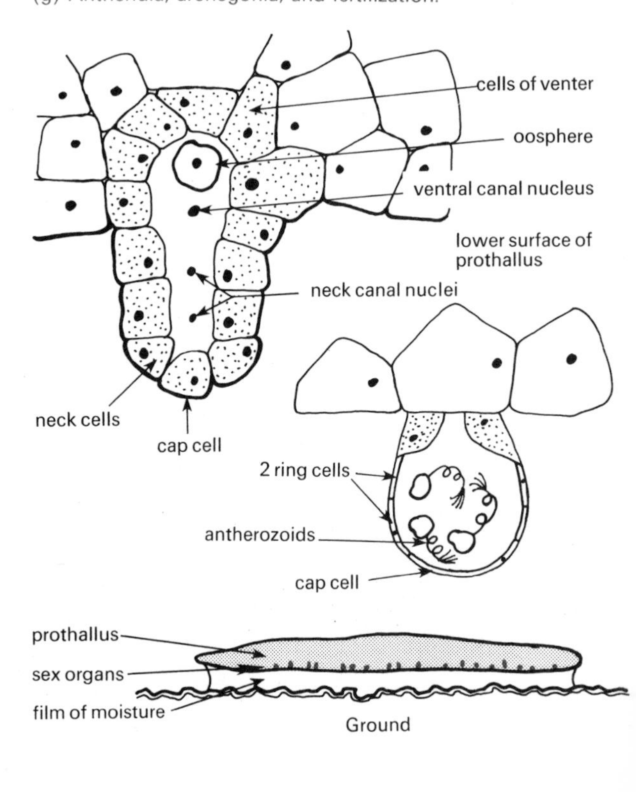

(h) Young sporophyte.

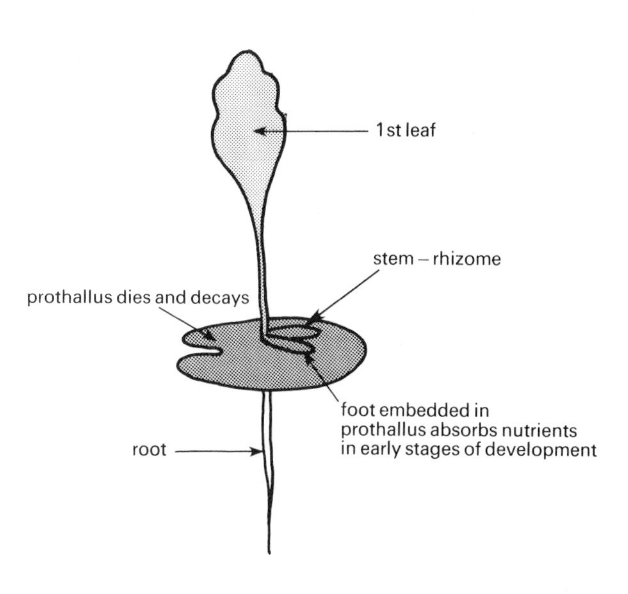

(i) Outline life cycles of *Selaginella* (Class Lycopodiatae).

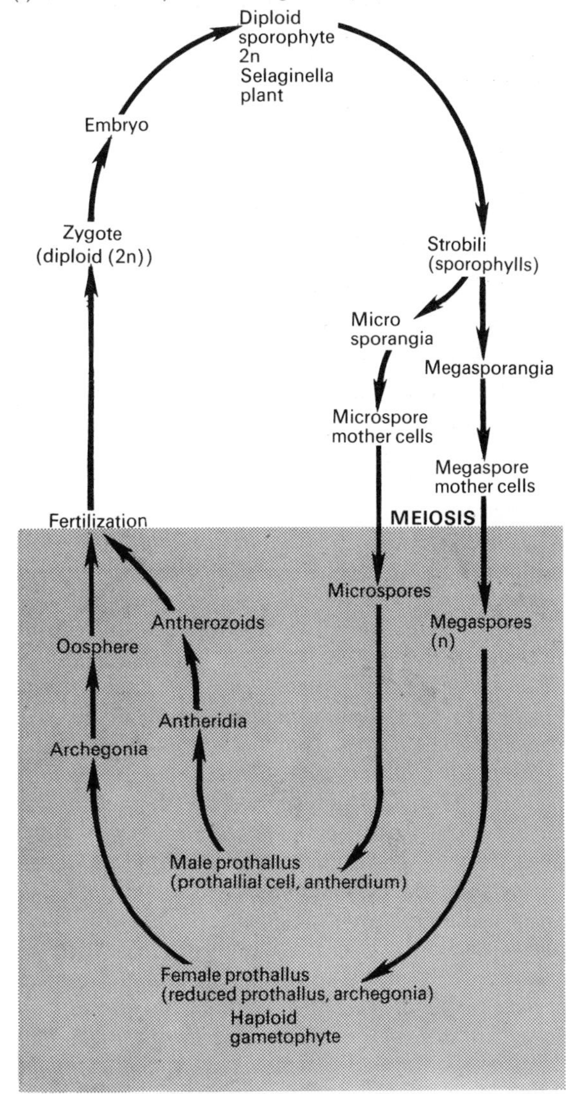

Tree ferns

Selaginella

DEVELOPMENT OF THE SEED HABIT

Some Pteridophytes, for example *Selaginella* in the Class Lycopodiatae,[350] are **heterosporous**, that is they produce two different types of spores: the larger **megaspores** and the smaller **microspores**. The megaspores give rise to female gametophytes that only bear archegonia, and the microspores give rise to male gametophytes that only bear antheridia. Both the megaspores and microspores begin their development into gametophytes whilst retained in the sporangia, which are borne on sporophylls grouped together to form primitive cones known as **strobili**. This development is dependent upon food reserves stored in the spores by the sporophyte plant, so that the gametophyte is dependent upon the sporophyte.

In *Selaginella* both gametophytes are eventually released from the sporophyte, and fertilization occurs by the motile antherozoids swimming in a film of moisture to the oospheres in the archegonia of the female gametophyte. This process is known as **zoidogamous** fertilization.

The seed-bearing plants, or Spermatophyta, which are the dominant terrestrial group of plants, do not depend on the presence of a film of water for a motile male gamete to achieve fertilization. Their gametophytes are very reduced and develop within the walls of the spores. The megaspore of the female gametophyte is completely retained by its sporophyte, on which it is structurally and physiologically dependent. The microspore and its male gametophyte, which is reduced to a single nucleus, is transferred as **pollen** to the megasporophyll bearing the megasporangium which contains the megaspores, in the process of **pollination**. The microspore, or pollen grain, develops a pollen tube down which the male gametes pass to the oospheres in a process of **siphonogamous** fertilization. Thus fertilization is no longer dependent on the presence of free water and shows a greater adaptation to a terrestrial existence than is seen in the non-seed bearing plants. Siphonogamous fertilization is seen in most Gymnosperms[351] and all Angiosperms.[351] The growth of the pollen tube depends to a certain extent on nutrients derived from the megasporangium tissues.

The fertilized oospore develops into the embryo sporophyte of the next generation within the megasporangium walls which form the protective coverings or **integuments** of the structure that is now known as the **seed**. In Gymnosperms the seed remains exposed on the surface of the megasporophyll, but in Angiosperms the megasporophyll encloses the seed in an ovary. The ovary wall develops into the **fruit** after fertilization.

Figure 2.14 *Spermatophyte life cycles (alternation of generations Gymnosperm series)*

(a) Outline life cycle of *Pinus sylvestris* (Class Pinatae).

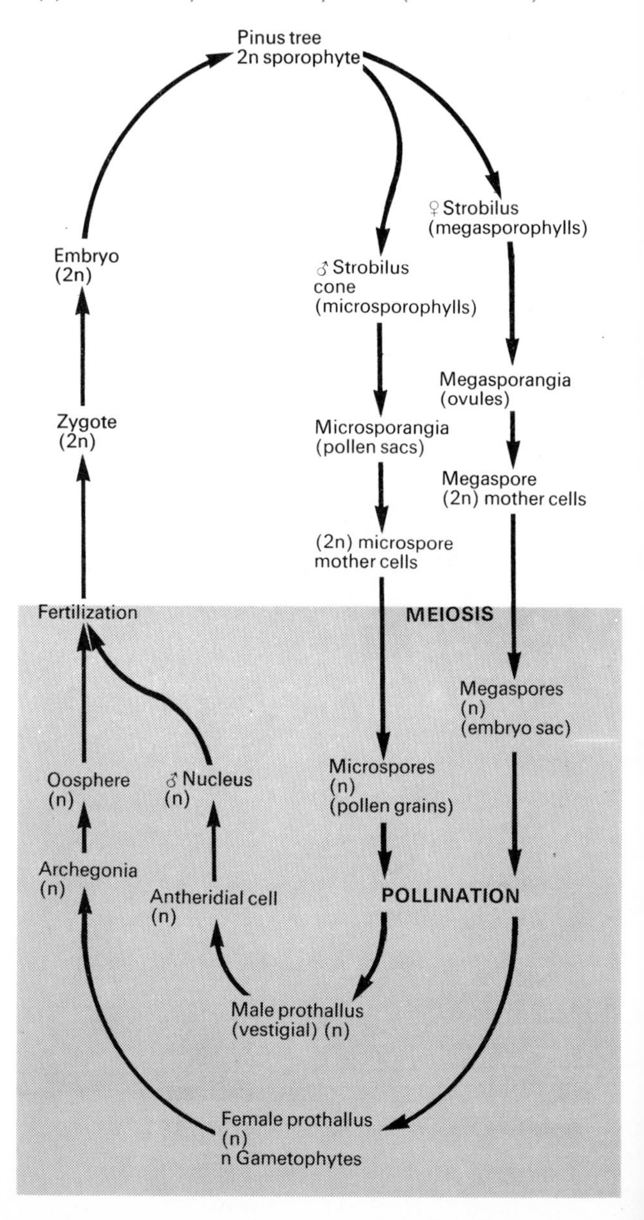

(b) Staminate (male) cone. Each cone is about 6 mm long, with spirally arranged microsporophylls bearing microsporangia or pollen sacs.

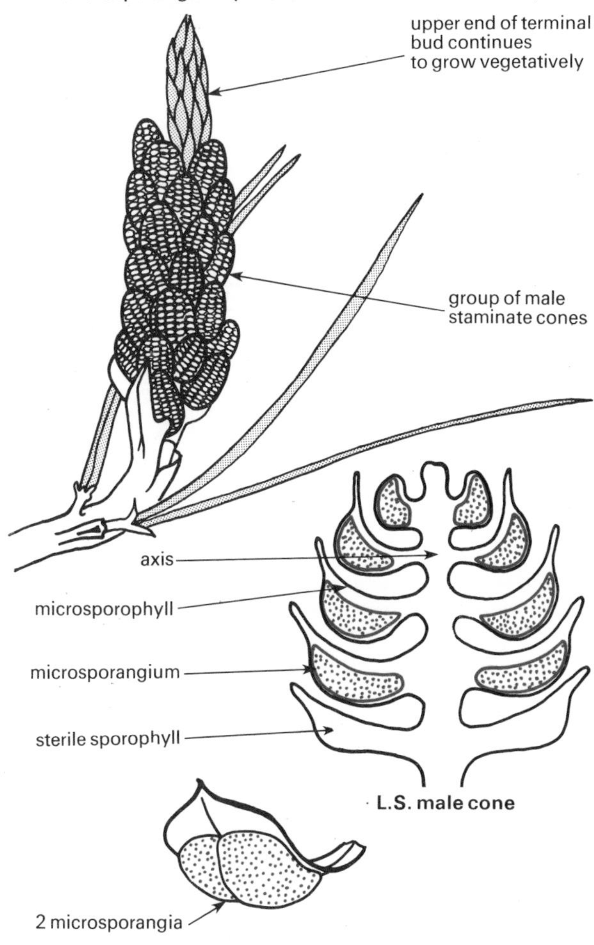

(d) Ovulate (female) cone. During the first year of development each ovuliferous scale develops two ovules.

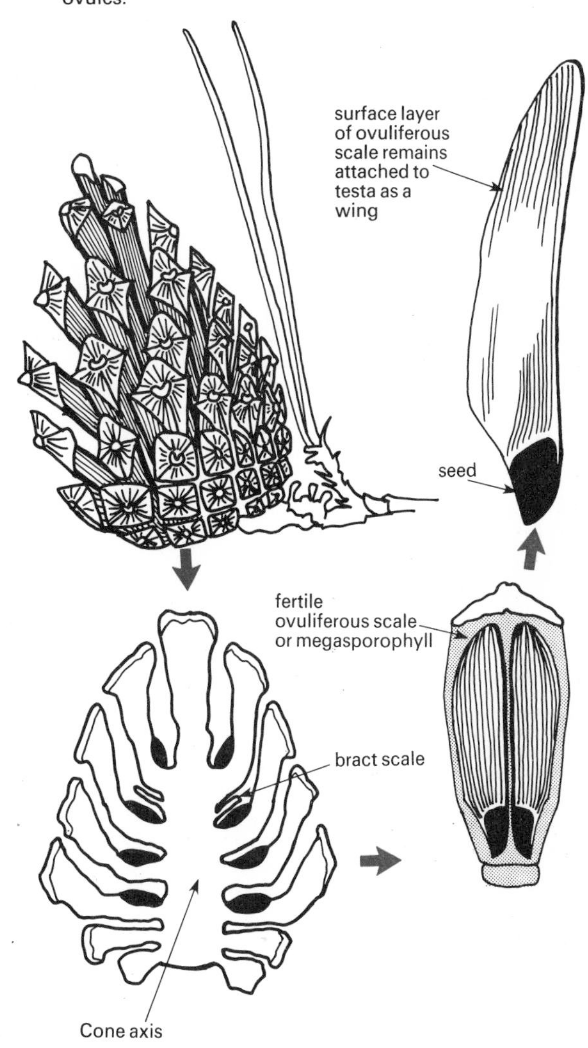

(c) Microspore or pollen grain. The vestigial male gametophyte begins to develop inside the pollen grain before being shed.

microspore or pollen grain

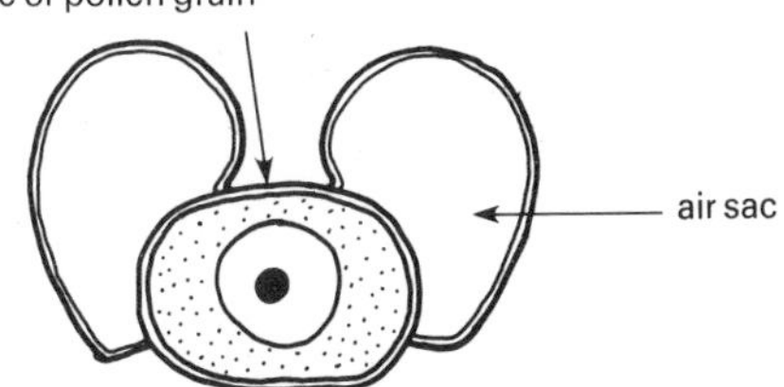

antheridial cell nucleus

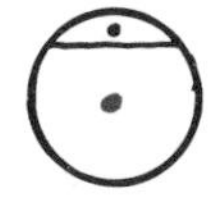

1 vegetative cell

2 vegetative prothallial cells

♂ gametophyte

pollen tube nucleus

Figure 2.14 (continued)

(e) L.S. ovule prior to pollination. This stage is reached between May and June, at which time the cones are about 2·05 cm long and green.

megasporangium wall
position of archegonial development
integument
micropyle
nucellus
Megasporophyll
wall of megaspore
♀ prothallus
bract scale

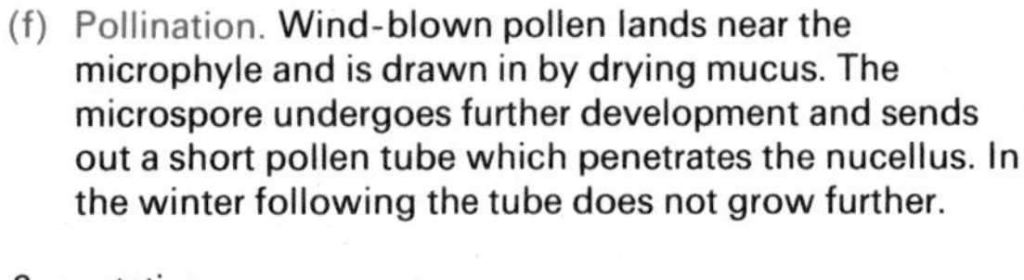

(f) Pollination. Wind-blown pollen lands near the microphyle and is drawn in by drying mucus. The microspore undergoes further development and sends out a short pollen tube which penetrates the nucellus. In the winter following the tube does not grow further.

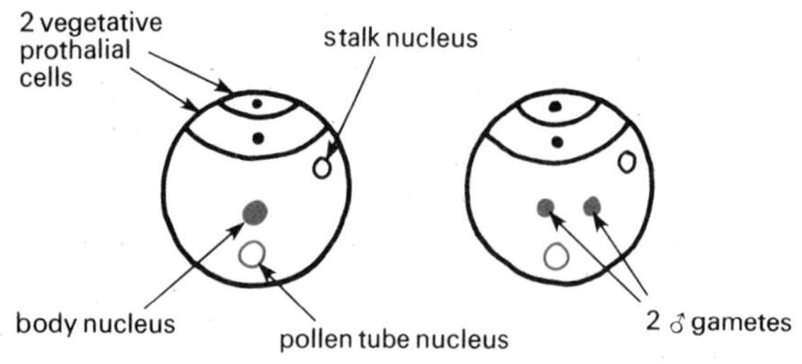

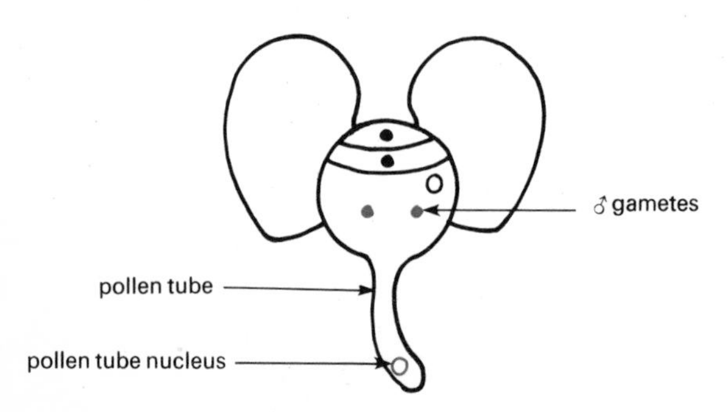

(g) Fertilization. Two or three large oospheres (0·5 mm across, which is 10 times the size of Angiosperm egg cell) develop in the short-necked archegonia. The following May the pollen tube continues growth and a male gamete fuses with an oosphere to form a diploid zygote.

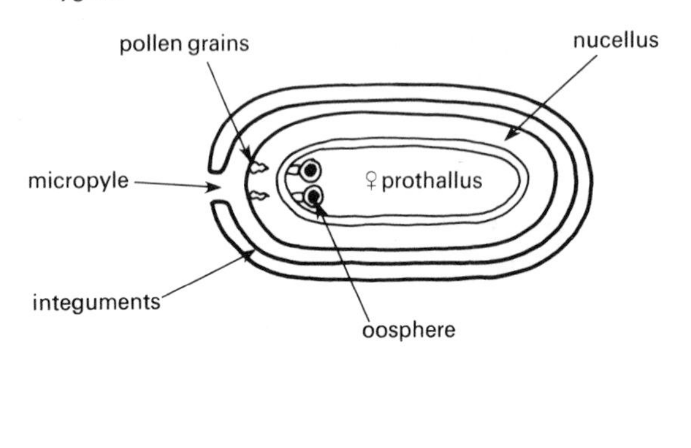

(h) L.S. seed. In dry weather, the cone scales open and the winged seed is carried on air currents, about two years after pollination.

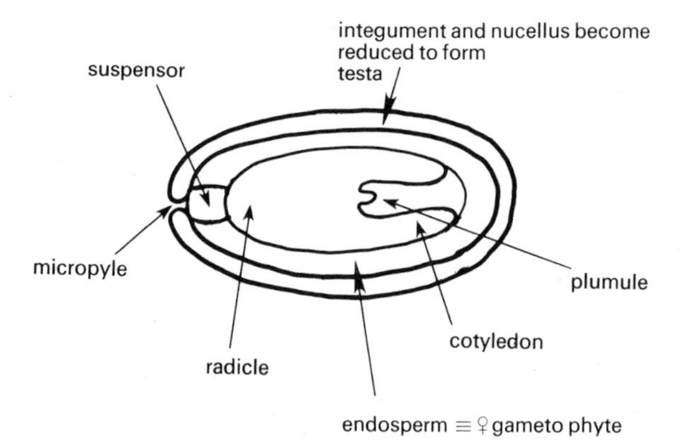

Figure 2.15 *Spermatophyte life cycles (Angiosperm series)*

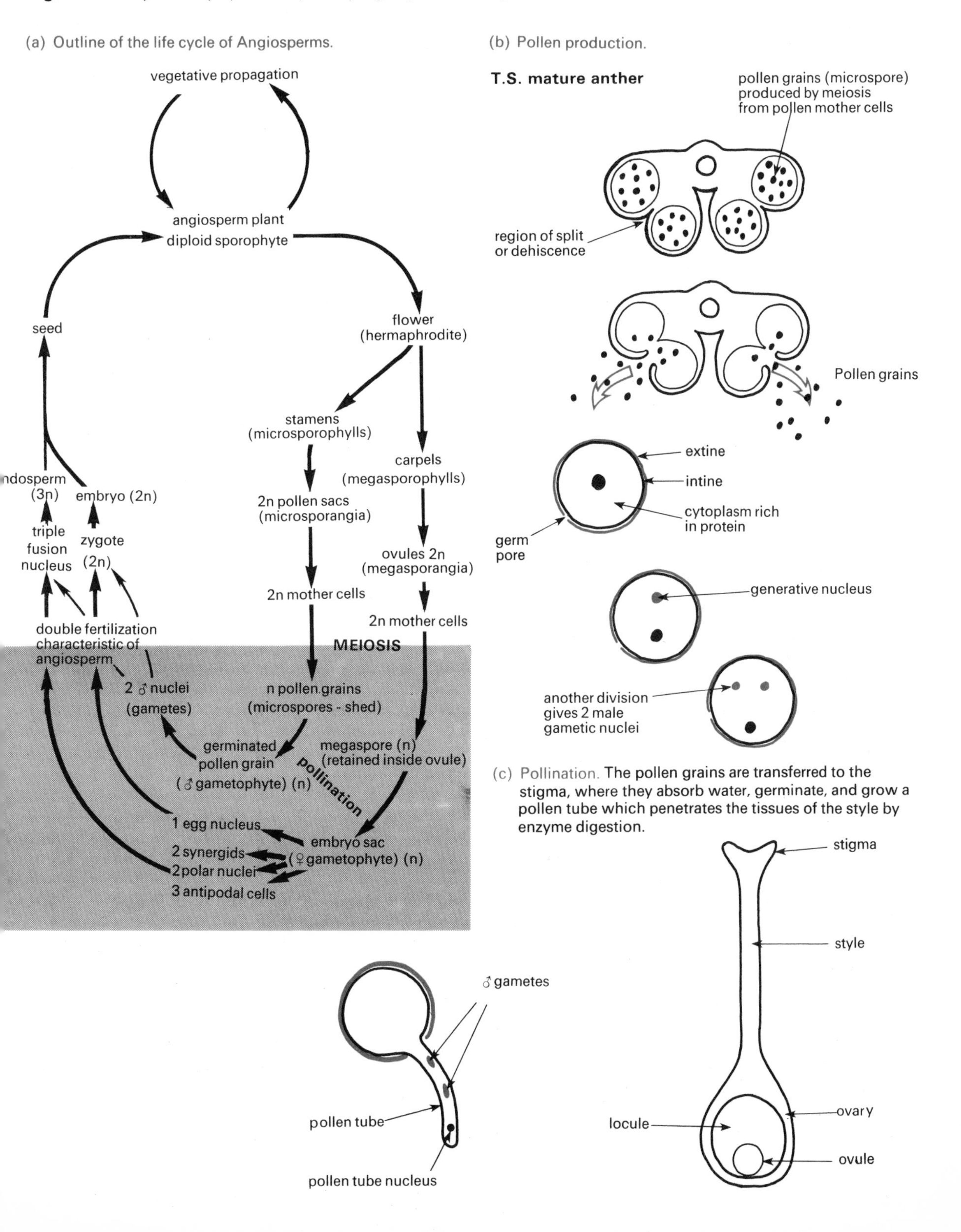

Figure 2.15 (continued)

(d) Development of the ovule.

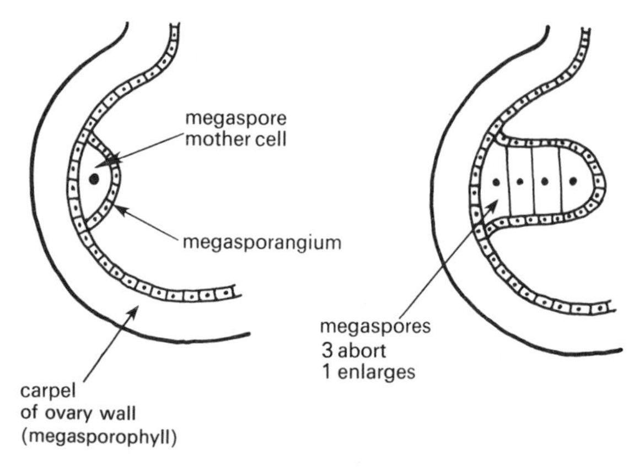

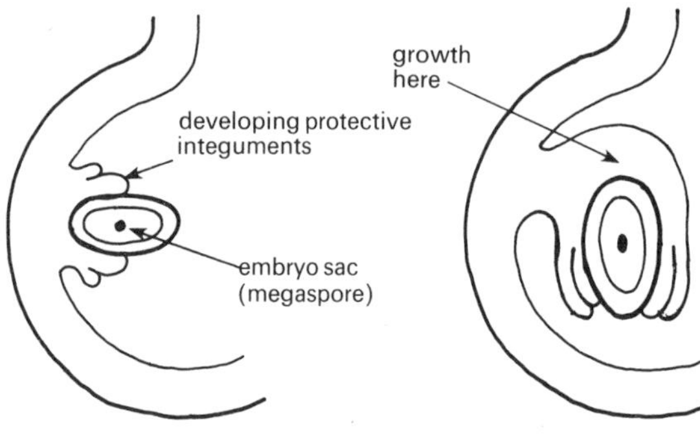

(e) Development of the embryo sac and fertilization. One male gametic nucleus fuses with the ovum nucleus, and the second male nucleus fuses with the two polar nuclei, in a process of **double fertilization** which is unique to the Angiosperms. The triploid 'triple fusion nucleus' can divide to give the endosperm food store tissue of endospermous seeds, or it can degenerate, as in non-endospermous seeds. The fertilized ovum develops into the embryo plant. The integuments develop into the seed coats or testa. The ovary wall develops into the pericarp or fruit wall.

Division of embryo sac nucleus

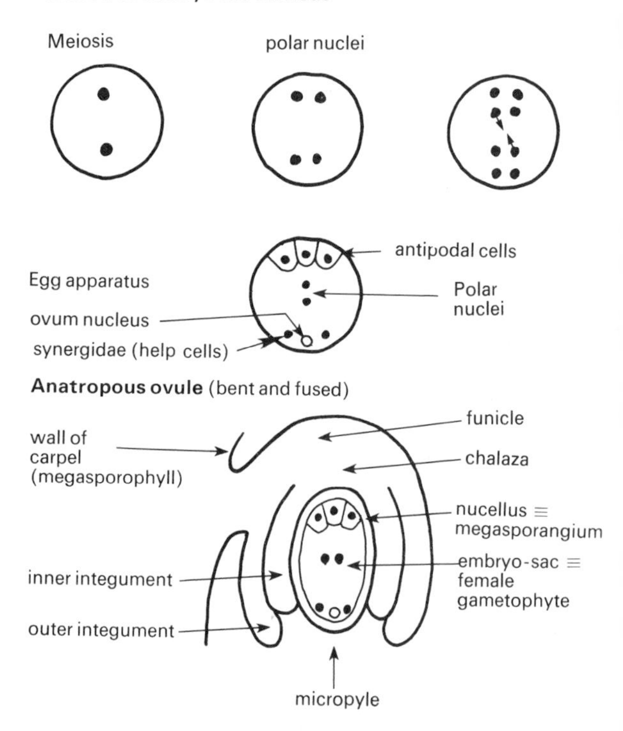

Pollination

STRUCTURE OF THE FLOWER

The flower is a part of the shoot modified for reproduction. It is usually borne on a stalk or **pedicel**. The flower parts are arranged on a swelling of this stalk, the **receptacle**. Generally the receptacle bears four sets of floral parts: the calyx of **sepals**, the corolla of **petals**, the androecium of **stamens**, and the gynoecium of **carpels**.

The **calyx** of sepals form the lowest and outermost whorl of floral parts. Typically their function is protection of the other floral parts during the bud stage. They are usually green, but may sometimes be coloured or petaloid.

The **corolla** of petals forms the next whorl of floral parts. These are usually brightly coloured and sometimes scented to attract insects, but in wind-pollinated flowers they may be green, reduced, or even absent. The whorls of sepals and petals are known together as the **perianth**.

The stamens forming the **androecium** are the male reproductive structures and may be arranged in a variety of ways in either purely male staminate flowers, or in hermaphrodite flowers. The carpels of the **gynoecium** are the female reproductive structures and may also be arranged in a variety of ways in either purely female carpellary (pistillate) flowers, or in hermaphrodite flowers.

Flowers may be solitary, but most plants have a special branching system, bearing a number of flowers, distinctly marked off from the vegetative region. This system is known as an **inflorescence**.

Figure 2.16 *Floral structure*
Flowers with separated or free floral parts are considered to be less highly developed than those with fused and often reduced floral parts. Dicotyledons usually have floral parts in fours or fives or multiples of these numbers. Monocotyledon flowers usually have floral parts in threes or multiples of threes. Flowers can be single-sexed or hermaphrodite.

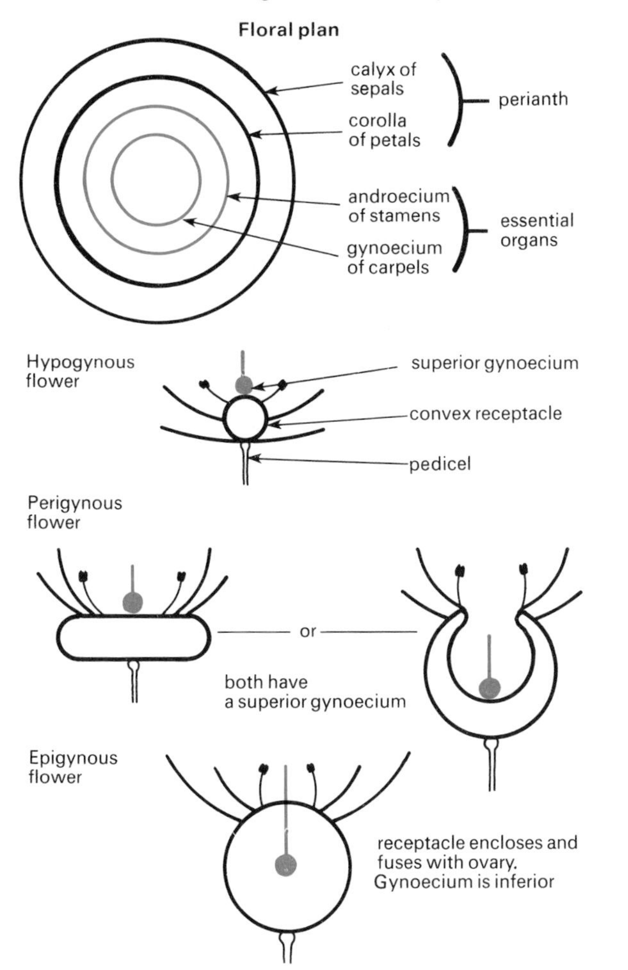

Figure 2.17 *Floral structure and floral formulae*
A floral diagram is a 'ground plan' of the flower showing the relative positions of the parts.

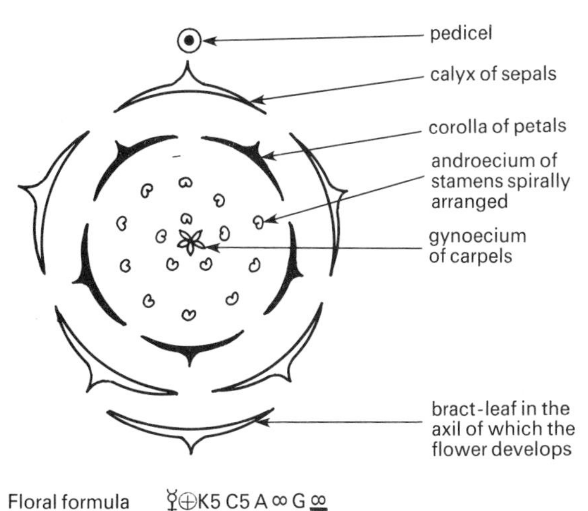

Floral formula ☿⊕K5 C5 A ∞ G $\underline{\infty}$

Derivation of floral formulae:

⊕—radially symmetrical (actinomorphic)
†—bilaterally symmetrical (zygomorphic)
♂—male flowers (staminate)
♀—female flowers (carpellary or pistillate)
☿—hermaphrodite ('perfect'—the rule in Angiosperms)
P—perianth (if sepals and petals not distinct)
K—calyx of sepals
C—corolla of petals
A—androecium of stamens (if fused to petals called epipetalous)
G—gynoecium of carpels
(5)—parts joined
K(5)C(5)—whorls joined (in this case parts also joined)
$\bar{G}^5$—superior ovary (in hypogynous and perigynous flowers)
$\bar{G}_5$—inferior ovary (in epigynous flowers)
∞—numerous parts

METHODS OF POLLINATION

Pollination is the process by which pollen is transferred from the anther lobes of the stamen to the stigma of the carpel. Flowers may be self-pollinating or cross-pollinating.

Self-pollination
Self-pollination occurs when the stigma receives pollen from the stamens of the same flower. This is only possible in hermaphrodite flowers when the stamens and carpels mature at the same time. The stamens are often arranged so that the pollen can fall on to the stigma(s). Many self-pollinating flowers, e.g. the garden pea, self-pollinate in the bud stage before the flower opens. Indeed, in some the flower never opens and is eventually destroyed by the development of the fruit.

Cross-pollination
In cross-pollination the stigma receives pollen from the stamens of a different flower, which may either be on the same or on a different plant. With single-sex flowers cross-pollination is inevitable; but with hermaphrodite flowers a variety of mechanisms are seen which favour cross-pollination and decrease the chances of self-pollination.

Stamens and carpels often mature at different times in the same flower, although overlapping frequently occurs. In some plants, e.g. dandelion, the anthers mature first (**protandry**), and in others, e.g. horse-chestnut, the carpels mature first (**protogyny**). Some flowers have special arrangements of their parts to prevent self-pollination. For example, the iris has a stigmatic flap on its petalloid style which prevents pollen on the back of a withdrawing insect from coming into contact with the stigma; the viola has an arrangement of its floral parts which ensures that the stigma is exposed to the incoming insect, but is covered as the insect leaves. Other plants, including many orchard fruits, have genetically controlled self-sterility incompatability mechanisms, by which the growth of the pollen tube of their pollen is slowed or stopped on their own stigma.

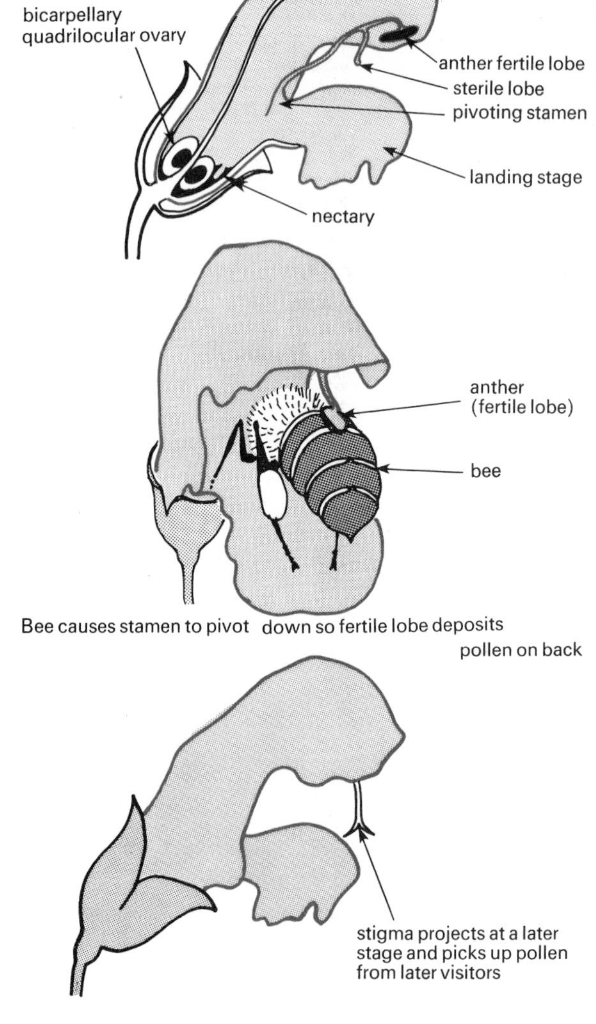

Figure 2.18 *Insect-pollinated (entomophilous) flower: garden sage*

POLLINATING AGENCIES

Cross-pollination can be carried out by a variety of agencies, including insects, wind, water, and even small birds, and mammals such as bats.

Insect-pollinated or **entomophilous** flowers have certain features in common. They are often large, or, if small, they are concentrated into large inflorescences. They are usually scented and possess nectaries. They are also usually highly coloured, especially with blue or purple pigments; however, insects are also attracted to white flowers due to the reflected ultra-violet light which they are able to detect. The arrangement of the floral parts is also designed to assist and to 'fit' certain insect visitors. For example, open white flowers with a food supply available to short-tongued insects are pollinated by flies. Blue or purple flowers with deep nectaries designed to be reached by insect mouthparts 7–12 mm long are pollinated by bees. Red flowers with even deeper nectaries are pollinated by butterflies, and night-flying moths are attracted to pale-coloured flowers with deep nectaries. So-called carrion flowers with dull brown colours and odours offensive to man are pollinated by insects, such as flies, that naturally feed on dung. The pollen grains of entomophilous flowers are sticky, spiny, and large, being adapted to cling to the insect visitors.

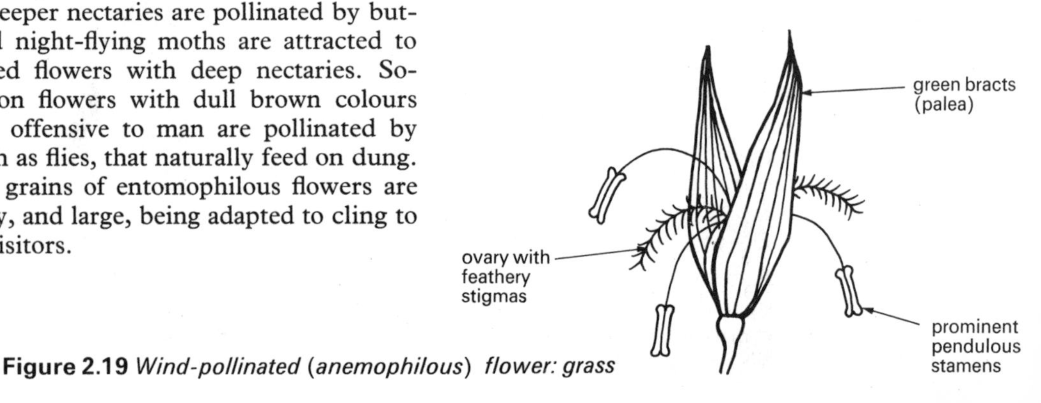

Figure 2.19 *Wind-pollinated (anemophilous) flower: grass*

In contrast, wind-pollinated or **anemophilous** flowers tend to be small and inconspicuous with the petals reduced, or absent, and no nectaries. They are often unisexual, with an excess of male flowers. Large quantities of light, dry, smooth pollen grains, sometimes with air sacs, are produced in hanging stamens that protrude from the flower. The stigmas also protrude and are large, sticky, and feathery to trap the wind-borne pollen. In temperate climates such flowers often develop before the emergence of the leaves in the spring.

Submerged flowering plants can show interesting pollination mechanisms. For example the Canadian pondweed (*Elodea canadensis*) produces a solitary female flower which is raised to the surface on a long stalk, so that the stigma is exposed to the air. At the same time, many male flowers are produced which break off and float to the surface, where they open and drift against the female flowers so that the anther lobes touch the stigma. In this way pollination occurs without the pollen getting wet. The stalk of the female flower then contracts and draws the fruit and seeds down under the water, where they germinate.

Fruits and seeds

After pollination and fertilization the ovule develops into the **seed** and the wall of the ovary forms the **pericarp** or wall of the fruit. Sometimes the receptacle is also involved in the formation of the fruit, which in these cases are known as **false fruits** or pseudocarps.

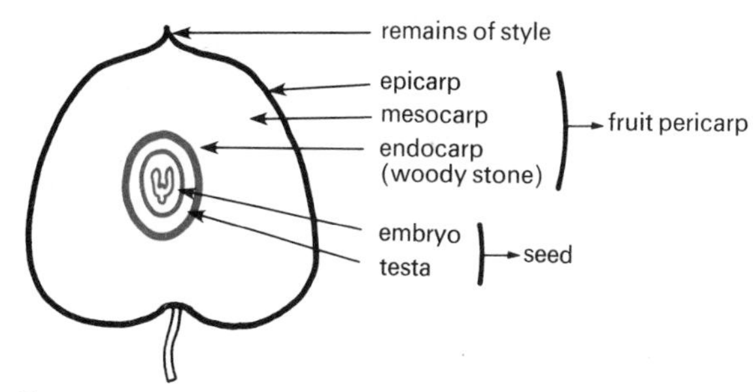

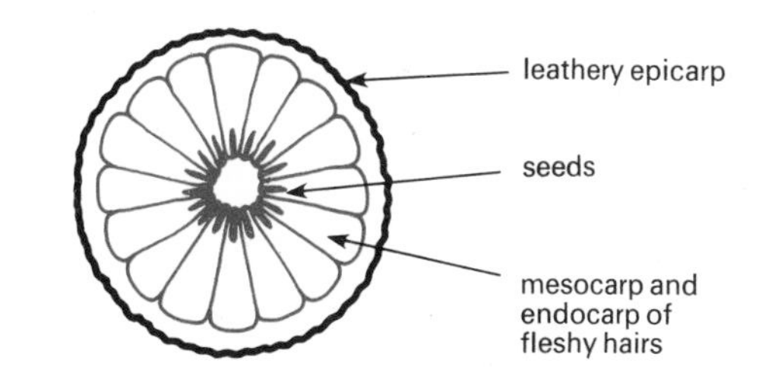

Figure 2.21 *Types of succulent fruit*
Drupe: fleshy fruit formed from monocarpellary gynoecium seed enclosed by a hard endocarp.
Berry: has no stony endocarp round seed.

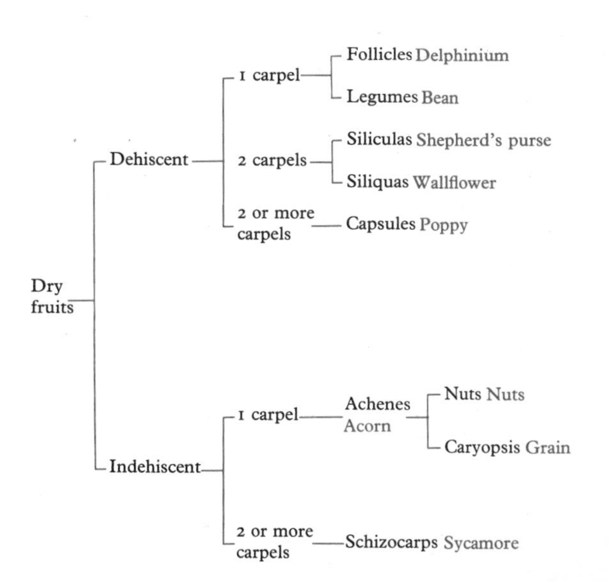

Figure 2.20 *Types of dry fruit*
Follicles: pod formed from a single carpel, splits down one edge. **Legumes**: pod formed from a single carpel, splits down both edges. **Siliqua**: special type of capsule, two carpels divided by central false septum. **Silicula**: similar structure to siliqua but short and broad. **Capsule**: many carpels, open by slits, pores, or teeth. **Achene**: one-seeded type of indehiscent follicle, almost always grouped. **Nut**: achene in which part of the fruit becomes woody. **Caryopsis**: achene in which the pericarp is fused to the testa. **Schizocarps**: many-seeded fruits which split into a number of one-seeded parts.

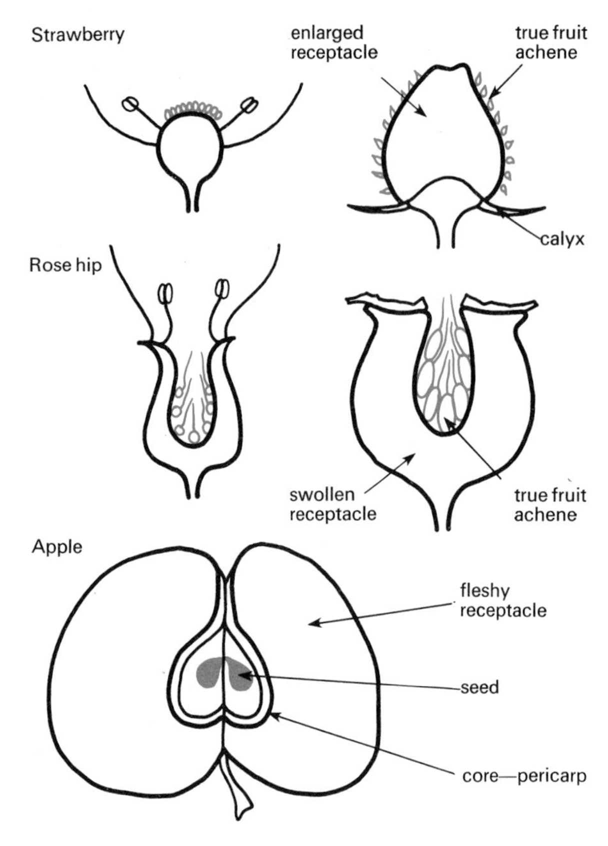

Figure 2.22 *False fruits*

DISPERSAL

There are a variety of dispersal methods of fruits and seeds, including mechanical dispersal, wind dispersal, and dispersal by animals.

Mechanical dispersal includes explosive fruits in which the drying of the fruit wall results in violent movements which disperse the seeds. An example is the pods of leguminous plants.

Wind-dispersed fruits are often winged, as in ash, elm, sycamore, and pine, or plumed, as in the dandelion, to aid their buoyancy. Wind-dispersed seeds are often small enough to be blown as dust. There are also censer mechanisms in which seeds are flung out of a perforated capsule which shakes in the wind, as in the poppy.

Water dispersal involves special water-resistant floating seeds and fruits which often have air cavities and special buoyant outgrowths.

Animal dispersal involves a wide variety of types of fruits and seeds. Many have hooked fruits and seeds designed to attach to animal coats. Succulent fruits attract animals, especially birds which, after feeding on the fruit, transport the seeds either stuck to their feet or beaks, or by swallowing the seeds and spreading them in their faeces.

SEED STRUCTURE

Seeds can be classified with respect to the presence or absence of endosperm food storage tissue, and with respect to the position of the cotyledons during germination. These either remain below ground (**hypogeal**) or are brought above the surface by the growing plumule (**epigeal**). Most dicotyledon seeds are non-endospermic, and most monocotyledon seeds are endospermic.

The embryo consists of the plumule, radicle, and cotyledon(s) or seed leaves. The **plumule** is the young shoot and is usually bent over so that the apical meristem at its tip is protected from damage as it pushes up through the soil. The **radicle** is the young root and is the first part of the embryo to start active growth, since it is necessary to exploit the soil solution for water and nutrients as soon as possible. Dicotyledonous plants have two **cotyledons** which act as the main food reserve, and monocotyledons have one which aids in the mobilization of the main food reserves in the endosperm. The **testa**, or seed coat, has several layers of thickened water-proofed cells which resist decay and the entry of water.

DORMANCY

When seeds fail to germinate in apparently suitable conditions they are said to be dormant. Dormant seeds have dense cytoplasm with a low water content and very low metabolic rate. Some cultivated seeds germinate immediately if conditions are right, but others are unable to germinate, even in optimum conditions, until after several weeks of dry storage. Weed seeds in particular show intermittent germination in at least two separate waves, thus some seeds germinate immediately whilst others remain dormant even under optimum conditions. In shepherd's purse (*Capsella bursa-pastoris*) there is an interval of several months between the two waves. Such intermittent germination contributes to the success of weeds, as it protects against the destruction of all seedlings at any one time.

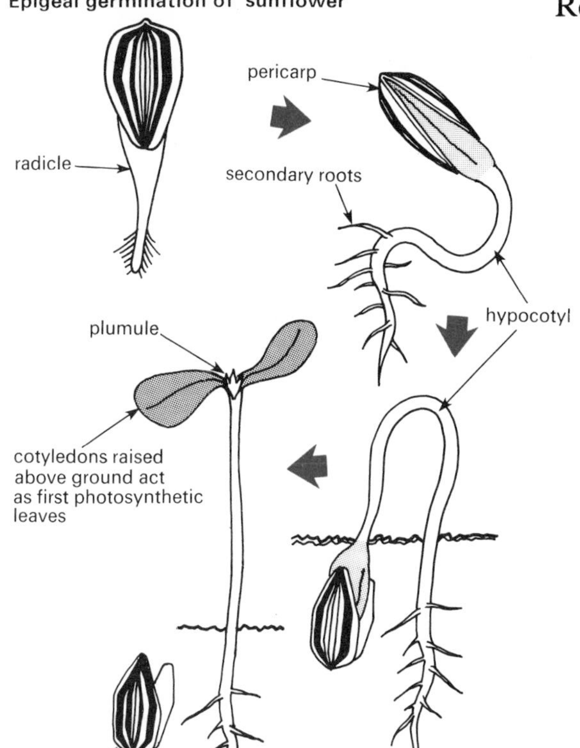

Figure 2.23 *Hypogeal and epigeal germination*

Dormancy allows for the dispersal of the seeds in distance and in time. Under certain conditions seeds can lie dormant for many years. Periods of up to ten years are common, wheat can be dormant for fifty years, and figures up to 2000 years are quoted for lotus seeds. However, with long periods of dormancy the percentage of seeds that germinate decreases. For example, although shepherd's purse seeds can remain dormant for thirty-five years, there is only a 47 per cent viability after sixteen years.

Germination

All seeds require moisture, oxygen, and a suitable temperature for germination. Some seeds also require exposure to orange-red light before they will germinate. This is thought to trigger the formation of some growth substance, possibly gibberellic acid.[223] In others, however, light inhibits germination. Fruits can contain germination inhibitors which prevent seeds germinating in the fruit, and sometimes the testa itself contains an inhibitor which must be dissolved out by water before the seed will germinate. In other species, mechanical damage of the testa is required before germination can occur. Once germination begins there are two major phases: the hydration phase and the metabolic phase.

The **hydration phase** involves the entry of water and the subsequent hydration of the dense dehydrated cytoplasm. As a result, the cytoplasm regains its activity, and germination enters the metabolic phase. In the **metabolic phase**, enzymes are activated and food reserves are rendered soluble and available by hydrolysis.[364] The metabolic phase requires a suitable temperature, the elimination of carbon dioxide, and, in some cases, red light. The phase is characterized by high respiration rates and the synthesis of proteins in the growing tips.

Fats, oils, and starch are used as respiratory substrates to provide energy for the synthetic reactions. Seeds have complex respiratory quotients[161] (RQ) as various reserves are mobilized and inter-converted. Proteins are broken down into amino acids which are translocated by diffusion to the active growing region where they are resynthesized into cell proteins. Glucose is used in the synthesis of cellulose for the new cell walls. The hormones gibberellin and cytokinin[223] are involved in the complex control of this metabolic phase.

The food reserves laid down by the parent plant for the developing embryo carry the young plant through its first surge of growth, prior to the formation of the roots and the first green leaves. As the plant grows it develops its characteristic structure.

HERBACEOUS PLANTS

Herbaceous plants generally lack secondary thickening, although a certain amount may occur. All annuals and biennials are herbaceous, as are some perennials such as grasses, lupins, etc.

SHRUBS

Shrubs are usually larger than the herbaceous plants and have woody secondary thickened aerial parts that last for several years. They are generally distinguished from trees by being smaller and branching close to the ground, although there is no exact distinction between the two; both shrubs and trees are woody perennials.

TREES

Trees are taken as being larger than shrubs, with a woody main stem or trunk which does not typically branch close to the ground in the mature form.

Length of life of plants

EPHEMERALS

Ephemerals can complete their life cycle within a period as short as 5–6 weeks, therefore several generations can be produced in one year, e.g. groundsel (*Senecio vulgare*).

ANNUALS

Annuals complete their life cycle within a year and pass the non-growing season in the form of dormant seeds, e.g. maize.

BIENNIALS

Biennials can complete their life cycle within two years. They are characteristic of temperate zones, where the vegetative growth and food storage completed in the first year give them an early advantage in their second year, so that they can grow rapidly to produce flowers and seeds ahead of many of the annuals. The underground food storage organs of biennials are often exploited as a food source by man, e.g. carrot, beet, turnip, etc.

PERENNIALS

Perennials live for many years and most flower and set seeds each year, although this may not occur until a certain amount of vegetative growth has occurred over a period of years.

Growth of dicotyledons

The stem

The stem apex consists of the apical meristem which is divided into two regions, the tunica and the corpus. Tunica cells give growth in area. Corpus cells give growth in volume.

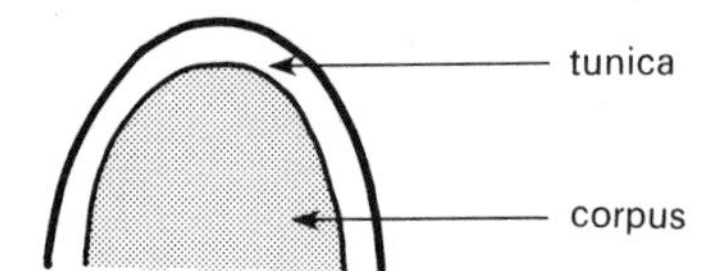

Figure 2.24 *L.S. apical meristem of shoot*

Immediately behind the apex, before any differentiation of tissues occurs, swellings arise which develop into leaves and lateral buds.

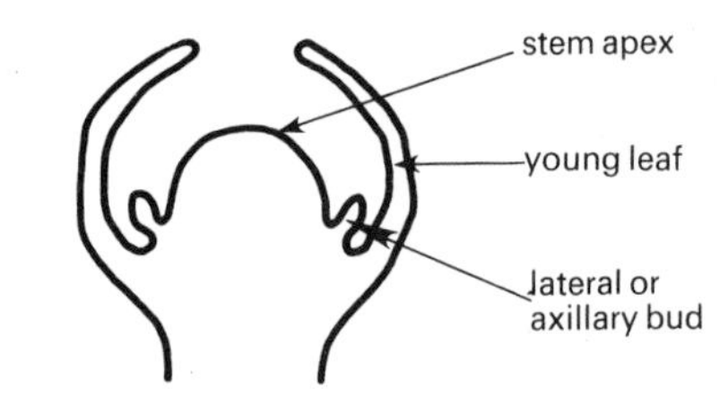

Figure 2.25 *L.S. shoot apex*

In the sub-apical region the cells fall into two regions, with the ground meristem making up most of the volume, and the procambium meristem arranged in procambial strands.

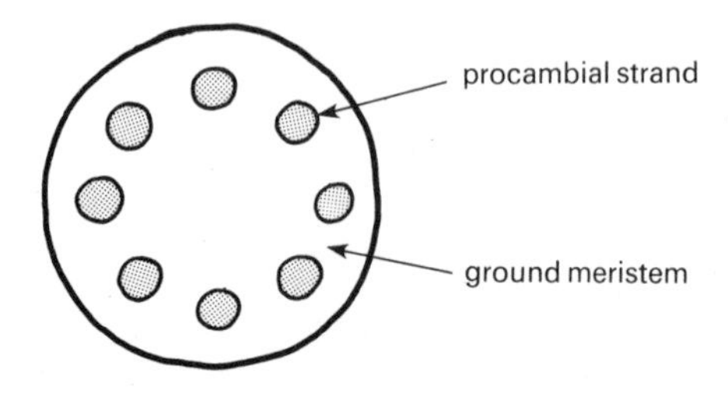

Figure 2.26 *T.S. sub-apical region of shoot*

The ground meristem cells enlarge by expansion of their vacuoles and the cell walls become separated at their corners to form inter-cellular spaces. After more divisions by transverse walls, division then stops and the cells differentiate into epidermis, collenchyma, and parenchyma. The procambium cells have little vacuolation and no inter-cellular spaces. They continue to divide by longitudinal walls and to elongate; they eventually give rise to vascular tissue which is predominantly longitudinal. Soon all the divisions are not only longitudinal but also periclinal (tangential to the surface).

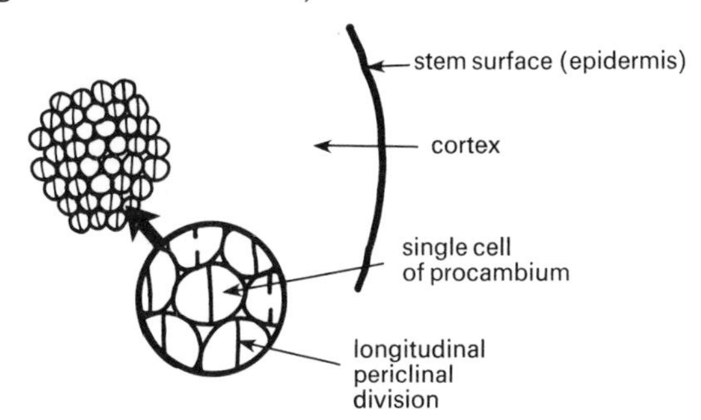

Figure 2.27 *Periclinal divisions*

These divisions produce cells in radial files.

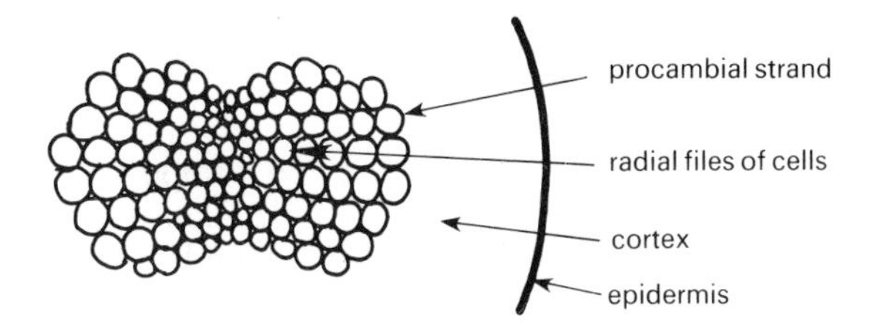

Figure 2.28 *Radial files*

Further down the stem the procambial cells become differentiated. Differentiation is from the edges of the strands inwards towards the centre of the strand. Whilst the stem is still elongating, annular and spiral vessels of the **protoxylem** form on the inner side. These patterns of thickening allow for extension in length. On the outer side thin, long cells of **protophloem** form whilst the stem is still elongating. The **metaxylem** of vessels with pitted walls which are not capable of further extension and the **metaphloem** of normal sieve tubes, companion cells, and phloem parenchyma are laid down in the region where stem elongation has stopped.

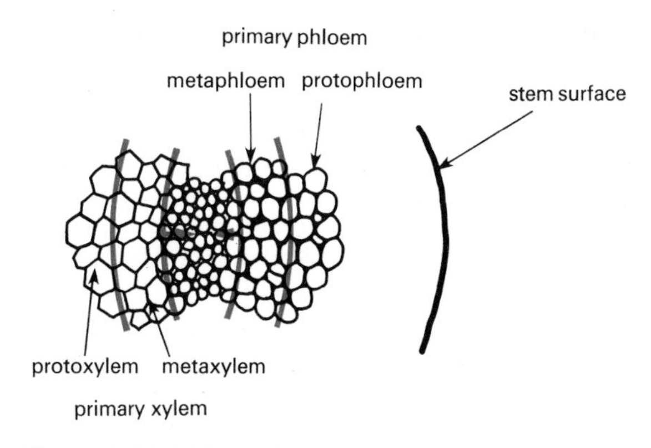

Figure 2.29 *Differentiation of procambial strand into vascular bundle*

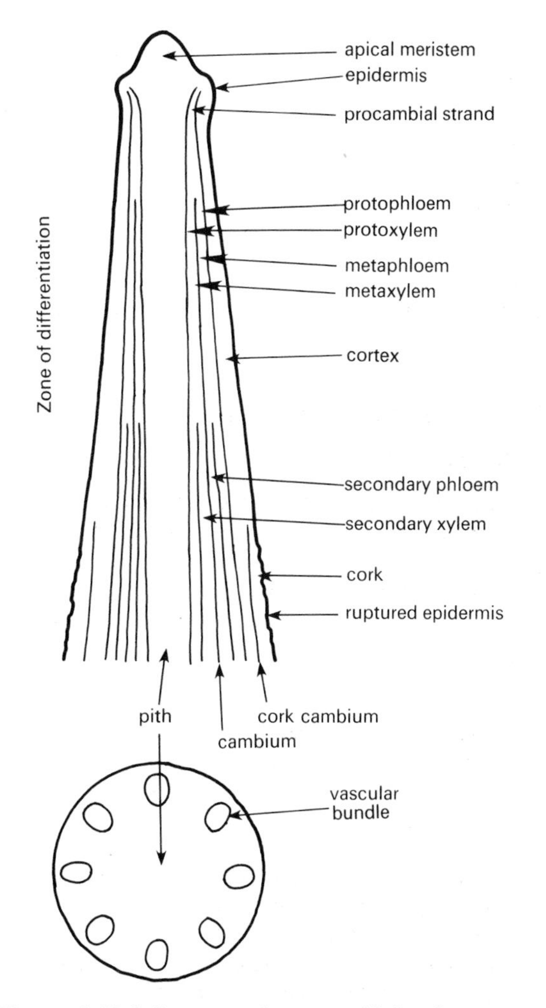

Figure 2.30 *L.S. young shoot, e.g. Helianthus*

In monocotyledons metaxylem and metaphloem meet, meristematic activity ceases, and the plant has no provisions for further growth. In this case the vascular bundles are said to be closed. In dicotyledons a remnant of the original procambium remains between the metaxylem and the metaphloem as the fascicular or vascular cambium, and here the vascular bundles are said to be open.

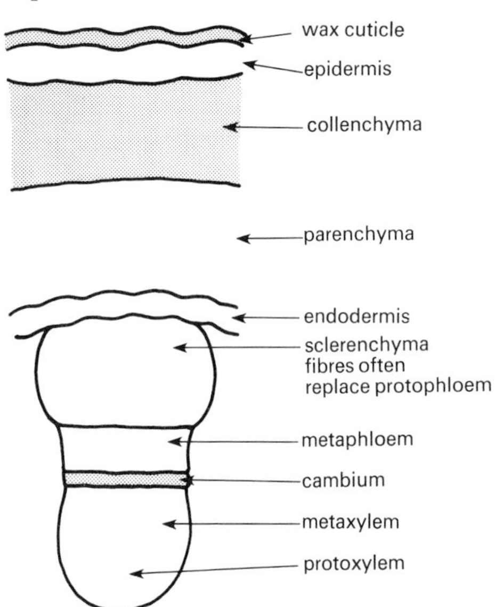

Figure 2.31 *T.S. detail of vascular bundle*

T.S. dicotyledon (*Helianthus*) vascular bundle

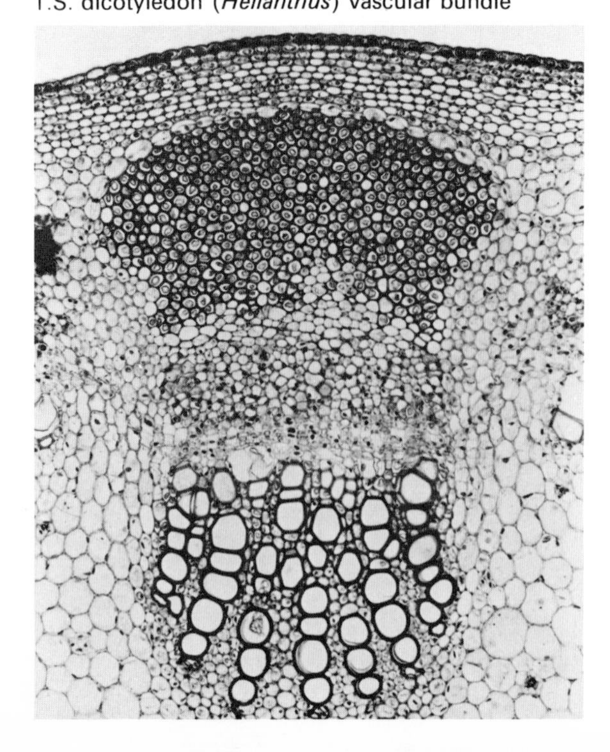

DISTRIBUTION OF VASCULAR TISSUE

The vascular and other supporting tissues are distributed in the various parts of the plant in patterns which are related to the resistance of the various forces acting upon them. In land plants the stem is exposed mainly to bending stresses due to the action of wind.

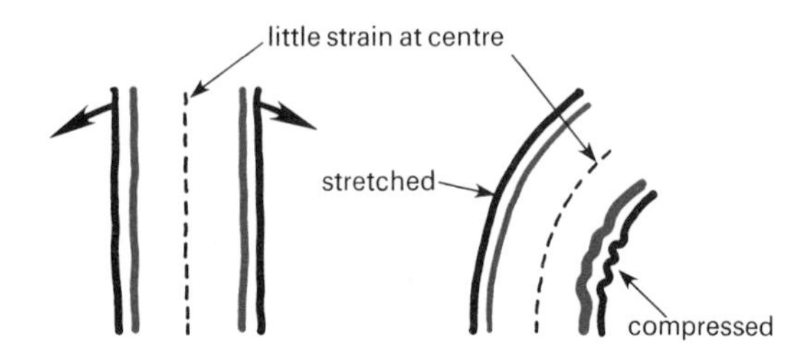

Figure 2.32 *Distribution of tissues in the herbaceous dicotyledonous stem with respect to bending strains*
The vascular bundles are found where the strains are.

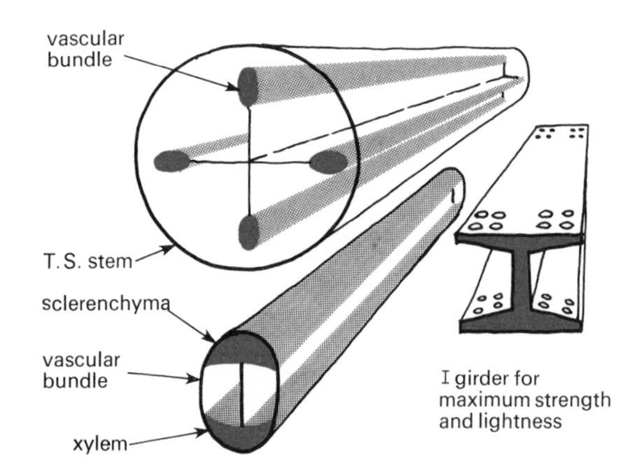

Figure 2.33 *Girder effect of the arrangement of bundles and tissues*

Roots which undergo a 'pulling' stress have the vascular tissue in the centre where it can best resist the forces involved.

LATERAL SHOOT GROWTH

Lateral shoots are formed from surface primordia in the axils of leaves. Since they are formed from the surface of the stem their development is described as **exogenous**. The initiation of the process of lateral shoot growth depends on the interaction of many factors including the level of hormones (auxin[223] from the shoot apex, and the auxin/cytokinin[223] balance). In fact, most lateral shoot buds remain dormant, hence the pattern of branching of a shoot system does not directly depend on the number and pattern of lateral shoot buds produced. This suppression of development of lateral buds, due mainly to the level of auxin diffusing back from the shoot apex, is referred to as **apical dominance**. If the apex is removed there is usually a vigorous growth of lateral branches.

MAIN FEATURES OF MONOCOTYLEDON STEM ANATOMY

The vascular bundles are scattered with no regular arrangement. They are closed, that is they have no cambium; and the metaxylem forms a deep V surrounding the phloem.

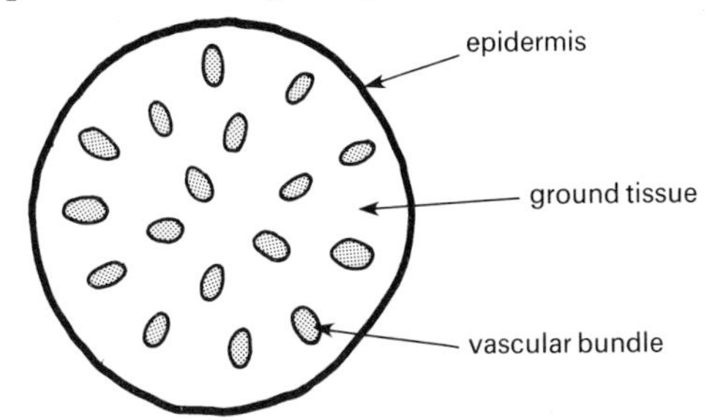

Figure 2.34 *T.S. monocotyledon stem*

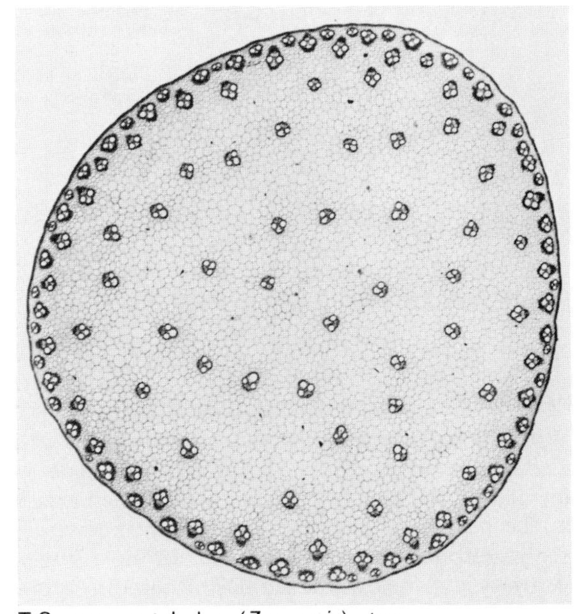

T.S. monocotyledon (*Zea mais*) stem

Palms, which are monocotyledons, attain tree-like dimensions from massive apical meristems, but not from the development of cambial strands.

SECONDARY THICKENING IN A SIMPLE WOODY STEM

This is produced by the division of cambial cells, both fascicular, and interfascicular (which arise by rejuvenation of parenchyma cells in the medullary rays between the vascular bundles).

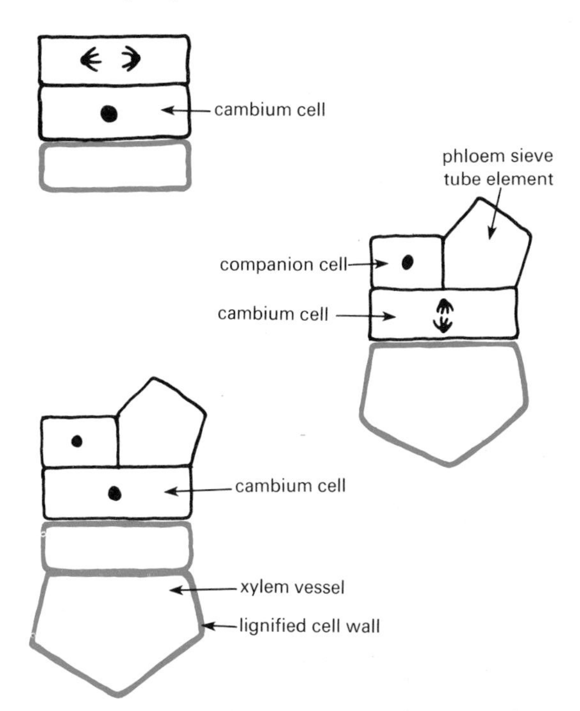

Figure 2.35 *Cambial cell division*

Fusiform initials give vascular tissue

Ray initial gives parenchyma of medullary ray

Figure 2.36 *Tangential longitudinal section (T.L.S.) to show ray and fusiform initials of cambium*

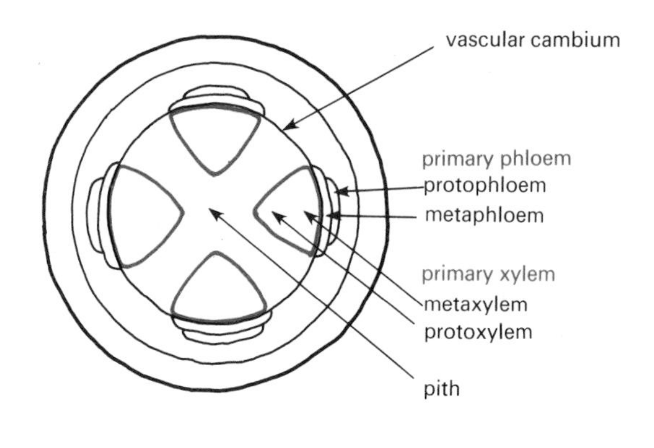

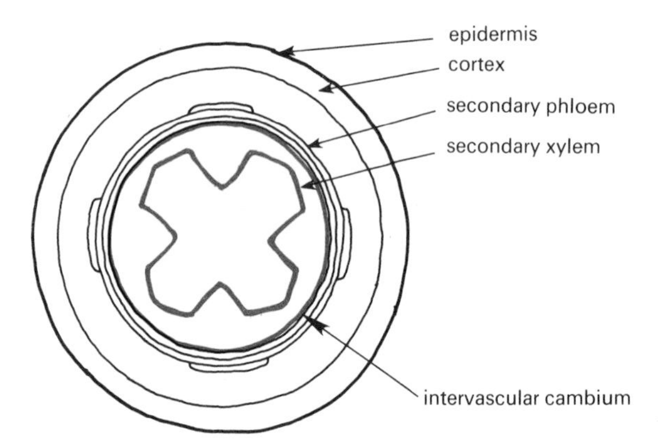

Figure 2.37 *Secondary thickening in a woody dicotyledonous stem*

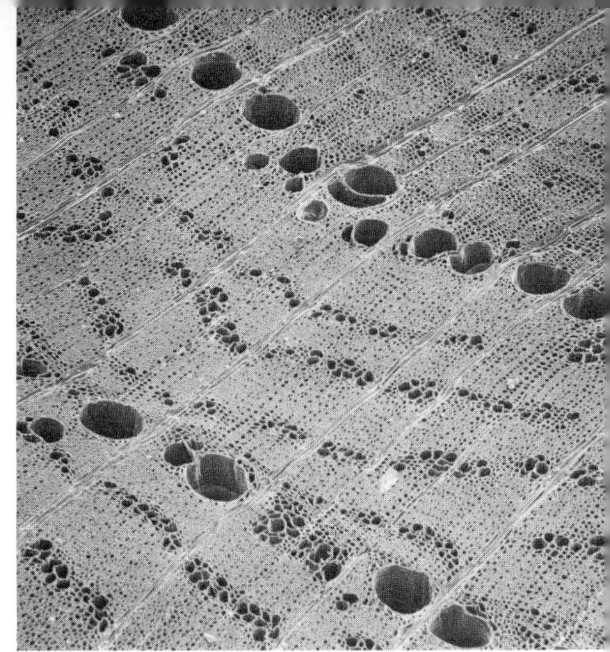

Scanning electron micrograph of growth rings in secondary xylem

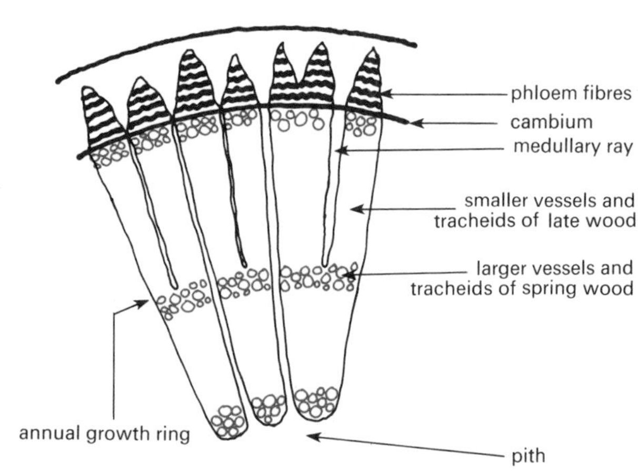

Figure 2.38 *T.S. woody stem to show growth rings*

Growth rings

Annual rhythms of cambium activity result in a zonation of rings in the secondary xylem. Any such zonation of the secondary phloem is obliterated because the unlignified phloem is stretched and broken as the diameter of the stem increases. The secondary phloem thus remains as a thin layer outside the cambium. In the xylem, growth rings are most clearly seen in woody plants of temperate zones, where the cambium activity is greatest in the spring and least in the winter. In the spring larger, thinner walled vessels with larger cavities are produced, while in the summer and autumn smaller, thicker walled vessels and tracheids are produced.

Disease, periods of drought, and cold spells can produce false annual rings, where the cambium activity is reduced more than once a year. The study of such rings can provide interesting evidence about the climate and conditions for tree growth in the past.

PERIDERM

With secondary growth, the outer regions are stretched and ruptured. A protective tissue is formed by the action of a cambium, which arises in the peripheral regions of the stem, known as the **phellogen** or cork cambium. In most plants it arises just below the epidermis in the hypodermis, e.g. elder, although in some plants it can arise actually within the epidermis or deep inside the endodermis.

To allow the interchange of gases through the cork to the internal tissues, lenticels are developed.

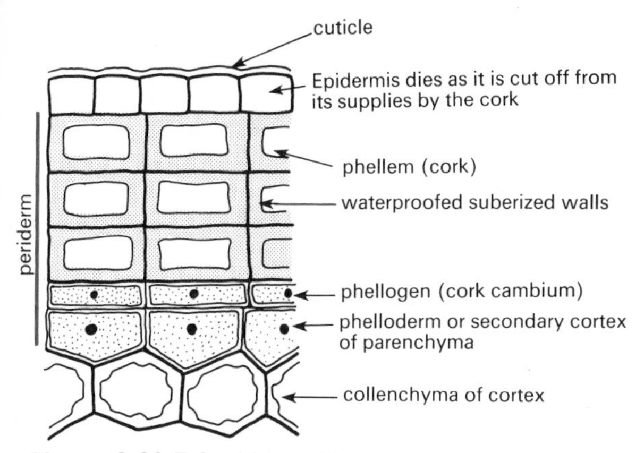

Figure 2.39 *T.S. elder periderm*

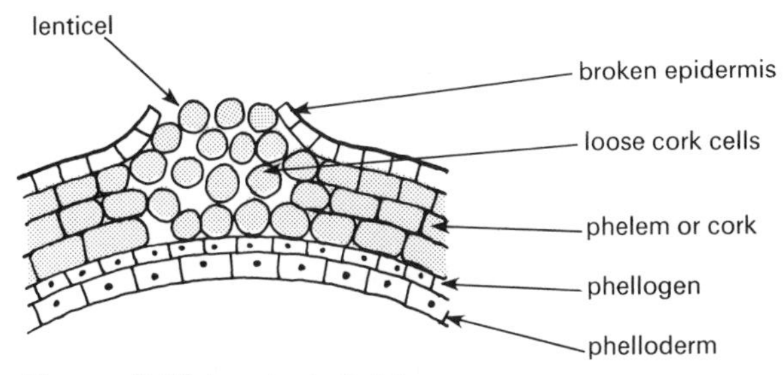

Figure 2.40 *Lenticel of elder*

In most cases the first-formed phellogen stops functioning after a time, and a new one forms in the tissues below. The shape of these successive phellogens gives rise to the different types of **bark**, a term which refers to anything outside the cambium.

The root

A group of meristematic cells just behind the tip give rise to four tissues, or **histogens**, which eventually differentiate into the adult tissues of the root.

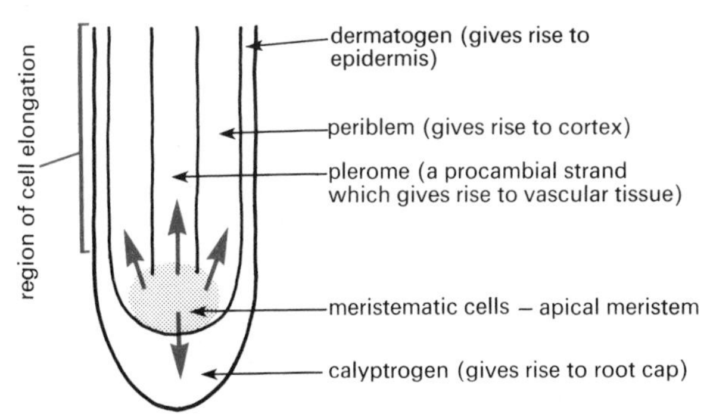

Figure 2.41 *L.S. root apex*

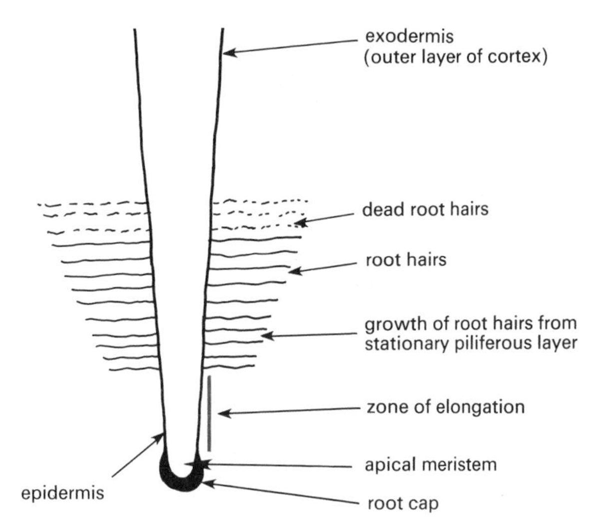

Figure 2.42 *Young root*

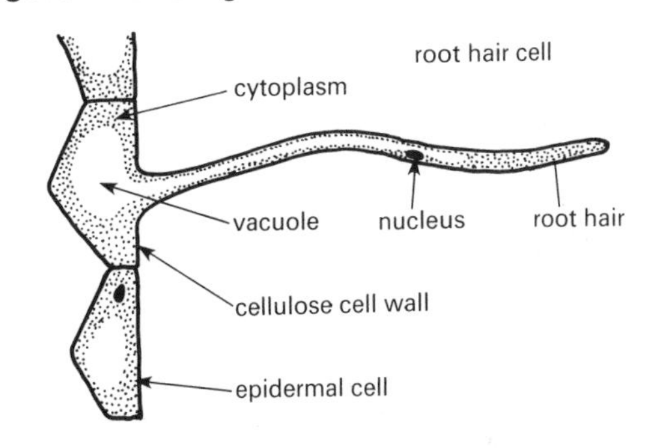

Figure 2.43 *Root hair*

VASCULAR DIFFERENTIATION

The zone of elongation in a root is short and therefore less stretching of the protoxylem and the protophloem occurs in roots than in stems. The differentiation of both xylem and phloem is from the outside of the strand to the inside.

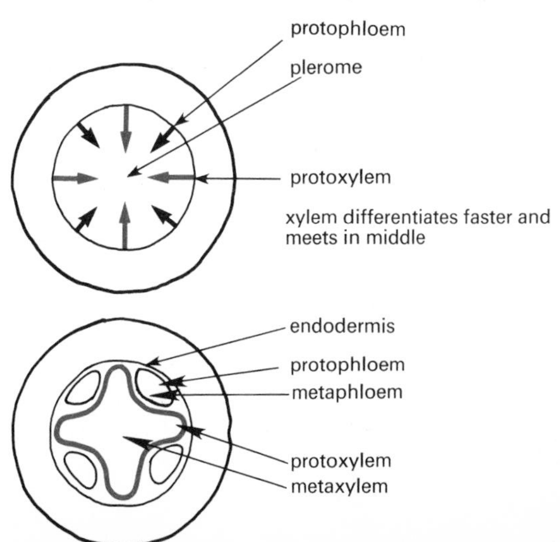

Figure 2.44 *T.S. differentiating plerome*

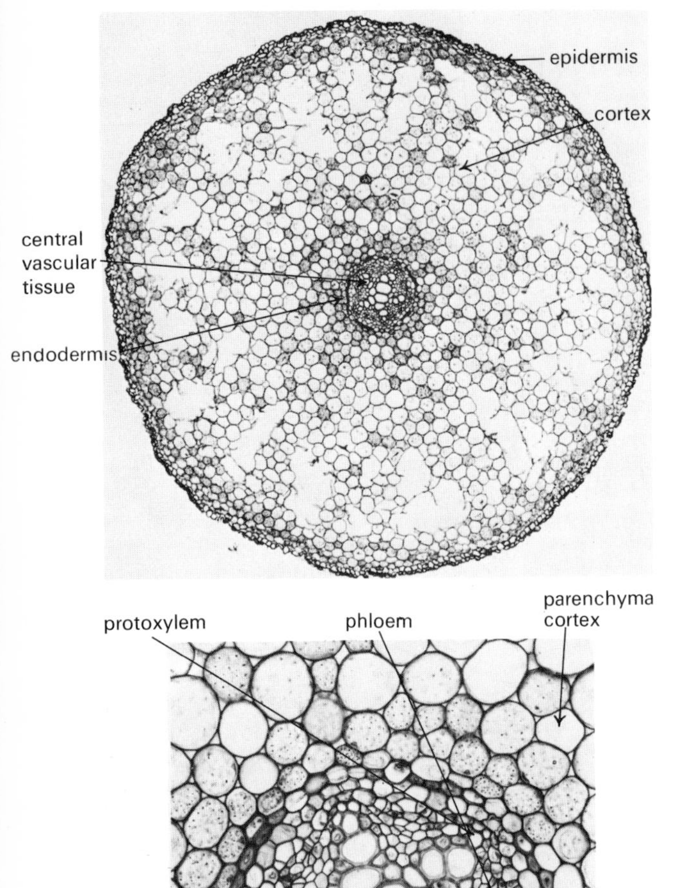

T.S. *Ranunculus* root, with enlargement of vascular tissue

Vascular development in roots normally proceeds to completion, that is no meristematic cells remain undifferentiated, and any cambium arises by a resumption of meristematic activity in mature differentiated tissue.

In Monocotyledon roots the central cells differentiate into parenchyma to form pith before vascular differentiation can reach it.

SECONDARY THICKENING OF A DICOTYLEDON ROOT

In those Dicotyledon roots which show secondary thickening the cambium produces xylem to the inside and phloem to the outside.

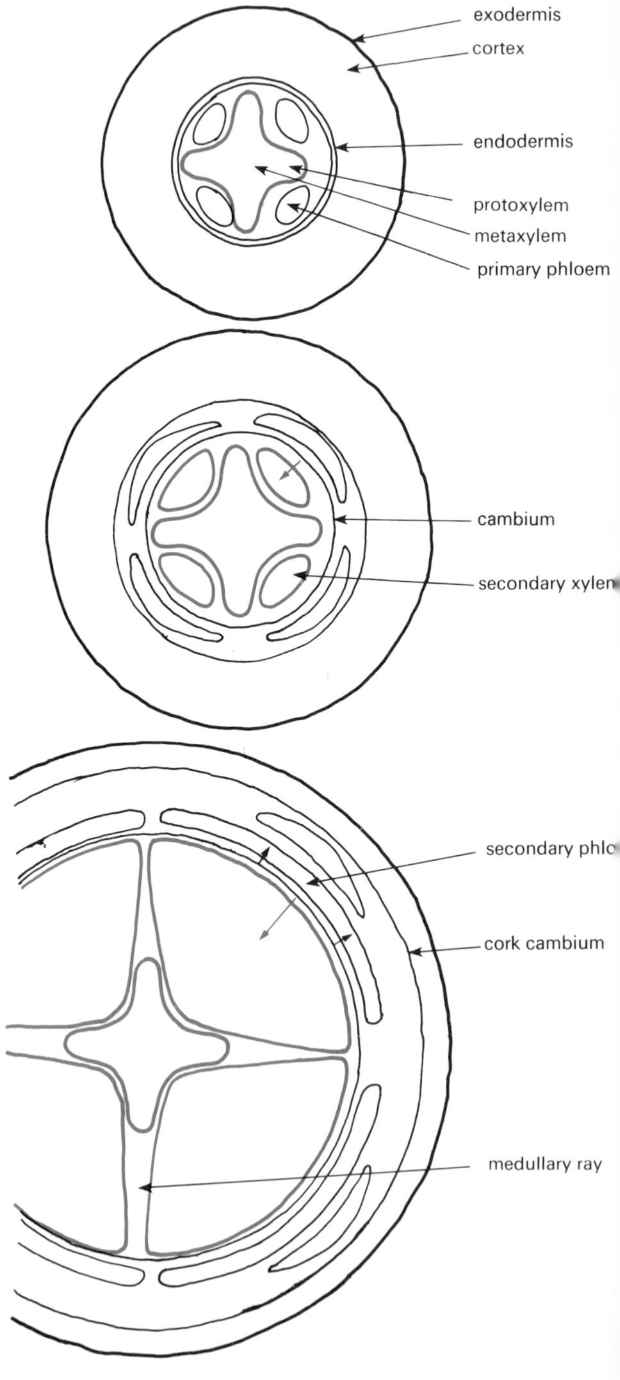

Figure 2.45 *T.S. root showing secondary thickening*

ROOT INITIATION

Lateral roots develop from primordia, or groups, of meristematic cells in the pericycle opposite the primary xylem groups. As they arise from deep inside the root tissues their development is described as **endogenous**. The initiation of the process of lateral root growth depends on the interaction of many factors including the level of auxin from the shoot apex and from the root apex. Since lateral roots only arise opposite protoxylem groups they will be vertically above each other and equal in number to the protoxylem groups, e.g. di, tri, and tetrarch.

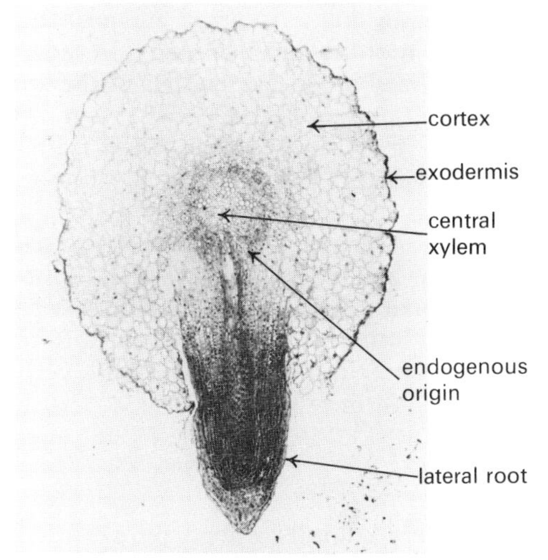

T.S. *Vicia* root showing the development of a lateral root

The Leaf

LEAF INITIATION AND DEVELOPMENT

Leaf primordia are initiated close to the apical meristem of the shoot under a complex of forces not yet understood. Leaves often show **heteroblastic** development, that is the first leaves formed are often of the juvenile form (which is different from those formed in later development). They arise exogenously as lateral protruberances of the tunica; at first all the cells of the leaf are meristematic, but meristematic activity later becomes restricted to the base of the growing leaf. The development of the leaf is completed in the bud, and the leaf eventually emerges by the osmotic expansion of cells already present, and not by cell division.

Dicotyledons typically have broad, dorsi-ventral leaf laminas, with reticulate venation. Monocotyledons typically have narrow, isolateral leaf laminas, with parallel venation.

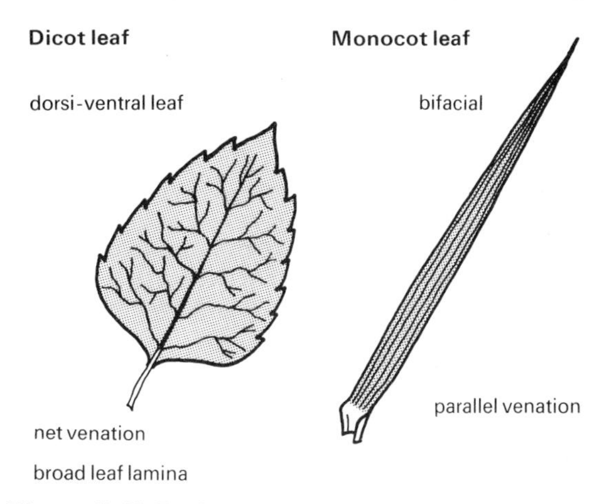

Figure 2.46 *Typical dicotyledon and monocotyledon leaves*

LEAF ABSCISION

Deciduous plants lose their leaves completely at one time of the year, usually the autumn in temperate zones, whilst so-called **evergreens** lose their leaves gradually at different times throughout the year. Prior to leaf fall the chlorophyll breaks down, photosynthesis stops, and the useful substances are absorbed back into the plant from the leaf. An abscision layer of parenchyma is formed across the base of the leaf stalk, or petiole, and on the stem side of this a layer of cork forms. The leaf breaks off due to the breakdown of the middle lamella of the cells of the cork layer. The process of abscision is under hormonal control.[224]

V.S. to show leaf abscision

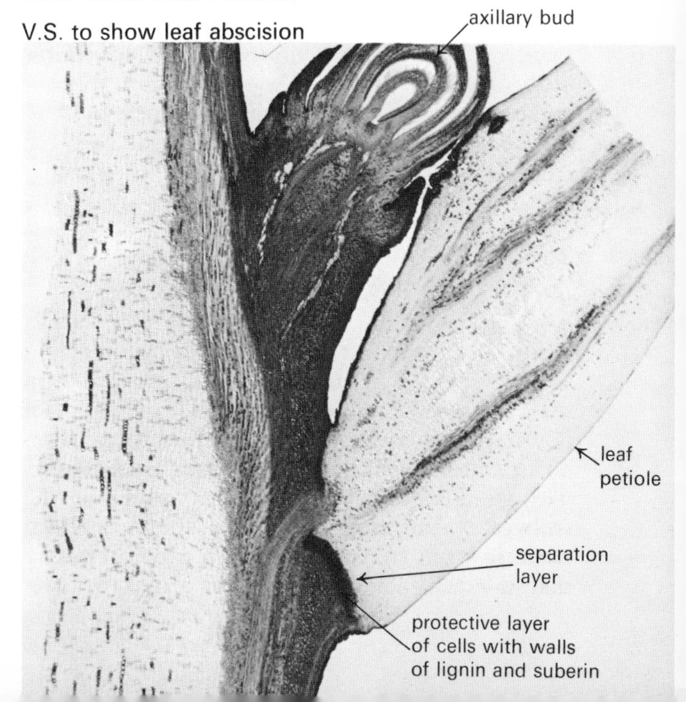

External factors affecting plant growth

Growth is the result of all the metabolic activities of all cells, therefore any factor that affects the plant's metabolism will affect growth. The complex interactions of these activities are such that it is sometimes impossible to isolate the effects of a single factor.

Climatic factors, including temperature, light, wind, etc; **edaphic factors**, such as the nature of the soil, availability of water, etc; and **biotic factors**, of interaction and competition with other living things, all interact with the genetic make-up or genotype of the plant to produce the actual plant or phenotype.[58]

The effect of temperature

The rate of all physiological processes increases with increase in temperature until the optimum temperature for enzyme-catalysed reactions is reached. The effects of temperature on growth are therefore very complex. Many plants require a period of exposure to low temperature before they can complete their life cycle by flowering and setting seeds. This effect is closely related to day-length requirements, and most plants that need cold treatment also require exposure to long days before they will flower. The time for which the plant must be exposed to low temperatures varies with different plants, but in all cases the receptor region for the stimulus is the stem apex.

Many biennials require a cold stimulus before flowering, and this usually occurs in the winter at the end of their first year of growth. If prematurely exposed to a cold stimulus they 'bolt' in the first year to produce flowers and seeds; if the cold stimulus is not encountered they often remain in the vegetative state indefinitely.

VERNALIZATION

The vernalization effect was first studied in wheats, of which there are both winter and spring varieties. **Winter varieties**, if planted in the spring, only grow vegetatively and will not flower in the same year since they require exposure to a period of cold before they flower. This stimulus is received by the apical bud of the embryo within the seed. They therefore need to be sown in the autumn of the year preceding that in which the crop is to be harvested. **Spring varieties**, if planted in the spring, do flower in the same year.

Winter varieties can be brought to flower in the season in which they are planted by means of vernalization. In this process the seed is moistened enough to allow the onset of germination, but not sufficiently to allow too much growth to occur. When the radicles have just emerged the seed is exposed to a temperature just above freezing for a few weeks. This seed, if planted in the spring, will flower in the summer of the same year. The advantages of this technique are that, since winter varieties yield more grain, by vernalizing and planting in the spring the losses that would normally have occurred in the soil over the winter months are avoided. In addition, when vernalized seeds are planted in the spring, they can be harvested before the true spring varieties, so that losses due to wind, rain, and fungal diseases of the damp late summer and early autumn can be avoided.

The name 'vernalin' has been suggested for the hormone involved in this process. Gibberellin[223] treatment can remove the need for a period of cold, and, although there are slight differences in the response, it is suggested that 'vernalin' may be some form of gibberellin.

The effect of light

Light has very complex effects on growth since the production of chlorophyll, photosynthesis, the opening and closing of the stomata, and tropisms depend on it, and it affects the anatomy of sun and shade leaves. Light tends to suppress growth by elongation, although it promotes leaf expansion in dicotyledons and can increase the rate of differentiation of tissues. In the dark there is a lack of development of chlorophyll, the internodes are long and thin and the leaves are reduced. A plant in this condition is said to be **etiolated**.

PHOTOPERIODISM

Some plants flower independently of the length of the light period to which they are exposed. They are called **day-neutral plants**, and some varieties can even complete their life cycle from seed to seed completely in the dark, if supplied with all the necessary nutrients. This group of day-neutral plants include tomato, cucumber, dandelion, and maize. Other types will only flower if they are exposed to more than about twelve hours of light in each twenty-four-hour period. These are called **long-day plants** and include onion, wheat, potato, radish, barley, and mint. Yet others will only flower if they are exposed to less than about twelve hours of light. These are called **short-day plants** and include primrose, chrysanthemum, strawberry, and salvia.

Not all plants fall into one of the above groups. In some flowering is only hastened by short days; some flower under a combination of day lengths; and some require a certain day length at one temperature but show no response to day length at another temperature. For the sake of simplicity, only those obligate short-day plants and obligate long-day plants will be considered here.

The photoperiodic stimulus is received by the leaves, which must be at a precise stage of development. Some 'influence' is then transmitted to the apical meristem(s) where the production of flowers is stimulated. In some plants the exposure of a single leaf to the necessary stimulus is enough to induce flowering. Some 'influence' must therefore be transmitted to the apical meristems where the production of flowers is stimulated. This 'influence' is assumed to be a hormone and has been called **florigen**. The concept of a universal flowering hormone is supported by grafting experiments in which a suitably induced plant will induce flowering in another plant on to which it is grafted. This occurs even when the graft is between plants of different genera. Furthermore, when an induced short-day plant is grafted on to a long-day plant, it will promote flowering in the latter.

The normal growth hormones of plants, auxin and gibberellins, influence flowering in some plants. For example, in long-day and short-day plants a raised level of auxin in the dark has effects similar to those of periods of low light intensity, that is it promotes flowering in long-day plants and inhibits it in short-day plants. Raised levels of gibberellins also promote flowering in many long-day plants. The relationships of these hormones to 'florigen' is not clear.

In trying to identify the exact nature of the stimulus it is necessary to distinguish between the high light requirements of photosynthesis and the low light requirements of photoperiodic induction. The red end of the spectrum is most effective in inducing flowering in photoperiodic plants, and the pigment responsible for detecting the stimulus appears to be the **phytochrome pigment** found in all plant cells. This pigment is a major receptor in all photomorphogenetic effects, including leaf enlargement, internode extension, germination, rhizome formation, and the formation of bulbs.

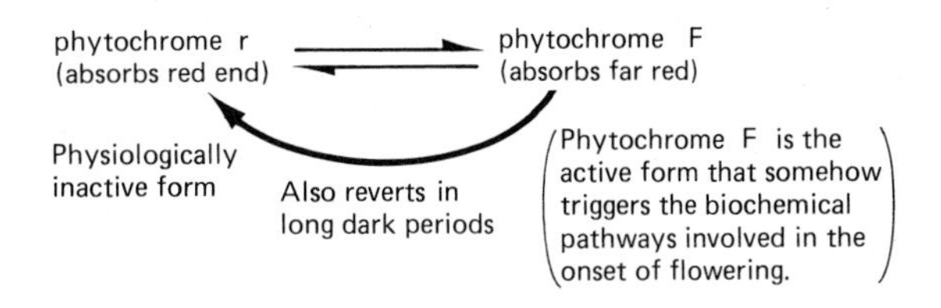

Experimental and applied aspects

Reproduction, growth, and development in both plants and animals is too wide a field for any experimental background central to all the issues involved to be summarized, although recent work on the cloning of individual organisms from the cells of a single individual (see under Genetics)[58] encompasses many fundamental processes involved in normal reproduction, growth, and development.

In **animal reproduction** much work has been, and is being, done on the control of fertility in humans, a problem of crucial importance as the limits of food, energy, and material resources are approached. The various mechanical methods of contraception, such as the rubber sheath, the diaphragm, and spermicidal creams, etc., are all less reliable than the intra-uterine device (IUD), and the oral contraceptive pill.

The mode of action of the **IUD** is still not fully understood, but it is thought to exploit a tendency of the female immune system to produce antibodies against the 'foreign' protein of the spermatozoa which are stored in the female for varying periods. During this time they may be attacked and destroyed by the leucocytes[180] produced by the female which migrate through the walls and into the lumen of the female tract. It is thought that the presence of an intra-uterine device stimulates a large influx of leucocytes from the blood stream. These subsequently attack the spermatozoa, and thus produce the contraceptive effect.

It is claimed that **oral contraceptives**, when used, are 100% effective, and although there are some circulatory side effects, these are minimal when compared to the risks associated with pregnancy. Substances known as progestagens which, unlike progesterone, remain active when taken orally, are combined with small amounts of synthetic oestrogen which have a stronger effect than naturally-occurring oestrogens. The usual procedure is to take the pill daily for twenty-one days, from the 5th to the 25th day of the menstrual cycle. The mode of action is not absolutely clear but it is thought that LH[22] section by the adenohypophysis of the pituitary is inhibited by some action on the hypothalamus,[22] and this in turn inhibits ovulation. Low dosage of progestagen throughout the whole cycle does not inhibit ovulation but controls fertility by some unknown action on the female tract.

Work on fertilization outside the mammalian female, and the subsequent reimplantation of the fertilized egg, has opened tremendous possibilities in the field of animal breeding. For example, fertilized eggs from good stock can be reimplanted into other females, thus increasing the reproductive potential of animals with good qualities.

With reference to the **growth and development of plants**, the production of secondary xylem in those spermatophytes with the tree habit is of great economic importance, and is of great interest in relating microscopic structure to macroscopic characteristics. Secondary xylem makes up the material that is known as wood. The lignified walls of the elements making up the xylem basically all have the same properties. Any differences, therefore, between different woods are due to variations in the thicknesses of the lignified walls, and in the distribution and pattern of the vessels, tracheids, sclerenchyma fibres, and parenchyma of which they are composed.

The so-called softwoods of the conifers[351] are composed entirely of tracheids; there are no vessels or fibres. In the spring wood the tracheids are larger than those of the later wood, with wider cavities and thinner walls. Good conditions result in wide rings made up mainly of thin-walled spring wood. Poor conditions limit the growth of the thin-walled spring wood and produce narrow rings with a greater proportion of the stronger late wood, which produces a better timber. The best softwoods are therefore grown on poor soils and in exposed situations. The so-called hardwoods of the angiosperm[351] trees have elements of all types in them. Spring wood has a greater proportion of larger, thin-walled vessels than the later wood. In contrast to the softwoods, good conditions result in wide rings consisting mainly of more durable late wood. The best hardwoods are therefore grown under good conditions which favour the production of wider growth rings. The terms softwood and hardwood can be misleading since not all hardwoods are actually harder than the softwoods. Hardwoods range from very hard, heavy timbers, which have a predominance of fibres, to very soft, light timbers, such as balsa wood, which have a predominance of parenchyma.

The **grain** of wood is the appearance conferred by the longitudinally-running xylem elements. The patterns in the grain are produced by irregular annual rings and medullary rays, usually seen in tangential longitudinal section.

In older, woody stems and branches the centre, or **heartwood**, loses its water-transporting function and becomes a depository for resins, tannins, and other waste products which give it a characteristic colour and feel. The transport of water in the transpiration stream occurs in the peripheral, young, secondary xylem, or **sapwood**. It contains living xylem parenchyma, and has a limited capacity for healing wounds by secreting gums and other substances.

'Timber' is a term covering a wide variety of woods, some with unique properties, but which generally have certain common characteristics determined by the microstructure of lignin-impregnated cellulose xylem tubules running longitudinally through the tree trunk. Timber is of relatively low density (about one-fourteenth that of steel), however, weight for weight its tensile strength is equivalent to four to five times that of the steel in common use. It is three or four times as strong in tension than it is in compression because the cell walls fold up in compression; but again, weight for weight, it has a compressive strength close to that of steel. In addition, since its planes of weakness are parallel to its strongest direction, that is between the tubes, it is strong and tough and therefore resists cracks. The good stiffness, low density, good tensile strength, resistance to stress concentrations, and availability in long pieces, renders timber suitable for use in beams and columns in building. However, the main draw-back to its use in beams is that it creeps slowly with time, and this sagging effect is seen in old timber beams in buildings. It is easily cut and, when fixed with nails or screws, does not split. This is due to its low lateral tensile and compressive strength, that is it is easily separated across the grain. Its porosity enables efficient bonding with glues and resins.

It has some disadvantages in its use, for example it comes into equilibrium with the relative humidity of the air and can absorb up to 25 per cent by weight of water. This causes swelling, particularly laterally (up to 10 per cent), and reduces its strength and stiffness by 66 per cent. This problem can be partially overcome by seasoning, in which the water content is reduced to that experienced during usage by controlled drying. Associated with the moisture content is the risk of rot due to fungal infection which sets in when the water content rises above 18 per cent by weight. These factors (swelling and rotting) are major problems in its use as a structural material.

Guided example

1 Why is the definition of growth of living organisms as an increase in size, volume, mass, or fresh weight not a particularly good one?
While such an increase in size, volume, mass, or fresh weight does occur when an organism is growing, any such increase could be due to an increase in water content. This is not restricted to living organisms. For example, a water-logged dead branch of a tree shows an increase in size, volume, mass, and fresh weight, but can hardly be said to be growing in the accepted sense.

2 The definition of growth as an increase in dry weight, that is the weight of the material after all the water has been driven off, is better since it avoids the problem of an increase in water content. Can you think of any objections to the use of this as a definition of growth?
An increase in dry weight could be due to an increase in non-living inclusions in the cells and tissues, for example, crystals of calcium oxalate which are not truly part of the cell structure. Moreover, it is possible for organisms to lose dry weight whilst growing, as is seen in germinating seedlings which deplete their food stores whilst undergoing very rapid and complex growth.

3 Can you think of any other definitions that would give a better description of the growth of a living organism?
Better definitions would include: an increase in the amount of cytoplasm, an increase in cell number (in multicellular organisms), and an irreversible increase in complexity.

4 The amount of growth of an organism per unit time is referred to as its **absolute growth rate**. For example, consider two plant stems, one initially 10 cm long and the other 20 cm long, which both grow 1 cm in a day. They therefore have the same absolute growth rate but this gives no idea of the amount of growth in relation to the size of the organism. A better expression is that of the **percentage growth rate**. Which of the two has the greater percentage growth rate?

$$\frac{1}{10} \times 100 = 10\%$$

$$\frac{1}{20} \times 100 = 5\%$$

The 10 cm stem has the greater percentage growth rate.

5 Different parts of a plant organism usually grow at different rates from each other, and from the overall growth rate of the body as a whole. This phenomenon is called **allometric growth**. Can you think of any examples in the growth of the human body?
The brain grows at a much greater rate than other parts of the body just before birth and in the early years of development. Another good example is the growth rate of the testes in the male, under the control of the Y chromosome, which is necessary to offset the influence of the maternal female sex hormones.

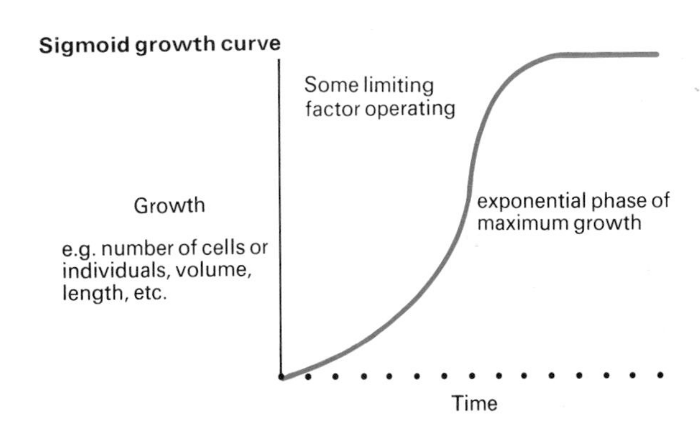

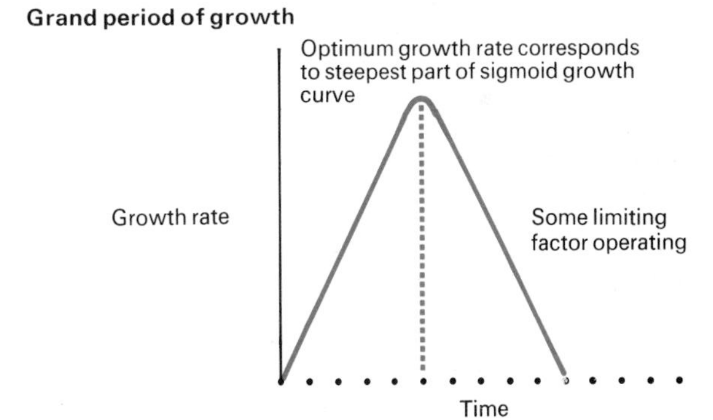

Figure 2.47 *Growth curves*
The grand period of growth of a cell, colony, organ, body, etc., occurs when the growth rate increases from zero to a maximum and then falls back to zero again.

Questions

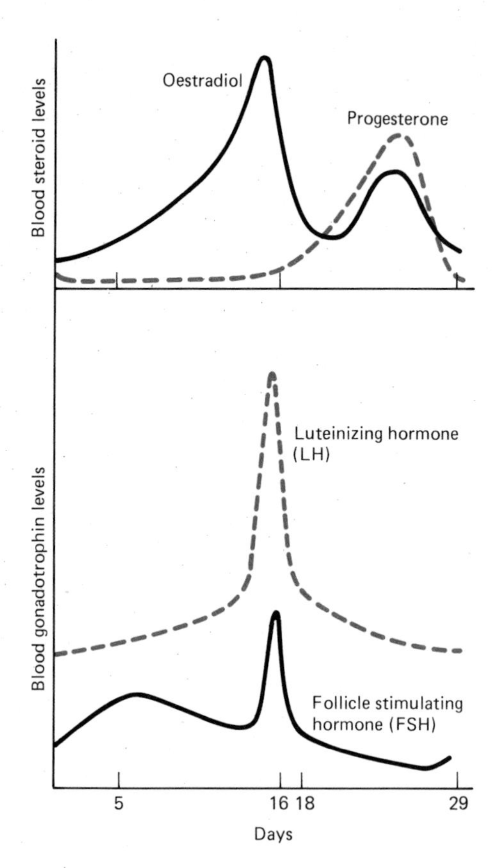

(Steroids and gonadotrophins are represented separately because levels are measured on different assay scales.)

1 The graph shows changes in the blood levels of four hormones during the course of the human menstrual cycle.

(a) Explain the part played by each hormone.
(b) Comment on any other regulatory aspects of the menstrual cycle not covered by the hormones represented in the graph.
(c) How would the graph have differed if pregnancy had been achieved on day 18 of the cycle?

(O & C)

2 Outline the events which occur between flower bud production and seed formation in an angiosperm.

What are the differences between an angiosperm and a pteridophyte (e.g. a fern) over the equivalent period of the life history? (O & C)

3 The two tables given below present some data on the effects of the growth factors abscisic acid (ABA), giberellic acid (GA) and kinetin (K) on the germination of two different kinds of seed.

Table 1 *The effect of ABA on the germination of seeds of tall fescue grass*

ABA concentration, ppm	Per cent germination after (days): 6	10	18
0	12	76	85
1	8	42	78
5	0	7	26
10	0	2	7

Table 2 *The effects of ABA and K in the presence of varying concentrations of GA on germination of seeds of lettuce*

GA concentration, mM	Germination, %: No addition	+kinetin, 0.05 mM	+ABA, 0.04 mM	+ABA and kinetin
0	10	15	0	0
0.05	21	27	0	17
0.5	66	69	0	57
5	95	97	0	73

Seeds of some plants require a period of moist low temperature treatment before their dormancy is broken. This is true of the seeds of sugar maple. The figure shows the actual changes that occur in levels of growth factors in seeds of sugar maple during this period of cold treatment. The seeds may start to germinate at any time after thirty days and all had germinated by fifty days.

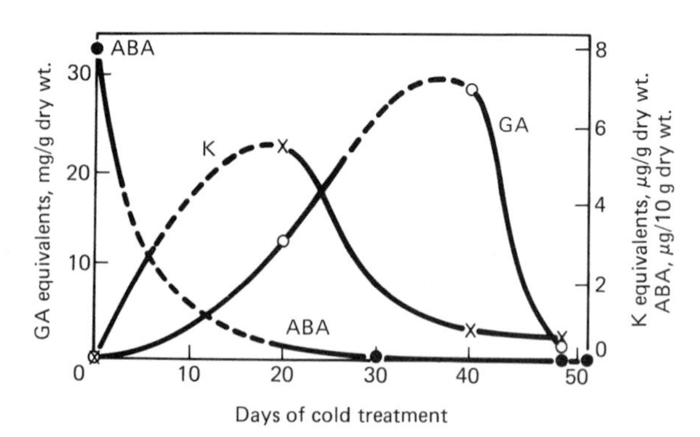

Changes in growth factors in seeds of sugar maple during cold treatment

(a) What is the effect of ABA on the germination of seeds of (i) tall fescue grass, (ii) lettuce?
(b) What are the effects of GA, and of GA in the presence of ABA, on the germination of seeds of lettuce?

(c) What is the role of K in the presence of GA and ABA in the germination of seeds of lettuce?

(d) After considering all the information given, discuss the role of the three growth factors in the dormancy and the breaking of dormancy of seeds of sugar maple. (C)

4 Explain concisely why a supply of water is necessary for seed germination to occur.

The following table is a record of changes in *dry weight*, in grams, of samples of maize seedlings and their various parts during germination in the dark at 25 °C.

Day	0	1	1	3	4	5
Whole seedling	225	210	208	206	175	155
Endosperm	200	189	188	155	115	84
Embryo (radicle and plumule)	2	3	5	15	23	36

Plot a graph of this data.

What causes whole seedlings to decrease in dry weight during germination?

How much greater is the weight loss of the endosperm than the weight loss of the whole seedling? What could account for this additional loss? (L)

5 By means of fully labelled drawings only, describe the structure of the vascular cambium of a woody plant. Briefly compare the *origins* of this meristem in the stem and root of dicotyledonous plants. What are the differences in structure and function between vascular cambium and cork cambium (phellogen)? (L)

Further reading

Fogg G. E., *The Growth of Plants* (Harmondsworth: Penguin Books, 1963).

Galston A. W. and Davies P. J., *Control Mechanisms in Plant Development*, Foundations of Developmental Biology Series (New Jersey: Prentice-Hall, 1970).

Gemmel A. R., *Developmental Plant Anatomy*, The Institute of Biology Studies in Biology No. 15 (London: Arnold, 1969).

Newth D. R., *Animal Growth and Development*, Institute of Biology Studies in Biology No. 24 (London: Arnold, 1970).

3 Genetics

Introduction

Genetics involves the study of heredity and is concerned with why offspring both resemble and differ from their parents. It also investigates how the inherited information is expressed in the developing organism; thus it overlaps, or rather is an integral part of, the study of growth and development. Through its concern with how offspring differ from their parents, that is with **variation**, it forms an essential part of the study of the theory of evolution.[80]

The expression of the genetic message depends on an interaction between the genetic information, or **genotype**, and the environment, to produce the actual organism or **phenotype**:

$$\text{genotype} + \text{environment} \longrightarrow \text{phenotype}$$

It is difficult, if not impossible, to separate completely the two components contributing to the phenotype, as a genotype must necessarily express itself in an environment. However, attempts can be made, and for convenience's sake they are often considered as separate influences on development.

In Nature it is difficult to determine whether a particular variation between individuals is genetic or environmental in origin. In order to make the distinction, the offspring of such individuals are allowed to develop under constant conditions. If the variation remains unchanged, then it is almost certainly genetic in origin. If the variation alters, then it was almost certainly environmental in origin. Thus, in order to study the effects of the genotype, environmental factors should be kept constant; and to study the effects of the environment the genotype should be kept constant. The environmental factors can be kept constant under controlled laboratory conditions, while the genetic factors, although they can never be completely standardized due to continual changes known as mutations,[94] can be kept relatively constant in a variety of ways.

One method is to use **clones**, which are groups of organisms produced by asexual reproduction

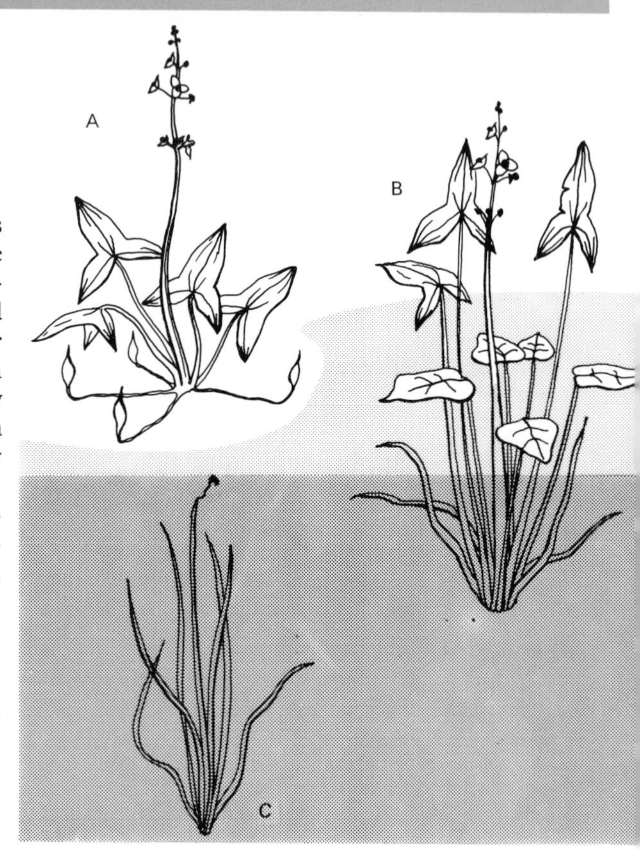

Figure 3.1 *Phenotypes of arrowleaf (Sagittaria sagittifolia)*
Three arrowleaf plants grown from a single root (A) in damp soil, (B) in 18 cm water with a gentle current, (C) submerged in deeper water with a fairly swift current.

from one parent. For example, only one King Edward potato plant has ever been produced from seed; the rest have all been obtained by asexual vegetative reproduction. They thus all have the same genotype and any difference observed between individuals must be environmental in origin. Another way is to use organisms produced by the splitting of a single fertilized egg; for example armadilloes habitually produce monozygotic quads, derived from a single fertilized egg, which therefore share the same genotype. The study of identical twins provides some information on the influence of the environment on human development in the so-called 'Nature or nurture' controversy.

Identical or monozygotic twins develop from a

single fertilized egg, or zygote, and are therefore genetically identical (except for any somatic mutations that may occur independently in each one). Such identical twins occur in about 0.3 per cent of all births. Since they are genetically identical, any observed differences between them will be due to environmental factors influencing the expression of their genotype. An interesting situation for a researcher occurs when such twins have been separated very early in life and reared apart. In this situation it should be possible to attribute any similarities to their Nature, that is their genes; and any differences to their nurture or environment. However, it is sometimes difficult to distinguish monozygotic twins from dizygotic twins, that is twins from the fertilization of two eggs, which are not genetically identical.

Table 3.1 *Average differences in selected physical characteristics between pairs of twins*

Difference in	50 pairs of identical twins reared together	50 pairs of non-identical twins reared together	19 pairs of identical twins reared apart
Height (cm)	1.7	4.4	1.8
Weight (lb)	4.1	10.0	9.9

These findings tend to indicate that height is mainly genetically controlled while weight is mainly environmentally controlled.

Table 3.2 *Corrected average differences in IQ tests for 50 pairs of identical twins*

Identical reared together (50 pairs)	Non-identical (52 pairs)	Identical reared apart (19 pairs)
3.1	8.5	6.0

In the absence of fingerprinting and analysis of blood proteins the identification of identical monozygotic twins can be difficult, and misidentification sometimes occurs. Studies on separated identical twins have been claimed to show that about 80 per cent of the ability measured by IQ tests is inherited. However, re-examination of the data shows that such a conclusion is highly suspect and no reliable conclusions can be drawn about the inheritance of IQ from these studies.

The nucleus

In the non-dividing cell, the so-called 'resting stage' or interphase nucleus contains scattered 'chromatin' material in the nucleoplasm or nuclear matrix. The term 'chromatin' was originally coined to include all the substances in the nucleus that stained preferentially with basic stains. Later, when the Feulgen reaction was developed for the selective staining of deoxyribonucleic acid[73] (DNA), the term 'chromatin' came to refer only to the DNA. However, as free DNA does not occur in the nucleus, the term also includes the associated nucleo-proteins, histones.

DNA is the carrier of the genetic code by which hereditary information is passed from generation to generation via the nuclei of the gametes and from cell to cell via the nuclei of the cells produced by division of the fertilized egg.

When the DNA is in the dispersed, lightly staining form seen in the interphase nucleus, it is genetically active, sending out information to the cytoplasm to control the complex of cell processes. This dispersed form of chromatin is referred to as **euchromatin**. When the cell is about to divide the euchromatin concentrates or condenses into an inactive state to form the **chromosomes**. In this state the DNA is more resistant to damage during the processes of nuclear and cell division. Moreover, since it is in discrete bodies like the chromosomes, the division of the genetic material between the nuclei can be carefully controlled, reducing the danger of any being lost. However, some chromatin remains permanently condensed and deeply staining during interphase, and is termed **heterochromatin**. An example of this is seen in female mammals where one of the X sex chromosomes[70] remains permanently condensed as a so-called Barr body; the basis of the sex test carried out on International female athletes concerns the presence of this body. Another example is seen in the mature mammalian spermatozoa where all the chromatin is heterochromatin and the DNA is thus protected for its passage to the fertilization of the female gamete.

Chromosome structure

The term chromosome means 'coloured body' and refers to the fact that they appear as dense darkly staining bodies when the nucleus is treated with various stains such as basic dyes and Feulgen reagent. Along their length are darker staining bands known as chromomeres.

Chromosomes are composed of DNA and a special class of nucleoproteins, known as histones, which form aggregates with DNA. The exact physical relationship between DNA and the structure of the chromosome is still not known, but one suggestion is that the chromosome consists of a tightly coiled DNA molecule wrapped around the histones.

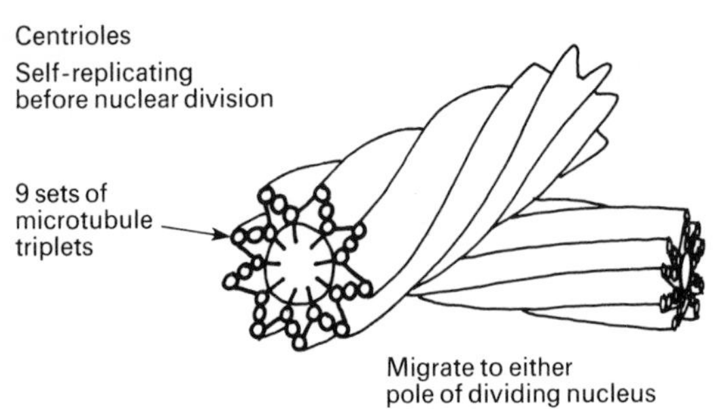

Chromosomes are not visible during *interphase* when the nucleus is not dividing and the DNA is active. This occupies about 2/3 of the cell cycle.

Three main divisions of interphase are G_1—gap period
S—synthesis of DNA
G_2—gap period

Prophase
Each chromosome appears as two sister chromatids joined at the centromere (DNA content is twice the normal diploid content due to replication)

nucleolus disappears by late prophase

Metaphase
Chromosomes move to the equator and attach to fibres at centromere

microtubules form spindle fibres

nuclear membrane replaced by spindle apparatus

Anaphase
Chromatids separate to each pole, thus becoming known as chromosomes

centromeres lead the way

Telophase
Reverse prophase.
Nucleus reformed, chromosomes de-condense

Figure 3.2 *Mitosis*
The centrioles occur in an area of clear cytoplasm, known as the centrosome, at one pole of the animal cell nucleus (their presence is suspected in plant cells). The centrioles show similarities in structure to cilia and flagella, and somehow control the movement of chromosomes during mitosis (and meiosis).

Mitosis

Mitosis is the process of nuclear division, seen in the dividing somatic or body cells, by which genetically identical daughter nuclei are produced. During interphase in these cells the DNA replicates itself so that, with the onset of mitosis, there is double the normal amount of DNA present in the nucleus. The chromatin, comprising the DNA and the nucleoproteins, condenses to form the chromosomes which then pass through a continuous series of movements resulting in the production of two diploid daughter nuclei, genetically identical to the original nucleus. This process is described by reference to so-called 'stages', but it must be remembered that it is essentially a continuous process.

Meiosis (reduction division)

Meiosis is the process of nuclear division, seen in gametogenesis in animals (and some plants, e.g. *Fucus*) and in spore production in most plants, in which the chromosome number is halved from diploid to haploid. The nucleus of each cell of a diploid organism contains two sets of chromosomes; one maternal from the female gamete and one paternal from the male gamete. As a result, each chromosome has a 'partner' in the other set which carries genes for the same characteristics. Two such chromosomes are said to form a **homologous pair**. The genes on the homologous partner may either be identical, or they may code for variants of the characteristics, in which case they are known as **alleles** or allelomorphic pairs of genes.

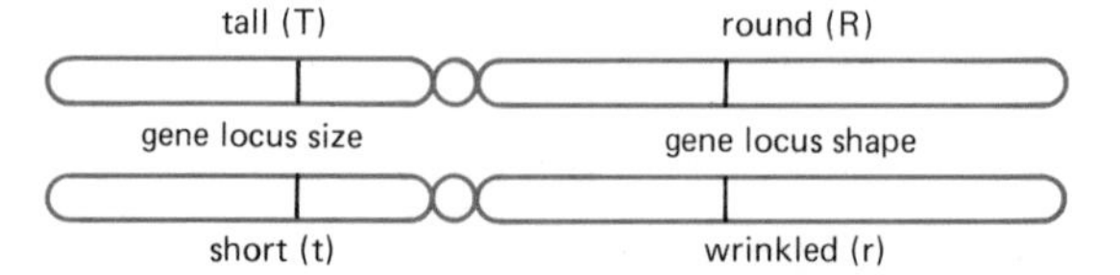

During meiosis such homologous chromosomes come together to form a pair, or **bivalent**. They then exchange genetic material by crossing over, and then repel each other into separate daughter nuclei. In this way the chromosome number is halved to produce haploid gametes; the diploid number being restored when the nuclei of two haploid gametes fuse in the process of fertilization.

First prophase

(1) Leptotene

Chromosomes condense as mass of long coiled threads

(2) Zygotene

Homologous chromosomes pair to form bivalents

(3) Pachytene

Homologous chromosomes shorten and thicken, and coil around each other

(4) Diplotene

Each chromosome now appears as 2 sister chromatids (not seen before due to their close attraction)

Homologous chromosomes now begin to repel each other

Breaks occur in opposing non-sister chromatids and crossing-over occurs

(5) Diakinesis

Point of cross-over known as a chiasma

Homologous chromosomes unwind as they continue to repel each other

Centromeres are centres of repulsion

Held together by attraction

Prevents complete separation occurring too early

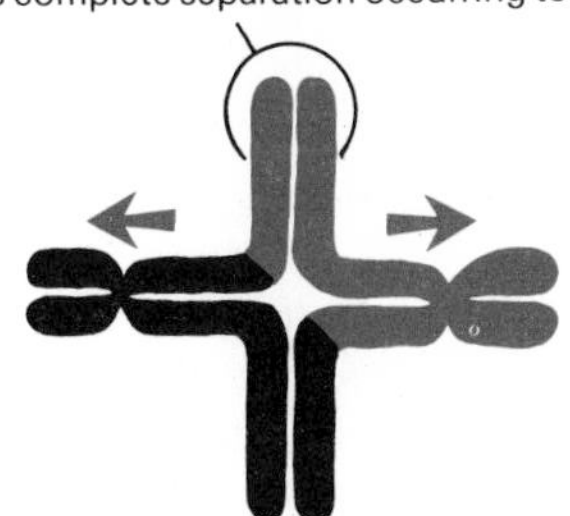

Chromosomes held together until they align on the equator

First metaphase

Chiasma (ta) align on equator

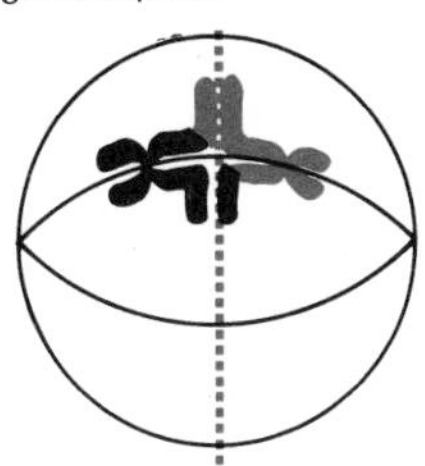

First anaphase

Chiasma (ta) are terminalized as chromosomes move to poles

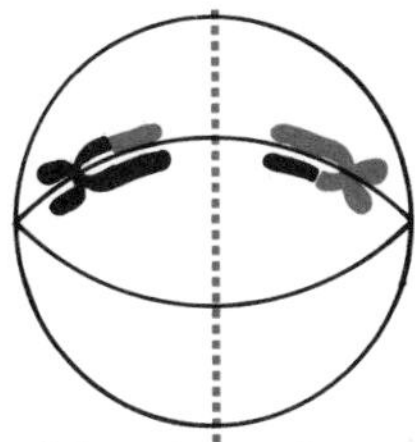

First telophase blends into **Second prophase**

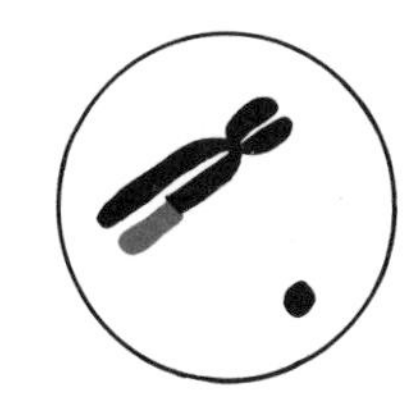

Second metaphase

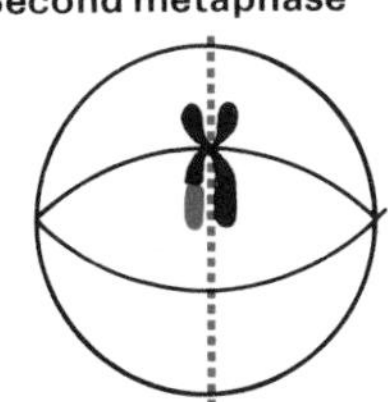

Second anaphase

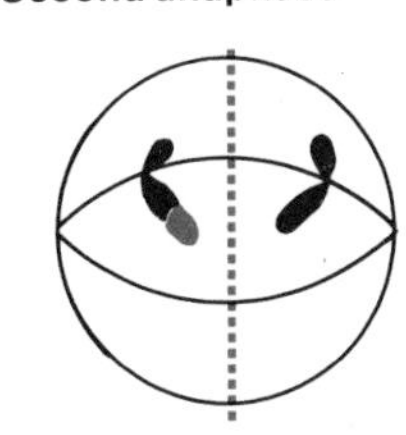

Second telophase

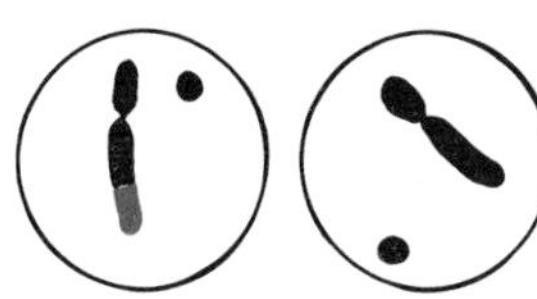

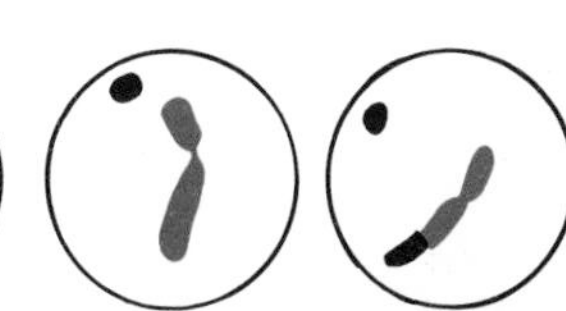

Figure 3.3 *Meiosis*

Four haploid daughter nuclei are formed with chromosomes of mixed paternal and maternal genes, due to crossing over, and with varying numbers of maternal and paternal chromosomes, due to independent assortment.

Meiosis results in a mixing or **genetic recombination** of the original maternal and paternal sets of chromosomes. This is brought about by the crossing over[69] that always occurs between homologous chromosomes, and by the independent assortment[64] of maternal and paternal chromosomes to each new gametic nucleus. It is this meiotic genetic recombination in gametogenesis and the subsequent random fusion of gametes that produces variation in the offspring of sexual reproduction.

GERM LINE CONCEPT

As an animal embryo grows and differentiates, certain cells are set aside as potential gamete-forming tissues. These so-called germ line cells are usually set aside at a very early stage of development, and are the only cells in which meiosis occurs. For example, in human embryos the primordial germ cells appear twenty days after fertilization. In the female they enter meiosis almost immediately and by the seventh month of development all reach the diplotene stage of meiosis, at which they remain until the ova are released from the mature ovary (a period of time that may exceed forty years).

Mendelian genetics

Gregor Mendel (1822–84), a monk at an Augustinian monastery in what is now Brno in Czechoslovakia, carried out a series of breeding experiments with the garden pea (*Pisum sativum*) over a period of eight years, which led him to formulate two laws of inheritance. The results of these experiments were read before the local Natural History Society in 1865 and published by that society in 1866. Although this publication was received by scientific libraries throughout Europe, Mendel and his work remained unknown to scientists until 1900, sixteen years after his death in 1884. In 1900, one of the laws governing inheritance was simultaneously discovered by three scientists working independently on similar problems, and these developments led to the rediscovery of Mendels results of thirty-five years earlier.

Mendel's work differed from that which had gone before in that he avoided earlier confusions by choosing simple, easily identified single features or traits, and by actually counting and reducing results to a measurable quantitative basis.

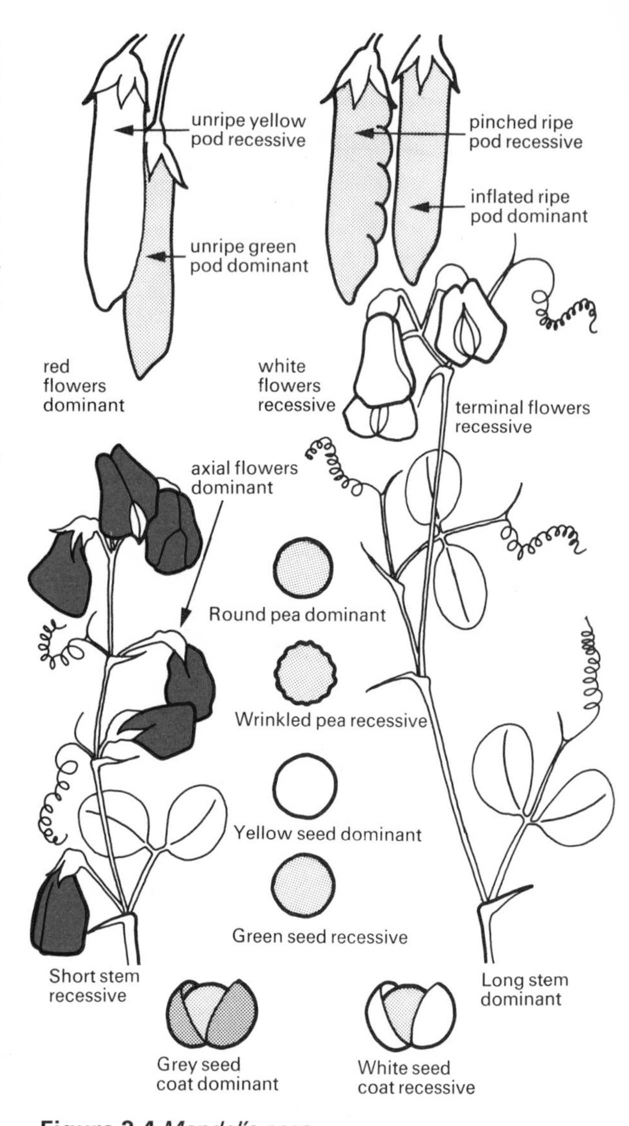

Figure 3.4 *Mendel's peas*
Some contrasting characteristics in the garden pea chosen by Mendel.

The garden pea was ideal for these studies. It is habitually self-pollinating in the flower bud stage, and thus has **pure lines** of unchanging inherited traits. There are many varieties with easily recognized differences. The varieties are inter-fertile and produce fertile progeny. The floral structure allows easy artificial cross-pollination and the plants are easily cultivated.

Mendel artificially crossed different pure lines, **the parentals**, to produce seeds which grew into the first filial or **F1 generation**. These were allowed to self-fertilize to produce seeds which grew into the second or **F2 generation**. These in turn were again allowed to self-fertilize to produce seeds which grew into the third filial or **F3 generation**. He noticed that with all the characters he chose to investigate, the F1 generation were all the same as each other and always resembled one of the parents.

Parentals: Tall × Dwarf
F1: all tall (hybrids)
In fact these were taller than the tall parent, a phenomenon known as **hybrid vigour**.

When the F1 generation was allowed to self-fertilize he found that some were like one parent and some were like the other, in a ratio of 3:1.

F1 : All tall
self-fertilization

F2 : 3 tall : 1 dwarf

When the F2 generation was allowed to self-fertilize he found that the dwarf ones bred true; but that, of the tall ones, some bred true but some gave rise to further 3:1 ratios of tall to dwarf.

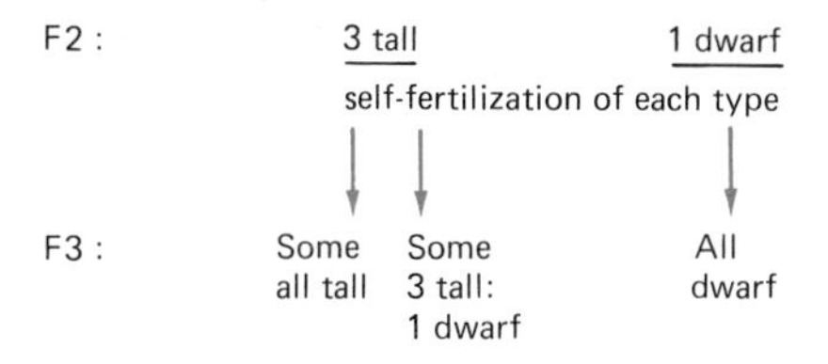

From these results he deduced that some characters were dominant to others (in this case tall being dominant to dwarf) and that each individual carried two 'doses' of the character, only one of which was carried in the gamete.

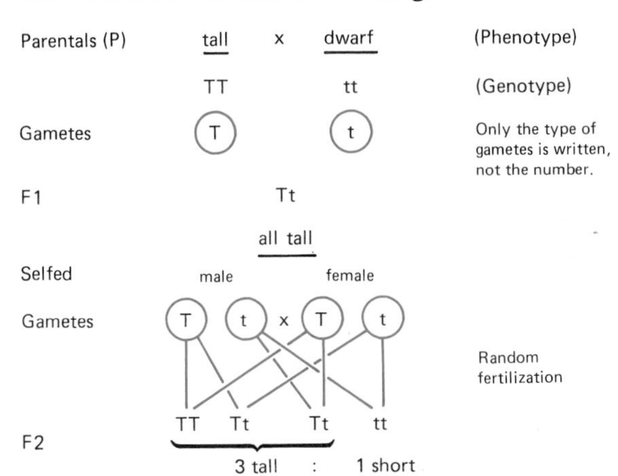

From these 3:1 monohybrid ratios he derived his first law of the **Segregation of Germinal Units** which states that, of a pair of contrasting characters (e.g. tall and dwarf), only one can be carried in a single gamete.

Tt — A pair of contrasting characters.

Gametes (T) (t) — Only one can be carried in each gamete, that is they segregate.

Random fusion subsequently results in the production of the **3:1 monohybrid ratio.**

During Mendel's time there was no knowledge of the importance of chromosomes and their behaviour in carrying the genetic message. It is now known that chromosomes occur in pairs and carry the genetic message in the form of genes. The members of a pair of these genes can differ from each other, in which case they are known as

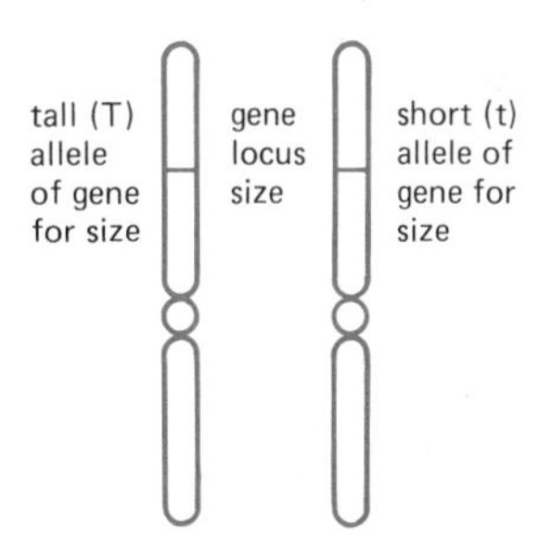

Pair of homologous chromosomes
(carry genes for the same characters)

alleles. During gamete formation these homologous chromosomes repel each other so that only one of each pair can enter a single gamete. This is achieved through a sequence of chromosomal movements called meiosis. Thus Mendel accurately described both the behaviour of chromosomes during meiosis and the significance of this behaviour, long before there was any direct visual evidence available. The segregation of genes into the gametes can be directly observed in maize which has a dominant gene for starch production and a recessive gene for non-starch production so that sugar accumulates. Both of these express themselves in the gamete in the pollen grains.

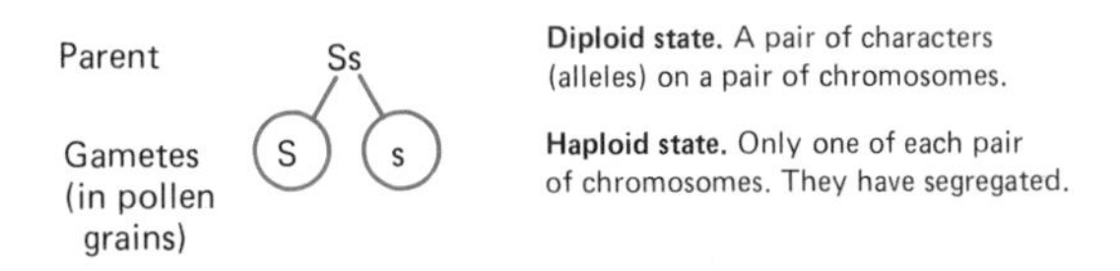

When pollen is put into iodine solution half the grains turn blue and half do not. This provides direct visible evidence of 'segregation'.

Mendel also considered the inheritance of two pairs of contrasting characters and, on obtaining a **9:3:3:1 dihybrid Mendelian ratio** in the F2 generation, formulated his second law of **Independent Assortment**. This states that, of a pair of contrasting characters, each one may be combined with either of another pair in a gamete. In terms of chromosome behaviour this means that each one of a pair of homologous chromosomes may combine or 'go with' either of another pair to form a gamete. In other words, chromosomes assort or segregate independently of each other.

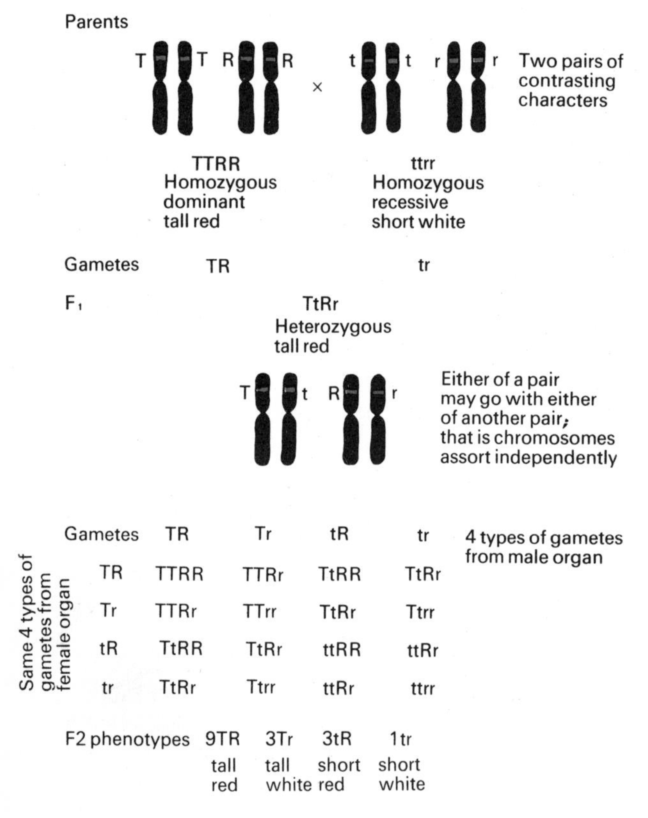

Figure 3.5 *Mendelian dihybrid inheritance*

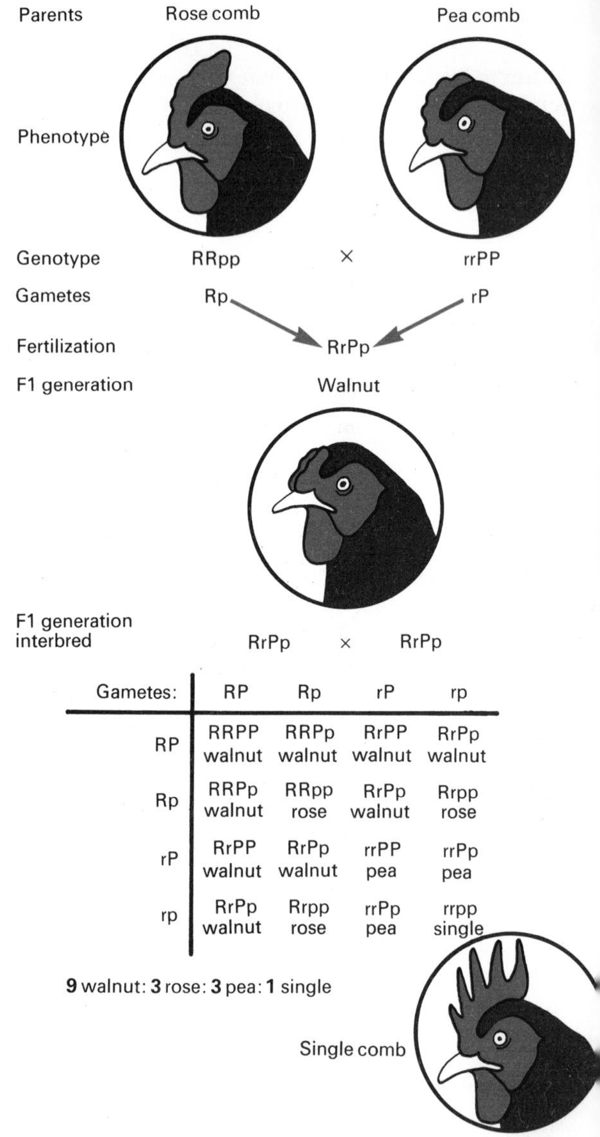

Figure 3.6 *Genetic interaction in the control of comb shape in chickens*
There are several varieties of poultry, each with a characteristic comb shape which is controlled by the interaction of two genes.

The first case of Mendelian inheritance described in animals was rather more complicated, due to interaction between the genes producing unexpected phenotypes. It involved the comb shape in four breeds of poultry: the rose comb, the pea comb, the walnut comb, and the single comb.

An example of more than one gene affecting a single character and thus obscuring the basic Mendelian ratios is seen in the Plant Kingdom in the inheritance of seed (caryopsis) colour in wheat.

Parentals $R_1R_1R_2R_2$ × $r_1r_1r_2r_2$
Deep red, each dominant allele adds a 'dose' of red — white

Gametes R_1R_2 — r_1r_2

F_1 $R_1r_1R_2r_2$
Intermediate red

Gametes	R_1R_2	R_1r_2	r_1R_2	r_1r_2
R_1R_2	$R_1R_1R_2R_2$	$R_1R_1R_2r_2$	$R_1r_1R_2R_2$	$R_1r_1R_2r_2$
R_1r_2	$R_1R_1R_2r_2$	$R_1R_1r_2r_2$	$R_1R_2r_1r_2$	$R_1r_1r_2r_2$
r_1R_2	$R_1r_1R_2R_2$	$R_1r_1R_2r_2$	$r_1r_1R_2R_2$	$r_1r_1R_2r_2$
r_1r_2	$R_1r_1R_2r_2$	$R_1r_1r_2r_2$	$r_1r_1R_2r_2$	$r_1r_1r_2r_2$

1 ////	4 ///	6 //	4 /	1
Deep red	Red	Pale red	Pink	White

Figure 3.7 *Duplicate dominant genes in wheat*
When strains of wheat with red caryopses are crossed with those with white caryopses, the ratio of red to white is sometimes 15:1 in the F2 generation. However, the red types in the latter do not all have the same colour density, there being a range of intermediate reds. The 15:1 ratio suggests dihybrid inheritance, as the numbers involved in the F2 ratio add up to 16 (e.g. 9:3:3:1 = 16, 15:1 = 16). In this case there are two genes contributing to the red colour development.

Test or back cross

A convenient method of determining the genotype of an organism is to cross it with a homozygous recessive, as genes from the homozygous recessive will not mask any genes from the organism under test.

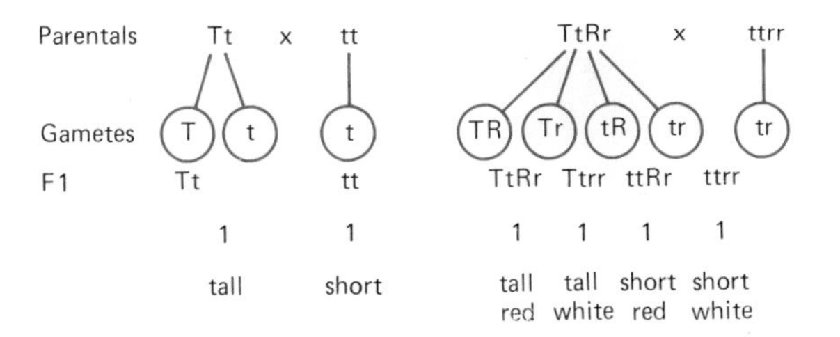

Incomplete dominance

Alleles do not always have a simple dominant/recessive relationship to each other; sometimes co-dominance or incomplete dominance occurs. For example

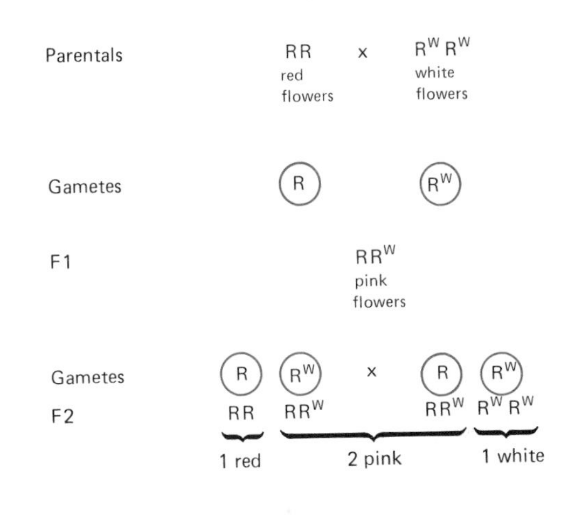

Non-Mendelian patterns of inheritance

The so-called 'classic' Mendelian traits are easily identified as distinct phenotypes and behave as discrete 'particles' of inheritance which affect only one character. However, many, if not most, inherited traits are affected by various genes and by interaction with the environment. This obscures their pattern of inheritance and prevents their description in simple terms of Mendel's first and second laws. Also, of course, the genotypes of the parents are not always known in outbreeding or cross-fertilizing organisms with no pure lines. Therefore many economically important traits in crop and livestock production under such multiple-factor or polygenic control, defy analysis in simple Mendelian terms, and the traditional system of breeding 'best to best' is still widely used in attempting to improve stock.

Mendel's concentration on a few discrete inherited traits in plants of known genetic composition enabled him to formulate his fundamental laws of inheritance. These formed the basis of modern genetics after many thousands of years of accumulated wisdom in plant and animal breeding had failed to indicate any underlying laws of inheritance.

The genetics of some widely studied organisms

Amongst the organisms most commonly studied in genetics are the fruit fly, the mouse, and maize.

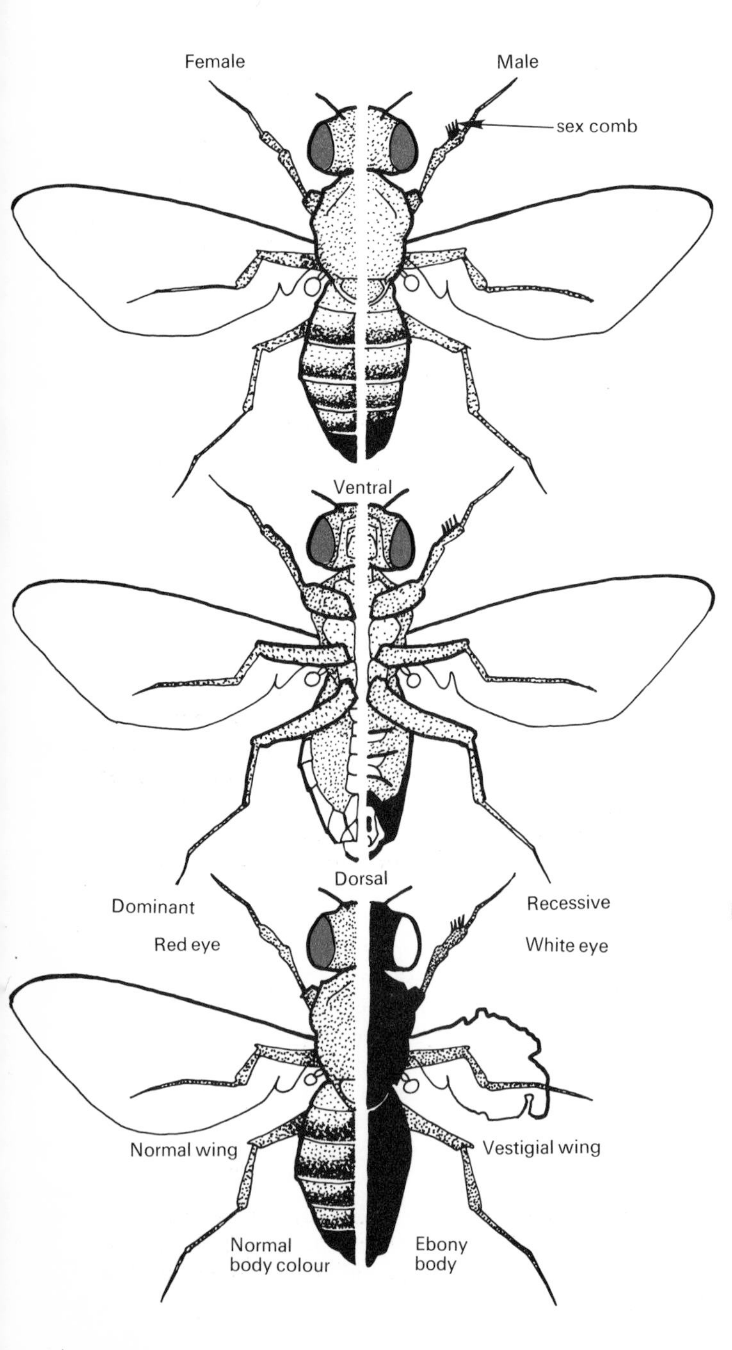

Figure 3.8 *Some inheritance characteristics of Drosophila melanogaster (the fruit or vinegar fly)*
Drosophila is ideal for practical genetic breeding experiments due to its short life cycle, small size, high rate of reproduction, low number of giant chromosomes, and the many easily-identified mutants.

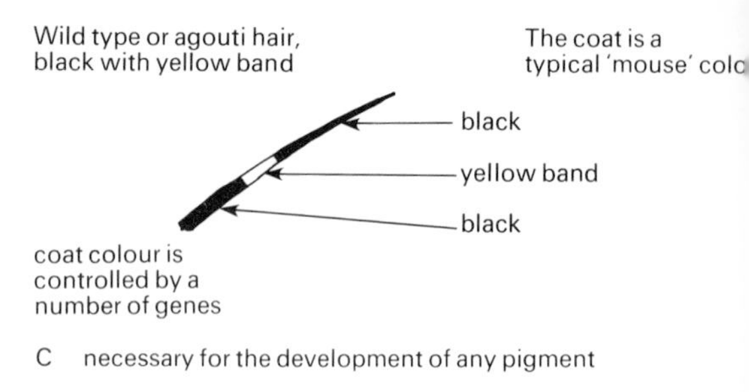

C necessary for the development of any pigment

A necessary for yellow band

B necessary for black pigment

c no pigment

a no yellow band

b produces brown pigment

(In fact even more genes are involved.)

C_B_A_ Agouti

C_B_aa Black: no yellow band

C_bbaa Brown

C_bbA_ Cinnamon: brown and yellow band

cc____ Albino

In breeding experiments only the heterozygous genes need be considered

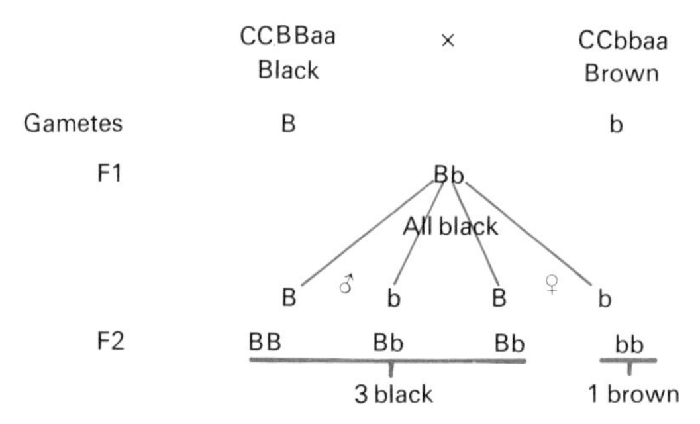

Figure 3.9 *The genetics of mice coat colours*

L. S. maize fruit

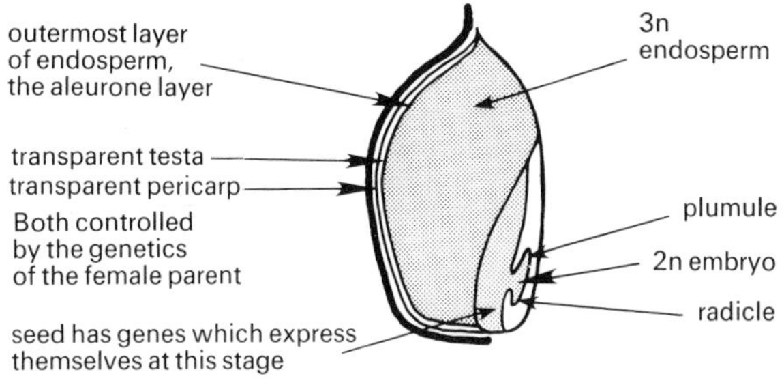

The colour of the aleurone is controlled by several genes

AA CC RR Pr Pr = purple

AA CC RR pr pr = red

AA CC rr pr pr = white

AA cc rr pr pr = white

aa cc rr pr pr = white

e.g. AA CC RR pr pr × AA CC rr pr pr

(can ignore all 'constant' genes i.e. AA CC pr pr)

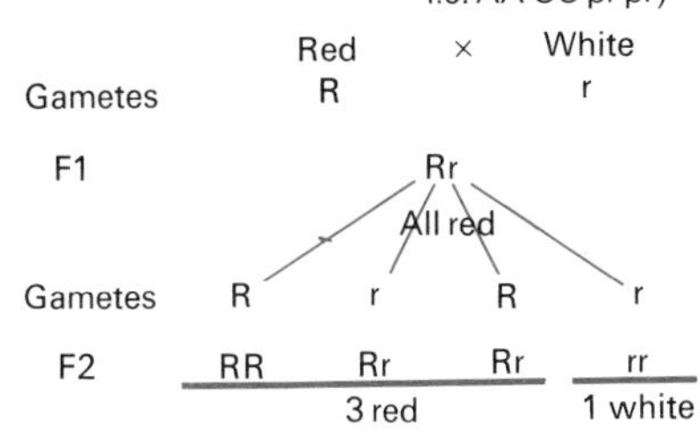

Figure 3.11 *The genetics of maize cobs*
Each maize cob is a collection of fruits since each 'grain' is a fruit developed from a single female flower. The fruit is a special type of achene with the fruit wall, or pericarp, fused with the seed coat, or testa. It is known as a caryopsis. The genes A, C, and R bring about different stages in the production of anthocyanin pigments. They are therefore complementary and, if any one is in the homozygous double-recessive condition, no pigment is produced. The gene Pr produces purple when A, C, and R are present. The triploid nature of the endosperm can be ignored as far as genetic calculations are concerned.

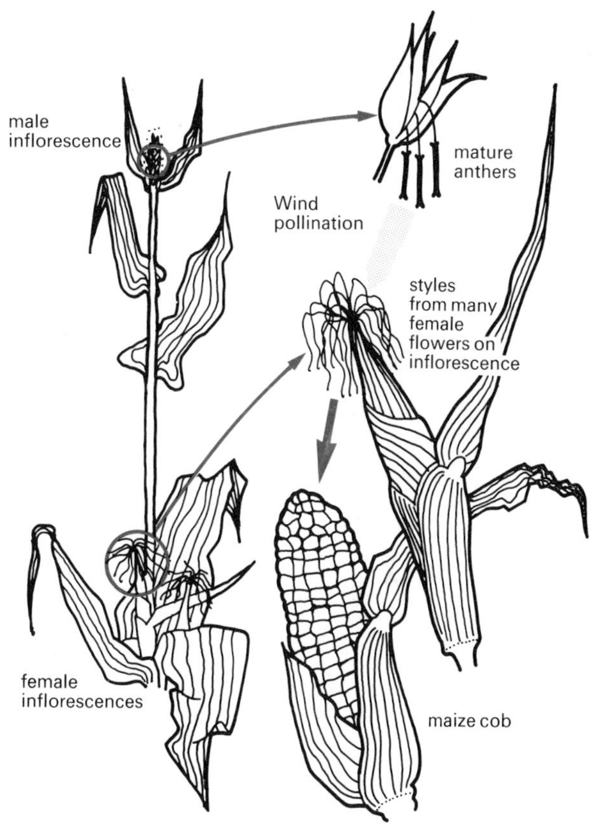

Figure 3.10 *The life cycle of maize*

Gene notation

The less common allele is usually referred to as the mutant allele, while the more common allele is usually referred to as the normal or 'wild' type allele. The 'case' of the letter used for a gene is determined by the mutant. If the mutant is recessive a small letter is used for both alleles, and the symbol $^+$ designates the dominant normal gene.

a	a^+
recessive mutant	dominant normal gene

If the mutant is dominant then a large letter is used for both.

A	A^+
dominant mutant	recessive normal gene

The simpler style of using a capital letter for the dominant gene and a small letter for the recessive, e.g. A, a, does not indicate which allele is the mutant.

Linkage

All organisms have more genes than they have chromosomes, therefore each chromosome must carry a number of genes, each in a specific position. Those genes which occur on the same chromosome are said to be **linked** and belong to the same **linkage group**. The number of linkage groups is equal to the haploid (not the diploid) number of chromosomes since each chromosome carries the same genes as its homologous partner. For example, assume that body colour and wing length of an insect belong to the same linkage group, so they are carried on the same chromosome. The fact that there are two chromosomes with different alleles (for example grey body, long wing; and black body, vestigial wing) does not mean that there are two linkage groups. They are different versions of the same linkage group of body colour and wing length. Theoretically, linked genes should not assort independently. For example, if AB, ab, are linked,

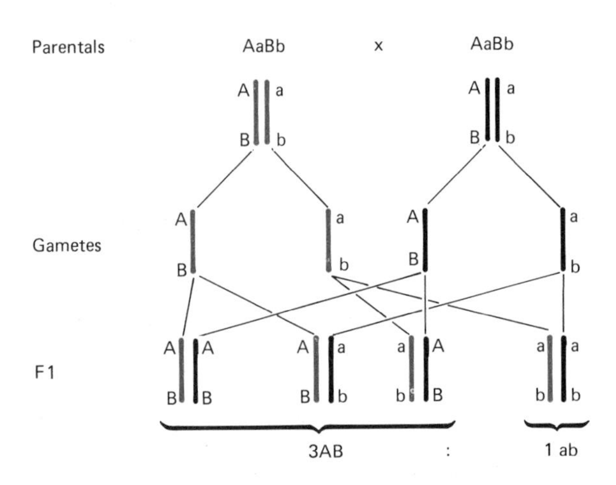

This gives the Mendelian monohybrid ratio, whereas normally, when considering two pairs of characters, the expected ratio would be 9:3:3:1. In other words, when dealing with a single pair of chromosomes the 3:1 ratio is expected. Again, with the test cross a 1:1 ratio should be obtained, and not the expected 1:1:1:1 ratio of two pairs of unlinked genes.

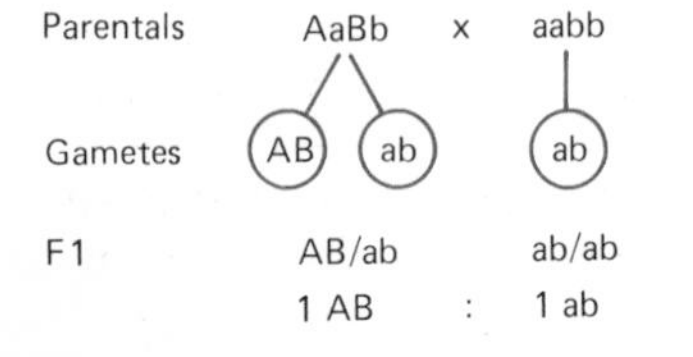

(Although when linked the convention is to show them as AB/ab x ab/ab.)

However the expected results for linked genes are seldom obtained. For example, in maize the genes for colour of aleurone and nature of endosperm are linked.

Dominant	Recessive
C: coloured aleurone	c: colourless
S: full endosperm	s: shrunken endosperm

As only one pair of chromosomes is involved, a test cross of a plant heterozygous for these genes with a homozygous recessive would be expected to produce a 1:1 ratio in the offspring.

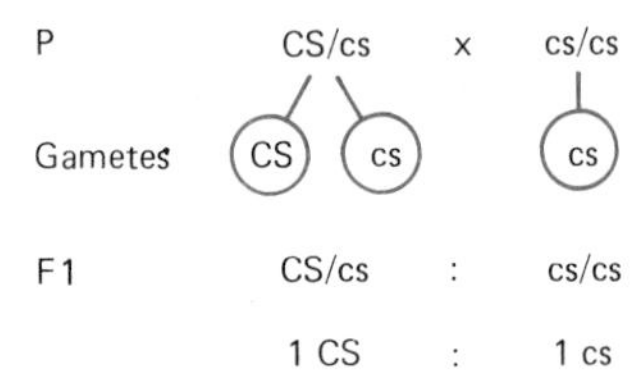

However, this result is not actually obtained. Some offspring with new combinations arise.

F1			
	CS/cs	4000	Expected 'parental' type
	Cs/cs	150	These 'unexpected' combinations have arisen by crossing-over
	cS/cs	150	
	cs/cs	4000	Expected 'parental' type

From these results a percentage crossing-over can be calculated.

$$\text{percentage cross-over} = \frac{\text{cross-over types}}{\text{total}} \times 100$$

$$= \frac{300}{8300} \times 100 = 3.6\%$$

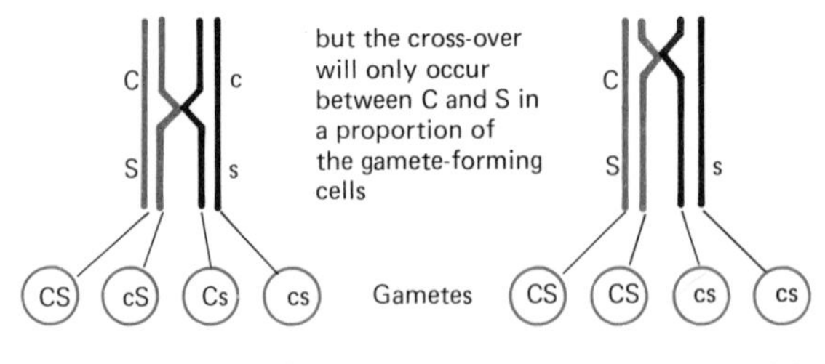

In a test cross these genotypes are expressed in the phenotype of the next generation, giving rise to the appearance of the so-called cross-over types.

In 1911 Morgan suggested that the strength of linkage between two genes is related to their distance apart on the chromosome. For example, if two gene positions or loci are far apart on the chromosome there is a greater chance of a cross-over occurring between them than if they are close together. This relationship enables chromosome maps to be constructed with genes in the correct linear order and with the correct distance between them. The map is considered to be 100 centimorgans long.

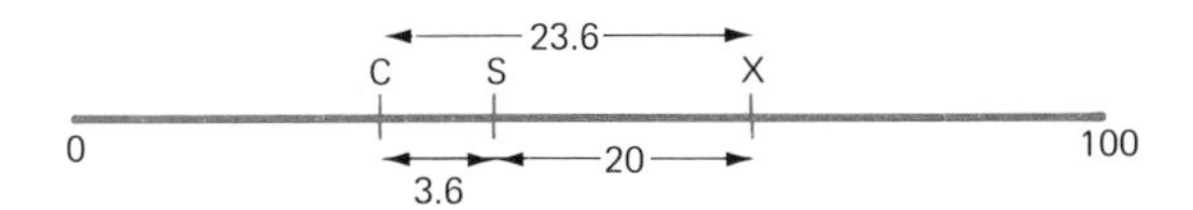

CROSSING-OVER

Crossing-over refers to the breakage of chromatids and the subsequent rejoining of non-sister

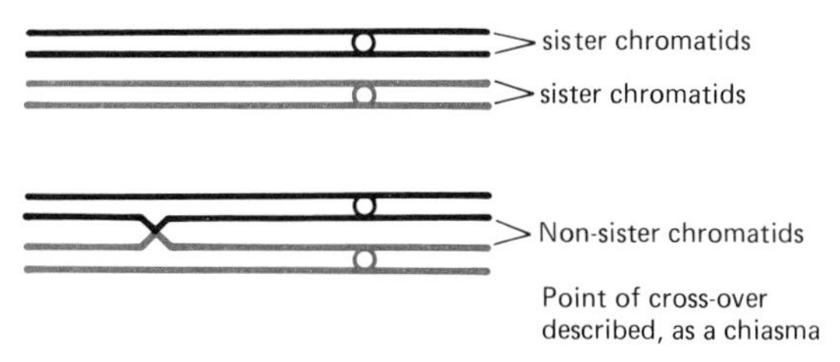

chromatids. It always occurs during meiotic prophase, the breaks coincide, and there is no loss of genetic material. There is a minimum 'separation distance' between cross-overs in the same pair of homologus chromosomes, as the occurrence of one somehow **interferes** with the formation of others in that region. Some regions of chromosomes are more prone to crossing-over than others, for example there are fewer at the ends and around the centromere. The frequency of crossing-over varies, but there is at least one occurrence on all pairs of homologus chromosomes and double cross-overs are common. The highest number of cross-overs recorded in a single pair is twelve in the broad bean (*Vicia faba*); the highest number seen in a single pair in man is about 7.

Crossing-over results in the exchange of genetic material between maternal and paternal chromosomes and leads to variation in the offspring. It also ensures that each pair of chromosomes remains a single unit until the end of first metaphase of meiosis.

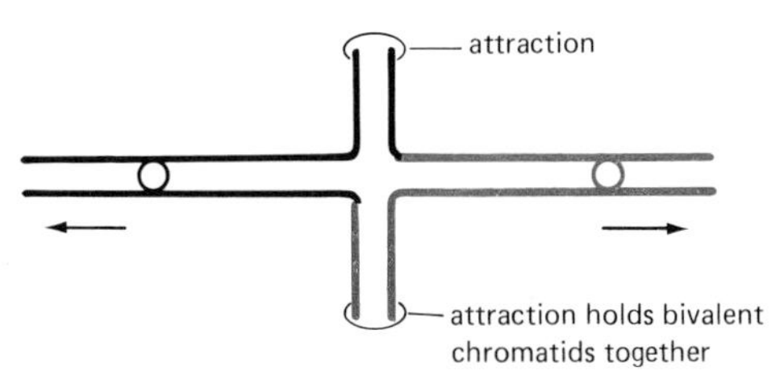

Without this, the homologous chromosomes would repel each other before lining up on the equator of the spindle in the first metaphase of meiosis and segregation would therefore be random, with little chance of the exact halving of the chromosomes into two haploid sets.

Various structural alterations to chromosomes, especially inversions, act to reduce the products of crossing-over and thus serve to keep sets of advantageous alleles together in blocks known as 'super genes'. In this way alterations to chromosome structure tend to reduce variation.

LINKAGE MAPS AND CYTOLOGICAL MAPS

It is expected that a linkage map constructed from the results of breeding experiments can be matched up with an actual chromosome and that the two are compatible.

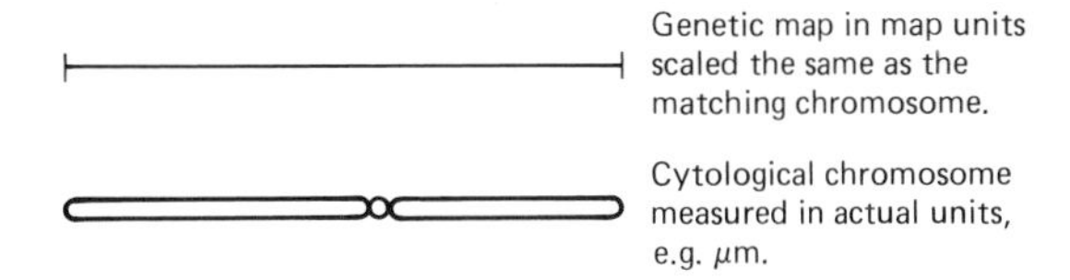

The first difficulty is to match a particular genetic map of a linkage group to a particular chromosome, as often individual chromosomes are indistinguishable from each other. One technique for identifying the correct chromosome is

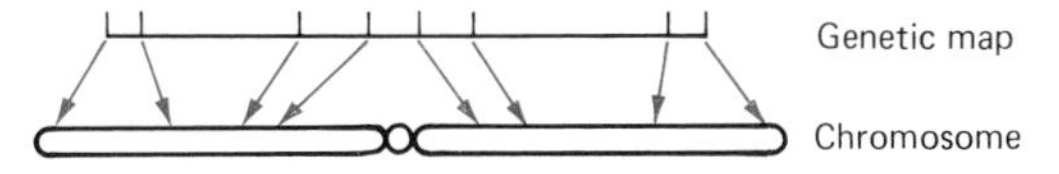

to watch for a translocation, in which a piece of one chromosome becomes detached and rejoins a completely different chromosome. If such a translocation occurs the genes carried on that segment of the chromosome appear in another linkage group in breeding experiments. This enables a linkage group to be matched to a particular chromosome.

When this matching has been achieved it is found that the genetic map does not match up with the actual chromosome. This is mainly due to the fact that cross-overs do not (as the mapping concept implies) occur with equal random frequency along the length of the chromosome. For example, the centromere prevents or interferes with crossing-over within a certain distance of it, and this so-called **centromeric inteference** causes this region to appear shorter on the linkage map than it is in reality. Also, double cross-overs between genes may not be detected since they restore the parental combination, and will tend to reduce the apparent distance between genes.

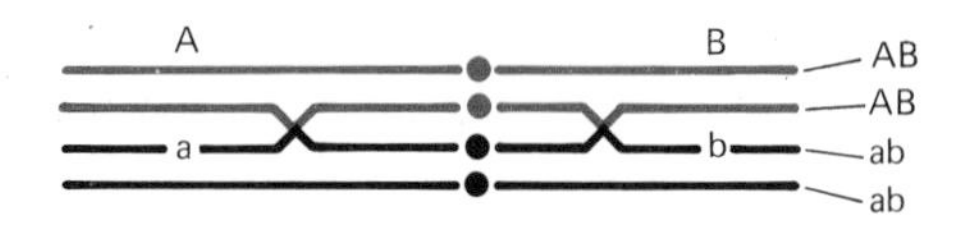

Sex determination

The control of the development of male and female reproductive structures in hermaphrodite organisms, such as many lower animals and most higher plants, is not fully understood.

In organisms with separate sexes a series of increasing numbers of genes are involved in the determination of sex, until a pair of distinct sex chromosomes can be recognized. Even in those organisms with sex chromosomes, however, there are sex-determining genes on the non-sex chromosomes or **autosomes**, and the sex-determining mechanism often depends on a balance between the sex chromosomes and the autosomes.

The sex with a pair of identical sex chromosomes is known as the **homogametic** sex and is usually the female. The sex with an unequal pair of sex chromosomes is known as the **heterogametic** sex and is usually the male. However, there are many exceptions in which the homogametic sex is male and the heterogametic sex is female, for example the Lepidoptera (butterflies and moths) some fish, and birds.

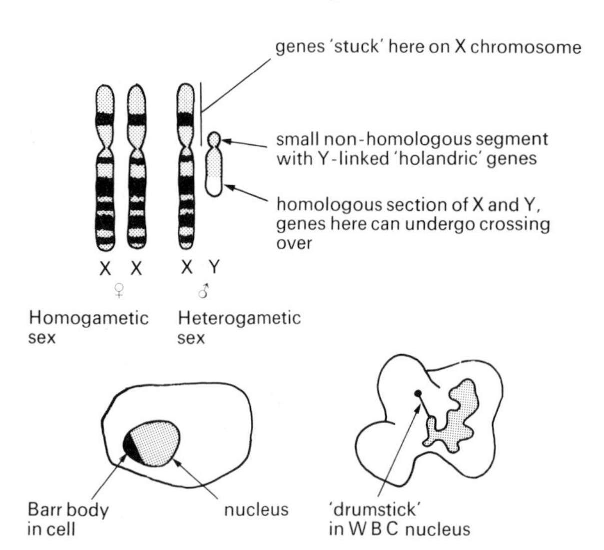

Figure 3.12 *Sex chromosomes and sex-linkage in man*
Genes 'stuck' on the X chromosome are inherited in a pattern related to sex, and are therefore known as completely sex-linked, e.g. RGCB, haemophilia, etc. (About sixty traits are known to be sex-linked.) *Holandric* genes are transmitted from father to son, e.g. hairy ears. Genes on the homologous sections behave like normal autosomal genes as they can cross-over from X to Y chromosomes. These are said to be incompletely or partially sex-linked, e.g. total colour blindness. The Y chromosome is strongly male-determining in mammals. One of the X chromosomes in the nucleus of all female cells remains permantly condensed throughout all phases of the cell cycle, and is seen as the Barr body, and drumstick chromosome in the white blood cells. This phenomenon affects the expression of X-linked genes.

Sex linkage

Sex chromosomes also carry genes which are not involved in sex determination, and are said to be **sex-linked**. In most cases sex-linked genes are X-linked, that is they are carried on the X chromosome. Such genes show an absence of male-to-male inheritance because the X chromosome of the male passes to the female offspring:

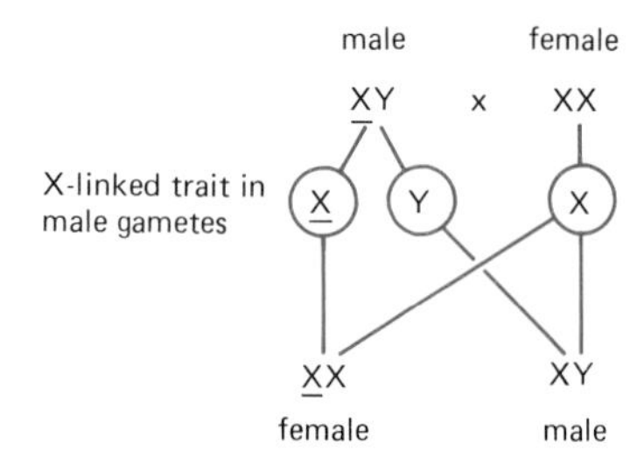

Sex-linked genes also show different results in **reciprocal crosses**, in which crosses are first carried out with the particular trait in one sex and later in the other, for example in *Drosophila*:

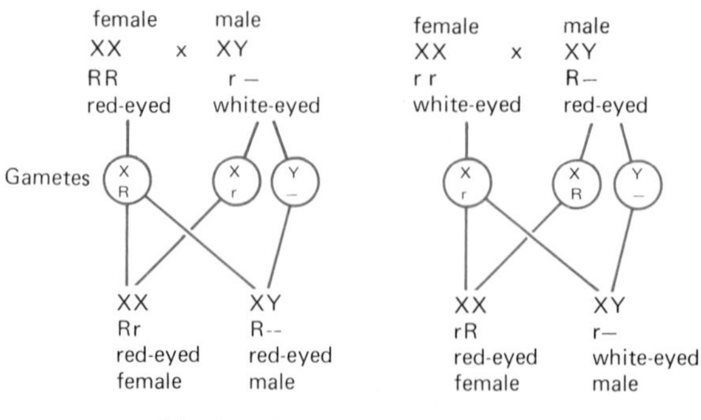

This was the first case in which a specific gene was associated with a specific chromosome (by Morgan in 1910).

A well-known example in humans is provided by **haemophilia**. This is an X-linked recessive gene, which is **semi-lethal**, that is, it is lethal in the double dose but not in the single dose. Thus females with two X chromosomes carrying the gene do not survive longer than about three to four months of foetal life. Females with one X chromosome carrying the gene do not develop the condition, but act as carriers by passing the gene to the next generation. Males with only one X chromosome can only have one dose of the gene but, as there is no dominant normal gene on the Y chromosome, this single recessive gene expresses itself and males suffer from the disease. The main symptom of haemophilia is an inability of the blood to clot normally and is due to a deficiency of blood-clotting factor VIII. Thus there is persistent bleeding following injury, and also recurrent bleeding into joints where movement continually ruptures small blood vessels.

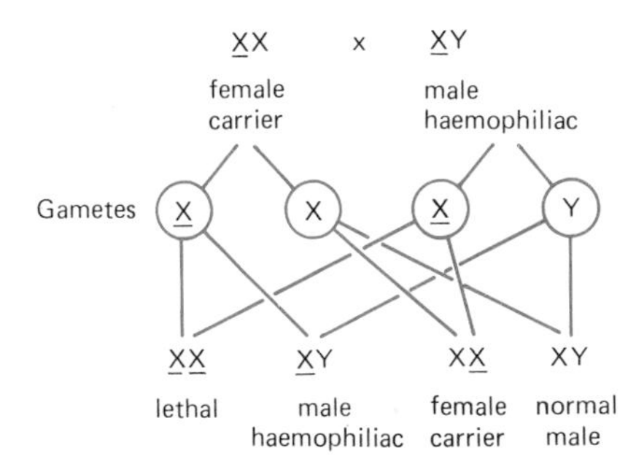

The net fertility of haemophiliacs is about 25 per cent of the normal level. As with any lethal or semi-lethal gene, it would be expected that the frequency of the gene in the population would gradually decrease. However, the frequency of the haemophilia gene in the population remains remarkably constant due to recurrent mutation, by which the normal gene spontaneously mutates to the haemophilia form at about the same rate as it is lost by selection against that trait.

The expression of sex-linked genes in the female is affected by the fact that one of the X chromosomes remains permanently condensed as a heterochromatic Barr body, in which form it is genetically inactive. For example, the **tortoiseshell cat** is heterozygotic for the X linked gene controlling coat colour (Y = black, y = yellow). If the X chromosome carrying the recessive gene is active, then the recessive yellow gene will be expressed; in those cells where the X chromosome carrying the dominant gene is active then the 'black' gene will be expressed. This gives rise to the phenotypic mosaic of the black and yellow tortoiseshell pattern.

In female carriers of haemophilia, the number of cells in which the normal dominant gene is switched off on a heterochromatic X chromosome will determine the degree of prolongation of blood clotting time which is seen in female carriers.

Sex limitation

In mice the gene for short ears is **pleiotropic**, which means that it affects another unrelated character, in this case the kidneys. The kidneys are enlarged and known as hydronephrotic. However, this is only seen in some offspring, so that the gene is said to have **incomplete penetrance**. Furthermore, a higher proportion of hydronephrotic kidneys are found in males (about 50 per cent) than in females (about 10 per cent). Thus this trait is also said to be **sex-limited** since its expression is limited more to one sex than another, normally as a result of hormonal influences.

Chromosome number

Different organisms have different numbers of chromosomes; some have very many small ones and some have a few large ones. The number of chromosomes is usually constant within a species. The number and size of the chromosomes does not appear to be directly related to the amount of genetic information carried, but it does have an effect on genetic recombination by crossing-over. Every pair of homologous chromosomes develops at least one cross-over during meiosis, but there is an upper limit to the number of cross-overs between chromosomes of each pair, due to so-called interference which prevents a cross-over from occurring too close to another. Long chromosomes therefore have a lower cross-over frequency per unit length, and shorter ones have a higher frequency per unit length. Thus if the chromosomes are long but few in number the genetic recombination due to crossing-over (a major source of variation in the offspring) is reduced when compared to organisms with larger numbers of smaller chromosomes.

Karyotyping

The morphology of the chromosome set as a whole, that is the number, shape, and size of the different chromosomes, is described as its **karyotype**. In any one species the karyotype is constant, except for a variety of chromosomal mutations that may occur. In some organisms each chromosome in a set has an identifiable size and shape; maize, for example, has ten such chromosomes in each set.

In other organisms it is impossible to identify all the individual chromosomes. Human beings, for example, have twenty-three chromosomes in each set, only a few of which can be identified on the basis of their appearance under the light microscope. However, specially treated chromosomes, when viewed with long wave ultra-violet light, show unique and identifiable patterns of fluorescent banding which enable individual chromosomes to be identified.

The fluorescent ends of the human Y chromosome are also visible as a small bright dot in the nuclei of non-dividing cells. Thus spermatozoa can be 'sexed' into X or Y types. (It is interesting to note that the fluorescent Y chromosome is only found in one other mammal, the gorilla.) The identification of all the chromosomes by their fluorescent patterns enables chromosomal abnormalities to be more easily identified, and similarities in the pattern of banding between organisms may suggest evolutionary relationships, for example man, chimpanzee, and gorilla show certain similarities in banding pattern not seen in other apes.

The structure and behaviour of chromosomes indicate that they are not merely strings of genetic material, but are highly differentiated cell organelles.

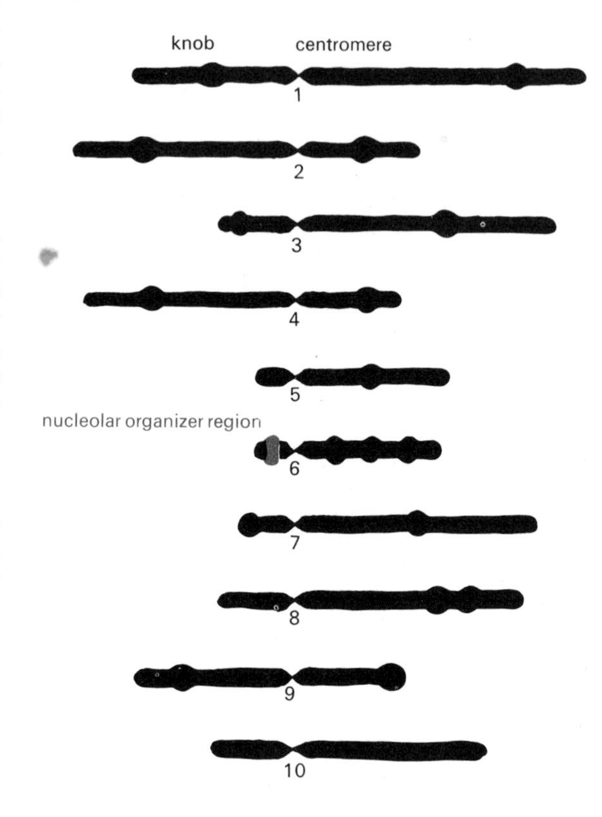

Figure 3.13 *Maize chromosome set*

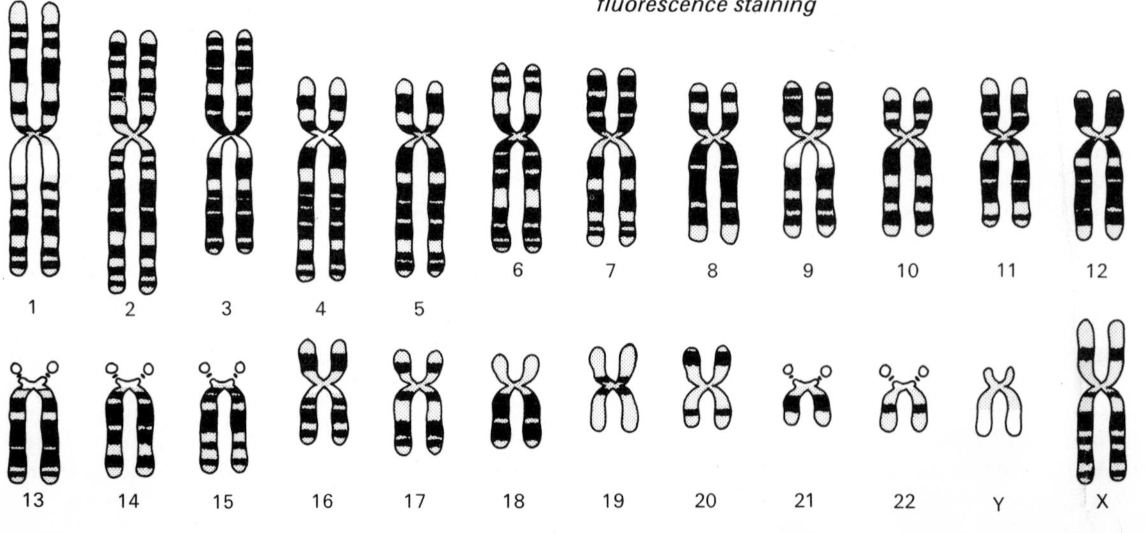

Figure 3.14 *Human chromosome set with u.v. fluorescence staining*

The genetic material

The molecular basis of inheritance is **deoxyribonucleic acid (DNA)**. Cells of organisms contain both DNA and a substance known as **ribonucleic acid (RNA)**. The DNA transmits the genetic message from generation to generation, and from cell to cell in an individual; the RNA is involved in the **transcription** and **translation** of this genetic information into the reality of cell proteins, mostly enzymes, which in turn control the structure and function of the rest of the cell.

Much of the work on the structure and function of DNA and RNA has been done on bacteria, and is assumed to apply to all living organisms. However, there are some important differences between the genetic material of bacteria and that of the cells of higher organisms. For example, bacteria are **akaryotic**, that is they do not have a true nucleus, whereas the cells of higher organisms are **eukaryotic**, with the DNA contained in a true nucleus surrounded by a nuclear membrane. In addition, the amount of DNA in a nucleus in higher organisms appears to be many hundreds of times the amount needed to code for all the possible information.

Fig. 3.15 *Structure and replication of DNA*
Strands of a single molecule are held together by hydrogen bonding between the bases. In replication the double helix unwinds and each old strand acts as a template or blueprint for the formation of a new strand. The replication process requires energy (ATP) and the enzyme DNA polymerase. Replication is semi-conservative, i.e. each new molecule contains half of the old.

Deoxyribonucleic acid (DNA)

DNA is a giant macro-molecule known as a polynucleotide. A **polynucleotide** is a string of sub-units or nucleotides. A **nucleotide** is a compound formed from one molecule of a pentose sugar (with five carbon atoms), one molecule of a nitrogenous base (either a purine or a pyrimidine), and one molecule of phosphoric acid. Such nucleotides are found free in the cytoplasm, as for example in ATP,[367] or as the basic units from which the nucleic acids are built. The sugar and phosphoric acid form the 'back-bone' of a strand of DNA and the bases 'stick out' at the side.

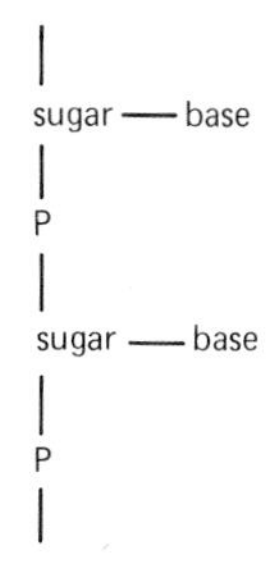

In the nucleic acids the sugar is always ribose. The purine bases are adenine (A) and guanine (G), and the pyrimidine bases are cytosine (C) and thymine (T). DNA is a double stranded molecule, made up of two such strands with the bases forming pairs. Adenine always pairs with thymine, and cytosine always pairs with guanine.

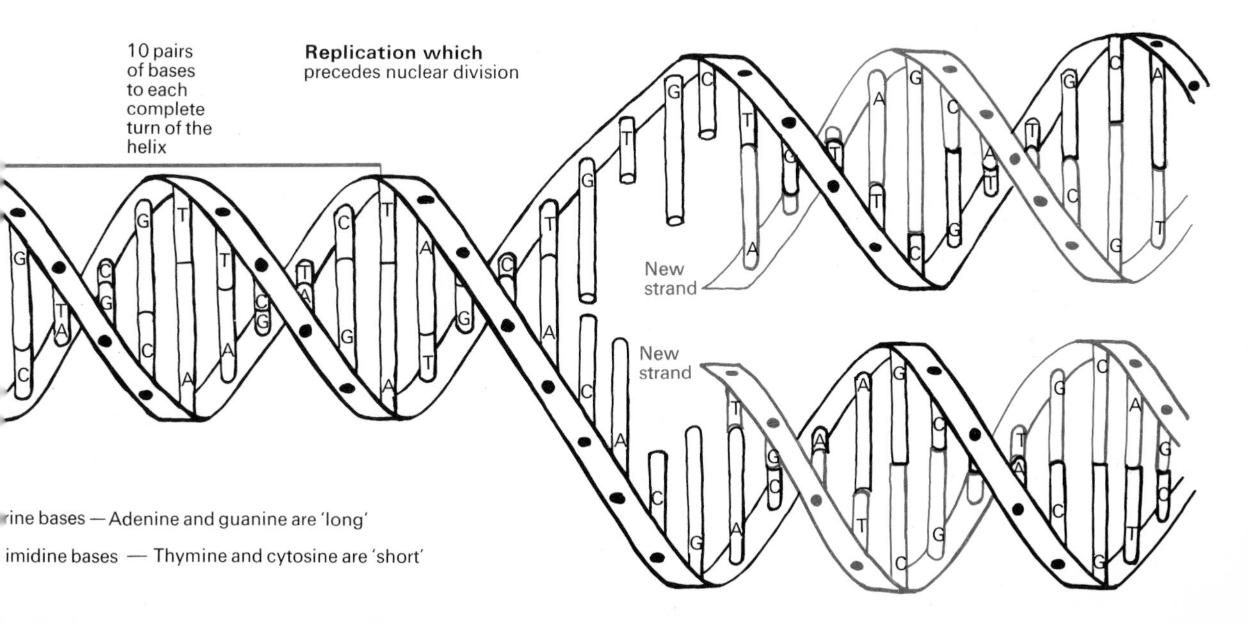

Ribonucleic acid (RNA)

RNA is similar in many features to DNA, but there are also some major differences. The sugar molecule has one more oxygen atom than that in DNA, and the base uracil is found instead of thymine. It is less polymerized (it is made up of fewer nucleotides) and is a single stranded helix with one strand doubled back and wound around itself.

There are many different types of RNA, each with a different structure and function.

Genetic RNA carries the genetic message in most plant viruses, some animal viruses, e.g. polio, and some bacteriophages.

Messenger RNA (mRNA) carries the genetic message from the DNA in the nucleus of cells to the cytoplasm, where the message is translated into protein in conjunction with the ribosomes.

Ribosomal RNA (rRNA) is part of the structure of ribosomes and is somehow involved in the translation of the message in mRNA.

Transfer RNA (tRNA) transfers or ferries amino acids from the surrounding cytoplasm to the mRNA attached to the ribosomes.

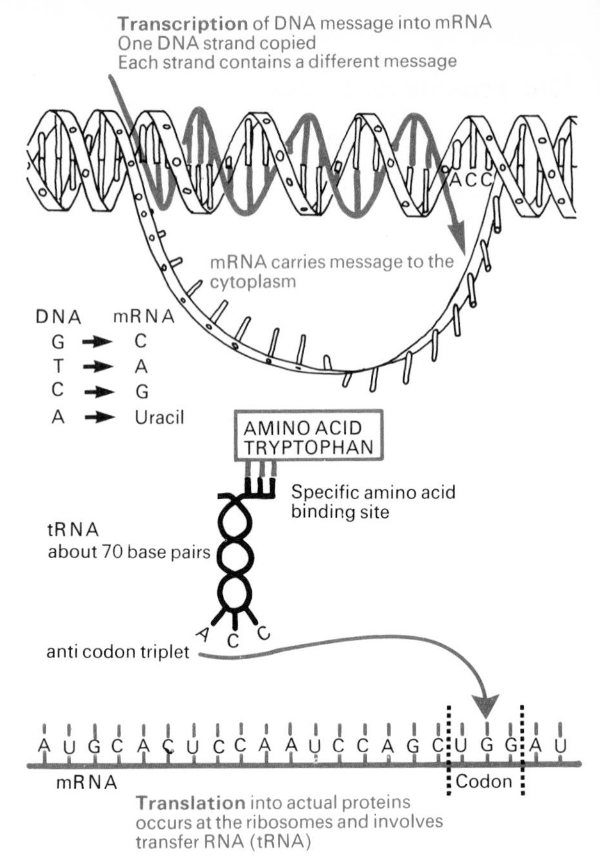

Figure 3.16 *Transcription and translation of the genetic message*

Figure 3.17 *Transcription and translation in context*

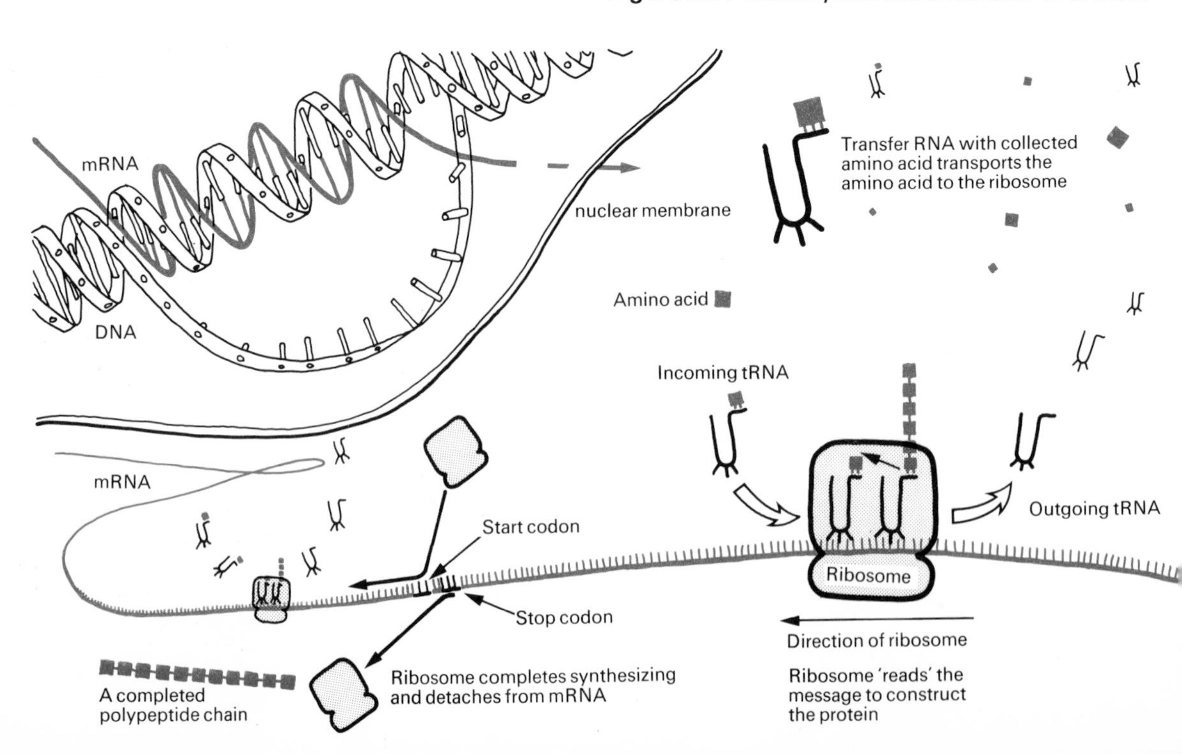

The genetic code

The **sequence of bases** on a strand of DNA contains a **coded message** for the construction of proteins. Proteins are the very basis of living matter; they form the structure of the cell and control all of its functions by means of those proteins which form enzymes.

All proteins are composed of combinations of about twenty different amino-acids joined together by peptide bonds to form long polypeptide chains. The sequence of bases on a particular section of a strand of DNA codes for the sequence of amino acids in a particular protein. Each amino acid is coded for by a 'word' of three bases known as a **triplet codon**, e.g. the base sequence ACC codes for the amino acid tryptophan. However, there are sixty-four different combinations of any three letters taken from four, and as there are only twenty different amino acids to code for, this implies that there are forty-four spare triplet codons.

In fact, more than one triplet codon can code for a particular amino acid, e.g. ACA, ACC, and ACG all code for the amino acid serine. Altogether sixty-one triplet codons have been shown to code for the twenty amino acids. The three spare are used as **punctuation marks**, for example stop and start signals between segments of DNA that code for different proteins. These are necessary because a strand of DNA will code for many protein molecules along its length. It is estimated that a sequence of about 1000 bases is necessary to code for a single polypeptide chain.

The complete genetic message for the construction of a living organism has been worked out. The full DNA code for the bacteriophage virus φX174 is now known, and serves to illustrate the problem of analysing the code for higher organisms. The code for φX174 is described on a computer print-out 15 m long. It is estimated that a similar print-out of the code for a human being would be at least 16 000 km long. Each human chromosome is estimated to contain 'DNA' with about 200 million base pairs, giving a total of about 10 000 million base pairs in the complete code for a human being.

The gene concept

The classic concept of the gene is as a unit of chromosome structure not sub-divisible by chromosome breakage, rather like a 'bead' in the 'necklace of the chromosome'. For many purposes this concept provides a good working model; however, for others it does not suffice. Many such 'structural' genes are found in fact to be sub-divisible by chromosomal breakages such as crossing-over. Thus the so-called **'one gene one enzyme' hypothesis**, by which a structural gene was considered as coding for the production of one protein enzyme, had to be modified. It was discovered that the gene for a protein, such as the globin part of haemoglobin, could be sub-divided by crossing-over, and that the protein globin consisted of two polypeptide chains. Thus, the original hypothesis was modified to the **'one gene one polypeptide' hypothesis**. However, in many cases the 'one gene, one enzyme' relationship still appears to exist between an enzyme and the gene that controls the final stage in the rendering of that enzyme active. In most cases the active enzyme is made by the 'dominant' gene, and not, due to some deficiency, by the recessive allele.

The gene concept is now very confused, and new terms have been introduced in an attempt to clarify the situation. The functional biochemical unit, that is the amount of genetic material that codes for a functional product such as an enzyme or a polypeptide, is known as a **cistron**. The smallest structural unit, which cannot be broken by crossing over, is known as a **recon**; and the smallest unit of genetic material that, when mutated, produces a phenotypic effect, is known as a **muton**. These terms are not clear-cut since a certain amount of genetic material could be a cistron, a recon, and a muton, all at the same time.

Further problems arise when attempting to relate these definitions of units of genetic material, derived from breeding experiments, to the structure of DNA. The exact relationship between the chromosomes and the DNA they contain is not yet understood. The length of a chromosome certainly bears no simple relationship to the number of genes carried. Indeed, the genetic material acts as an integrated system, with all the 'genes' interacting in a complex of ways with each other and with the surrounding cytoplasmic environment.

Experimental and applied aspects

Molecular basis of inheritance

Work on the **genetic transformation** of bacteria in the 1920s and 1930s, and on viruses in the 1950s, led to an understanding of the nature of the genetic material.

The bacterium *Diplococcus pneumoniae* has a virulent strain with smooth capsules (**S**) and a non-virulent strain with rough capsules (**R**). It was found that, from a mixture of heat-killed (**S**) and living (**R**) injected into mice, living (**S**) could be isolated, and that this transformation was stable and inherited. Later experiments showed that the only substance in the bacteria that could effect this transformation by passing from the heat-killed (**S**) into the living (**R**) types was a substance known as **deoxyribonucleic acid** (**DNA**).

Modification of the phenotype of the *Diplococcus pneumoniae* capsules is very common, for example, if it is grown on a sucrose-containing medium the capsules are smooth, and if on a glucose-containing medium the capsules are rough. However, the transformation by DNA is both stable and genetic, indicating its role as the genetic material.

Other evidence was gained from work on **viruses**.[288] These consist basically of a thread of DNA protected by a protein sheath. They infect cells by attaching to the cell surface and injecting their DNA into the cytoplasm of the host, leaving a 'protein ghost' outside the cell. The injected DNA then directs the host cytoplasm to construct complete viruses which are liberated when the host cell is destroyed. It was the investigation of this process of virus infection and reproduction that further demonstrated the role of DNA as the genetic material. Most plant viruses, however, do not contain DNA but contain RNA which acts in the same way.

Control of the expression of the genetic code

As the fertilized egg cell divides into the multicellular body, each cell receives a complete copy of the genetic message. Each cell is said to be **totipotent**. The totipotency of animal cells has been demonstrated by transplanting the nucleus from a cell in the intestine wall of *Xenopus* (the African clawed toad) tadpoles into a fertilized egg which has had its own nucleus destroyed, and observing its subsequent development into an individual identical to the one from which the intestinal cell nucleus was taken. This indicates that the cytoplasm of the egg cell is important in controlling the full expression of the genetic message.

Single isolated plant cells can give rise to whole plants if provided with the correct stimulus and conditions. For example, single carrot root phloem cells in suspension in a culture medium with coconut milk develop into embryoids which grow into plantlets and then into mature plants. Such cloning of single cells from good stock will be of considerable economic importance.

A single cell expresses its genetic totipotency as it is freed from the complex of spatial 'fields of influence' which are normally imposed upon it by adjacent cells. During normal development it is these three dimensional gradients of influence which control the differentiation of the cells, only allowing them to express that part of their genetic information which is relevant to the future role of that cell in the organism.

As the final role of each cell becomes determined, only that part of the genetic message which is relevant to its position and function in the body is expressed. Therefore in all cells of the adult body most of the genetic message is switched off and only those genes required are switched on. This control of the expression of the genetic code could theoretically occur by

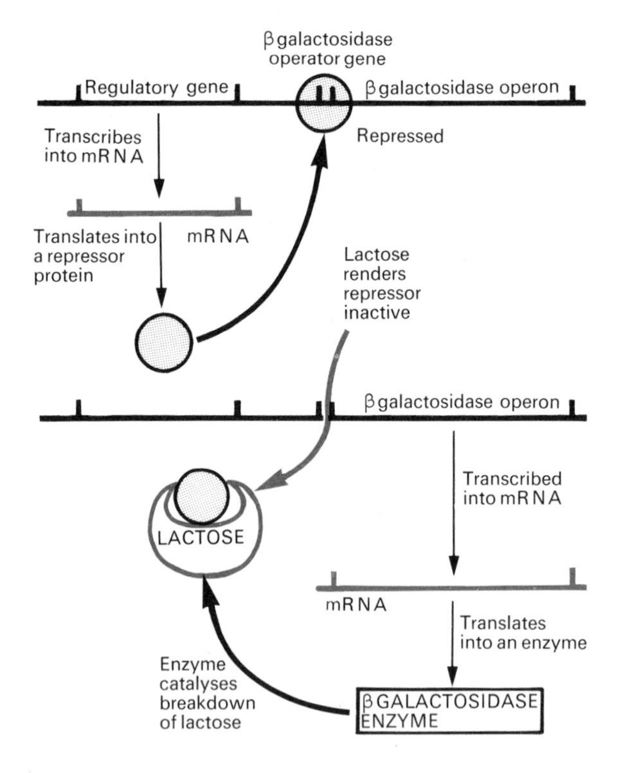

Figure 3.18 *Enzyme induction (Jacob–Monod model)*
In *E. coli* the enzyme *β* galactosidase is only produced in the presence of its substrate lactose (milk sugar). Thus the presence of the substrate induces the synthesis of the enzyme. The mRNA produced is unstable and short-lived (about two minutes) so that when the lactose disappears the gene is switched off and the mRNA does not remain to direct the synthesis of the enzyme. However, no similar process has yet been found in eukaryotic (nucleated) cells. Nevertheless in higher organisms there is some evidence that some vitamins and hormones exert their effect by regulating the transcription of certain segments of DNA.

regulation of either transcription or translation. Experimental evidence indicates that, in most cases at least, control is exerted by the regulation of transcription. Studies on the production of the enzyme β galactosidase by the bacterium *Escherichia coli* indicate one method by which this control can be achieved.

Observations of the special types of chromosomes, such as the salivary gland or **polytene** chromosomes of the Diptera, e.g. *Drosophila* (the fruit fly) and the lamp-brush chromosomes of the Amphibia, also indicate that control is exerted by the regulation of transcription.

In the salivary glands of fly larvae there are large cells and nuclei, which increase in size but do not divide. These contain very large chromosomes (which appear to be accumulations of many chromatids that have not separated) in a permanent mitotic metaphase which show **somatic pairing**; so that they are permanently condensed and paired in these body cells. Similar chromosomes are seen in the larvae of *Chironomus* (a midge, the larva of which is known as the 'common pond bloodworm').

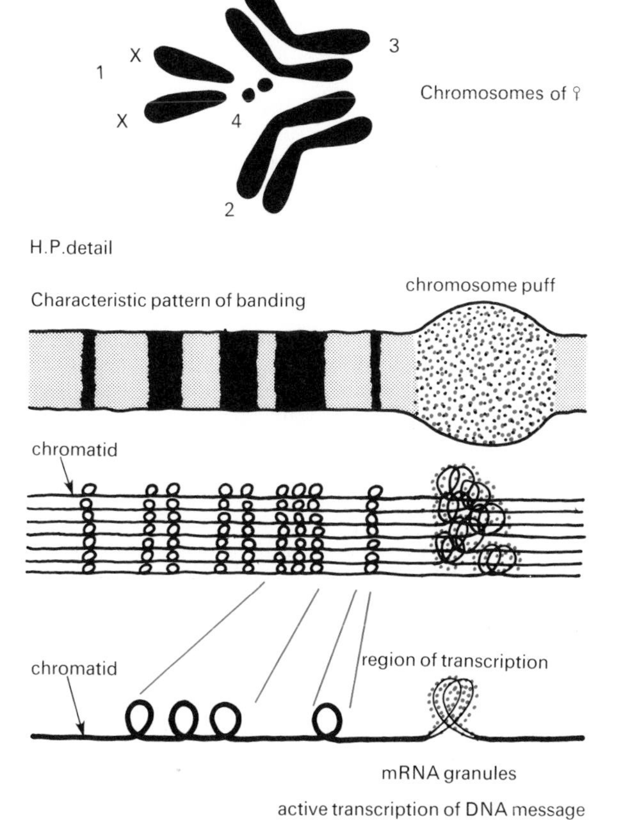

Figure 3.19 *Polytene chromosomes of Drosophila melanogaster*

At certain periods, puffed regions known as Balbiani rings can be seen on these polytene chromosomes. These are regions of active transcription of the genetic message into mRNA. In other words they are regions where the DNA or 'genes' can actually be seen in operation.

Lampbrush chromosomes occur in the primary oocytes of organisms with yolky eggs, but they are best seen in the Amphibia, particularly the Urodeles. The primary oocyte nucleus enters meiosis and reaches the diplotene stage, at which it remains for eight to nine months. During this time loops arise in the chromosomes which, like the Balbiani rings, are regions of active mRNA formation in the process of transcription.

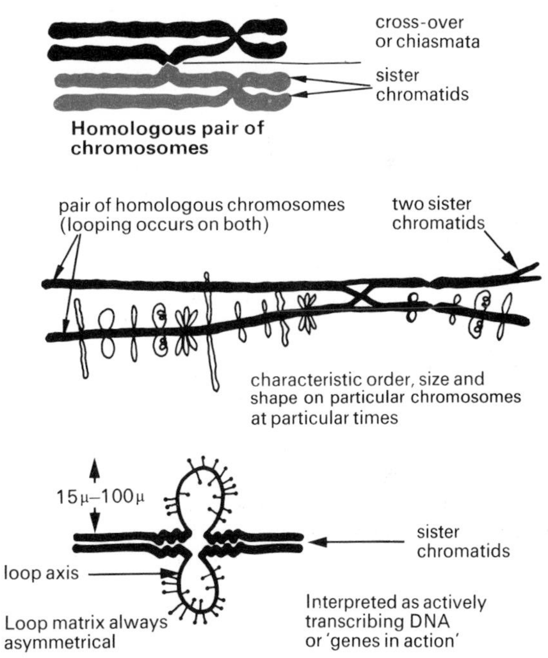

Figure 3.20 *Lampbrush chromosomes*
These occur in primary oöcytes of organisms with yolky eggs, but they are best seen in the Amphibia. In newts, the nucleus is 0.5 mm in diameter and the chromosomes can be up to 1000 μ long. They start meiosis, enter diplotene, and stay there for up to nine months, during which period they develop the changing pattern of loops which are thought to be important in the organization of the egg cytoplasm, and upon which development after fertilization depends.

Guided example

Cytological basis of inheritance

The knowledge that chromosomes are the structures that carry the genetic message is based on a variety of evidence, the main points of which are summarized below:

(a) The zygote nucleus is formed by the fusion of the nucleus of the female gamete and the nucleus of the male gamete. The male gamete contributes only nuclear material to the zygote, and characteristics from the male appear in the offspring. Therefore it would appear that the genetic message must be contained in the nucleus of cells. When the nucleus is examined the most obvious structures during nuclear division are the chromosomes.
(b) Mitotic nuclear division of the zygote ensures exact replication of the chromosomes to all the cells of the body. Meiotic nuclear division in gametogenesis ensures the balanced distribution of the chromosomes to the gametes and subsequently to the next generation. Thus chromosomes have complete continuity between cells and generations.
(c) At meiosis in gametogenesis the chromosomes behave in the same pattern as Mendel's genetic factors.
(d) The number of linkage groups (groups of linked genes as determined by breeding experiments) never exceeds the haploid number of chromosomes.
(e) As a result of linkage studies, genes can be mapped relative to each other in a linear sequence as if borne on an elongated structure such as a chromosome.
(f) Sex determination in many organisms is accounted for on the basis of the possession of observably different sex chromosomes.
(g) Studies of the inheritance of sex-linked characters indicate that the genes controlling these factors are carried on the sex chromosomes.
(h) Structural alterations to chromosomes, such as inversions, translocations, deletions, and crossing-overs, result in an alteration to the behaviour of genetic factors.

1 Which of these do you consider to be the more convincing pieces of evidence that chromosomes carry the genetic material? Rank them in order of their relative importance (in your opinion).
There is no absolutely correct rank order, but certainly some are more direct pieces of evidence than others.

(h) is strong evidence, particularly with translocations. Translocations between chromosomes can be observed cytologically and result in observable genetic effects because a certain number of genes change their linkage group. Moreover, deletions result in loss of both chromosomal and genetic material. (f) and (g) are helpful as they also involve observable chromosomal differences.

(b) looks convincing, but maternal cytoplasm also has complete continuity between cells and generations. For example, all the mitochondria in the body of an individual are derived from the cytoplasm of the egg, which in turn is derived from the cytoplasm of the maternal tissues. So an argument based on continuity alone is not strong evidence. (a) and (b) together are more convincing, since the male contribution to inheritance shows that the genetic material cannot be carried in the maternal cytoplasm alone. Note that (a) and (b) do not exclude that some genetic material is carried in the maternal cytoplasm.

(c) and (d) are similar in that the behaviour of the genes in breeding experiments is related to the behaviour and form of the set of chromosomes. (e) only relates genes to linear structures and there are many of these in cells other than chromosomes, such as, for example protein fibres.

*From the above, one suggested rank order could be (h), (f) and (g), (c) and (d), (a) and (b), (e). However, it is as a **body of evidence** that these points prove that chromosomes carry the genetic material because there are no other structures in the cell to which all the points apply.*

Questions

1 (a) Compare the prophase of mitosis with the prophase of the first division of meiosis.
(b) Comment on the significance of any differences between these two types of prophase.
(c) Where does meiosis occur in the following: (i) a fern, (ii) a mammal, (iii) a flowering plant? (L)

2 Distinguish between the terms genotype and phenotype, illustrating your answer by reference to Mendel's second law of independent assortment.

In mice, coat colour is controlled by several pairs of alleles, including the following:

A, agouti (wild-type colour), dominant over a, black

C, coloured (i.e. pigmented), dominant over c, albino

If a black individual, CCaa, is crossed with an albino, ccAA, what will be the appearance of (a) the F1 and (b) the F2 generations? Explain briefly your results. (L)

3 What is meant by 'sex-linkage'? A normal red-eyed male fruit fly was crossed with a female fly which was also red-eyed. All the female offspring had normal red eyes, but about half the males had white eyes and half normal eyes. How would you account for these results? (SUJB)

4 (a) Briefly describe, with the aid of *one* example in each case, what you understand by the following: (i) chromosome mutation, (ii) incomplete dominance, (iii) sex-linkage.

(b) Blood groups ABO in humans are under the control of genes (I) which occur in three forms, I^A, I^B, I^O, any two of which can occur together:

Blood type	*Genotype*
O	I^OI^O
A	I^AI^A or I^AI^O
B	I^BI^B or I^BI^O
AB	I^AI^B

A baby has blood type B, his mother has blood type A, his paternal grandfather has blood type A, and his paternal grandmother has blood type B. Determine (i) the genotype of the baby and (ii) the possible genotypes of the baby's father. Use symbols as above to show how you reached your conclusions. (L)

5 In a species of plant petal colour is determined by one pair of alleles and stem length by another. The following experimental crosses were carried out.

Experiment 1 A purple-flowered plant was crossed with several red-flowered plants. The progeny were all purple-flowered plants.

Experiment 2 A short-stemmed plant was crossed with several long-stemmed plants. The progeny were all short-stemmed.

Experiment 3 A different purple-flowered, short-stemmed plant of the same species was crossed with several red-flowered long-stemmed plants. The following progeny were obtained:

37 purple-flowered short-stemmed
34 red-flowered short-stemmed
41 red-flowered long-stemmed
35 purple-flowered long-stemmed

(a) What are the dominant alleles?

(b) What are the probable genotypes of the purple-flowered and short-stemmed plants used in Experiments 1 and 2?

(c) With the aid of diagrams, explain the results obtained in Experiments 1 and 2.

(d) With the aid of a diagram, explain the results of Experiment 3.

(e) If the purple-flowered, short-stemmed plant used in Experiment 3 had been self-fertilized, what proportion of the progeny would you have expected to be red-flowered and what proportion would you have expected to be short-stemmed?

(f) From your knowledge of reproduction in flowering plants explain (i) how the crosses required in Experiment 3 could be ensured, (ii) how self-fertilization could be ensured. (AEB)

6 (a) State Mendel's two principles of segregation.

(b) Explain why the 'back cross' is also termed the 'test cross'.

(c) In Mendel's notes it was found that he recorded that 258 plants yielded 8023 seeds of which 6022 were yellow and 2001 were green, and that each pod usually yielded both kinds of seeds. Explain these results. (L)

7 What is the *evidence* that deoxyribonucleic acid (DNA) is involved in the mechanism of heredity? Briefly describe the structure of DNA. How do the structure and the role of DNA differ from those of ribonucleic acid (RNA)? (L)

Further reading

Paterson D., *Applied Genetics* (London: Aldus Books, 1969).

Levine L., *Biology of the Gene* (Saint Louis: Mosby Co., 1973).

Penrose L. S., *Outline of Human Genetics* (London: Heinemann, 1973).

Watson J. D., *The Double Helix* (London: Weidenfeld and Nicolson, 1968).

4 Evolution

The basic concept of organic evolution is that today's organisms have evolved from ancestors which were of different form; in other words, all organisms are related to their ancestors by descent with modification.

Evidence for evolution

All the evidence for evolution above the level of the species is indirect and open to many alternative interpretations. However, when taken together, the various pieces of information provide a considerable amount of evidence in support of the theory of evolution. This in turn provides a unifying theme for a wide variety of often apparently unrelated phenomena.

Evidence from taxonomy [331]

The fact that organisms can be placed in a hierarchical system of classification, from the simple to the complex on the basis of structure, suggests that they could be related by descent with modification.

Evidence from comparative anatomy and morphology

Classification is based on the observation that certain structures show similarities, particularly during embryonic development, and these similarities indicate that these structures could have been derived from some common ancestral form.

A good example of such **homologous structures** is seen in the vertebrate pentadactyl limb, where it is suggested that adaptive radiation of a common structure, that is a diversity of form arising from different needs, has resulted in the **divergent evolution** of differently adapted structures. Indeed, the theory of evolution conceives of a series of major changes such as the development of the pentadactyl limb, being followed by their adaptive radiation into divergent forms. However, many such homologous structures are not homologous genetically, that is their development is not controlled by the same genes, and it is hard to see how such similar phenotypes with differing genotypes could be related by direct descent with modification.

Figure 4.1 *Vertebrate pentadactyl limbs*

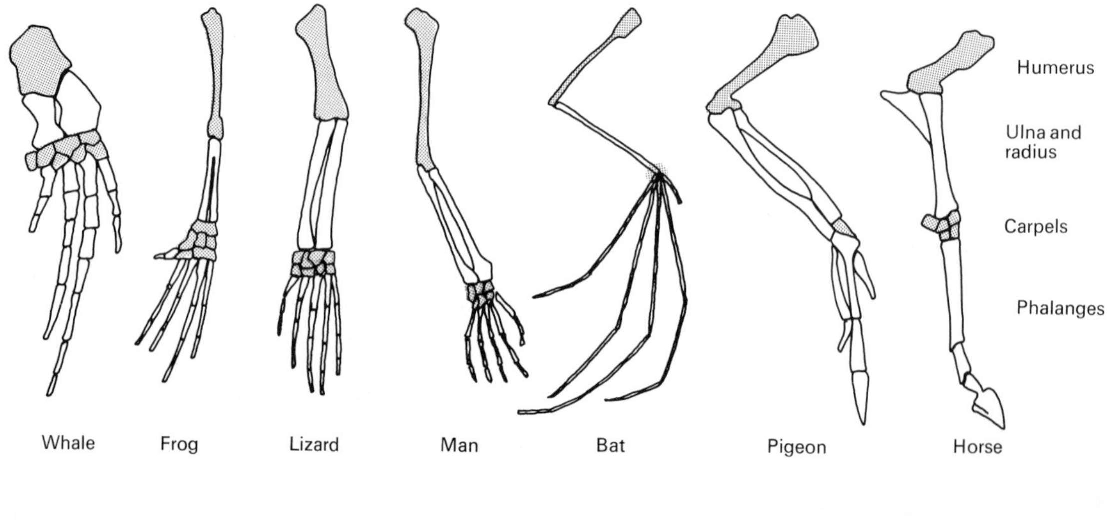

The possession of **vestigial structures**, such as reduced bones in the horse limb and stomata on the petals of flowering plants, is also taken as evidence of descent with modification.

Analogous structures are those that have the same functions but which are not homologous, for example the eye of the squid and the eye in vertebrates. The possession of analogous structures is not taken to imply a close evolutionary relationship between the organisms that possess them. It is regarded as an example of **convergent evolution**, whereby structures of different origins have become adapted to a common function.

Evidence from palaeontology

The study of fossils introduces a time-scale to evolutionary considerations. Fossil evidence indicates that, in most cases, simple forms appear in time before the complex; that within groups there is a trend towards increase in size of the largest members; and that the process of evolution is irreversible.

The simple explanation of fossilization is that fossils were formed by organisms which died in water and were rapidly buried in silt. As the silt formed **sedimentary rocks**, an impression of the hard structures of the organisms was left. Most fossils are found today in the layers of sedimentary rocks. It can be imagined that the chances of the occurrence of the correct conditions for fossilization would have been very rare. However, the fossil record is far from complete and prior to the appearance of the reptiles and mammals the fossil record is scanty and unreliable.

The age, and therefore the sequence, of the different strata in the **geological column** (see Table 4.1), has in the past been largely based on the fossils found in the layers of sedimentary rock. The justification for such dating assumed the correctness of the theory of evolution.

However, other dating techniques now exist, including those based on the rates of radioactive decay[361] of material found in the fossils. These use the fact that radioactive C^{14} is formed in the atmosphere by the activity of cosmic radiation on carbon dioxide in the upper atmosphere. It is an unstable radioactive isotope[361] which decays with a half life[361] of about 5600 years. Plants absorb some of this C^{14} during photosynthesis and it is incorporated into their organic compounds and into those of animals in the food chain founded on them. When the organism dies the accumulation of C^{14} stops and the amount present in its tissues decays. By assuming that the atmospheric level of C^{14} has remained constant over the period considered, and that it is incorporated into plant and animal tissues in the same proportion as it is in the atmosphere, it is possible to estimate the length of time elapsed since a plant died from the level of C^{14} remaining in the fossil.

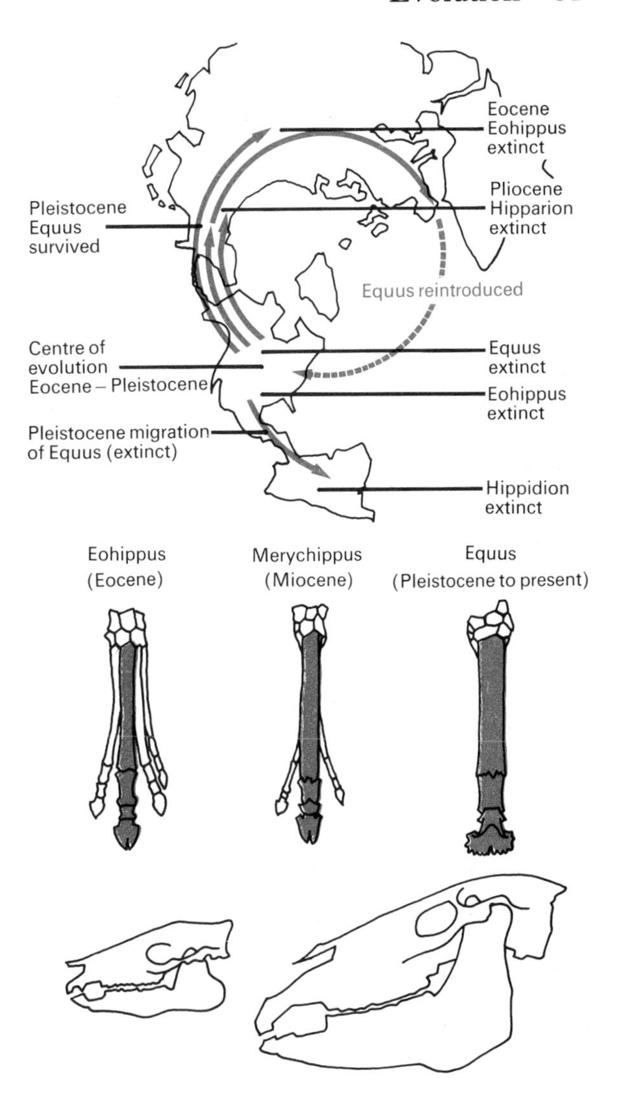

Figure 4.2 *Suggested pattern of horse evolution*
Eohippus: a small four-toed browser 25–50 cm tall, which fed on soft herbage 60 million years ago; *Merychippus*: a 100 cm tall three-toed grazer 26 million years ago; *Equus*: a large one-toed grazer with teeth and skull well adapted to chewing grass.

Table 4.1 *The geological column*

Era	*Period*	*Estimated time since start of period in millions of years*	*Some main features of the animal fossil record*	*Some main features of the plant fossil record*	*Era*
Cenozoic age of mammals	Holocene (recent) 10 000 years ago	0	If the time since life first appeared is equated to a 24-hour period then humans first appear at 23.59 hours and recorded human history begins at 23.59.75		Cenophytic
	Pleistocene	1 Ice ages	Man		
	Pliocene	10	Last of the modern mammalian groups established		
	Miocene	25 Alps and Himalayas	Giant sharks		
	Oligocene	40	Large number of modern mammalian families		
	Eocene	60	Main orders of mammals established, including a few modern families		
	Palaeocene	70	A few more mammalian orders established		
Mesozoic age of reptiles	Cretaceous	135	Large reptiles extinct. Most modern life forms	First woody angiosperms. Flowering plants begin to radiate	
	Jurassic	180 Continents separate	Marsupials and placental insectivore mammals appear. First bird fossils. Reptiles dominant. First Teleosts		Mesophytic
	Triassic	225	First Dinosaurs	Conifers dominant on land	
Palaeozoic	Permian	270 Continents join	Adaptive radiation of reptiles and bony fish	Great floral change between early and late Permian. Deciduous plants appear. Many plant extinctions	
	Carboniferous	350 GB on the equator	Adaptive radiation of insects and amphibia	Coal forests	Palaeophytic
	Devonian	400	First amphibian fossils. Age of fish	Vascular land plants. Trees and seed habit	
	Silurian	440	First insect fossils	First land plants	
	Ordovician	500	First vertebrate fossils. Molluscs dominant in sea	Eophytic	
	Cambrian	600	Age of Trilobites. All main types of invertebrates present in the sea	Complex green algae from Cambrian onwards. All periods based on sedimentary rock strata	
	Precambrian	4–5000 Age of Earth	Represents well over 75% of Earth's history	Some bacterial and blue-green algae fossils. Some unaltered Precambrian sedimentary rocks. Altered igneous and sedimentary meta-morphic igneous rocks	

There is still much controversy as to the degree of accuracy obtainable with these radioactive dating techniques, and the limits of accurate carbon dating may be only about 40 000 years.

Figure 4.3 *Carbon dating of fossil remains*
(Colours show proportions, not actual distribution within the structures.) The remaining C^{14} has a known half-life. Thus, assuming that the proportion of C^{14} in the atmosphere has not changed with time, the age of the fossil can be calculated.

From the fossil record two very dramatic examples of adaptive radiation, dominance, gigantism, and almost complete extinction are seen in the reptiles of the Mesozoic period, and in the vegetation of the Carboniferous period.

VEGETATION OF THE CARBONIFEROUS PERIOD

During this period, that lasted about 65 million years, 250 million years ago, there was further development and a wide adaptive radiation of the variety of plant types first established in the Devonian period. Several types showed the trend of the development of gigantism followed by extinction, as also occurred with the ruling reptiles of a slightly later period. The Carboniferous period saw widespread fluctuations in the levels of the Earth's crust which lead to land emergences and submergences and the production of large areas of brackish swamps. During the periods of submergence the lush vegetation of the low-lying swamplands was rapidly covered by sedimentation. This caused the formation of coal beds; a process which was repeated several times. At this time the land masses of which Britain was a part lay on the equator.

The Carboniferous flora was dominated by the Pteridophyta, the Pteridospermae (seed ferns) and the Gymnospermae. The major coal-forming vegetation was dominated by the tall tree-forms of *Lepidodendron* (the club-moss scale tree), which grew to over thirty metres, and *Calamites* (the horse-tail tree) which grew to over fifteen metres. In the ground flora beneath the forest canopy were *Lycopodium* and *Selaginella*, herbaceous club-mosses very similar to present day forms; the herbaceous horse-tail *Sphenophyllum*; and representatives of several families of Ferns that have persisted to the present day. The seed ferns reached their peak of development at this time, many growing to five metres in height. The early Gymnosperms developed, including members of the order Cordaitales which grew to thirty metres with strap-like leaves and winged seeds in loose cones; and members of the order Coniferales which had cones intermediate in structure to those of the Cordaitales and the modern conifers. There were no deciduous trees and no true flowering plants. Also during this period the reptiles became the first vertebrates to breed on land, and certain species of insects developed wings.

MESOZOIC REPTILES

The first traces of the reptiles are found in the Carboniferous period. During the Permian period the reptiles began their very wide and successful adaptive radiation which led to their dominance in the Jurassic; this was followed by their almost complete extinction in the Cretaceous period, modern reptiles being a declining remnant of this once dominant group. Of particular interest are the Dinosaurs, a loosely-defined group of ruling reptiles which dominated nearly the whole of the Mesozoic period.

By the end of the Cretaceous period all of the Dinosaurs had disappeared completely, so that no man ever saw a dinosaur. The cause of this extinction is not understood. The carnivorous types would have died out in the wake of the extinction of the herbivorous types. The herbivorous types may have died out due to a change in climate and in land levels resulting in a change of vegetation.

Evidence from geographical distribution

The pattern of distribution of plants and animals can be interpreted as providing information about possible evolutionary relationships. Plants and animals are often discontinuously distributed over widely separated areas. In plants and animals that have efficient dispersal structures this discontinuous distribution can arise as a result of the long range transport of these structures to other suitable environments. However, in those organisms without such structures a widely discontinuous distribution is more difficult to explain and involves consideration of both ecology and evolution.

ANIMAL GEOGRAPHY

The present-day distribution of **mammals** could be explained in terms of their migration across known land bridges, between permanent continents, which have undergone a series of breaks and rejoins. However, to explain the biological similarities of the southern continents, especially the occurrence of side-neck turtles, lungfish, freshwater fish, and invertebrates, it is necessary to take account of alterations in the formation of the land masses of the world. At one time massive continental-sized land bridges were postulated, but there is now an increasing acceptance of the theory of **continental drift**.

A more recent influence on the distribution of plants and animals, particularly animals in the northern latitudes, has been the series of Ice Ages.

Although the geographical distribution of birds, which are highly mobile and have a great range of movement, is usually more difficult to interpret in terms of evolutionary change than that of other types of animals, at least two classic cases exist. One example is that of Darwin's finches in the Galapagos[91] Islands, and another is provided by the woodpeckers and cockatoos in S.E. Asia, first described by Wallace.[86]

A recent dramatic change in the geographical distribution of a bird species (the significance of which is still not understood) is seen in the collared dove (*Streptopelia decaocto*). In a few decades this species has spread from Asia Minor and southern Asia across Europe, reaching Great Britain in 1956. This has been described as perhaps the most dramatic change in distribution known in any vertebrate.

As a result of all these interacting influences, the world can be divided into six major regions on the basis of the distribution of animals. These are the Neotropical, Ethiopian, Palearctic, Nearctic, Oriental, and Australian regions.

Figure 4.4 *Discontinuous distribution of tapirs*

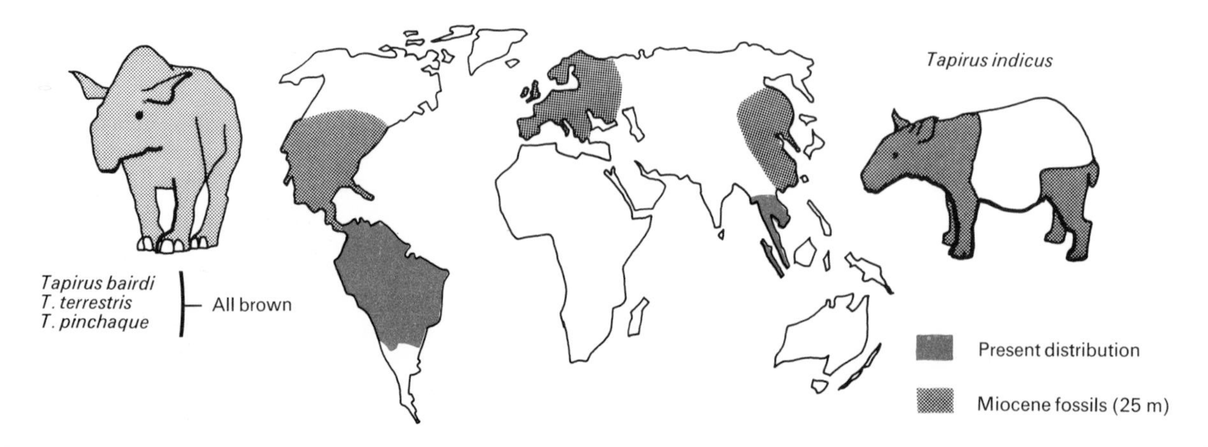

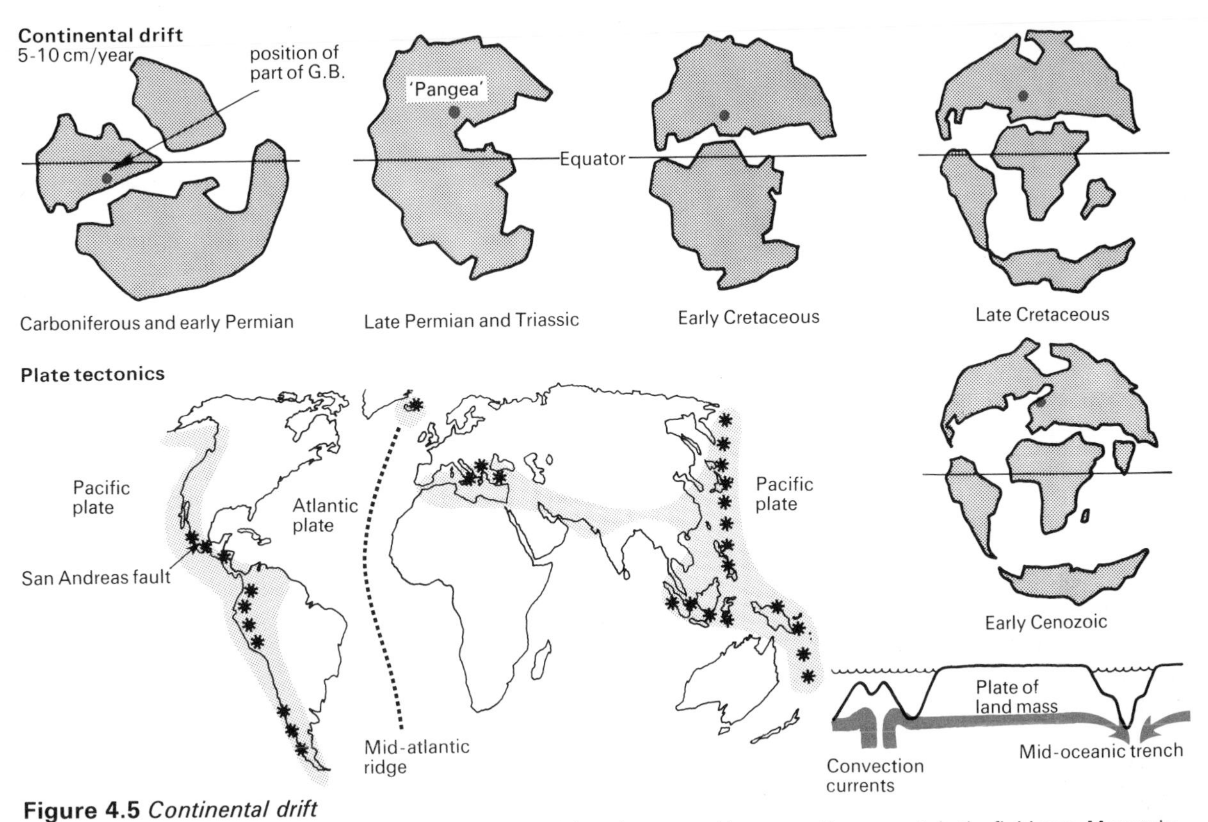

Figure 4.5 *Continental drift*
'Plates' of the Earth's crust float on denser, more fluid material, and are moved by convection currents in the fluid core. Mountain ranges and volcanic areas are formed where plates meet.

Figure 4.6 *The Ice Ages*

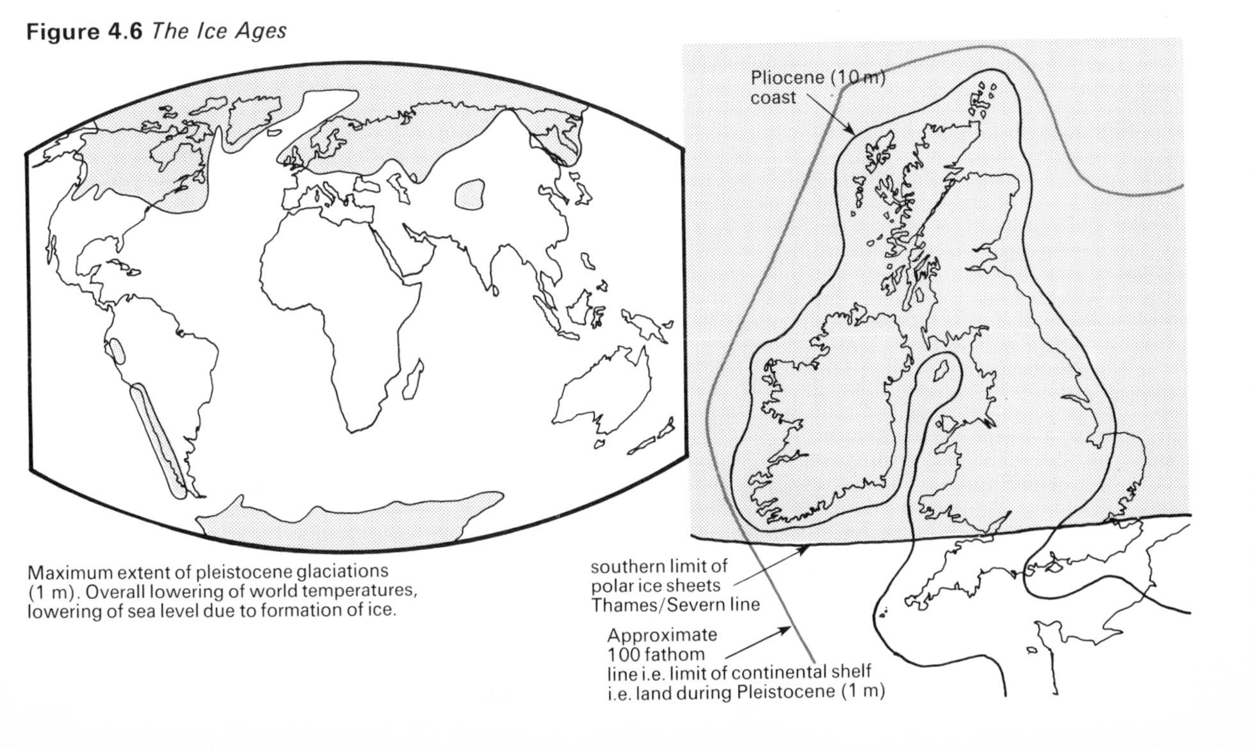

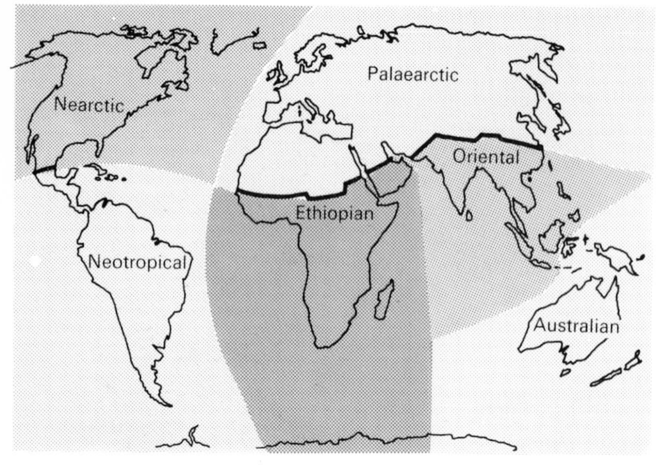

Figure 4.7 *Zoogeographical regions*

Neotropical region
This region is mostly tropical but also lies partly in the south temperate zone. It is connected to the Nearctic region by the intermittent land bridge of the Central American Isthmus, which has been broken and remade several times with the rise and fall of sea levels during and between the Ice Ages. It has the highest number of endemic families of any region (16). The tapirs are considered to belong to the same genus as those in the Oriental region despite the separating distance. It is the only region with a large endemic Order of mammals, the Edentata (the sloths, anteaters, and armadilloes of South America). It has Platyrhine monkeys with prehensile tails and non-opposable thumbs, and two families of marsupials. Nearly all the amphibians are Anurans. The birds are very striking (South America is often called the 'Bird Continent').

Ethiopian region
The Sahara acts as a natural barrier between the Ethiopian and Palaearctic regions, preventing many exchanges. The most varied mammalian fauna of all regions is to be found in the Ethiopian region, including Catarrhine monkeys, with no prehensile tails, opposable thumbs, clavicle, pectoral mammary glands, and cheek pouches (as the fore limbs are used in locomotion).

Palearctic region
This is not rich in varieties of animals when compared to other regions. Very few families are found only in this region (endemic). The Camelidae is the only family with a discontinuous distribution in the Palearctic, being found in Arabia and Asia. The region has certain similarities with the Nearctic, the two together sometimes being known as the **Holarctic**. Both are north temperate regions with land connections to tropical regions, and both have been connected by intermittent land bridges across the Bering Strait.

Nearctic region
This region shows similarities with the Palaearctic, although it has more reptiles and more endemic families. Four mammalian families are shared with the Neotropical region; all are recent immigrants from the Neotropical region. The modern horse was reintroduced to the Nearctic by man after originating there, migrating to the Palaearctic, and becoming extinct in the Nearctic. Both reptiles and Urodele amphibia are abundant. The Anuran amphibian *Ascaphus* shares peculiarities with only one other species, the *Liopelma* of New Zealand. The two are in-

Figure 4.8 *Distribution of woodpeckers and cockatoos*

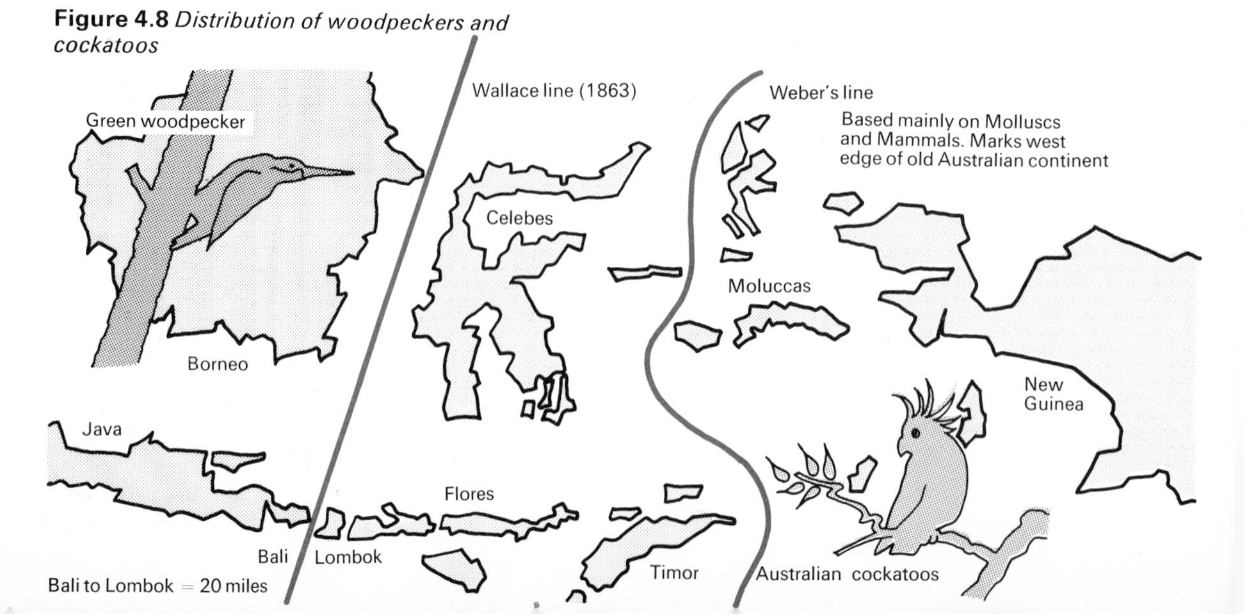

cluded in the family Liopelmidae. The genus *Liopelma* contains the only three species of amphibia found in New Zealand.

Oriental region
This region is mainly tropical; the fauna resembles that of the Ethiopian region. It has 50 per cent of all the primate families. Reptiles are abundant.

Australian region
This region is unique in having no recent land connections with adjacent regions. Vertebrate fauna are sparse, lacking variety and numbers, but a high proportion are endemic. Six families of marsupials show adaptive radiation parallel to that of placental Eutheria in other parts. There are two families of Monotremes, the duck-billed platypus and spiny anteater. Parrots are abundant. There are remarkably few fresh water fish, amphibia, and reptiles, with no Urodele amphibia. Many species in this region have been introduced by man.

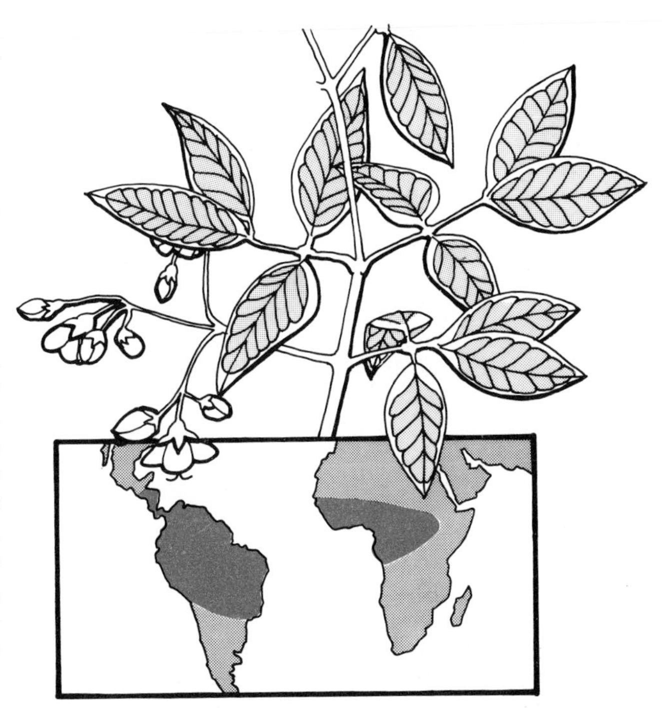

Figure 4.9 *Discontinuous distribution of Symphonia globulifera*

PLANT GEOGRAPHY

On a global scale, plant distribution is primarily controlled by **climatic conditions** and as a result different types of plant are distributed mainly according to **latitude** and **altitude**. Climate also has an effect on **edaphic factors** (soil factors), which in turn contribute towards determining the pattern of plant distribution. Generally it can be said that climate determines the extent of a plant's distribution, and that the edaphic factors determine the density of that distribution. In other words, the climate determines whether a plant type can colonize an area and the edaphic factors determine whether it actually will, and, if so, in what abundance. There have been great variations in climate since the emergence of the Angiosperms, and these, coupled with changes in the distribution of land and sea, are reflected in the modern distribution of plants.

Generally the dispersal of Angiosperms is a short-range process, with sea passages forming effective barriers to their spread. It is therefore difficult to explain the patterns of distribution of those types which are separated by oceans, other than by means of Continental Drift. In historical times man has been a major distributor of plant species, either accidentally or, more typically, as a result of the quest for new crop plants.

Figure 4.10 *Distribution of some major crop plants by man*

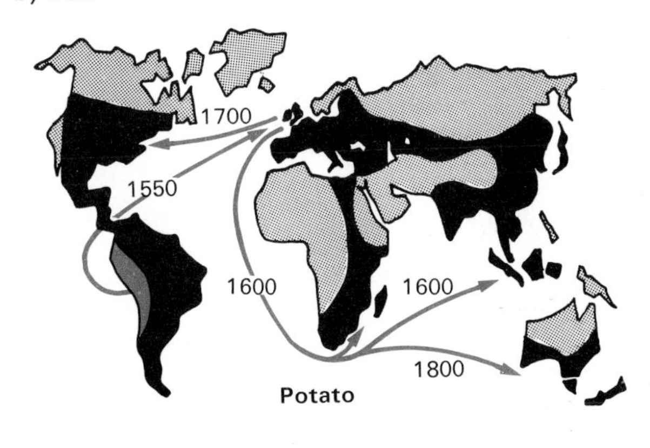

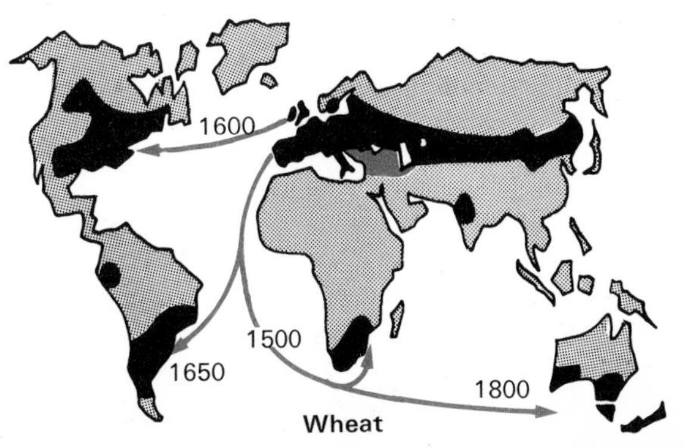

Evidence from embryology

Embryological and larval stages of some animals show striking similarities and it is suggested that this demonstrates an evolutionary relationship between them. The homology of structures is particularly derived from embryological studies, where fundamental similarities of structure are observed prior to their divergence in the adult. Relatively minor changes in an embryonic stage could lead to larger changes in the adult forms that might confer an advantage under certain circumstances. **Neoteny**, in which sexual maturity is gained in the larval or juvenile form and the fully adult stage is abolished, can produce offspring that develop similarly, and is also suggested as a means by which new forms could arise.

Evidence from life cycles

The life cycles of plants show an underlying pattern, known as the **alternation of generations,** which is found throughout the plant kingdom. This remarkable uniformity of basic life cycle is taken as evidence that all such plants are related by descent.

Evidence from comparative cytology and biochemistry

There is a remarkable uniformity of cell structure and function in all living organisms. This also is taken as evidence that present-day organisms are related by descent from some common ancestor.

Less widely found similarities, such as the sharing of certain blood groups in man and the apes, are taken as evidence for a closer evolutionary relationship. For example, serum immunized against human blood gave an agglutination of 64 per cent with gorilla blood, 42 per cent with orang utang blood, 29 per cent with baboon blood, and 10 per cent with ox blood. Differences in the sequence of amino acids in haemoglobins[156] and cytochromes[160] are used in attempts to estimate the time of divergence of related organisms from a common ancestor. It is assumed that the different types arose by individual gene mutations affecting single amino acids in chronological sequence.

The mechanism of evolution

If all the points discussed in the preceding sections are taken as evidence in support of the theory of evolution, and it is assumed that organisms are related by descent with modification, there remains the problem of explaining *how* such changes could have occurred. In other words we must ask, 'What is the mechanism of evolution?'

Lamarck (1744–1829) observed that organs varied with use and disuse, and suggested that such acquired characteristics could be inherited if they fulfilled a 'need'. Many people did, and still do, find this an attractive proposition.

Charles Darwin (1809–82) was led to formulate his theory, based on the idea of **variation and natural selection**, as a result of observations of the flora and fauna of all parts of the world made during the five-year voyage of HMS Beagle from 1831–6. He spent the next twenty-two years developing this idea and amassing as much material in support of the theory as possible. Another naturalist, **Alfred Russell Wallace** (1823–1913), working in the East Indies, independently formulated the same theory in a short essay in 1855. In 1858 Darwin and Wallace had a joint communication read to the Linnaean Society and, in 1859, Darwin's **'Origin of Species by Means of Natural Selection'** was published.

Darwin based his **theory of natural selection** on a series of observations and inferences. He noticed that a far greater number of offspring are produced by organisms than actually survive, and that there is a struggle for existence. He also observed that individual members of a species show small variations and therefore differ slightly from one another. From these two observations he inferred that some of these small variations may have enabled some individuals to survive more easily than others. Hence there is survival of the fittest in the struggle for existence, a process he called natural selection. He also assumed that many of these advantageous variations were inherited by the offspring of those that survived, and that these small differences could gradually accumulate over periods of time to result in the appearance of a new species.

However, the 'weak' point of the theory was its lack of explanation of the mechanism by which the advantageous variations arose and were inherited. Indeed, the scarcity of ideas on inheritance at that time led Darwin to accept the theory of the inheritance of acquired characteristics.

The first major rejection of Lamark's theory came from Weismann (1834–1914), who produced the concept of the continuity of the 'germ plasm', in which the gamete-producing 'germ-line' cells are conceived as being 'isolated' from the body cells or 'soma', and hence from any environmental influence. This concept became known as the 'Weismann Barrier' and, although it was not at the time based on full experimental evidence, there followed a general rejection of Lamarkism by orthodox biologists. The barrier concept was reinforced on the rediscovery and development of Mendelian genetics, and later by the Watson–Crick model of the structure and function of DNA.

There thus came about an almost complete rejection of Lamarkism amongst biologists since there was no apparent mechanism by which such acquired characteristics could be inherited.

More recently, however, it has been shown that there is a mechanism by which the structure of DNA *can* be modified by influences from the cytoplasm which is itself exposed to environmental influences. In addition there is evidence for a certain degree of cytoplasmic inheritance, and this again could be affected by environmental factors. Such ideas have resulted in the 'reappearance' of a form of Neo-Lamarkism, in which the possibility of environmental influences altering the genetic material is not rejected out of hand.

Neo-Darwinism

A satisfactory genetic explanation of the origin and inheritance of the variations on which natural selection acted awaited the work of De Vries on discontinuous or 'sudden' mutations, and the rediscovery of Mendel's work in the early 1900s. The rapid development of the study of genetics by Bateson (who coined the term 'genetics') in England and Morgan in America firmly established the so-called 'mutation theory', by which the variations on which natural selection acted were considered as arising by means of these random 'discontinuous' variations, such as tallness and shortness of stem length in Mendel's peas. It was then realised that the 'smaller', less obvious mutations which give rise to so-called 'continuous' variations, as, for example, small variations in the length of a limb, could also contribute towards evolution. This removed the dependence of the theory of evolution on the major discontinuous mutations as the only source of inheritable variation.

Thus Neo-Darwinism retained Darwin's original concept of natural selection but embraced the developments in modern genetics, in particular those aspects that became known as **population genetics.**[92]

The origin of variation, in individuals within a population, is initiated by the gene and chromosomal mutations, and the genetic recombination that results from sexual reproduction. Neo-Darwinism, however, considers that it is the breeding population, rather than the individual, which is the evolutionary unit on which natural selection operates. Such population thinking allows the possibility of gradual evolution, and reduces the theory's dependence upon major mutations as a source of variation. In almost all species there are **reproductive isolating mechanisms** separating groups of individuals into breeding communities which thus act as **'gene-pools'** in which the genes of the members can be mixed by random breeding. Examples of such isolating mechanisms include both prezygotic and post-zygotic types.

Prezygotic reproductive isolating mechanisms are those that prevent the fusion of gametes in the process of fertilization and zygote formation. This type includes ecological, geographical, ethological, temporal, mechanical, and gametic isolation mechanisms.

Ecological isolation occurs as a result of organisms occupying different habitats so that, apart from a few chance migrants, they never meet to carry out mating. This can be seen in practically all natural populations.

Geographical isolation is similar to ecological isolation but is usually defined by various geographical barriers, such as seas, rivers, mountains, deserts, etc. An example can be seen in the Galapagos Islands.[91]

Ethological isolation occurs in groups which, although they may share the same habitat have incompatible behaviour patterns. An example is provided by the herring gull and lesser black backed gull.[91]

Seasonal isolation occurs when there are differences in the timing of breeding mechanisms.

Mechanical isolation is due to structural incompatibilities preventing interbreeding; and **gametic isolation** is due to genetic incompatibilities preventing the successful fusion of gametes, as are seen for example in many flowering plants.

Postzygotic reproductive isolating mechanisms are those that prevent the successful perpetuation of the result of fertilization, should it occur. This group includes the so-called **hybrid inviability**, in which zygotes fail to develop; **hybrid breakdown**, in which the offspring have a reduced reproductive viability; and **hybrid sterility**, in which the offspring are sterile.

These isolation mechanisms maintain the integrity of the gene pool and ensure the rapid spread of favourable variations throughout that population. They also prevent the 'loss' of these genes to other populations. In the absence of such isolating mechanisms the local group, facing local environmental demands, would be unable to maintain any genetic advantages specific to its habitat. However, a certain amount of genetic exchange is necessary from time to time between breeding populations, so as to introduce fresh genetic variations via hybridization. It is suggested that the correct balance between isolation and migration will result in **speciation** (the formation of a new species).

Some people maintain that Neo-Darwinism is trivial, that it cannot explain the really important events of evolution, and that it only explains how species maintain themselves in a changing environment, rather than explaining how species arise.

The central dogma of orthodox Darwinism is that the raw materials of evolution are the random chance mutations that occur, and that the driving force in the evolution of organisms is the operation of natural selection leading to the survival of the fittest (or the best adapted).

If this is believed, the question then arises as to whether sufficient time has passed since the formation of the Earth for the laws of chance to have operated through random mutation and natural selection to produce the present fantastic beauty and complexity of living organisms.

In an analogy that is often drawn in relation to the origin of life, evolution, and the laws of chance, it is stated that if enough monkeys were sitting at typewriters typing letters at random for long enough, they would eventually produce one of Shakespeare's sonnets by pure chance. To put this analogy into perspective it has been calculated that, to type the first verse of Genesis ('In the beginning God created the heaven and the earth') by pure chance, it would take one million monkeys on one million typewriters, typing at a rate of twelve and a half keys per second for twenty-four hours a day, the same amount of time as is needed to erode away 400 rocks, each of four light years in diameter, by the removal of one grain of sand once every 1000 years! That the complexity of life could have arisen by pure chance is rejected by many, indeed many convinced evolutionists believe that natural selection only plays a minor role in the evolutionary process.

There is a need to loosen the bonds of the orthodoxy of Darwinism so that fresh ideas may be introduced in trying to understand the forces at work. Indeed, once it is postulated that to shorten the odds the 'laws of chance' must work within certain constraints, there is no limit to the suggestions that could be made. Perhaps it is for this reason that many biologists cling so tenaciously to the idea of random chance and natural selection, assuming that a bad law is better than no law at all.

It certainly must be stated that the theory of natural selection, despite strenuous claims to the contrary, remains just a theory; views which conceive of evolution as a goal-directed process are equally valid.

Species and speciation

Phyletic evolution is conceived of as occurring in a single lineage of descent and results in a particular organism becoming better adapted to its environment. **Speciation** is the process by which a lineage of descent splits into two or more new lines of descent to produce new species.

There is no single definition of a species; a species is characterized rather than defined. In practice, **similarities in appearance** are taken as the main criteria for placing individuals together in a species, and a species is thus the smallest taxonomic grouping commonly used. Members of a species generally do not differ from one another more than the offspring of a single pair may do. In addition, members of a species are usually distinctly different from members of other species, there being no intermediate forms. However, some species are **polymorphic**[100] and the offspring can differ considerably from the parents, be distinctly different from each other, and show no intermediate forms.

Another concept of the species is that of the **breeding unit**. Members of a species are capable of interbreeding and producing fertile offspring. If hybridization does occur with a member of another species the offspring are

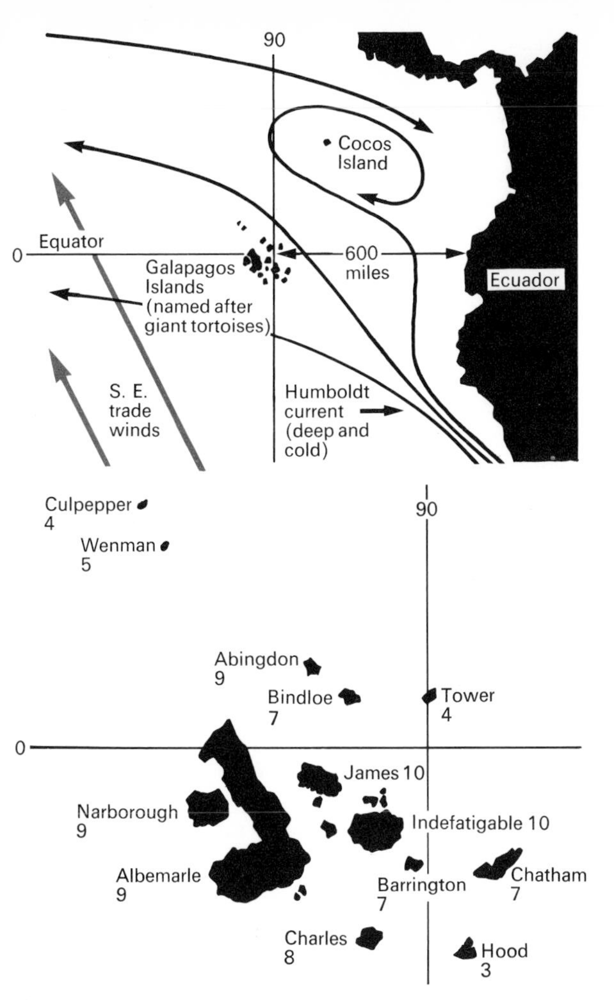

Figure 4.11 *Galapagos Islands*
The land area, about 3000 square miles, is spread over 23 000 square miles. It is volcanic in origin, about three million years old. The Islands were first discovered in 1535. Darwin visited four islands in 1835, spending five weeks there. Now 5000 humans, and feral (escaped) goats, rats, and dogs are destroying the natural flora and fauna. (The figures indicate the number of finch species that occur on each island.)

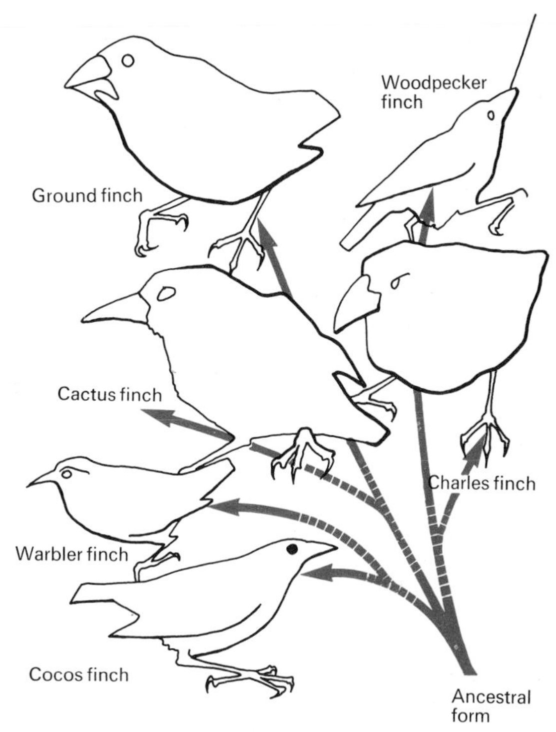

Figure 4.12 *Darwin's finches (a selection)*
'Seeing this gradation and diversity of structure in one small, intimately related group of birds one might really fancy that . . . one species had been taken and modified for different ends.' (Charles Darwin)

Figure 4.13 *Ring species and races of gull*
A and B do not mix, do not form intermediate populations, and behave as reproductive isolated species. A breeds inland and is migratory in winter (as far as North Africa), B breeds on cliffs and is non-migratory. However, if eggs are swapped between A and B the resultant offspring ('changelings') will mate with the 'wrong' type. A can be regarded as belonging to every species in the chain towards the east, but by the time the chain reaches A from the west, the two ends are separate species.

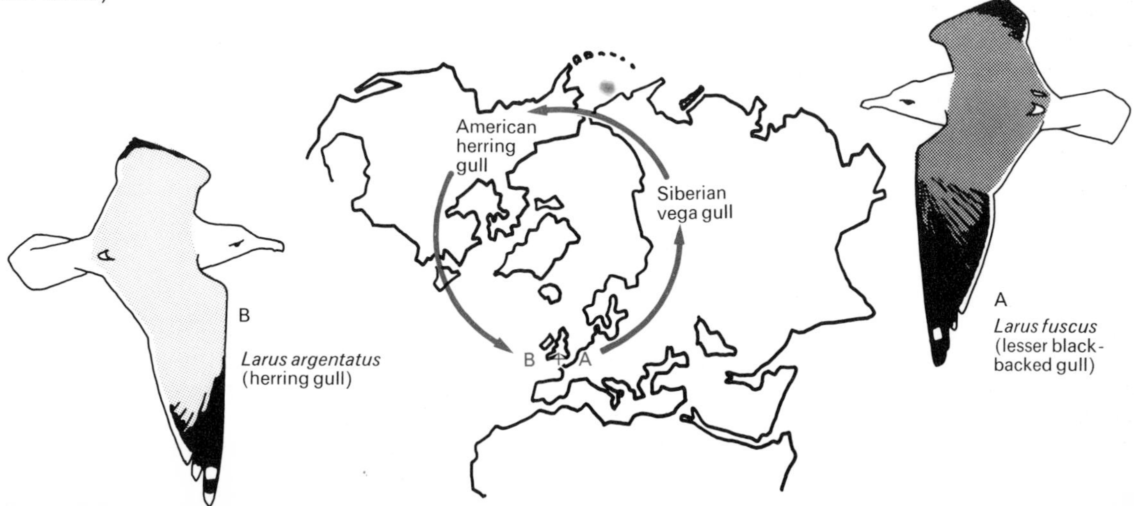

usually sterile. However, this definition cannot apply to organisms where sexual reproduction is not strongly developed, e.g. bacteria and some plants. Neither, of course, can it apply to fossil species.

The formation of species or speciation depends on a complex of factors, among which reproductive isolation is considered critical. This often, but not always, involves geographical isolation. The classic example of geographical isolation leading to speciation is that of **Darwin's finches** of the Galapagos Islands. Such speciation caused by geographical isolation is referred to as **allopatric speciation**, and is considered to involve several steps: the initial invasion of an area by a random group of members of the original species; the subsequent isolation of these invaders so that they are cut off from the other members; and the gradual accumulation of genetic differences via the **'founder effect',**[94] **genetic drift,**[94] **mutation,**[94] and **selection**.

Genetic isolation caused by barriers such as differences in breeding season, behavioural differences, and genetic incompatibility within a single population results in what is known as **sympatric speciation**. This occurs when there are strong selection pressures for different **ecotypes** which lead to the formation of subpopulations, or **demes**, which eventually accumulate sufficient differences to prevent interbreeding.

Population genetics

Evolution can be regarded as changes in populations. It can therefore be considered as a problem in population genetics. In natural populations it is difficult either to analyse the genotypes of members of the population, or to predict the genotype and phenotypes of the offspring of random mating within the population.

In 1908 **Hardy** (an English mathematician) and **Weinberg** (a German physician) analysed the behaviour of genes in natural populations. They derived a law which states that in the absence of selection or other changing factors such as genetic mutation, the proportion of alleles in a large population will remain the same, generation after generation, regardless of whether the alleles are dominant or recessive.

Each diploid individual is considered as having just two copies of a gene (in actual fact each cell of that organism will have two copies). In a population of 100 individuals, there are considered, therefore, to be 200 of these genes, for example either of the alleles T or **t**. The proportion of these alleles in the population is known as the **allele frequency** or, less accurately, as the gene frequency.

The members of the population can have one of three genotypes: **TT**, **Tt**, or **tt**. To calculate what the proportion of these possible genotypes will be in a sexually reproducing population, the following assumptions must be made.

That there is **no mutation of T to t** or vice versa; that there is **true random mating**, as if all the genes were shed into a single gene pool; that all three genotypes are **equally fertile**; that **no selection** for or against any particular type is operating; that the population is **closed** with no immigration or emigration; that **meiosis is normal**; and that the population is **'large'** in a statistical sense.

If these conditions are considered as existing then the following result can be derived.

Let p = the frequency of one allele—the dominant (T).
q = the frequency of the other allele—the recessive (t).
$p + q = 1$ (i.e. the total is taken as being unity because either T or t is present).

The frequency of either of the alleles which may be contributed by the male to the next generation will be $(p + q)$; similarly for those contributed by the female. For example, let there be 200 alleles in a population of 100 diploid organisms. Assume there are 80 dominant alleles and 120 recessive alleles. Therefore the frequencies of these alleles are given by:

$$p = \frac{80}{200} = 0.4$$

$$q = \frac{120}{200} = 0.6$$

$(p + q = 1: \quad 0.4 + 0.6 = 1)$

Half the population will be males and half females and the alleles will be randomly distributed between them.

50 males	50 females
100 alleles	100 alleles

The alleles will occur in the same relative proportions or frequencies in the two sexes.

50 males	50 females
100 alleles:	100 alleles:
40T + 60t	40T + 60t

For both males and females, $p + q = 0.4 + 0.6 = 1$.

Assuming that these gametes will fuse at random, the expected frequency of the genotypes which result will be:

$$(p + q)^2 = p^2 + 2pq + q^2 = 1 \text{ (since } p + q = 1)$$

Now p is the frequency of the dominant gene T, and q is the frequency of the recessive gene t.

$$p^2 + 2pq + q^2 = 1$$

p^2 = expected frequency of homozygous dominants (TT)
$= 0.4 \times 0.4$
$= 0.16$

$2pq$ = expected frequency of heterozygous genotype (Tt)
$= 2 \times 0.4 \times 0.6$
$= 0.48$

q^2 = expected frequency of homozygous recessives (tt)
$= 0.6 \times 0.6$
$= 0.36$

In a natural population the only *known* value is the frequency of the homozygous recessives. However, it is possible to substitute values in the equation to obtain the frequency of all the genotypes. Consider a population of plants, 70 per cent of which are tall and 30 per cent of which are short. It is known that tall is dominant to short.

Then

Phenotype	Tall	Short
Genotype	TT or Tt	tt
Quantity (%)	70	30

$$p^2 + 2pq + q^2 = 1$$

$\therefore \quad q^2 = \frac{30}{100} = 0.3$
$q = 0.55$
$p = 1 - q = 0.45$
$\therefore \quad p^2 = 0.2$
$p^2 + 2pq + q^2 = 1$
$0.2 + 0.5 + 0.3 = 1$
TT Tt tt

In other words 0.2 of the total population are homozygous dominants and 0.5 of the total are heterozygous.

Returning to our original population of 100, this means that there are 20 homozygous dominant, 50 heterozygous, and, as was already known, 30 homozygous recessive alleles. When this is known, predictions can be made about the proportion of genotypes and phenotypes that should appear in the next generation as a result of random mating. When these predictions are incorrect it is an indication that one or more of the assumptions made are not operating, for example, mating may not be random or selection may be favouring a particular phenotype. It is the establishment of a theoretical ideal genetic behaviour for a population that enables any forces acting on that population, for example selection pressures, to be identified.

It is interesting to note that Mendel's laws are a special case of the Hardy–Weinberg Law, where p and q are equal. Indeed, Mendel expressed his results in terms of the binomial equation $(a + b)^2 = a^2 + 2ab + b^2$.

Factors altering expected allele frequencies in populations

Allele frequencies within populations can change under a variety of influences.

Mutation[94] will change the relative frequencies of two alleles; for example, any dominant allele can mutate to the recessive form and vice versa.

Selection, both artificial and natural, will tend to favour some alleles rather than others. An example in which mutation and selection balance the effect of each other is seen in the X (sex-) linked recessive allele for **haemophilia**.[71] This is lethal in the homozygous state, so that females with two of the haemophilia alleles die before birth. Theoretically, this should gradually reduce the numbers of the allele in the population because 30 per cent are lost in each generation. However, the rate of recurrent mutation of this gene (about 30 000 mutations per 1 000 000 000 cells) almost exactly balances the rate of its disappearance due to selection. Thus the frequency of the haemophilia allele remains fairly constant in each generation.

If the population is small **random fluctuation** can occur as the laws of chance and probability do not operate ideally. Indeed, due to territorial behaviour, the size of breeding groups in natural populations can be relatively small, as not all members will be successful in winning and defending territory—often a prerequisite for breeding. Due to these small breeding groups and the resultant non-random breeding, some alleles may not be contributed to the 'gene pool' in the same proportions as they exist in the wider population, and some may even be absent altogether. Thus there is a chance of random fluctuations due to this so-called **genetic drift**. This is also seen when new areas are colonized by small numbers of 'pioneers', in which case the genetic drift that occurs is referred to as the **founder effect** as the founder members will not possess all the alleles in the same proportions as in the parent population.

A form of unbalanced meiosis, known as **meiotic drive**, can also occur, by which one member of a bivalent or pair of homologous chromosomes may be retained in an egg while the other is lost via the non-functional polar bodies.[21] In this way the different types of gametes are not produced in the expected ratios.

Genetics of change

Mutation

Mutation is a sudden inherited change in the structure of the genetic material, and can be due to changes in individual genes, changes in chromosomal structure, or changes in chromosome number. Most mutations are harmful, or at least of no immediate advantage, as the organism is already adapted to its environment and any change would tend to upset this adaptation. However, the variation produced by mutation may be beneficial, should the environment change.

Mutations can occur in body or somatic cells, in which case any effect they exert is restricted to the mitotic descendants of the original cell in which the mutation occurred. This can give rise to patches of mutated cells or **mosaics**. If the somatic mutation occurs in a bud primordia in plants, the whole of the resultant branch will be mutant. The nectarine was one such 'bud sport' of a peach tree. If it occurs in one of the two daughter cells of the first cleavage of the zygote, then approximately half of the adult organism will show the mutation.

Mutations can also occur in the germ-line cells of the gamete-producing tissues, in which case the mutation will pass to all of the cells of the offspring.

GENE OR POINT MUTATIONS

These occur due to an error in the copying of a sequence of bases during the replication of a new DNA strand. Once the error is made it is perpetuated by further DNA replications. The altered gene is said to be **allelomorphic** to the original, and mutation is the origin of all such alleles of genes. Without such alleles the existence of the genes themselves could never be demonstrated as it is only when a gene changes its effect that its original role can be recognized.

Dominant gene mutations express themselves immediately, whereas recessive mutations may be masked or hidden for several generations before expressing themselves in a homozygous recessive individual. It is difficult to detect the 'sudden' appearance of a recessive mutation unless the organism is haploid, as in the male drone honey bee and the gametophyte generation of plants.

The frequency of gene mutation varies, but is of the same order in a wide variety of organisms, of about one mutation at a given locus per million cell generations. This indicates that the same forces are at work in all types, as is expected considering the universal occurrence of DNA. However, some genes do mutate far more frequently than others and form the so-called 'genetic hot spots' on certain chromosomes.

Mutations can revert back to the original form, but this usually occurs at a different rate from the original mutation. For example, A is mutated to a at rate x, and then reverts back to A at rate y.

$$\mathrm{A} \underset{y}{\overset{x}{\rightleftharpoons}} \mathrm{a}$$

Some striking changes occur in *Drosophila* which could be reversions of earlier mutations, for example mutations are seen whereby wings are produced instead of halteres,[331] legs instead of antennae, and primitive chewing mouthparts instead of the proboscis.

The far-reaching effects of some gene mutations are well seen in the example of **sickle-cell anaemia**. In this condition, newly-formed red blood cells are normal in shape but, when the oxygen in their haemoglobin is released to the body tissues, most of them change to an abnormal sickle shape. These sickle cells are destroyed in the spleen, thereby reducing the number of red blood cells. This cuts down the amount of oxygen carried. Also, due to their shape and

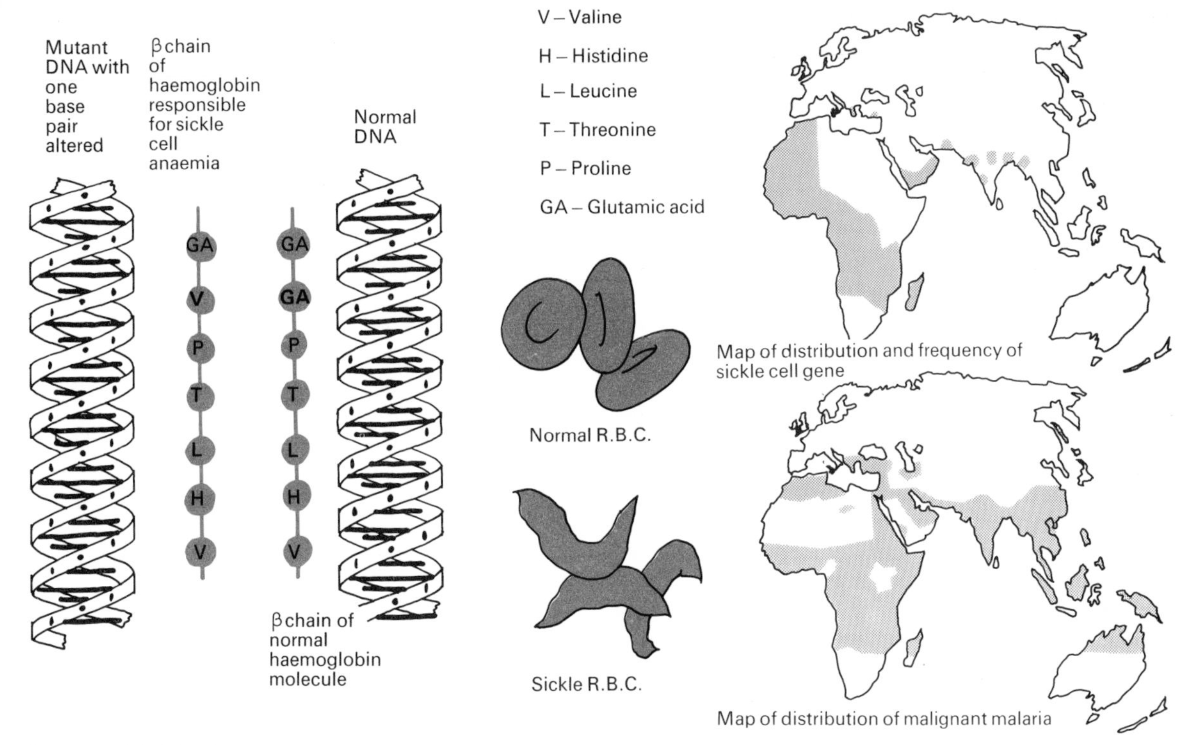

Figure 4.14 *Distribution of sickle-cell anaemia and malaria*

lower flexibility, they do not pass through capillaries easily, and so reduce the efficiency of circulation. Sickle-cell anaemia is usually fatal in childhood. In less severe cases, sickle red blood cells are only produced when the supply of oxygen is low, e.g. at high altitudes or when the need for oxygen increases (such as during exercise). People affected in this way are said to 'have the sickle-cell trait'.

The sickling is a characteristic not of the red blood cells themselves but of the haemoglobin in the cells; the haemoglobin molecule differs from the normal molecule in just one of the 557 amino acids that make up the protein part of the molecule. It is inherited as a recessive gene.

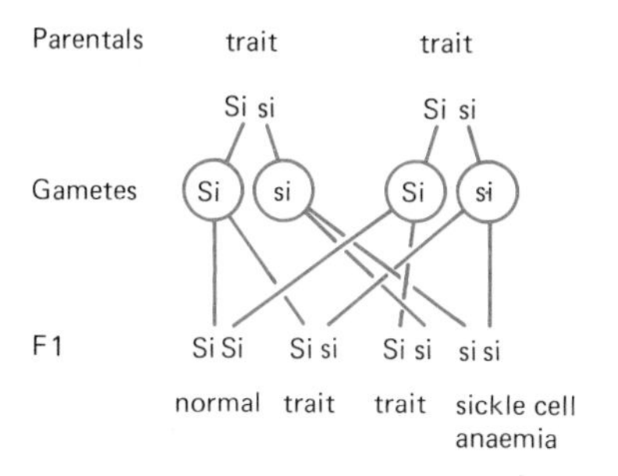

The si gene is rare in most populations (as would be expected with a semi-lethal gene). However, in some parts of Africa the trait is found in as much as 40 per cent of the population. Its occurrence is related to the distribution of malaria, as the heterozygote carriers of the trait are more resistant to the disease.

Red blood corpuscles from a sickle-cell anaemia sufferer (× 6000)

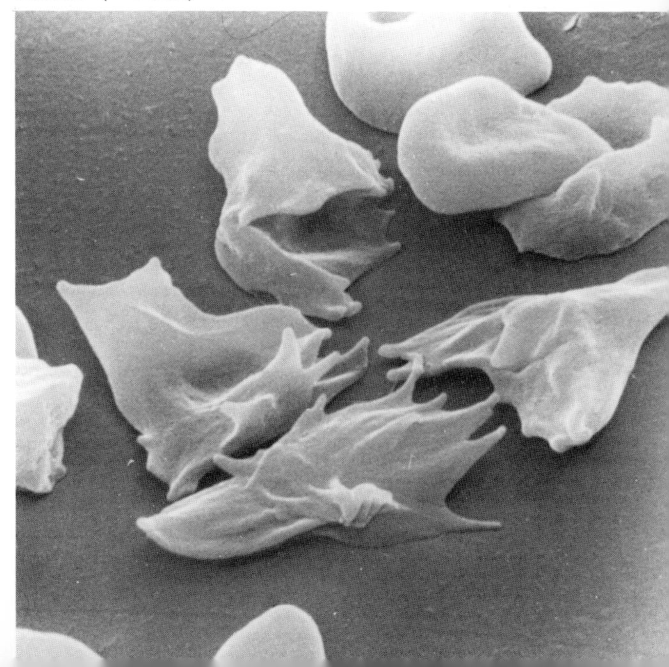

CHROMOSOMAL MUTATIONS

Structural changes in chromosomes

There is a variety of structural changes that chromosomes may undergo. Although these changes do not result in the gain of new genetic material, the new gene arrangements which result can have far-reaching effects. Indeed, chromosome changes, for example translocations in which a piece of one chromosome can break off and become attached to another, can produce a genetic situation from which the development of species may be the ultimate outcome. Changes in chromosome structure include deletions, inversions, and translocations.

A **deletion** is the loss of a segment of chromosome. Normally chromosomes have only one centre of movement, the centromere. Therefore if a chromosome gets broken only the centromeric segment will move to the pole of the spindle apparatus in meiosis and the acentromeric segment is lost.

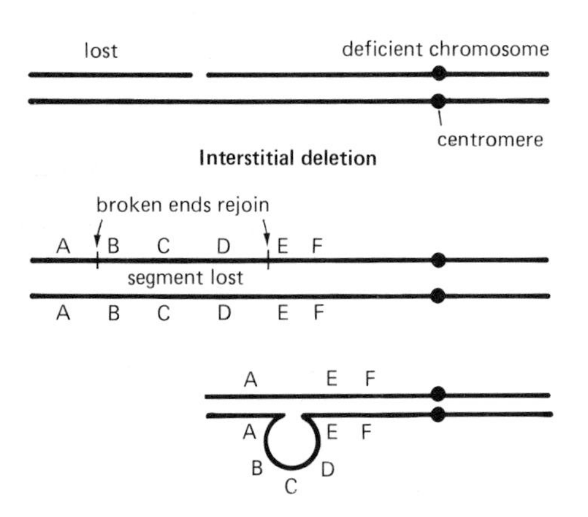

results in a loop being formed in whole chromosome at meiosis, as alleles pair

The effect of these deletions depends on the amount and the nature of the genetic material lost.

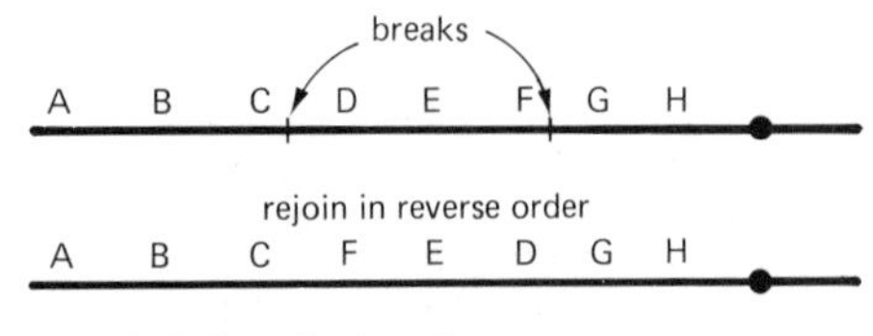

results in inversion loop formation at meiosis

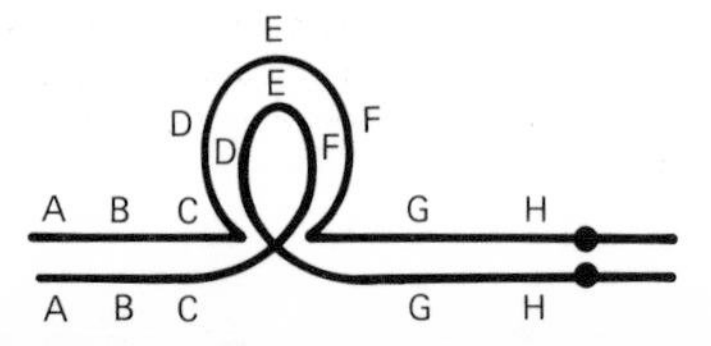

An **inversion** occurs when a segment of a chromosome breaks off and rejoins the 'wrong way round' so that the gene sequence is reversed or inverted.

If a cross-over occurs within the loop the products are usually lost as a result of the complicated chromosomal formation and movements that occur at first anaphase. Thus inversions decrease variability in the offspring and act to keep advantageous combinations of genes together. These advantageous combinations or co-adapted gene loci can be considered as **super-genes** which are inherited as a unit.

A **translocation** occurs when a segment of chromosome breaks off and rejoins to a non-homologous chromosome. There is usually a mutual exchange of material.

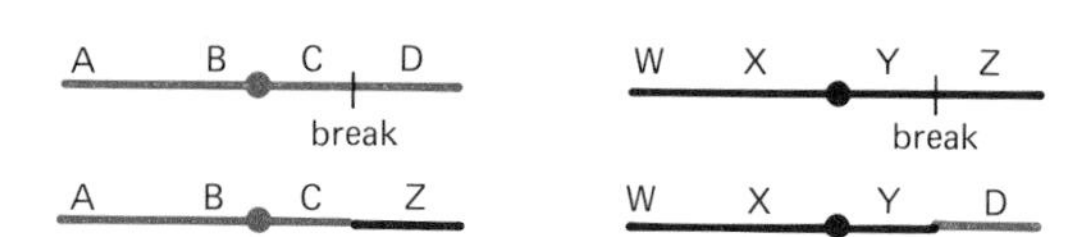

this causes the formation of tetravalents at meiosis instead of bivalents

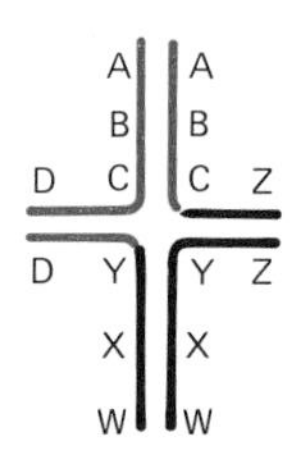

The tetravalents formed lead to complications at anaphase of meiosis.

Chromosomal mutations are found quite frequently in populations. Usually there are no visible distinguishing features of organisms with such mutations, so it is difficult to understand why they occur so frequently or what their selective advantage could be.

In terms of chromosome structure it is possible to have a **basic structural homozygote** with normal pairs of homologous chromosomes, a **structural heterozygote** with one or more pairs of chromosomes where one member has a structural alteration, or a **structural homozygote** with one or more pairs of chromosomes where both members have a structural alteration.

Populations of organisms can be polymorphic for chromosome structure, so that individuals with different chromosome structures are selected for under certain circumstances, and selected against in others. Wild populations of *Drosophila pseudobscura* are polymorphic for inversions of chromosome number III, and these flies undergo a seasonal cycle of changes in frequency as a result of differing selection pressures favouring one type and then another. For example, in Southern California those carrying the so-called 'ST' inversion regularly decrease from representing 53 per cent of the population in March, to 28 per cent in June, and then increase in frequency to 50 per cent in the autumn.

Variations in chromosome number

The basic chromosome number of a species can change for a variety of reasons. A loss or gain of individual chromosomes of a set is known as **aneuploidy**. A gain of whole sets of chromosomes is known as **euploidy**, of which there are two types: autopolyploidy, and allopolyploidy. Such polyploid organisms are usually sterile as the increased number of chromosome sets causes complications at meiosis so that fertile gametes cannot be formed. As a result polyploidy is much more common in plants than in animals because plants often have strong powers of asexual vegetative reproduction which enable them to survive without sexual reproduction. Animals also generally have separate sexes, the differentiation of which depends on a delicate balance of sex chromosomes and non-sex chromosomes; this does not permit a variation in chromosome number to the same extent as is possible in hermaphrodite plants. The only animal groups where polyploidy is common are those with powers of asexual reproduction, although there may be isolated polyploid tissues (for example, liver cells are generally tetraploid).

Autopolyploids have multiple sets of their own chromosomes. This situation usually arises by misdivision at mitosis; the chromosomes replicate into chromatids but do not separate to the poles of the spindles. Thus all the chromatids are included in one nucleus, and a diploid (2n) nucleus would become tetraploid (4n). This process can be repeated until large numbers of sets of chromosomes are built up. If it occurs in the zygote then the whole organism will be polyploid, but if it occurs later then only those cells produced by mitosis from the affected cell will be polyploid. Fertile diploid gametes are sometimes formed, and when one of these fuses with a normal haploid gamete an autotriploid is formed. Many garden plants have arisen like this, including varieties of hyacinths, tulips, and daffodils, all of which perennate and reproduce asexually by bulbs. Autopolyploid plants are often larger and more vigorous.

Allopolyploids are formed by the hybridization of two organisms, usually plants, with different chromosome sets. Usually the organisms must be reasonably closely related; thus it can occur between members of different species, but is very rare between members of different genera. The F1 hybrid is usually sterile, due to problems at meiosis, as the chromosomes will not form homologous pairs.

Somatic doubling of the chromosome number must occur before fertile gametes can be produced. It occurs by non-disjunction, as in the formation of autopolyploids, thus there is no real hard-and-fast division between these two so-called different types of polyploidy. Hybridization between species to form fertile **amphidiploids** would appear to have occurred time and time again in nature, although the only recorded example is that of **Spartina**.

Table 4.2 *Hybridization of* Spartina *(cord grass)*

Spartina maritima (stricta) (uncommon European species) × *Spartina alterniflora* (an American species which reached GB in 1829 via shipping in the river Itchen near Southampton. Now very rare, possibly extinct)

↓

Spartina townsendii (male sterile F1 hybrid first seen in Southampton area in 1870, spread along whole of south coast by vegetative reproduction. Builds mud flats. Sometimes fertile)

↓ Somatic doubling of chromosomes (chromosome numbers are not given as the situation is very complex and controversial)

Spartina anglica (fertile amphidiploid first seen in 1892. Spread from E. Dorset to Sussex by 1907, covering many thousands of acres of tidal mud flats. Stabilizes mud, shuts out tide, outcompetes salt marsh plants.

The occurrence of hybridization can raise questions about the validity of the classification, as one of the criteria used in the definition of a species is that members will not interbreed with non-members.

Table 4.3 *Artificial hybridization between different species*

Primula

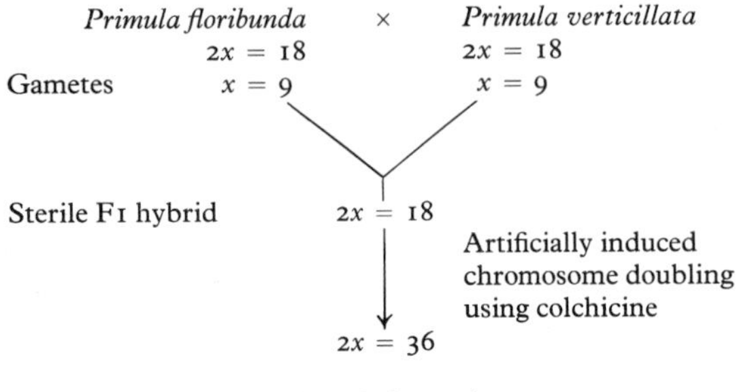

Primula kewensis
(fertile new species)

***Galeopsis* (hemp)**

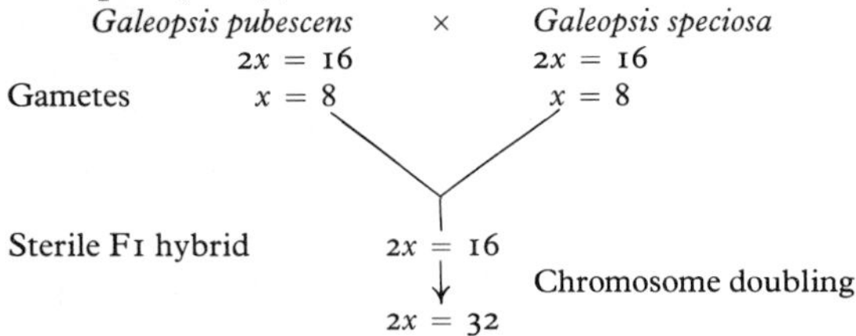

Similar to and fertile with the natural species *G. tetrahit* (common hemp nettle). This gives an explanation of the origin of *G. tetrahit* in nature.

Table 4.4 *Artificial hybridization between different Genera*

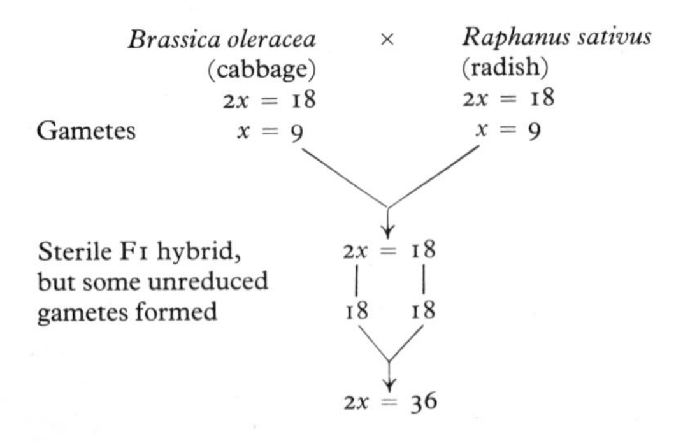

Raphanobrassica
(fertile and robust)

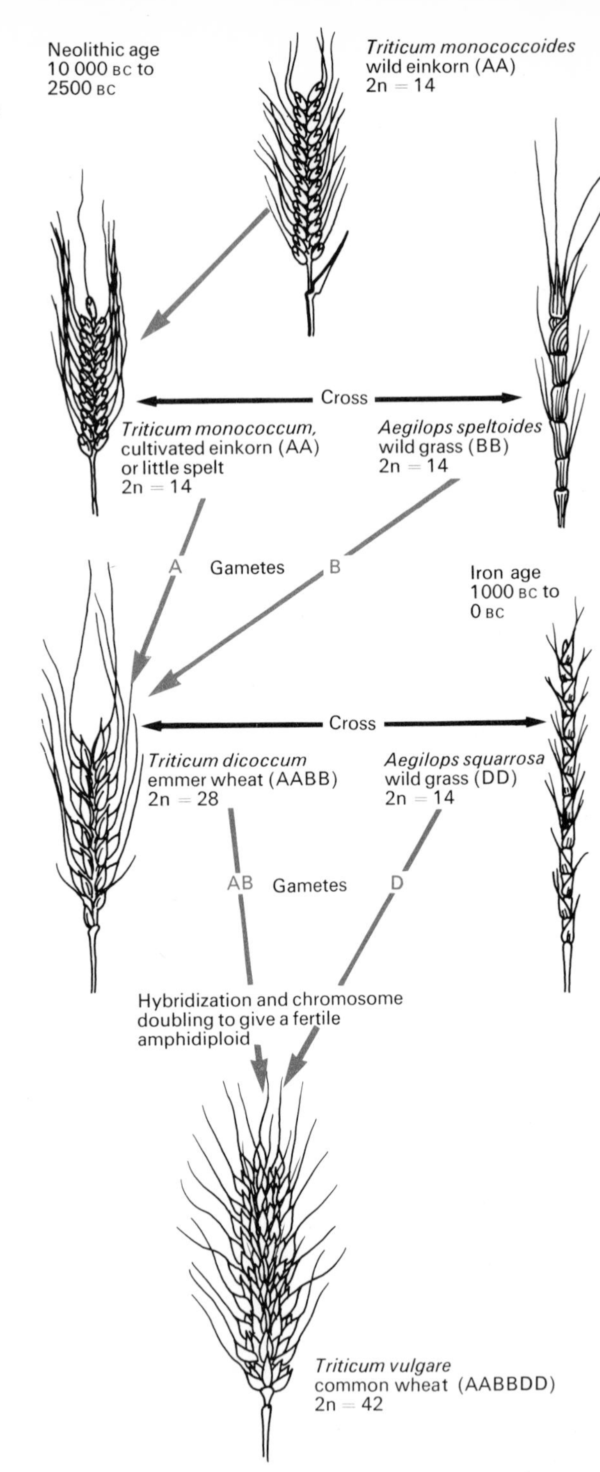

Figure 4.15 *Evolution of wheat—a suggested scheme*
Triticum monococcoides is a wild wheat still found in S.W. Asia. It is not free threshing. *Aegilops speltoides* is widely distributed in the middle east. *T. monococcum* is still cultivated in some areas, but is not free threshing. *A. squarrosa* grows in Asia. *T. dicoccum* cultivated in Egypt since 5000 B.C. *T. vulgare* has some 3000 cultivars. It is easily threshed as the glumes do not enclose seeds.

The importance of polyploidy is demonstrated by its frequency of occurrence in the Angiosperms[351] which are the dominant plant group. About 30 per cent of all Angiosperm species are polyploid, and as many as 75 per cent of the grasses. It is thought that polyploidy has played a vital role in the success of the Angiosperms as the dominant flora of the world. Fossil evidence indicates that there were four major periods of Angiosperm evolution: the Cretaceous, Tertiary, Pleistocene, and Recent. One idea is that during these times there were climatic and edaphic changes that favoured hybridization and polyploidy. Hybridization followed by somatic chromosome doubling could lead to the almost immediate appearance of new species. Diploids and tetraploids are considered to have been the major sources of new genera and families, whereas higher polyploids are considered to have given rise to species. However, beyond a certain point, polyploidy acts as a conservative force in evolution by reducing genetic variability, due to the complexities of the chromosome arrangements. Thus in polyploids the expression of gene mutations and the production of genetic recombination by crossing-over are both inhibited.

Because polyploids both develop new types rapidly, and accumulate different chromosome sets or **genomes**, they tend to have the ability to rapidly colonize new areas. Some believe that they tend to be involved in the rapid exploitation of trends rather than in the origination of these trends.

MUTAGENIC AGENTS

Mutagenic agents increase the natural rate of mutation. They include ionizing radiations and many chemicals.

Ionizing radiations include those emitted by radioactive materials[361] and ultraviolet light (u.v.). U.v. is one type of electromagnetic ionizing radiation. The wavelengths of around 260 nm are absorbed by DNA, and these are also the wavelengths at which little energy is absorbed by proteins, so that most of this u.v. light reaches the DNA in the nuclei in the cells. One effect of these radiations is to cause bonding between thymine bases in the same strand of DNA. This produces so-called thymine dimers, which alters the behaviour of the DNA and results in mutations.

The range of **mutagenic chemicals** is very wide, as is the nature of their effect on the DNA. Nitrous acid reacts chemically with DNA to convert adenine bases to hypoxanthine and cytosine to uracil. This disrupts normal DNA structure and is therefore mutagenic. Base analogues which are similar to the bases in DNA can be incorporated into the genetic material by mistake and this will also result in mutations. Chemicals in the mustard gas group are mutagenic, and even caffeine has been demonstrated to increase the mutation rate in bacteria.

Selection

Selection is not always an agent for change: it can also act as a stabilizing conservative force by eliminating unfavourable genetic variations. This **stabilizing selection** occurs when the mean phenotype of a population coincides with the optimum phenotype for that character in any specific set of environmental conditions.

Directional selection results in a shift of the mean phenotype towards the optimum if the two do not coincide.

Disruptive selection occurs when two extreme phenotypes are favoured. It can lead to balanced polymorphism[100] maintained by differences in the environment, each type or morph being adapted to the different conditions. In this way species with a wide ecological range develop different **ecotypes**, for example the grass *Agrostis tenuis* has ecotypes which thrive variously on high calcium soils, low calcium soils, and soils contaminated with heavy metals from mine workings. When the ecology of a region changes gradually over some distance, the phenotypic characters show all forms of intermediate conditions along what is known as a **cline**.

Features of apparently neutral survival value are often found to be maintained at a definite frequency in the population. In many cases this is due to the fact that although the character may be neutral (e.g. the number of bristles in *Drosophila*), the gene controlling that character may have multiple effects that are not neutral and which are positively selected for.

With favourable conditions a population may increase in size as selection against change is less strong. As a result variation among the offspring can increase. If adverse conditions set in, selection becomes stronger and only the new, better-adapted types may survive.

Figure 4.16 *Selection pressures*
In the early 1960s in Bangkok DDT at 0.005 ppm killed 50% of mosquitoes. In 1967 in Bangkok DDT at 5.0 ppm killed 50% of mosquitoes. In 1969 in Bangkok DDT at 15.0 ppm killed 50% of mosquitoes. (O.P. = optimum phenotype.)

Stabilizing selection

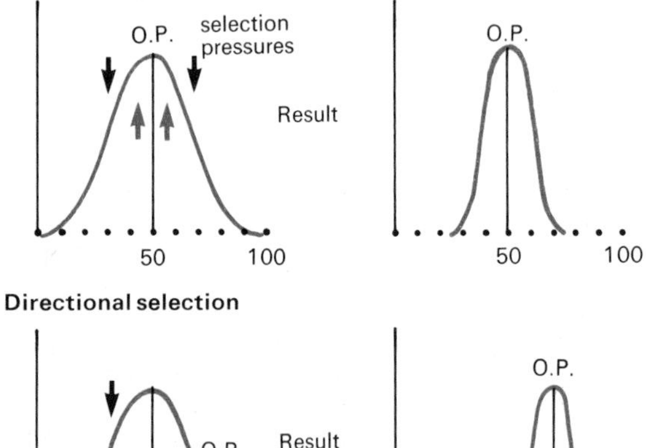

Directional selection

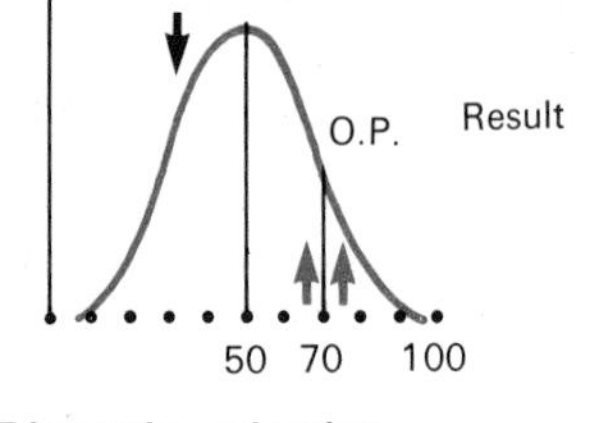

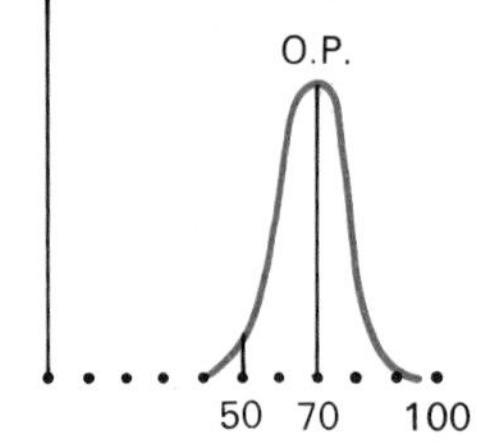

Disruptive selection

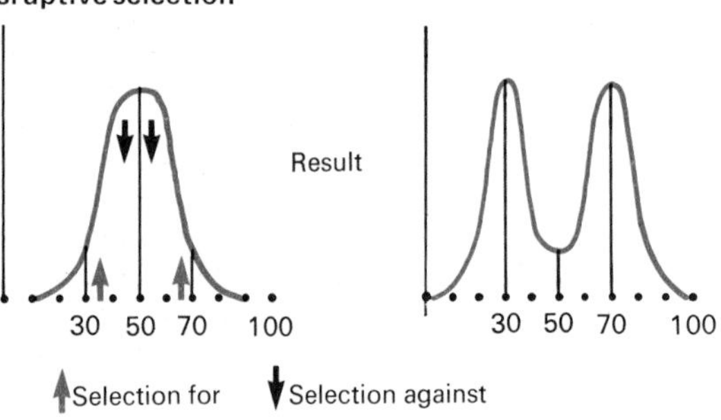

Polymorphism

Polymorphism is a type of variation in which members of a freely interbreeding population have clearly distinct characteristics, and there is no continuous range of intermediates. Species which have two or more different forms or **morphs** occurring together in the same habitat in such proportions that the rarest of them cannot be maintained at the observed frequency merely by continuing mutation are said to be **polymorphic**. Such polymorphism is typically under the control of **super-genes**. A super-gene consists of two or more genes that are so tightly linked that they are rarely separated by crossing-over. They are therefore inherited as a unit and behave like normal single genes. The tight linkage is due to a complex of factors, including the genes being very close together and being on an inverted segment of a chromosome.

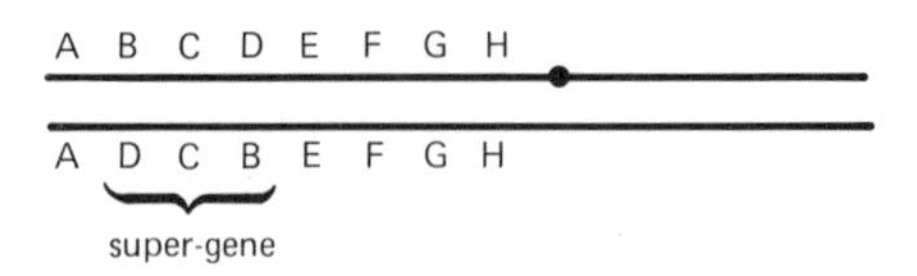

The genes making up the super-gene are co-adapted, that is they represent some favourable combination of characters that it is advantageous to keep together. The development of super-genes is considered to be one of the great principles of genetics as it shows how favourable genetic combinations can be preserved.

Super-genes exist in different allelic forms, as do single genes.

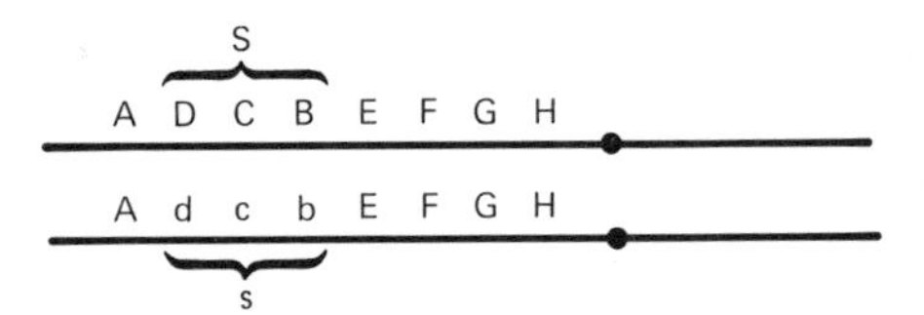

It is an advantage to keep such different alleles in a population as it allows a population to adapt to changing circumstances, should they arise. In a stable environment there is usually a stable equilibrium between the different morphs in a population, and this is referred to as **balanced polymorphism**. Examples are red–green colour blindness; human A, B, and O blood groups; and cyanogenesis[101] in clover leaves. However, the balance between different morphs is sensitive to environmental change and the polymorphic system can respond rapidly to changes in environmental conditions. When this occurs the condition is referred to as **transient polymorphism**.

An everyday example of polymorphism is provided by the existence of distinct male and female forms in a species. The super-genes controlling the differences between females and males are often on the X and Y chromosomes.

Experimental and applied aspects

Balanced polymorphism

RED–GREEN COLOUR BLINDNESS

This is an X-linked recessive 'gene' occurring in about 8 per cent of the male population and about 0.5 per cent of females. It is in fact a super-gene of two very tightly linked genes, one concerned with red perception and the other with green perception.

This type of colour blindness is maintained at a remarkably high frequency in European countries, and is slightly less frequent in less industrialized centres. This indicates the operation of selective pressures, but the adaptive significance of this condition is not clear.

Red–green colour-blind people have a greater ability to detect differences in shades, and this confers a greater ability to detect cryptic colouration camouflage patterns. It has been suggested that this could be an advantage under certain circumstances, but one would imagine it to be of greater selective advantage in hunting communities than in industrialized societies. It is interesting to note that red-green colour-blind females have been preferentially employed in the wool-dye industry for matching different batches of the same colour.

WHITE CLOVER

In white clover (*Trifolium repens*) there are two dominant genes which control the ability of the leaves, when damaged, to produce hydrogen cyanide in a process known as **cyanogenesis**. One gene (S) controls the presence of the glucoside substrate and the other (E) controls the presence of the enzyme which catalyses the hydrolysis of the substrate to produce hydrogen cyanide.

If both enzyme and substrate are present in the cells of a single plant, then hydrogen cyanide is produced when the two come together as the leaf is chewed, crushed, or disturbed by frost. In a population there is a polymorphic situation with plants of four possible phenotypes namely:

S_E_	S_ee	ssE_	ssee
Strongly cyanogenic	Very slightly cyanogenic due to spontaneous breakdown of substrate	Acyanogenic	Acyanogenic

In frost regions the extreme cold activates the process of cyanogenesis whilst the leaf is still on the plant, respiration is inhibited, and therefore these cyanogenic types are selected against and their frequency is reduced. In warmer regions small animals avoid the cyanogenic plants and graze on the acyanogenic ones, therefore the former increase in frequency.

These differing selective pressures result in a balanced polymorphism with the acyanogenic morphs being favoured in frost zones (as determined by the mean January temperature), and the cyanogenic morphs being favoured in warmer regions.

Transient polymorphism

PEPPERED MOTH (*Biston betularia*)

The peppered moth has two distinct morphs: the light coloured 'natural' form and the dark coloured or 'melanic' form. They are active at night, and during the day they rest on vertical surfaces, usually tree trunks, particularly birch.

In 1848 a census of the population near Manchester showed 99 per cent of the population were of the light form and 1 per cent of the melanic form. In 1894 another census showed that 1 per cent were of the light form and 99 per cent of the melanic form. It was also observed that there had been a corresponding darkening of the bark of trees due to atmospheric pollution from smoke emissions. Today the proportions of light and melanic forms vary in different parts of the country. In rural areas the light form predominates and in industrial areas the melanic form predominates.

Experiments, including direct observation, have shown that the less well camouflaged types, the light form on a dark background and the dark form on a light background, are preyed upon more by birds. Thus natural selection favours the dark form in industrial areas and the light form in rural areas. In this way the frequency of the dominant melanic gene increases in industrial areas and decreases in rural areas and the frequency of the recessive light gene decreases in industrial areas and increases in rural areas.

Such industrial melanism is found in a wide variety of organisms, including mammalian species.

Figure 4.17 *Distribution of dark and light forms of Biston betularia in GB.*

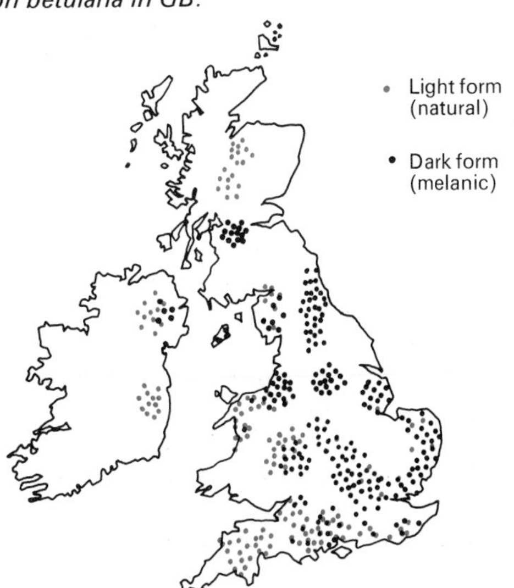

Guided example

Examine the information on the Galapagos Islands given in Tables 4.5 and 4.6 and answer the following questions.

Table 4.5 *Comparison of prevalence of finch species on the Galapagos Islands*

	No. of species	*Percentage of unique species*
Outer islands		
Cocos	1	100
Culpepper	4	75
Wenman	5	75
Hood	3	67
Tower	4	67
Chatham	7	36
Abingdon	9	33
Bindloe	7	33
Central islands		
Charles	8	25
Narborough	9	20
Albermarle	9	20
Barrington	7	14
James	10	5
Indefatigable	10	0

A seed-eating ancestral form is postulated, with wide adaptive radiation due to isolation, adaptation to different conditions, and lack of competition. Flightless or ground-dwelling birds are characteristic of the small islands due to the danger of being blown off and to the lack of predators. (The islands also have unique marine-feeding iguanas, a unique type of flightless cormorant, and one species of penguin which is the only one to be found in the tropics.)

Table 4.6 *Comparison of species in the Galapagos Islands and the British Isles*

	Land snails		*Insects*		*Reptiles*		*Land birds*		*Mammals*	
	GI	*GB*	*GI*	*GB*	*GI*	*GB*	*GI*	*GB*	*GI*	*GB*
Unique species	15	4	35	149	17	0	28	1	0	0
Species also found elsewhere	0	83	0	12 551	0	13	1	130	0	40
Approximate percentage of unique species	100	4.8	100	1.2	100	0	97	0.8	—	0

GI... Galapagos Islands; *GB*... *Great Britain*

1 What suggestions could be put forward to explain the presence of the finches in the Galapagos Islands?
One suggestion could be that they arose there; the more accepted view is that insufficient time has passed since the emergence of the islands (about two million years) for this to have occurred, even if the 'necessary' conditions existed, and that they have been colonized by finches blown there from the mainland of South America 600 miles away.

2 What alternative suggestions could be made to explain the variety of finches on the Islands?
Again there are two main suggestions. One is that the islands were colonized by a mixed flock of species from the mainland; the other is that they were colonized by a flock of a single type, perhaps only once. The basic similarities between the different species suggest that the second alternative is the more acceptable.

3 If it is accepted that colonization was by a flock of a single type, what explanation could be put forward to account for the variety of species that are seen today?
The explanation, according to the Theory of Natural Selection, would be that the original type has undergone adaptive radiation to the different environmental niches on the islands, and that isolation on the different islands has led to the origin of new species.

4 What explanation could be put forward to account for the observation that the central islands have the greater number of species, but that the islands towards the edge have a greater percentage of unique species?
The accepted explanation is that, since the central islands are closer together, a certain amount of mixing has occurred between the species of the different islands so that there are a wide variety of types, but few which are unique; whereas the islands to the edge of the group have less variety but more unique species, as their isolation prevents any inter-mixing with other types from other islands.

5 As is seen in Table 4.6, the British Isles have very few unique species compared to those of the Galapagos. What differences between the islands could account for this?
These observations can be accounted for by their different geographical origins. The Galapagos Islands are small oceanic islands, volcanic in origin, which have always been isolated from the nearest land mass, whereas the British Isles are large continental islands, only recently separated from the continent by a narrow sea channel. This means that there are many species common to both the British Isles and the continent as there has been less geographical isolation than in the Galapagos.

Questions

1 Outline the evidence for evolution.
Show how various types of evidence have been used to deduce the evolutionary history of a *named* group of animals. (L)

2 Define a species. Describe, with examples, the processes which take place during the evolution of a new species. (L)

3 A groundsel plant may have one of three types of flower. The flower type is controlled by a pair of alleles, neither of which shows dominance. Each flower type and its corresponding genotype is shown in the table below:

Genotype	*Phenotype*
RR	long ray florets
Rr	short ray florets
rr	no ray florets

In a population of 600 groundsel plants there were:

350	with	long ray florets
100	with	short ray florets
150	with	no ray florets

(a) Calculate the frequency of the R allele.
(b) Calculate the frequency of the r allele.
(c) Using the Hardy–Weinberg formula calculate the *expected* number of the Rr genotypes.
(d) Assume that there was no selection against any of the three types. Suggest *one* reason which would account for the fact that the actual number of Rr genotypes is less than the expected number. (JMB)

4 In a large population two alleles B and b occur, and the genotype frequencies are

BB	Bb	bb
49	42	9

Assuming that mating is random, that all individuals produce roughly equal numbers of gametes and that genes B and b do not mutate, show by calculation what the genotype frequencies are in the next two generations and comment upon the rate of evolution of such a population. Discuss the effects upon this rate of evolution if (a) non-random mating occurs, (b) mutations occur, and (c) the population is small. (O & C)

5 Two closely related species of finch (*Geospiza fortis* and *G. fuliginosa*) are found on some of the isolated Galapagos Islands. The frequency distributions of beak lengths of these finches on two of the islands are shown in Figures 1 and 2. Both finches feed on seeds. Assume that beak length can be taken to indicate the type of seed on which each species feeds. The same range of seed size occurs on both islands.

(a) The frequency distribution in Figure 1 is for *G. fortis* on Daphne Island. It is the only one of the two species to live there. To what type of distribution does this frequency distribution approximate?

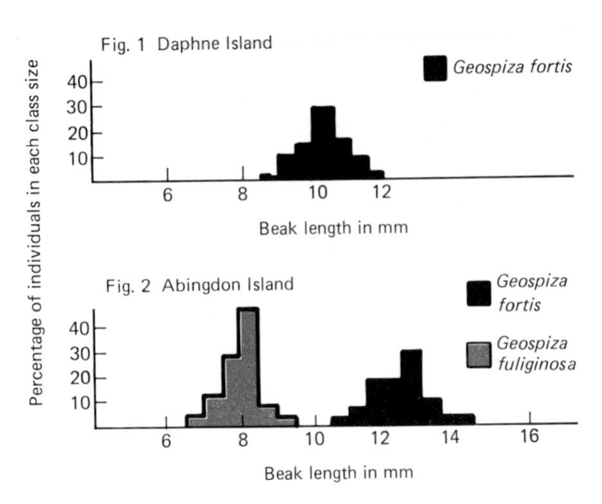

(b) On an island not referred to in the above figures (which has the same range of seed size as the first two islands) *G. fuliginosa* is found, but not *G. fortis*. The distribution of beak lengths of *G. fuliginosa* on this island is almost identical to that of *G. fortis* on Daphne.

On Abingdon Island both species occur. The frequency distributions in Figure 2 show their beak lengths.

Suggest an explanation for the fact that most frequent beak length for each of the two species when they occur on Abingdon differs from that of each species when they occur on the other islands. (JMB)

Further reading

ASE Lab Books (compiled by G. W. Shaw), *Cytology, Genetics, and Evolution* (London: John Murray, 1973).

de Beer Sir G., *Atlas of Evolution* (London: Nelson, 1964).

de Beer Sir G., *Homology, an Unsolved Problem*, Oxford Biology Readers No. 11 (Oxford: Oxford University Press, 1971).

Dowdeswell W. H., *The Mechanism of Evolution* (London: Heinemann, 1975).

Edwards K. J. R., *Evolution in Modern Biology*, The Institute of Biology Studies in Biology No. 87 (London: Edward Arnold, 1977).

Lack D., *Darwin's Finches* (Cambridge: Cambridge University Press, 1947).

Savage J. M., *Evolution* (New York: Holt, Reinhart and Winston, 1900).

5 Ecology

Introduction

Ecology has been defined as the search for the causes of the patterns in the distribution of animals and plants which arise as a result of their interactions. All organisms live in a complex web of inter-relationships with both their particular environment and each other. The inter-relationships between members of the same animal species and with the environment are expressed in their behaviour.[232] Population genetics[92] is also involved in this web of inter-relationships, where a variety of forces act on members of a population to modify or maintain the state of the gene pool.[89]

Some of these aspects have been outlined in other chapters, and an attempt will be made here to deal with further ways in which organisms interact with each other and with their environment, particularly with regard to the flow of energy which provides the unifying theme within their environment.

The study of a particular species in relation to its environment throughout its life history is referred to as its **autecology**. It can reveal interesting results of interaction with the environment, for instance seasonal changes, or even the formation of a range of variations between members of the same species in different habitats. These ecological races are known as **ecotypes**. However, the autoecological approach reveals little, if anything about the organism's interaction with other living organisms in the same environment.

Synecology is the name given to the study of organisms in communities. It provides a more accurate picture of individual species which may be derived from an autoecological approach. There is a considerable overlap between the two approaches, and the distinction is not as clear-cut as the use of these two terms may imply.

The ecosystem

A fairly self-contained system of interacting organisms and non-living material can be referred to as an **ecosystem**. The size of an ecosystem depends on the limits set by the investigator, as all so-called ecosystems are ultimately part of the biosphere or whole global ecosystem. Although the concept of an ecosystem is therefore rather artificial, it does enable regions of manageable size, for example a wood or pond, or even a puddle, which illustrate particular themes, to be 'isolated' for study.

Habitat

A habitat is a particular place with fairly distinctive environmental features which enable it to be defined as a region within an ecosystem. It has been defined as an organism's 'address' within the ecosystem. An example is the leaf litter in a wood.

Ecological niche

An organism's ecological niche is defined by its functional role within the community or ecosystem. If the habitat is an organism's 'address', then its niche is its 'profession'. It describes, amongst other things, its trophic position in the complex of food-webs that exist in an ecosystem; for example, the ecological niche of an earthworm in a woodland soil is that of a decomposer of leaf litter and a soil conditioner. Two or more species rarely have the same niche within a habitat, indeed specializations for a particular niche are amongst the most frequently observed characteristics of a species. However, a taxonomic species is not usually a single ecological unit; within most species there are local populations which occupy slightly different niches.

Populations

Population ecology is concerned particularly with the fluctuations in numbers of particular populations within a community, and with the factors that cause them. Animal populations show more rapid numerical changes than most plant populations because animals are characteristically mobile and respond rapidly to various forces. As a result of their immobility plants generally interact more slowly with their environment and produce fairly stable patterns of distribution which are more easily studied than those of animal populations.

During the colonization of a new area, or the re-establishment of a population in an area where there has been some destructive agent such as fire on heathland, a population usually shows a typical population growth curve. At first there is a time of establishment, individual growth, and maturation, in which the numbers do not increase significantly. Once established, however, if conditions are right, there can be a period of exponential[357] or very rapid growth during which numbers increase dramatically. As the numbers increase, various limiting factors which prevent the further exponential growth of the population come into force, and the population then levels off at its optimum size.

For dynamic equilibrium to be obtained, the gains due to reproduction and immigration must balance the losses due to mortality and emigration. Dynamic equilibrium of population size can exist in natural communities, but catastrophic falls and explosive increases in number also occur. The environmental factors causing such dramatic fluctuations can often be apparently identified, such as seasonal depletion of food supply or unfavourable temperature. Sometimes, however, there is no clear-cut explanation and the cause remains obscure.

A particular example of dramatic fluctuation is found in the predator/prey relationship between the Canadian lynx and the snowshoe hare in the Hudson Bay area of Canada. Estimates of population trends of both these animals were made for the period from 1845 to 1935. They showed that there was a cycle of peak abundance and a population crash about every decade. A simple explanation is that the predators increase in number as the prey increases and eventually destroy the prey, so that their own numbers then drop. However, care is needed in drawing such conclusions as it is almost certain that this cycle of abundance is not a simple cause and effect relationship between the predator and prey.

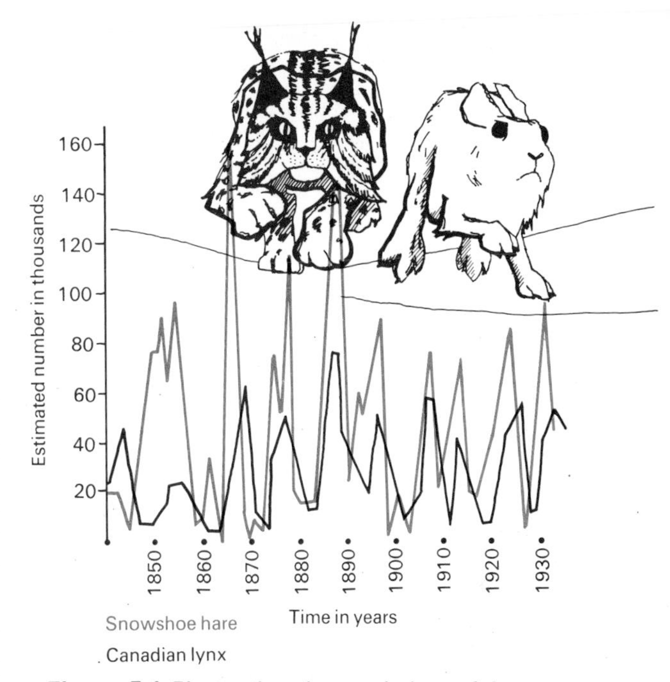

Figure 5.1 *Fluctuations in populations of the Canadian lynx and snowshoe hare*

Interactions

The factors affecting organisms in their environment mainly fall into three broad categories: climatic, edaphic, and biotic.

Climatic factors include light, rainfall, temperature, wind, and humidity.

Edaphic factors include the topography (slope and aspect) and nature of the soil.

Biotic factors include competition for light, space, food, shelter, etc., both within species and between different species; and many other interactions such as feeding, reproduction, and dispersal.

Examples of these factors are given in the description of particular habitats.

The complexity of some biotic interactions is illustrated by those plants that can prevent the development and breeding of the insects that feed upon them. Several species of plants contain ecdysone[28] and other hormone-like substances which have a strong juvenile hormone action on a variety of insects. When the insects feed on the plants these substances are absorbed in their gut and interfere with their normal development, thus reducing further attacks on the plants by reducing the numbers of insects.

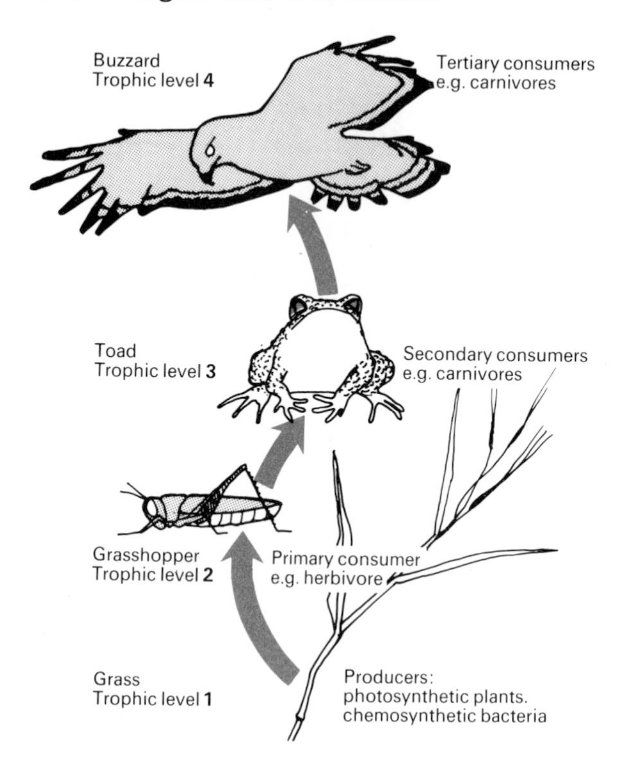

Figure 5.2 *The tropic levels of a simple food-chain*

Figure 5.3 *Energy flow through an ecosystem*

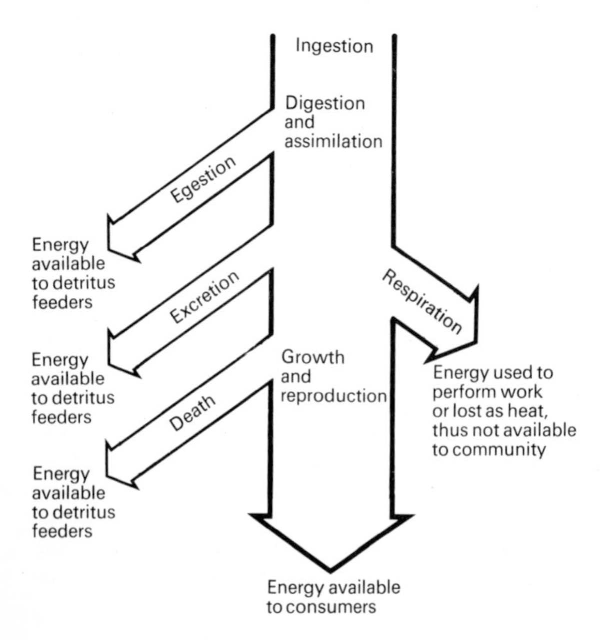

Food-chains and webs

Feeding relationships are one of the most important ways in which living organisms interact. Heterotrophic[134] organisms, such as animals, fungi, and some bacteria, obtain their food either directly or indirectly from autotrophic[127] organisms, especially photosynthetic plants. The photosynthetic plants are therefore known as **producers** and the heterotrophic organisms as **consumers**. The organic material consumed by heterotrophic organisms is required for the synthesis of their body matter, for the energy necessary for that synthesis, and for the energy needed for general metabolism and movement. A sequence of organisms, from the producers, to first order consumers, to second order consumers, etc., at different feeding or **trophic levels**, is known as a **food-chain**. However, few simple chains exist as there are many inter-connections between different feeding relationships which together produce complex **food-webs**.

ENERGY FLOW

Energy from the Sun is incorporated into organic compounds by the process of photosynthesis. These organic compounds pass along a food-chain or web from the producers to the consumers. At each transfer of materials between members of different feeding or trophic levels, about 80–90 per cent of the potential energy is lost and is therefore not available to the consumers of the next trophic level. Much of the food material remains undigested and passes through an animal to form one of the first links in the food-chains, based on the activity of **decomposers**. Energy is also lost as heat liberated during digestion and respiration.

These figures show that when plants are eaten by herbivores only about 10 per cent of the plant material is converted to animal material, and when herbivores are eaten by carnivores only about 10 per cent of the herbivore material is converted to carnivore material. The amount of photosynthesis carried out by the producers determines the amount of energy entering a particular food-chain. Due to the conversion inefficiencies at each trophic level, more energy will be available to those at lower trophic levels than to those at higher levels. For this reason, the number of trophic levels in a food-chain is normally limited to about four. However, food-chains with more than four levels can exist because most carnivores can feed at more than one trophic level and do not rely entirely on the animals immediately before them in the chain to provide all the food they require.

BIOMASS

The biomass is defined as the mass of living material in an ecosystem. Due to the loss of material and energy at each trophic level, the biomass of the producers will be greater than the primary consumers, and that of the primary consumers will be greater than that of the secondary consumers, and so on. This gives rise to the concept of a **pyramid of biomass**. It is important to study an ecosystem throughout the year when estimating the biomass at each level, as seasonal fluctuations in populations occur. The pyramid of biomass gives a more accurate picture of the flow of energy and materials through a food chain than the pyramid of numbers because it is not always true that the number of organisms decreases with increase in trophic level.

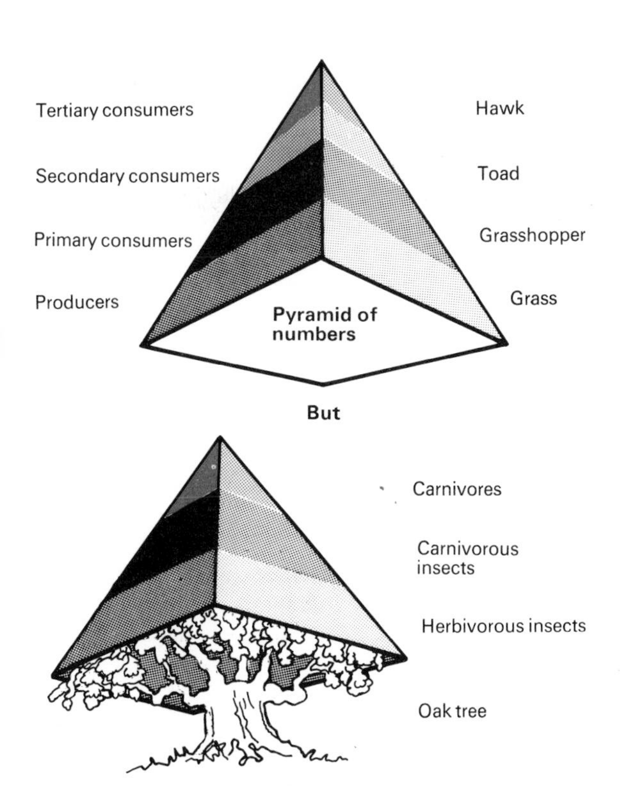

Figure 5.4 *Pyramid of numbers*

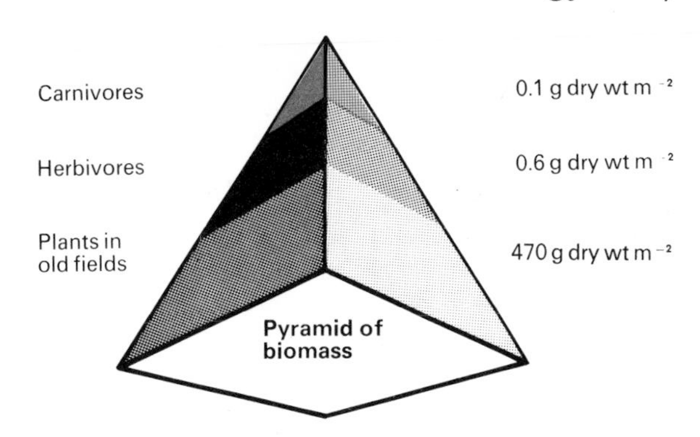

Figure 5.5 *Pyramid of biomass*

Special interactions

Certain very close associations can occur between organisms in an environment, for example symbiosis and parasitism.

Symbiosis

This is an association of organisms of different species which causes harm to neither but brings benefit to both. There is a wide range of interdependence of symbionts, from those bordering on **commensalism**, where each has virtually no effect on the other, to those bordering on parasitism, where the benefit is very one-sided. Examples of symbiotic associations are seen in the mycorrhizal[114] association of certain soil fungi with the roots of some higher plants; the association of certain bacteria with the tissues of higher plants to form nodules; the association between algae and fungi in the lichens;[108] and the association between algae and animal cells as seen in the endodermal cells of Coelenterates.[332]

NODULE FORMATION

An extremely important symbiotic association occurs between some nitrogen-fixing bacteria and the tissues of many Angiosperm and a few Gymnosperm species, to produce root, or even leaf and stem, nodules. Root nodules are characteristic of the Leguminosae and are formed with the bacterium *Rhizobium*. In this association the host plant obtains its nitrogen requirements from the bacteria and the bacteria obtain organic compounds, including carbohydrates and growth substances, from the host.

LICHENS

Lichens are composite organisms consisting of a symbiotic association between **algae** and **fungi** (mostly Ascomycetes). The algae gain protection and mineral nutrients, while the fungi gain organic nutrients, including vitamins. Primitive **homoiomerous** types have a plant body or **thallus** composed of an undifferentiated mixture of algal cells and fungal hyphae. **Heteromerous** types have the algal cells confined to a specific region of the thallus, usually in a sub-cortical layer beneath the upper surface, which produces an appearance rather like the structure of a dicotyledonous leaf with its photosynthetic palisade mesophyll. There is a continuous series of increasing complexity of structure from flat **crustose** types, to leafy **foliose** types, to shrubby **fruticose** types. Along with this increasing complexity goes an increasing sensitivity to SO_2[316] pollution; thus lichens can be used as monitors for air pollution. However, lichens resist ionizing radiations[361] better than any other plants.

Vegetative reproduction occurs by fragmentation and dispersal of the thallus, and in some cases by the production of special structures known as **soredia**, which consist of a few algal cells with fungal hyphae.

Homoiomerous type
Fungi
Algae
Heteromerous type
Algae
Fungi
Algae
Fungi
T.S. dicot leaf
Photosynthetic palisade mesophyll layer

Figure 5.6 *Structure of lichens*

The algal partners are usually members of either the Cyanophyceae (the blue-green algae) or the Chlorophyceae (the green algae) and can live a separate existence. The fungi are known as **obligate symbionts** because they are not found as free-living individuals.

Lichens are often highly pigmented (the pH-sensitive pigment litmus is extracted from the lichen *Rocella tinctoria*).

ALGAE/ANIMAL CELL ASSOCIATION

Cyanophora paradoxa is a unicellular animal which contains two or more symbiotic blue-green algal cells. The algal cells have no cell wall and function like the chloroplasts of the green plant cell, there being complete physiological inter-dependence between the members of the association. Their reproduction is simultaneous so that the association is maintained. The animal cell gains nutrients and oxygen from the photosynthesizing algal 'cells', and the algal 'cells' gain carbon dioxide and inorganic nutrients from the animal cell.

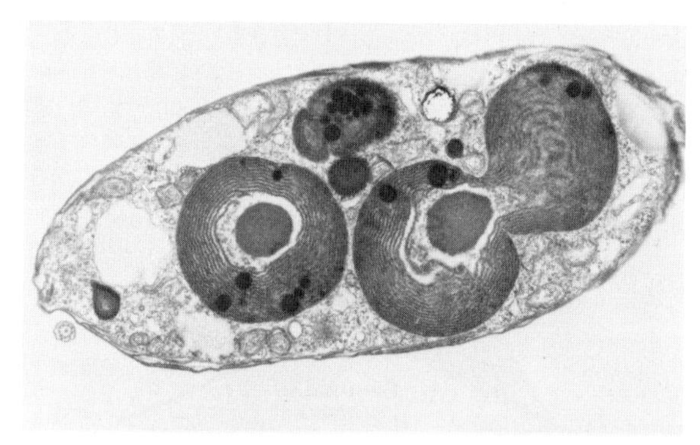

Cyanophora paradoxa showing two blue-green algal symbionts, one in the process of division

Parasitism

A parasitic relationship exists when an organism (the **parasite**) lives on or in another organism of a different species (the **host**) from which it obtains its food, protection, and optimum conditions for its survival. Some parasites do not apparently harm the host, but others do in a variety of ways, for example by depriving the host of nutrients, by the production of toxins, by irritation and inflammation, by introducing other disease-causing pathogens, etc.

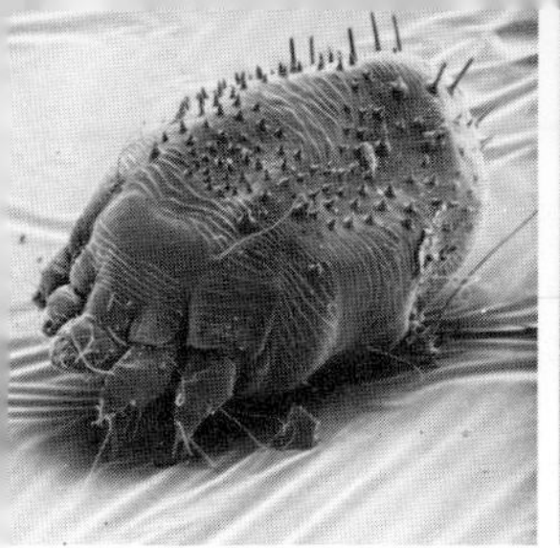
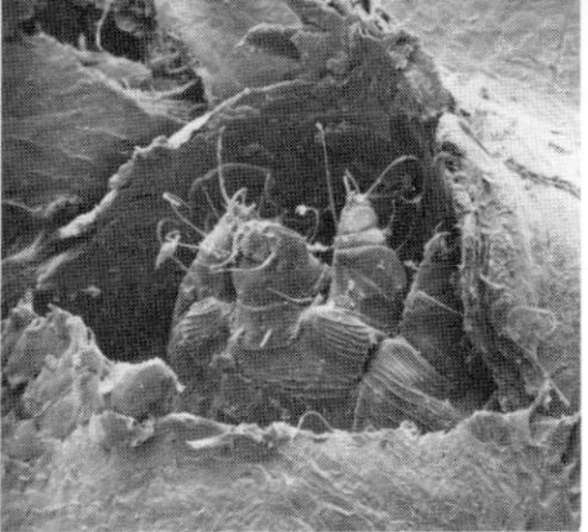

Photomicrographs of a scabies mite (*Sarcoptes scabei*) (× 150)

ANIMAL–ANIMAL PARASITES

There are two major groups of animal parasites, those that live on the surface of their host, the **ectoparasites**, and those that live inside their host, the **endoparasites**.

Ectoparasites
These live exposed to the external environment and thus do not show great structural adaptations when compared to their free-living relations. Adaptations most commonly seen are related to attachment, whereby special structures such as hooks and suckers are developed to maintain their position on the host. In addition, the body shape is usually flattened to lessen the danger of mechanical damage due to scratching or rubbing by the host. Mouthparts are well developed since, being on the outside of the host, feeding is still by biting, sucking, or chewing. Transference to new hosts is relatively simple, and so cause no major changes in their life history. Examples of such animal ectoparasites include fleas, lice, mites, ticks, and leeches.

Endoparasites
These live in constant optimum conditions, surrounded by a plentiful supply of readily available food. There thus tends to be a reduction in those structures involved in locomotion, food capture, perception, and coordination. However, protection against the host's defensive mechanisms is required, so special developments such as the secretion of mucus and of anti-enzymes is seen in many species. Since there is transference to a new host is a risky business, there is often an elaboration of the life cycle involving one or more secondary hosts. When the secondary host actively transports the parasite to the primary host, it is known as a **vector**. Resistant stages during the life cycle often occur so that the organism can withstand adverse conditions outside the host.

Gut parasites are, strictly speaking, attached to the outside of their host, as the lumen of the gut is not actually within the body of the host. However, they are usually considered as endoparasites. Although surrounded by digested food at a constant temperature, and protected from environmental fluctuations and predators, they still occupy a hazardous position. The host's digestive juices threaten to digest parasitic tissues, peristalsis[145] threatens to eliminate them with the faeces, and conditions are virtually anaerobic. Gut parasites show a variety of adaptations to these conditions.

Blood parasites can show a remarkable circadian rhythm[235] of movements in the body which increase the chances of their being picked up by a blood-sucking vector such as the mosquito. For example, the larvae of the filarial Nematodes, which parasitize the blood of vertebrates producing elephantiasis in man, move into the blood vessels of the skin at times when the mosquito feeds on the host. At other times they are not found in these vessels. Such rhythms of blood parasites are thought to be in response to some aspects of the host's circadian rhythms.

Even with modern drugs and medicines, and control methods based on a detailed understanding of the life cycles of the parasites, the death, disease, poverty, and misery arising as a result of infection by animal parasites is unimaginable. Amongst the most devastating are the protozoal diseases, for example malaria (*Plasmodium* spp.),

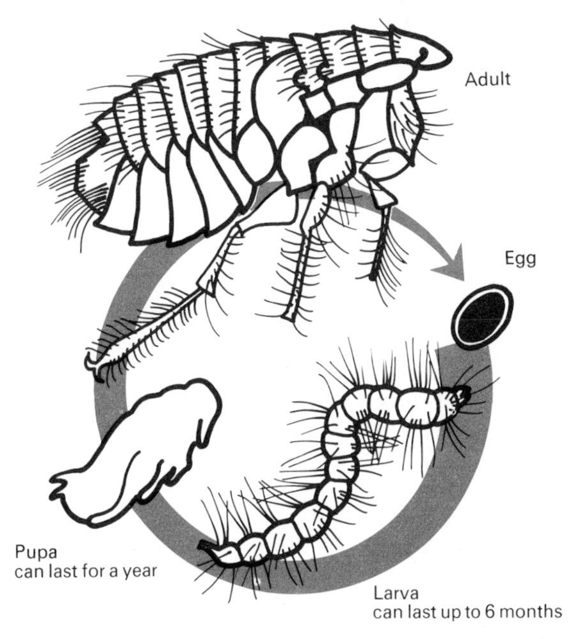

Figure 5.7 *Flea life cycle*
The cat flea (*Ctenocephalides felis*)[311]

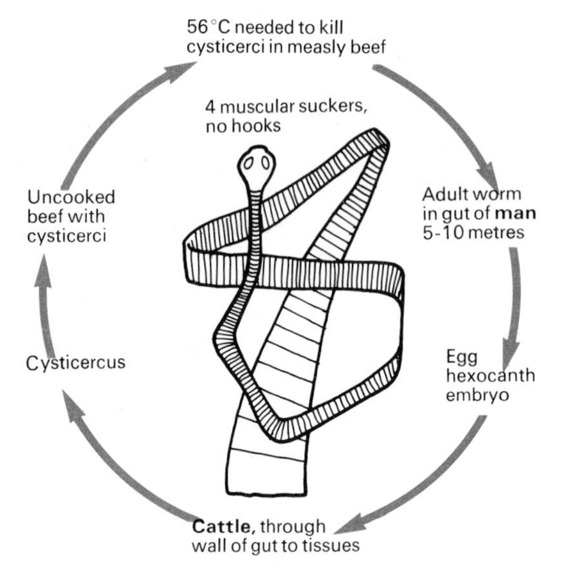

Figure 5.8 *Tapeworm life cycle (Taenia saginata, the beef tapeworm)*
The adult can reach a length of 15 m and live for about 25 years. Each segment can contain 80 000 eggs, and nine such segments can be released each day (750 000 eggs per day).

sleeping sickness (*Trypanosomonas* spp.); the Platyhelminth diseases, for example schistosomiasis or bilharzia caused by blood flukes; and the Nematode diseases, for example hookworm (*Ancylostoma duodenale* and *Necator americanus*).

ANIMAL–PLANT PARASITISM

Animals parasitize plants both externally and internally.

Aphids (greenfly and blackfly) penetrate soft regions of plants with their tubular, piercing mouthparts and suck the sugary contents of the phloem sieve tube elements. (Certain types of aphids are 'cultured' by certain species of ants to act as sources of 'sugary honeydew'.)

The nematode *Heterodera rostochiensis* (the potato root eelworm) infests the tissues of the potato root. After mating, the fertilized eggs overwinter in the dead body of the female, and lie dormant in the soil until stimulated to emerge by potato root secretions. The larvae are attracted to the roots and penetrate the root tissues by using a sharp stylet. The cycle is thus repeated. The roots of a single potato plant can contain up to 50 000 eelworms.

PLANT–ANIMAL PARASITISM

Plant parasites of animals are all fungi. In contrast to the many viral and bacterial diseases, there are very few fungi which parasitize animals and man. Of those that do, there are three main groups: the dermatophytes, the yeasts, and the systemic mycoses.

The dermatophytes infect the skin, hair, and nails, and include the so-called 'ringworm' and 'athletes foot' fungal infections.

The yeast-like infections include *Candida albicans*. This is a normal saprophyte found on the skin and other surfaces of the body which forms part of the natural ecosystem of the skin. However, if the resistance of the body is lowered they can invade the tissues, producing the condition known as 'thrush', which usually appears as fluffy white patches of growth on inflamed red patches.

The systemic mycoses are caused by those fungi which penetrate the body and infect the internal organs. These are very rare in temperate climates. One example of this group is *Cryptococcus neoformans* which causes cryptococcosis by invading the lungs and the cerebrospinal fluid. The symptoms appear as a form of meningitis.

PLANT–PLANT PARASITISM

Plants can parasitize plants both externally and internally.

The colourless flowering plant dodder (*Cuscuta* sp.) parasitizes clover, nettles, heather, and gorse. On germination in the ground the seedling grows rapidly and if it attaches to a host its roots die away. If no host is reached the seedling soon dies.

The fungus *Phytophthora infestans* infects the tissues of potato plants, especially the leaves, causing the disease known as 'potato blight'.[304]

Biosphere

This refers to the region of the Earth and its atmosphere in which life exists. It extends from about 600 m above and to about 10 000 m below sea level. There is a continual energy input from the Sun, but as far as materials are concerned the Earth represents a closed, finite system, within which materials necessary for life must be recycled in various biogeological cycles.

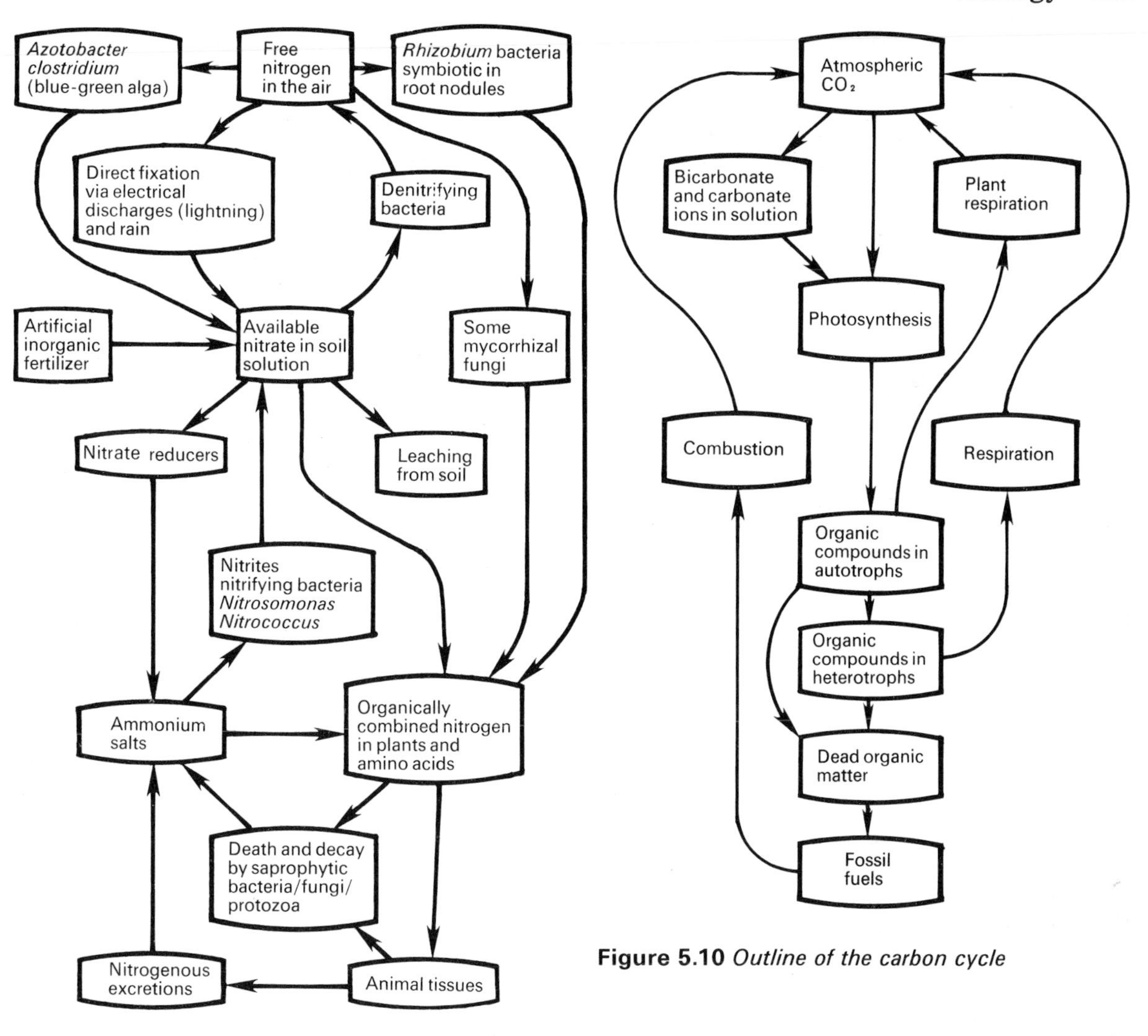

Figure 5.9 *Outline of the nitrogen cycle*

Figure 5.10 *Outline of the carbon cycle*

The two great dimensions of the biosphere are the **hydrosphere**, of rivers, lakes, and oceans; and the Earth's surface, or **lithosphere**, particularly the soil, which is sometimes referred to as the **pedosphere**.

Hydrosphere

The aquatic environment provides ideal conditions for life, but in large bodies of water such as fresh water lakes, and oceans, there are limitations to its productivity of biomass due to problems of available light and nutrients for the photosynthetic producers at the base of the food-chains. Furthermore, there are problems of osmoregulation,[183] especially for animals.

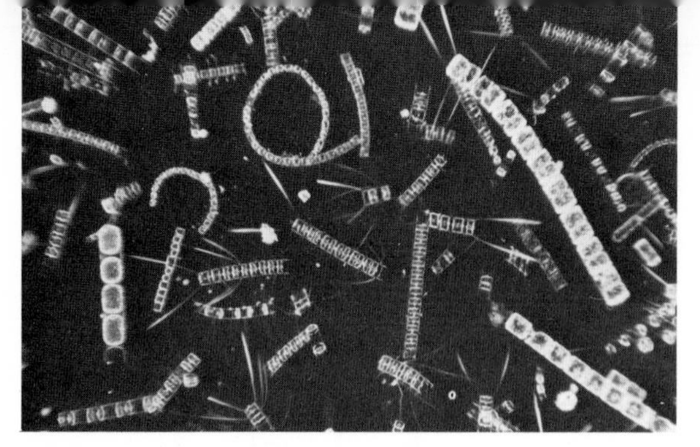
Phytoplankton (× 60)

PHYTOPLANKTON

This consists of small photosynthesizing plants that float in the surface layers of the fresh water lakes and the sea. These organisms are of central importance to the ecology of these environments as they form the basis of all the food-chains. Phytoplankton includes dinoflagellates, microflagellates, diatoms, green algae (Chlorophyceae) and the blue-green algae (Cyanophyceae). The most important, with regard to the amount of photosynthetic production, are the microscopic **nannoplankton**, especially the green flagellates from 2–25 μm in size.

The limiting factor affecting the productivity of the sea is the availability of nutrients to the photosynthesizing phytoplankton. Nitrates and phosphates, which are particularly important to the productivity of the phytoplankton, are at a relatively low concentration in the sea. Nutrients get lost by sedimentation to the depths where light is limiting, and only in areas of strong upwelling currents, which bring the nutrients back into the photosynthetic surface regions, are they available in amounts large enough not to limit productivity. One cause of such upwellings is off-shore winds which, by moving the surface waters away from the coast, bring cold water rich in nutrients to the surface. The fertility of coastal waters is further increased by the flow of river water from estuaries which is rich in nutrients leached from the soil. Most phytoplankton and its associated fauna are therefore found in the coastal waters of the continental shelf region, which are much more fertile than the open sea.

In temperate regions there is a seasonal rhythm of planktonic growth which is related to the formation and breakdown of a fairly sharp division or **thermocline** between the warmer surface waters and the colder deeper waters. In the winter the water has the same temperature and mineral content at all depths. In the spring and early summer the surface waters gradually warm up and become less dense than the deeper colder waters. By midsummer there is a fairly distinct thermocline established at about 15 m which effectively separates the surface waters and the deeper waters into two distinct layers.

With the increased light and temperature in the spring and early summer the phytoplankton increase their rate of photosynthesis and consequently their rate of growth and reproduction. In so doing they deplete the nutrients in the surface layers and, as a result, begin to die off. In the autumn the thermocline is destroyed by storms and the cooling of the surface waters so that nutrients from the deeper layers replenish the surface waters. As a result there is an autumnal 'pulse' of phytoplankton growth before the light and temperature drop below that required for photosynthesis. The phytoplankton die off, the nutrients re-accumulate, and the process is repeated the following year.

Figure 5.11 *Seasonal 'pulses' of plankton*

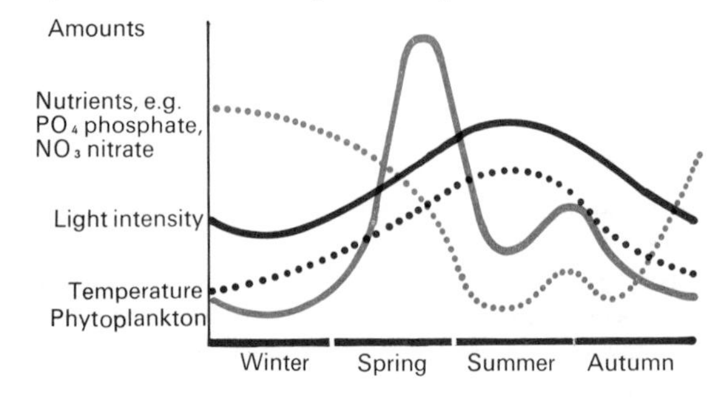

ZOOPLANKTON

This consists of small animals that float in the surface layers of both fresh-water lakes and the sea, and which feed on or 'graze' the phytoplankton. The zooplankton contains both permanent members, the **holoplankton**, and temporary members, the **meroplankton**. The permanent members include protozoa, tiny jelly-fish, some

Zooplankton (× 16)

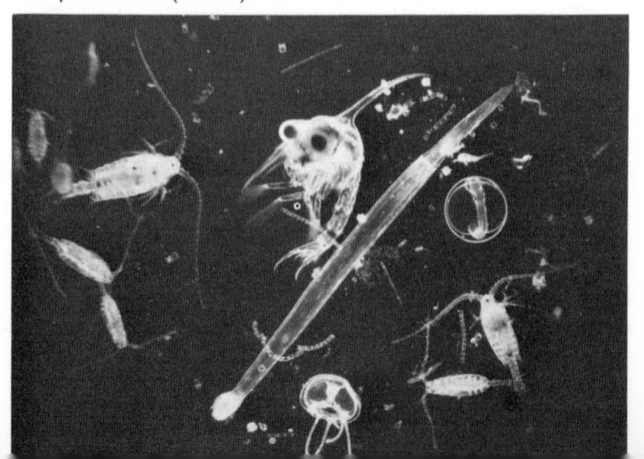

molluscs, Copepoda, some larger Crustacea known collectively as 'krill', free-floating Polychaetae worms, and predatory arrow worms. The temporary members include the planktonic larvae of many Polychaetes, echinoderms, crustaceans, and fish. These feed and grow in the surface waters, and then metamorphose into the adult form which is no longer planktonic.

Lithosphere

That part of the Earth's crust in which most life occurs, and on which most life depends either directly or indirectly, is the soil.

SOIL

The soil has an **inorganic component** composed of the products of weathering of parental rock material, and an **organic component** composed of living organisms and dead organic matter or **humus**.

Inorganic components

Weathering can be caused by physical processes, for example the expansion and contraction of rock due to changes in temperature; the expansion of water on freezing can open cracks in which water is trapped; wind and water erosion; and the grinding action of large bodies of moving ice, as in glaciation. It can also be caused by chemical processes. For example, rain is often weakly acidic with carbonic acid or, in areas polluted with sulphur dioxide, more strongly acidic with sulphuric acid. These acids will dissolve out certain minerals from the parent rock, contributing to its further weathering by physical processes. Thus weathering produces the inorganic or mineral soil structure.

The properties of soils depend to a large extent on the nature of the parent rock and the degree of weathering that it has undergone. These properties, and various biotic factors, in turn affect the nutrients available to the plants growing in the soil. The size of soil particles ranges from stones and gravel to sand, silt, and clay. Sand particles are relatively large and angular. Therefore sandy soils have large air spaces between their particles, lose water rapidly by evaporation and drainage, and, due to the nature of the parent material, tend to be poor in nutrients. Clay particles are relatively small and rounded. Therefore clay soils have small air spaces between their particles, hold much water by capillary action and other retentive forces, and, due to the nature of the parent material, tend to be rich in nutrients.

An equal mixture of sand and clay particles produces a fertile **loam** soil with good air spaces, good water retention, and good structure. Such soils are neither too loose, as with sand, nor too sticky and compacted, as with clay, but have what is known as good 'crumb and block' structure. Block structure is seen if the soil forms good blocks when turned with a spade. Crumb structure is seen if it forms good friable crumbs when raked or crumbled in the hand.

Organic components

The invasion and colonization of the inorganic component by living organisms completes the process of soil formation. Once formed, however, the process of soil development continues under the influence of a complex of weathering and biotic factors.

A fertile soil contains large numbers of living organisms, both plant and animal, which, by their various activities, contribute to soil fertility. The majority are either saprophytic, photosynthetic, chemosynthetic, or holozoic; but some are parasitic pathogens that can cause disease in plants and, to a lesser extent, in animals. Soil organisms can be classified according to size into macrobiota, mesobiota, and microbiota.

The **microbiota** are the micro-organisms found in the soil. They are particularly involved in recycling organic matter and rendering it available again for use by the higher plants (and indirectly the animals for which the vegetation forms the basis of a food-chain).

Bacteria are particularly important to soil fertility: saprophytic bacteria break down complex organic compounds and chemosynthetic bacteria are important in the nitrogen cycle. Actinomycetes,[343] a form of filamentous bacteria, are also important saprophytes in the soil; as are the so-called 'slime' bacteria (Myxobacteriales)[343] which are particularly active in the decomposition of cellulose.

Protozoa may number as many as one million in a gram of soil. *Amoebae*, colourless flagellates, and ciliates may all be found. Being holozoic they feed on other organisms and, by their excretions and eventual death and decay, can liberate substances useful to both other soil micro-organisms and higher plants. In the encysted form many protozoa can survive in dry soil for up to fifty years.

Fungi are also very important soil saprophytes, and they tend to dominate the microbiota in acid forest soils where they are the main decomposers of the cellulose and lignin of fallen leaves, etc., known as the 'soil litter'. In neutral soils they are also the first organisms to attack cellulose.

Many of the microbiota, particularly the fungi, produce antibiotics which play a key role in competition between these soil organisms in their micro-habitat.

The **mesobiota** include the relatively larger animals found in the soil. Nematode round worms are particularly numerous: figures of up to twenty million per square metre of soil surface are sometimes quoted. Certainly their biomass can equal that of earthworms in a good fertile soil. Most live on fluid decomposition products and the associated bacteria, although some are serious plant pests, e.g. the potato eelworm. They form an important food source for larger animals and even for a type of fungus that captures them. Annelid worms smaller than earthworms are found, and play an important part in the maintenance of soil fertility.

Micro arthropods such as springtails and soil mites and smaller insect larvae are also numerous. Springtails are one of the most abundant of all soil arthropods. Some are predatory, but most feed on fungi and plant material, and some can digest totally undecomposed wood. Soil mites are perhaps the most important group of all soil animals. There are a tremendous variety: up to 17 000 species have been described. They have an important influence on the complex interrelationships of the soil fauna and flora.

The **macrobiota** are the larger organisms found in the soil, for instance the larger insects and their larvae, various molluscs such as slugs and snails, earthworms, and the roots of higher plants.

Earthworms are particularly important to soil fertility and often occur in large numbers in good soils, for example up to about 500 per square metre of soil surface. By their burrowing they increase aeration and drainage of soils. Their feeding method entails continually swallowing the soil to digest out the organic content, and they thus gradually turn over the soil. One estimate, made by Darwin, of the amount of soil turnover by earthworms in a fertile soil is as high as 40 tonnes per hectare per year; over a period of 65 years this would produce a surface layer up to two metres deep. *Lumbricus terrestris* does not form worm casts on the soil surface but it does draw leaves from the litter down into its burrow, which greatly speeds up their decomposition.

Millipedes are important in organic decomposition as most feed on decaying vegetation. Centipedes are mainly active carnivores. They are of two types: the **lithobiomorphs** have poorly waterproofed cuticles and live in the upper layers of damp soils, while the **geophilomorphs** are better waterproofed, less liable to desiccation, and actively burrow throughout the soil.

The **roots** of higher plants growing in the soil also contribute to soil activity. They affect the soil by removing water and minerals and, by channelling through the soil, they increase the aeration and drainage. Root growth is continuous at variable rates throughout the life of a plant, stopping only when the soil is near freezing. This continuous growth is necessary for the continual exploitation of new soil regions, and also because it is only the young growing root tips which absorb materials from the soil. Most feeding roots of most trees are within 15 cm of the surface of the soil and tend to be most active under the 'drip-line' where rain dripping off the edge of the leaf canopy provides a plentiful supply of water and contains nutrients washed from the leaves. Roots do not branch much when nutrients are low and therefore grow through poor soil more rapidly. They only 'slow-up' and branch when nutrients are available.

In addition to these factors, certain conditions prevail immediately around roots giving rise to an especially active region of the soil known as the **rhizosphere**. Root secretions, mainly sugars and amino acids, attract and stimulate a wide variety of soil organisms, including the nodule-forming bacteria *Rhizobium* near legume roots, and fungi which can surround and invade the root tissues to form a symbiotic relationship known as **mycorrhiza**.

Mycorrhizal associations occur mainly in plants growing in acid soils. In such soils bacterial activity is suppressed and, as a result, mineral nutrients, particularly nitrates, are not readily available to the plants since they remain locked in the raw, undecomposed organic matter. However, with the aid of mycorrhiza, nitrates are rendered available so that plants with such associations can grow vigorously in very poor soils. Alkaline soils are fatal to many plants as their mycorrhiza are killed under these conditions, and the plants themselves cannot survive without them. Mycorrhizal fungi render nutrients, particularly nitrates, available to the roots in a form that can be taken up. In addition their felt-like, spongy mass retains water around the root. In most cases root hair development is suppressed and the fungal hyphae act as organs of absorption for the root, transporting water and mineral salts into the plant tissues. The fungi gain organic compounds for respiration and synthesis, growth factors, and a certain degree of protection.

There are different degrees of association between the roots of the plant and the fungi. **Peritrophic** mycorrhiza form felt-like spongy masses around roots but do not actually penetrate the epidermis. **Ectotrophic** mycorrhizal fungi surround the root, but also invade the epidermis and root cortex. They are mostly Basidiomycetes and are found in many trees including pine, birch, and beech. Many of the familiar toadstools of woods are the sporing bodies of these mycorrhizal fungi, for example *Amanita*, *Coprinus*, and *Boletus*. **Endotrophic** mycorrhizal fungi actually penetrate the cells of the roots. They are found in all members of the Orchidaceae and Ericaceae. Many of these plants cannot survive the seedling stage without their fungal partner. *Calluna vulgaris* (ling), the dominant plant of the dry heath community, owes its success to the endotrophic fungus which penetrates all tissues of the plant. Even the seed coats are invaded, thus ensuring the establishment of the association in the next generation.

Humus is the product of all the activities of the soil flora and fauna upon dead organic matter. It is a complex of colloidal[370] and fibrous organic substances. It colours the soil its typical dark colour and in this way affects the heat absorption and radiation of soils. Its colloidal nature results in the retention of water and nutrients, and its fibrous nature contributes to the formation of good soil structure. In alkaline or neutral soils there is rapid decomposition of dead organic matter to form the type of humus known as **mull**. In acidic soils the plant litter itself is acidic and usually heavily lignified, bacterial activity is suppressed, and decomposition is slow. This leads to the accumulation of layers of raw humus or **mor**.

SOIL PROFILES

Under the influence of a complex of factors, including particularly the nature of the parental material, the climate, and topographical factors such as the degree and direction of any slope, a soil develops a characteristic sequence of layers known as a profile, described in Table 5.1. This develops particularly on flat, poor, easily drained soils under heavy rainfall which washes or leaches down materials from surface layers and redeposits them lower down. Two examples of such profiles are seen in the **podzol** and the **rendzina** soils.

Table 5.1 *Soil profiles*

		L	Litter—ecological sub-system
A	Top soil	A_0	Humus
		A_1	Dark with high organic content
		A_2	Light-coloured leached region low in organic and inorganic content
		A_3	Transitional
B	Mineral soil	B_1	Transitional
		B_2	Dark zone of maximum deposition of transported materials
		B_3	Transitional
B	Parent material	C	Parent material present if soil formed *in situ* (if soil has been transported by wind, water, etc., it may not overlay its original parent material)

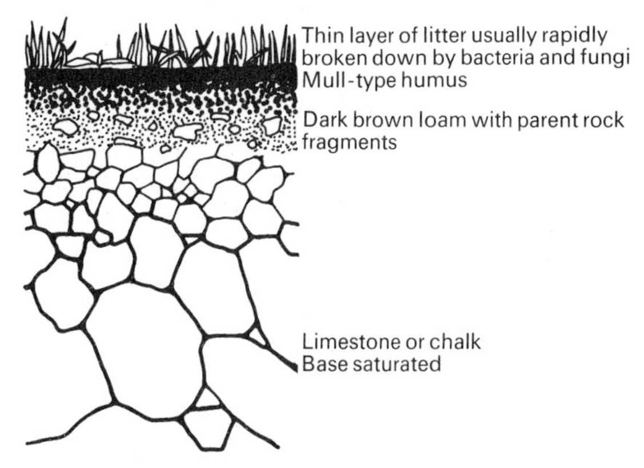

Figure 5.12 *Rendzina profile*
The rendzina soil is an extreme case of the parent rock controlling the nature of the soil. It is a shallow, well-drained soil with fragments of the parental rock reaching the surface. Beech and ash are typical trees supported by this type of soil, with most conifers absent. Surface leaching may sometimes occur, giving rise to acid surface conditions, supporting a heath vegetation, immediately over strongly alkaline deeper layers which support deeper rooting, base-loving plants.

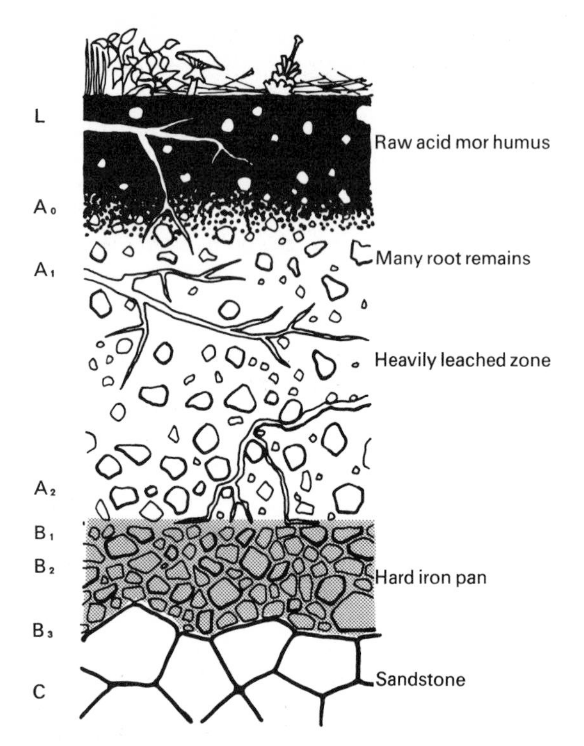

Figure 5.13 *Podzol profile*
There are few bacteria, and the fungi present only attack cellulose and not lignin, so that the tough acidic litter accumulates on the surface as a raw acid mor-type humus. Such acidic soils have a low calcium and phosphate content but high concentrations of iron and aluminium which are dissolved under acid conditions and leached down to the B layers. Humus-enriched leached substances, particularly iron, are deposited as a hard iron pan. This pan acts as a barrier to most roots, although a young pine can penetrate this layer with a tap root and recycle the materials to some extent via its leaf litter.

Succession

Once plants have colonized an area, they and their associated fauna modify the conditions, especially that of the soil, so that other plants are able to establish themselves. These compete with the established plants and, if they become established, in turn modify the habitat. In this way biotic modification of the habitat leads to a succession of plant communities which in turn leads to a relatively stable community, the nature of which is mainly determined by the climate, so that it is referred to as the **climatic climax**. A succession of communities, each occurring in a sequence leading to a climax, is called a **sere**.

Once established, the dominant canopy-forming plants of the climatic climax are adapted to their environment. They create modified climatic conditions of light, shade, humidity, temperature, etc., in which characteristic shrub, field, and ground layer plants survive. These in turn support a characteristic fauna.

Sometimes edaphic features (those relating to the nature of the soil) prevent the succession progressing to the climatic climax so that a local **sub-climax** community develops.

Grazing also is a powerful force in interrupting natural successions. Only those plants with features that enable them to survive the removal of large parts of their photosynthetic surfaces, or those that are not grazed, can survive. Features which enable plants to survive grazing include strong powers of vegetative reproduction; the rosette form of growth which keeps the leaves and growing point tight against the soil; and leaves and stems with intercalary meristematic growing regions at their base, as in the grasses. Young trees are particularly vulnerable to destruction by grazing and browsing, and this often prevents the reafforestation of many heathlands.

Plant communities

SAND DUNE SUCCESSION

A good example of how plants can modify their own environment is seen in the sand dune succession. Sand is continually moved by waves and wind, and plants living on it must be able to survive periodic exposure of their roots. They must also be able to withstand high levels of salinity from the sea water and its spray. Such plants need deep roots or underground stems and the ability repeatedly to send shoots to the surface. They must also be able to withstand both drought and floods of sea water.

A succession occurs from new dunes with coastal plants to old dunes with plants of the limestone grassland if the soil has many shells; to heath if the soil is derived from almost pure sand; and to woodland if conditions permit (this rarely occurs in the British Isles). The main pioneer plant of this succession is marram grass (*Ammophila arenaria*) which can become established on a sandy sea-shore above the high water mark. Sand carried by wind blowing up the beach is deposited initially just behind each tuft of grass. Marram grass spreads rapidly by means of its underground stem or rhizome. It can also withstand repeated burial by the blown sand because when it is buried the rhizome grows up towards the surface, and it has stiff erect leaves. The grass thus helps the sand pile up increasingly to form the first line of dunes. Little else can survive on the seaward side of these dunes as they are exposed to much salt spray, and wind and wave action. Behind the first dunes a second, lower line of dunes is built up from the sand that escapes from the first line. They are more protected, more stable, and more rapidly colonized than the first line.

Sand dunes are deficient in humus and minerals, especially nitrates, although they can contain some dead sea-weed and bird droppings which are nitrate-rich. There are few micro-organisms as initially there is little on which they can feed. However, air is moved through the dunes by the movement of the tides, and this promotes rapid root growth if sufficient nutrients are available.

Early pioneers of the second line of dunes include leguminous plants which can obtain their own supply of nitrates from their symbiotic root nodule bacteria. The early flora of the dunes is very varied, including seasonal 'migrations' from inland of annual garden weeds, and many plants with the rosette habit which protects them from harsh conditions and grazing. Mosses and lichens are also important early invaders. As these plants become established the dunes become more stable. Surface stability is crucial in preventing scouring by the wind, so plants such as sand sedge and sand fescue are important here. The accumulation of dead organic matter and micro-organisms increases the fertility of the sand, allowing plants that could not have survived the original conditions to invade the area. Eventually a heath or even woodland community may become established.

Under the correct conditions fresh dunes can be formed to the seaward of the first line of dunes. If this occurs the original first line of dunes will mature and, if the process continues, a gradual reclamation of land may take place.

Animal communities

The study of animal communities does not always show the same clear-cut relationships that can be seen in plant communities. This is mainly due to their mobility. Animals can usually move around their habitat to search for food, shelter, and suitable conditions of light and humidity; and to avoid predators. This greater complexity of behaviour and their interaction with the plants of the habitat makes the study of animal communities very complex. This can be seen in the brief outline description of the fauna of the heathland.[118] However, if a habitat with very special conditions and a mainly sessile animal population is chosen then a clearer picture of an animal community's ecology can be gained.

SANDY SEA SHORE

The sandy sea shore is a distinctive habitat that makes special demands on animals if they are to establish themselves successfully. As a result the fauna of the sandy sea shore form a very distinct group with relatively few species.

The sandy shore does not provide any surfaces for the attachment of seaweeds, so that they and their associated fauna are absent. The surface is subject to wave action; variation in temperature, humidity, and salinity when exposed between tides; and movement caused by the wind. Animals that have adapted to this environment come from a wide variety of Phyla, including Annelida, Crustacea, Mollusca, and Echinodermata. However, most show a common adaptation to this environment by the adoption of a burrowing habit. The fauna is thus rather less motile than others, and forms a community in which clearer relationships exist between them and their environment. Indeed, under the influence of a complex of environmental factors (such as the period of exposure between tides and the moisture content of the sand) the fauna of the sandy sea shore shows a zonation between the low and high water marks whereby each type of group of related animals is restricted to a particular part of the beach. For example, Echinodermata are generally only found close to the low water mark, whereas Crustacea can be found in the spray zone above the high water mark, as well as on the mid- and lower-shores.

The burrowing habit provides protection from fluctuations in temperature and desiccation when the sand is exposed between high tides. However, it also poses particular problems in feeding and reproduction. Many of the animals are **detritivorous**, feeding by swallowing the sand and digesting the organic detritus content. Others are filter feeders, filtering suspended food particles from the water, while others siphon up organic particles from the surface. Both of these types only feed when the sand is covered by the tide. Others are carnivorous.

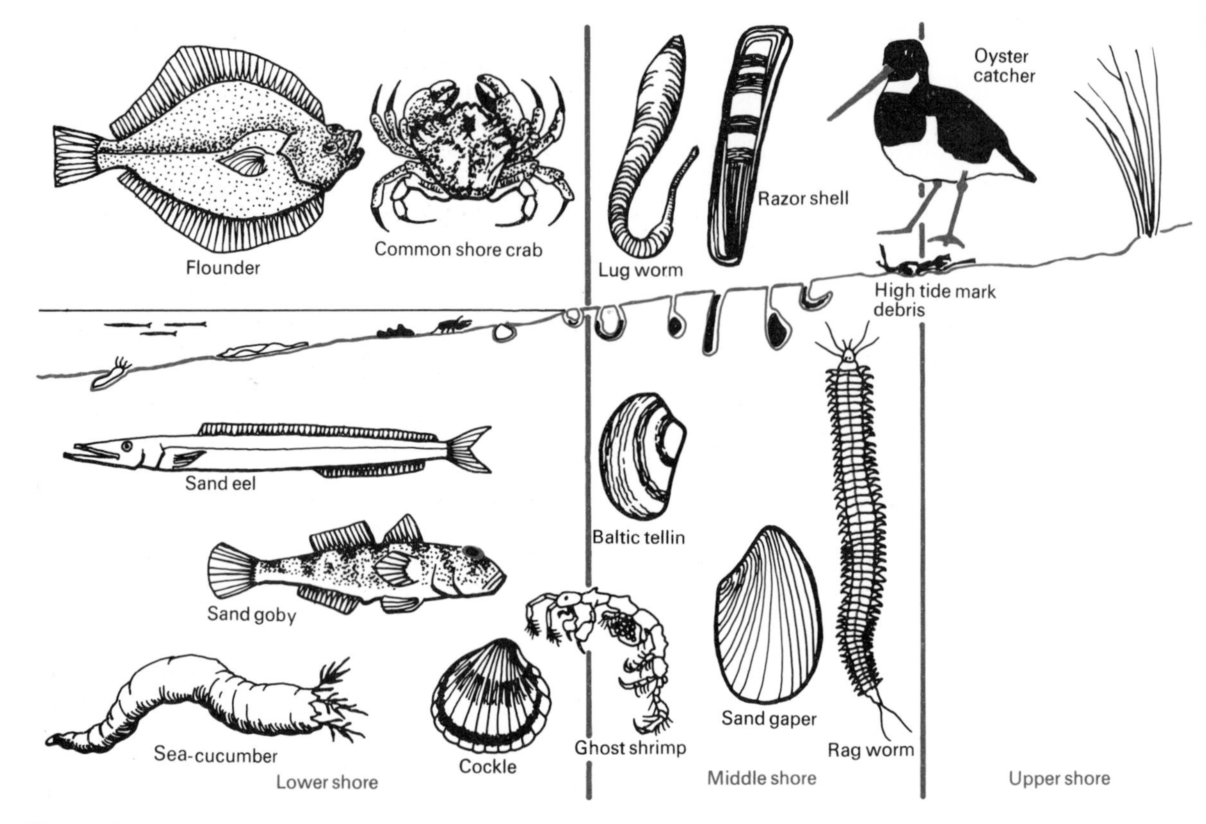

Figure 5.14 *Sandy sea shore fauna*

The relative immobility of the burrowing forms cause problems of reproduction and dispersal. Fertilization is external and the shedding of the gametes is timed to coincide with immersion by the incoming tide. The majority have aquatic larval stages which, after feeding and growing as members of the zooplankton, eventually undergo metamorphosis and sink to the bottom to develop into the sessile adult.

Heathland

The heathland community provides good examples of the interaction of organisms with climatic, edaphic, and biotic factors.

Heathland is found where the soil is acid, infertile, and some factor prevents the establishment and growth of trees. The soil is almost always sandy. The climatic climax vegetation for many lowland areas on sandy soil with a high rainfall can be deciduous oakwood. In such an oakwood mineral nutrients are taken up by the roots and eventually returned to the soil via the falling leaves, resulting in a recycling of nutrients from the deeper layers to the surface layers of the soil. This recycling of nutrients, the upward pull of water from the soil by transpiration, and the protection from heavy rain provided by the canopy of the trees, prevent excessive downward washing or leaching of minerals and humus. However, if the trees are removed, the soil can degenerate into an acid **podzol** with a characteristic heathland community, dominated by common heather or ling (*Calluna vulgaris*), in which the number of plant species is small. Regeneration of woodland cover may occur if conditions are suitable, but is usually prevented by grazing. Thus in many cases the heathland community is a sub-climax maintained by grazing; but it is also in many ways maintained by the nature of the soil and known as an **edaphic climax**.

After the clearance of the trees the rain water washes through the acidic sandy soil leaching the minerals and humus down to deeper levels where the aluminium and iron compounds in particular are redeposited as a hard **iron-pan**. This continues even when the heath vegetation is established. Heath plant leaf litter is acidic in itself. There are few bacteria or litter-decomposing animals, and only a raw acid humus or mor is formed which accumulates on the surface. Thus acid water continues to drain down, liberating minerals and bases from the surface layers and leaching them down to deeper layers. Most of the heath plants such as ling, heather (*Erica cinerea*), gorse (*Ulex europaeus*), bracken (*Pteridium aquilinum*), broom (*Sarothamnus scoparius*), and lichens, all show **xeromorphic** adaptations to conditions of water shortage because, although rainfall is usually high, drainage is equally rapid and soil water is not always readily available.

In a poor soil the development and distribution of the root systems is important. The weight of root system under an area of heathland can be larger than under many areas of woodland. The root-to-shoot ratio is large because the roots have to exploit large volumes of poor soil. Moreover, the roots of different species stratify at different depths which reduces competition between them. *Calluna* has relatively shallow roots whilst the whortleberry (*Vaccinium myrtillus*) and bracken have deeper roots. Most heath plants have to supplement the limited supply of nutrients by means of special symbiotic relationships, such as root nodules and mycorrhizal associations.

Insectivorous plants, such as sundew (*Drosera* sp.), are also found on heathland and can supplement their nitrogen supply by trapping and digesting insects.

Pine trees can colonize acid podsols as they do not require much calcium, and their roots can penetrate and exploit the minerals in the deep lying hard iron pan. Birch trees may also establish themselves, and, if the young trees are not exposed to grazing, there can be a gradual return to the climatic climax of deciduous forest.

The **soil fauna** is very poor in heathland soils as important decomposer animals, such as earthworms, millipedes, and beetle-mites, are almost entirely absent, but there are numerous species of Collembola and Acari which are important decomposers of heathland soils. Typical invertebrates of the heathland are the ants and spiders, with the latter as the most numerous invertebrate carnivores. Vertebrates include lizards; and insectivorous, seed eating, and predatory birds.

Periodic **fires**, either man-made or naturally incinerated, are a particular feature of heathlands. They act as 'decomposers' by removing the excess accumulation of dead wood and litter, although too fierce a fire can cause greater acidity, lower nitrogen, and loss of soil structure and soil fauna. The re-establishment of the heathland community takes a long time. Mosses and lichens are pioneers of burnt soils, while gorse and heather sprout from burnt stubs. After about twelve years the plant canopy builds up sufficiently to produce changes in the micro-climate beneath it which affect the sub-canopy plants and animals. The maximum biomass of the vegetation is reached in about twenty-five years, although individual ling plants can live for as long as fifty years.

Figure 5.15 *Sandy heath and valley bog*

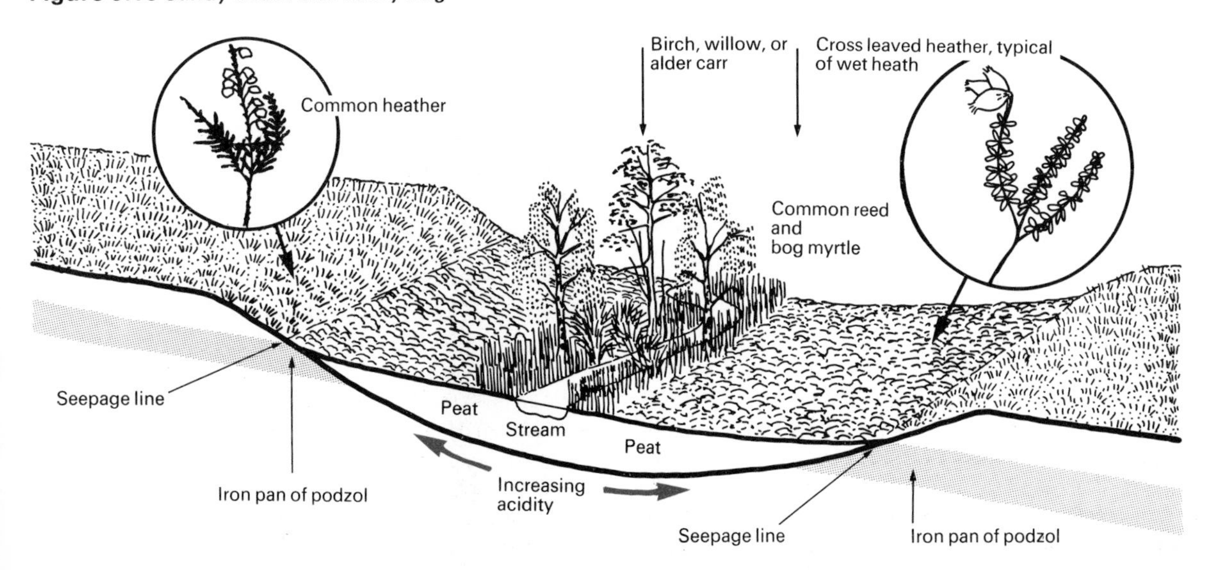

VALLEY BOGS

In certain areas acid water run-off from above the relatively impermeable iron-pan of a podsol can accumulate above impervious material, such as clay, in shallow valleys. *Sphagnum* moss grows well in such conditions, and it further acidifies the water. The acid conditions, the lack of oxygen caused by the permanent water-logging, and the absence of the decomposing bacteria, lead to the accumulation of the remains of dead *Sphagnum* plants to form layers of *Sphagnum* peat. This retains water and gradually, as layers of fresh sphagnum grow, a valley bog or mire develops.

In some valleys a small stream runs down the centre of the bog and produces an interesting zonation of the vegetation. It forms a central region in which drainage from the bog concentrates nutrients, the stream washes away some of the acidity, and the moving water results in better oxygenation. As a result tree species can become established and a thin strip of forest develops alongside the stream; such regions are referred to either as gallery forests or **carrs**. Because of the tendency to alkaline (fen-like) conditions, the dominant trees are usually willow and alder which can provide excellent conditions for a wide variety of animals. Towards the edge of the valley bog the conditions become more acid and *Sphagnum* moss dominates.

Historical ecology

By using a variety of techniques it is possible to gain some insight into the ecology of the vegetation of the past. One such technique is pollen analysis or palynology; another involves the study of hedges.

Palynology

There is always a continuous 'pollen rain' falling on the landscape from the vegetation. Pollen grains have coats which, although they are broken down by oxidation, are very resistant to decay in anaerobic and acidic conditions. The 'pollen rain' is trapped and preserved in acid bogs where conditions favour their preservation. The washing down of these pollen grains, and the annual incremental growth of the surface layers of the bog which trap the next year's pollen rain, result in the preservation of vertical stratification of layers of pollen grains. Study of this pollen grain profile provides information on past vegetation, and on how the vegetation has changed with time.

The process of pollen stratification also occurs in acid soils such as podzols, but not as clearly as in *Sphagnum* bogs. Soils do not show the same annual incremental surface growth and the stratification of the pollen grains is due entirely to a downward washing or leaching through the soil. Soils buried beneath ancient earthworks or burial mounds provide interesting pollen records of the vegetation that grew, and the stratification that existed, at the time when the soil was buried.

Hedges

There are an estimated 800 000 kilometres of hedgerow in Great Britain, covering an area of about 2000 square kilometres. Under the pressures of urbanization, road improvements, and recent agricultural practice, they have been disappearing at a rate of about 8000 kilometres per year for the last twenty-five years.

Hedges have a variety of origins; some were planted as such to mark boundaries, some were self-sown on strips of unused boundary land, and some arose from the border of mixed relict woodland. They provide conditions of humidity, light, and shelter which suit a distinctive flora, especially climbers such as ivy, bryony, and hop, although no wild flowers are restricted entirely to hedgerows. A wide variety of animals also find the conditions suitable, and hedges are particularly important in providing birds with

Bronze age barrow from which the pollen profile in Figure 5.17 was taken

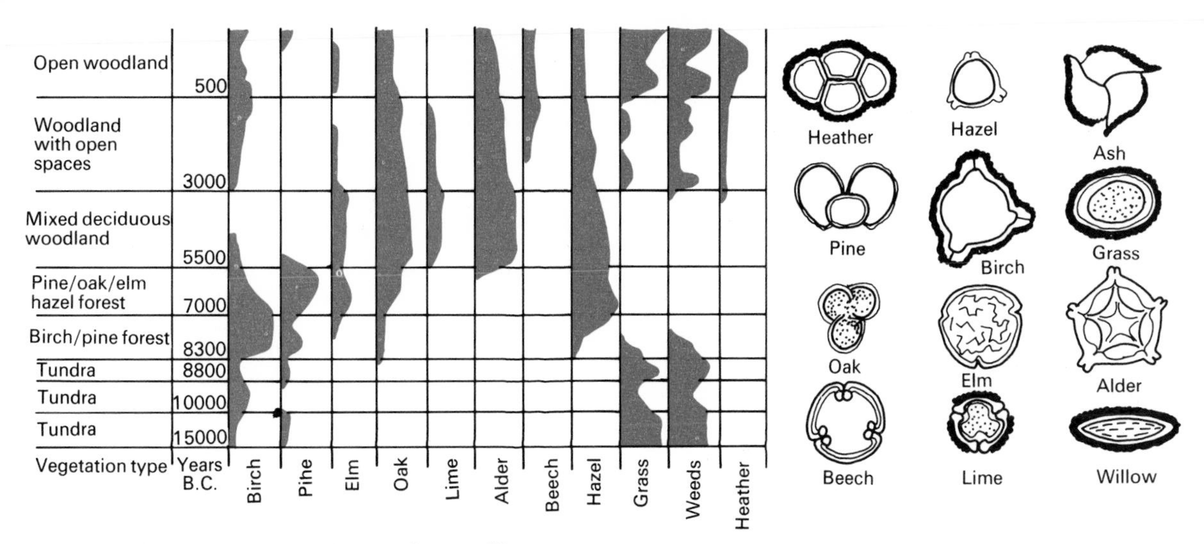

Figure 5.16 *Acid Sphagnum bog pollen profile*

Figure 5.17 *Pollen profile of late Bronze Age barrow at Berrywood, near Burley, New Forest* (SU/212052) (After Dimbleby G. W. (1962) and McGregor R. (1962).)

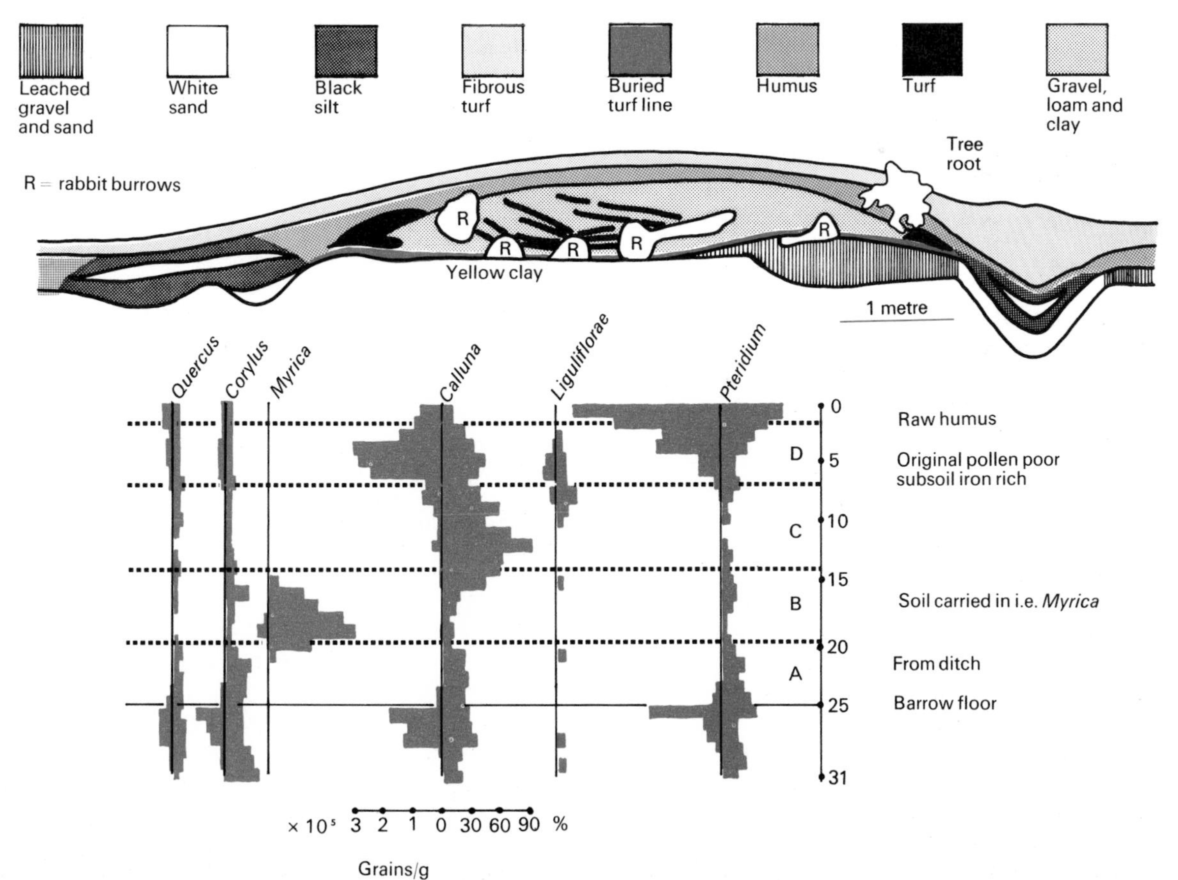

nesting sites. By forcing the air flow up and at the same time slowing it down, hedges are effective windbreaks, particularly on the leeward or down-wind side.

The trees most associated with hedges are the elms; these are not often dominant in woodlands but, until recently, flourished in hedgerows. However, these hedgerow elms have been almost entirely wiped out in many areas of Britain by Dutch elm disease.[306]

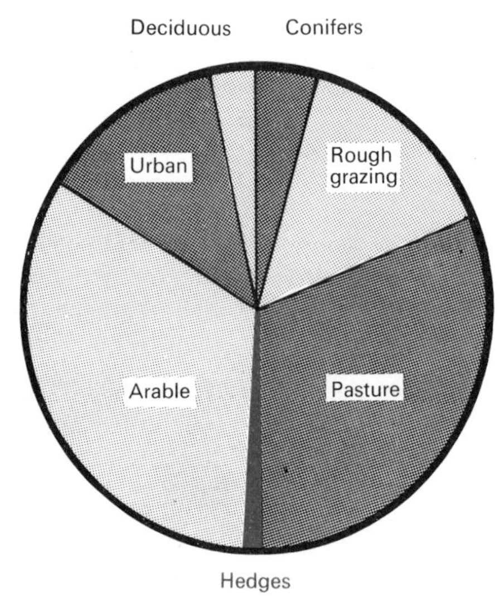

Figure 5.18 *'Pie' chart of land areas in Britain*
The land area under hedges is about one-third that of deciduous wood, but, as they form a network in farmland, hedges allow woodland edge plants and animals to live in agricultural areas.

A relationship exists between the age of the hedge and the number of species of woody shrubs and trees found in it. Generally, the older the hedge the more such species are found. If the hedge was planted as a mixed hedge then obviously this relationship does not hold true, but in any one area it has been calculated that 92 per cent of the variation in species content between hedges can be related to the age factor. Simply stated, the relationship is one species for every one hundred years, thus 'modern' hedges have but one species (usually the common hawthorn *Crataegus monogyna*), while Saxon hedges have ten or twelve species.

Certain species are used as extra guides in dating; thus any hedgerow with maple is considered to be at least 400 years old, with the spindle tree at least 600 years old, and with *Crataegus laevigata* (a hawthorn) at least 1000 years old. Non-woody plants can also provide information, for example, it appears probable that any hedge with woodland plants such as dog's mercury, bluebell, and wood anemone at the bottom is of great age, and was once the boundary of a medieval clearance of woodland for agricultural purposes.

Experimental and applied aspects

The study of ecology is essentially a practical subject. An attempt will be made to outline some of the basic techniques involved, but these must, by definition, be performed and understood in the field. Such techniques include sampling methods and the subsequent collection and analysis of the data so obtained, and are designed to reveal any patterns of distribution of organisms. These patterns can in turn reveal the nature of some of the interactions occurring within that particular ecosystem or habitat.

Sampling methods

These can be either random or selective, depending upon the particular type of investigation. In both cases the results are placed in the context of an overall map of the area under study.

Random sampling methods enable representative samples of a wide area to be made without 'bias' on the part of the observer. Thus, for example, if the investigation was of meadow grassland, random sampling would ensure that the investigator was not attracted to a particularly interesting, but atypical, area. The simplest and most common method of random sampling is that of **random quadrat sampling**, in which a square frame is thrown at random (usually over the shoulder) and the area where it falls subsequently examined. The size of the quadrat varies, but is often a metre square. This technique may miss small local deviations, but it does in general provide statistically valid data of a wide area so that a rough profile of the community structure can be constructed. It is better for plants than for animals, and is of limited use on rough terrain.

Table 5.2 *Tabulated results of random quadrat sampling*

Name of species	*Quadrat number*										*Average*	*Frequency*
A	1	2	3	4	5	6	7	8	9	10		
Density												
Cover												

Within each quadrat the number of individuals of each species can be recorded, and converted to a measure of their **density**. More quickly, the **cover** or percentage area covered by each species can be estimated by eye. Those species with the highest percentage cover are the **dominants** of the community. In both cases the average over a number of quadrats (usually ten) is taken. The percentage of the total number of quadrats in which a species occurs is referred to as its **frequency**, and provides an index to the type of distribution of a species. The results obtained can be tabulated for each species, as shown in Table 5.2.

A non-random or selective form of sampling is that of the **belt transect**. This is suitable when changes in the habitat or community are to be studied, as, for example, across a path in a wood, across a sea shore, etc. A continuous series of data is taken along a belt of variable length and usually of a metre width. (If a series of data is collected along a single line, this is termed a line transect.) It provides information on the frequency and distribution of organisms in relation to biotic, physical, and topographical changes along the transect, as shown in Figure 5.19.

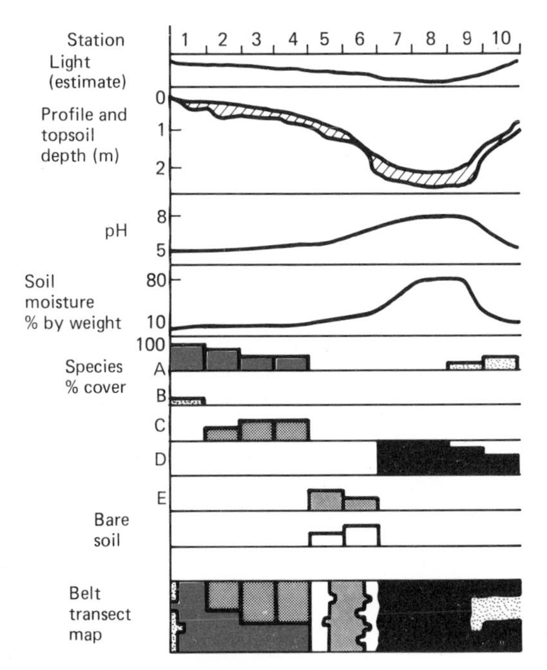

Figure 5.19 *A belt transect sample*

Guided example

1 When undertaking an ecological investigation of an area it is necessary to obtain data on the numbers of different organisms present. Generally it is easier to obtain such information about plants rather than animals. Why should this be so, and can you think of any exceptions?
Plants are usually fixed and, if care is taken, there is little danger of counting the same plant individual more than once. Animals, on the other hand, are usually mobile, and there is a danger of counting the same individual more than once, and collecting inaccurate information. Exceptions would occur in aquatic environments where larger plants may be detached and floating, and where microscopic plants, such as are found in the phytoplankton, are free-floating; and in environments such as the sandy sea shore where many animals can be relatively sessile.

2 In a typical terrestrial ecosystem, therefore, the numbers of animals present are usually estimated by using various sampling techniques. One of the simplest techniques is to count all the animals in a small sample area or volume and then multiply up, to obtain a figure for the whole area under investigation. If two woodlice were counted in an area of 100 cm^2, how many would be estimated as being in a square metre of that environment?
200, because there are 100 × 100 cm^2 in 1 m^2.

3 Can you think of any criticisms of this method?
It assumes a uniform distribution across a relatively wide area, which for most animals (particularly those such as woodlice with specific requirements of humidity, etc.) would not be true. This problem could be reduced by taking a random sample of several areas and calculating the average number of animals before multiplying up, but the problem then arises of counting the same animals more than once.

4 Another method is to capture a sample of animals from the particular area, mark them, return them to the population, and allow them to mix with the unmarked population, and then to capture a second sample. The total population is then estimated using the following relationship. If A is the total number of marked animals, B the total number of animals in the second sample, and C the number of marked animals in the second sample, then the total population number, X, is given by

$$\frac{X}{A} = \frac{B}{C}$$

What is the estimated total population number, if $A = 100$, $B = 80$, and $C = 40$?
Using the formula above,

$$\frac{80}{40} = \frac{X}{100}$$

$$X = \frac{80}{40} \times 100 = 200$$

Questions

1 'Plants destroy their own environment.' Explain and illustrate this statement by reference to examples of plant succession. (L)

2 Explain clearly the meaning of the terms 'habitat', 'environment', and 'niche'. From your own field observations, describe the range of physical factors which may affect the distribution of animals within a *named* habitat. (L)

3 What is meant by a 'soil profile'? Describe *one* such profile and give an account of the factors which may have led to its formation. (L)

4 The energy content of a grassland has been estimated for each trophic level as follows:

	Kilojoules per square metre*
Primary producers	5033
Herbivores	122
Omnivores	12
Carnivores	10

* 1 kilocalorie is equivalent to 4.2 kilojoules.

(a) Comment on the biological significance of these data.
(b) What is meant by 'primary producer' in this habitat?
(c) Discuss the factors that would affect the net production of the grassland's primary producers.
(d) The table below gives figures for two species of arthropod that live in the same grassland habitat. Plot graphs to show the numbers of each species against time.

Time (weeks)	*Numbers of individuals* *Species A*	*Species B*
1	0	50
2	5	100
3	50	500
4	100	1000
5	500	2000
6	2100	1000
7	800	400
8	400	80
9	10	50
10	50	500
11	150	1000
12	500	2000

(e) What deductions can be made from the graphs concerning the biological interrelationships of the two species?

Outline an experiment which might be carried out to provide further information in support of your deductions. (L)

5 The diagram below refers to a deciduous woodland in Britain. It shows the usual duration of the leafy condition and the time of flowering (where appropriate) for six of the species which grow in the woodland.

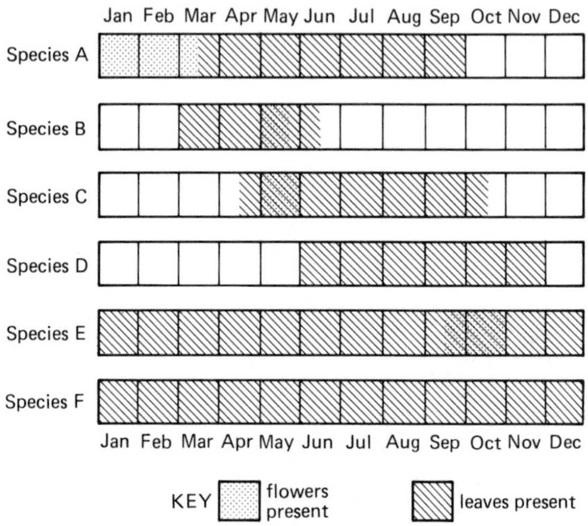

Jan Feb Mar Apr May Jun Jul Aug Sep Oct Nov Dec
Species A
Species B
Species C
Species D
Species E
Species F
Jan Feb Mar Apr May Jun Jul Aug Sep Oct Nov Dec
KEY flowers present
leaves present

(a) (i) Two of these species are deciduous trees. Give the letters which identify them.
(ii) Explain how you arrived at your answer.
(b) (i) Is Species A likely to be wind-pollinated or insect-pollinated?
(ii) Explain how you arrived at your answer.
(c) (i) Suggest *one* explanation for the short duration of the leafy condition in Species B.
(ii) Give *one* example of a plant or type of plant which could be Species B.
(d) Included in these species are a moss, bracken fern, and ivy. For each, give the letter which identifies it and explain how you arrived at your answer. (JMB)

Further reading

A.S.E. Lab Books, *Ecology*, compiled by A. Davies (London: John Murray, 1973).
Jenkins P. F., *School Grounds: Some Ecological Enquiries* (London, Heinemann Educational Books, 1976).
Kormondy E. J., *Concepts of Ecology*, Prentice-Hall Foundations of Modern Biology Series (New Jersey: Prentice-Hall, 1969).
Owen D. F., *What is Ecology?* (Oxford: Oxford University Press, 1974).

PART TWO

Comparative physiology

6 Nutrition

PLANT NUTRITION

Most plants and many bacteria are capable of synthesizing all their organic nutritional requirements from inorganic materials, using energy from sunlight, by the process of photosynthesis. Some bacteria are chemosynthetic, using energy from chemical reactions for their synthetic reactions. Both photosynthetic and chemosynthetic types are also called **autotrophic**. The fungi and some bacteria are **heterotrophic**, which means that they cannot synthesize their own organic compounds and so require an external supply.

Photosynthesis

The key factor of photosynthesis is that light energy is trapped as chemical energy in a wide variety of stable chemical compounds that do not spontaneously break down, and which can therefore be stored. In outline the process can be represented by the following equation:

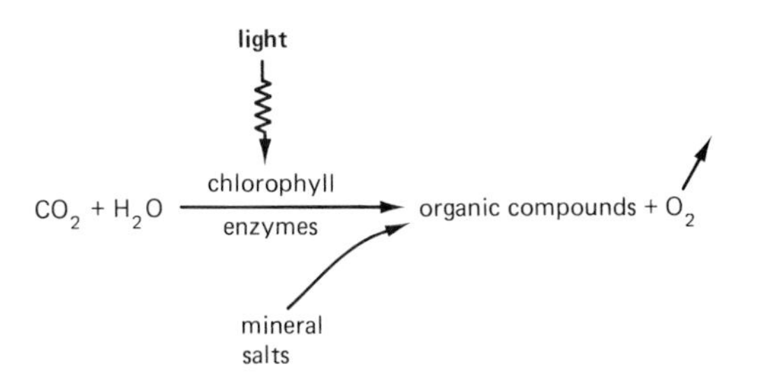

Photosynthesis occurs in two main stages:

(a) a light-dependent reaction; and
(b) a non-light-dependent reaction or so-called 'dark reaction'.

It is the main source of organic compounds in the world and the only source of atmospheric oxygen.

Conditions necessary for photosynthesis

CARBON DIOXIDE

On warm, sunny days the atmospheric level of 0.03 per cent carbon dioxide is the factor that slows up the reaction more than any other. Under these conditions the carbon dioxide concentration is known as the **limiting factor**. Carbon dioxide supplies the carbon for the organic compounds.

WATER

Water is never a direct limiting factor in photosynthesis as water shortage would produce other disturbances to plant metabolism before it could directly affect photosynthesis. It supplies the hydrogen for the reduction of the fixed[129] carbon dioxide into organic compounds; oxygen is released as a by-product.

LIGHT

Only about 1 per cent of the light falling on a plant is used in photosynthesis, and it usually only acts as a limiting factor at dawn and dusk, or in shaded positions (although shade plants have an optimum rate of photosynthesis at relatively low light intensities). It supplies the energy for the synthesis of carbon dioxide into organic compounds.

CHLOROPHYLL

This is a complex organic molecule containing magnesium. There are two main types, **chlorophyll a,** which is present in all photosynthetic organisms which evolve oxygen, and **chlorophyll b,** which is present in the higher plants and green algae only.

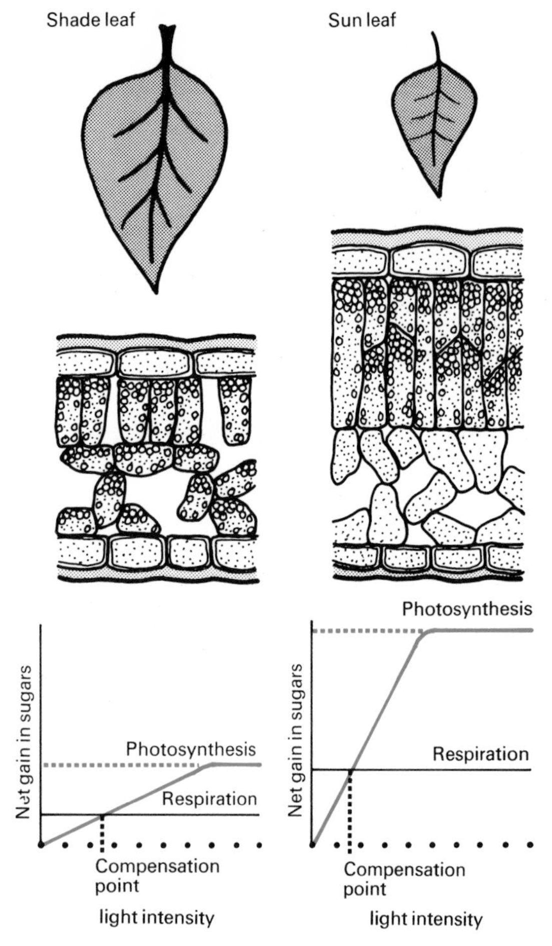

Figure 6.1 *Sun and shade leaves (with T.S.) and compensation points*
The compensation point is that point at which the rate of photosynthesis equals the rate of respiration.

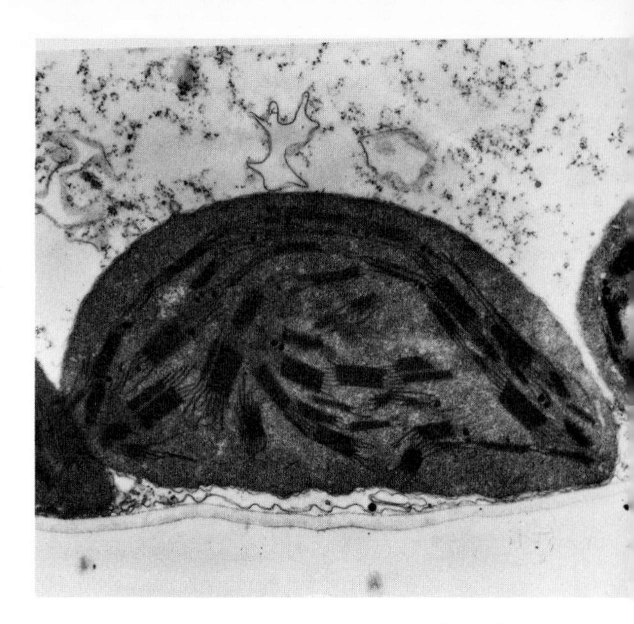

Chloroplast section of maize leaf (×12 175)

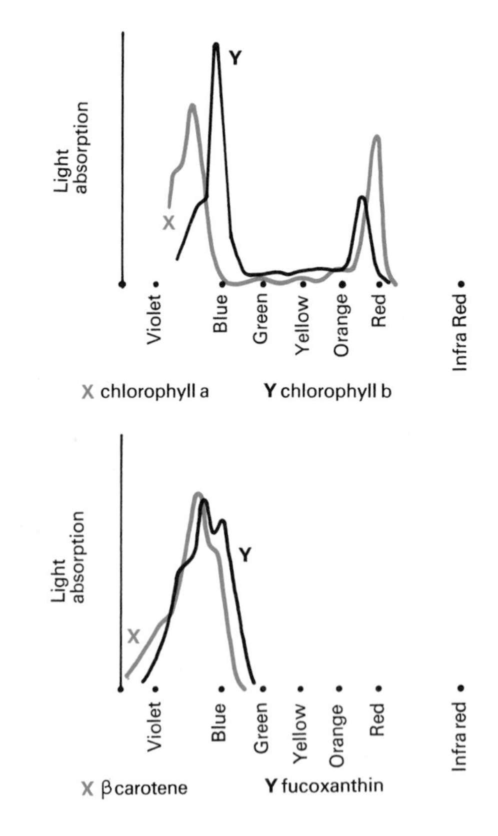

Figure 6.2 *Absorption spectra of photosynthetic pigments*
The wavelengths of light which are absorbed are those utilized in the process of photosynthesis.

The chlorophyll is localized within chloroplastids, and converts the electromagnetic energy[371] from the Sun into chemical energy of chemical bonds. It is usually associated with accessory pigments such as carotenoids which protect the chlorophyll from photo-oxidation and absorb light energy which is transferred to chlorophyll a.

Photosynthesis is rarely a limiting factor, except in extreme cases of **chlorosis**, when the plant loses chlorophyll and becomes yellowy-green in colour. This may happen when the plant grows in the dark, or is lacking in certain essential chemical elements.[132]

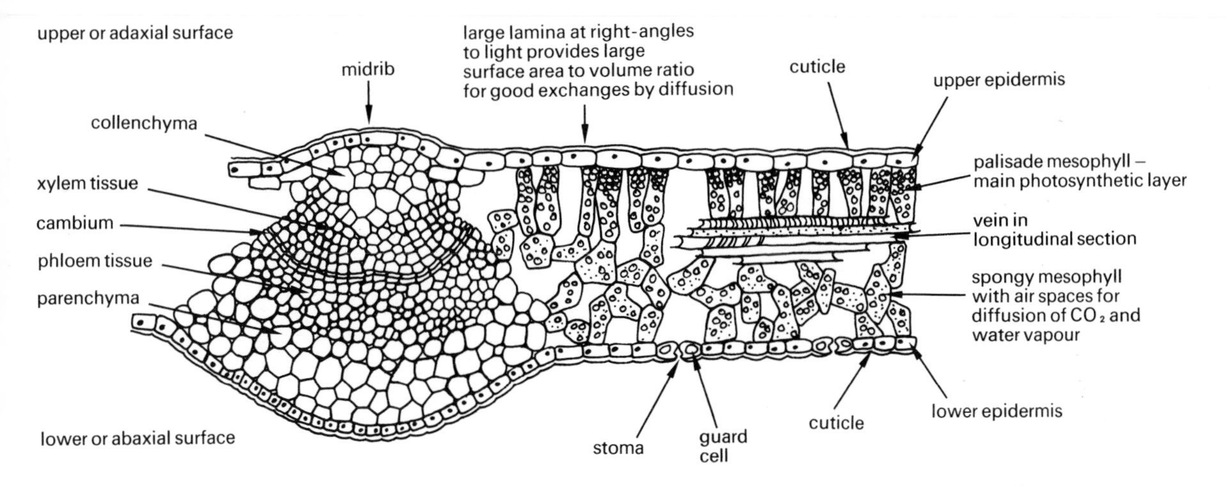

Figure 6.3 *T.S. dicotyledonous leaf (highly magnified)*
Palisade mesophyll cells are elongated with many chloroplastids at the edge for rapid CO_2 uptake from the extensive vertical intercellular spaces. Also light can penetrate without having to pass through many cross walls and air spaces. The spongy mesophyll has large air spaces for rapid diffusion of CO_2 to the palisade mesophyll. Their regular horizontal arrangement possibly aids in supporting the thin lamina due to the turgidity of cells acting against the non-elastic epidermis.

TEMPERATURE

Since photosynthesis is an enzyme-catalysed reaction there is an optimum temperature for the whole process of about 30 °C. However, only the non-light-dependent carbon assimilation process is affected by temperature. The light-dependent reaction is unaffected by temperature changes and thus at low light intensities temperature cannot be a limiting factor.

ENZYMES

All the steps in the complex biochemical pathways of photosynthesis are catalysed by enzymes, and thus any factor that affects enzyme activity will affect photosynthesis.

LEAVES AS PHOTOSYNTHETIC STRUCTURES

The structure and arrangement of the leaves are adapted to their function as the main centres of photosynthesis.

The outline biochemistry of photosynthesis

LIGHT REACTION

The light reaction results in the production of energy-rich ATP[367] (adenosine triphosphate) and so-called 'reducing power' which are used in the non-light-dependent reaction for the reduction and synthesis of carbon dioxide into organic compounds.

DARK REACTION

The dark reaction does not have to occur in the dark, it is simply not dependent upon light so it can, and does, proceed in the light.

The carbon dioxide in the air must first of all be trapped, or fixed, by an acceptor molecule into a form which can then be reduced and synthesized into organic compounds. There are two main pathways by which carbon dioxide is 'fixed' or trapped.

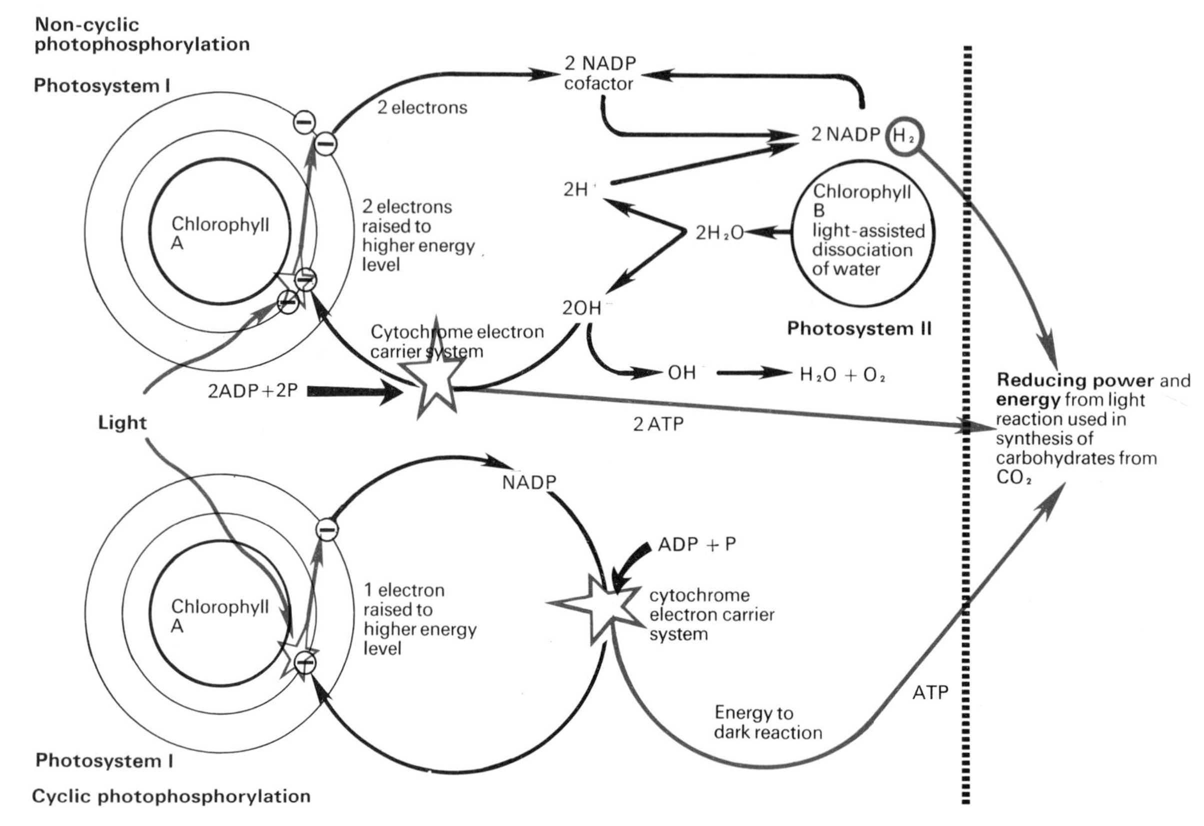

Figure 6.4 *Outline of the light reaction of photosynthesis*
In non-cyclic photophosphorylation the electrons that are returned to the chlorophyll are not the same ones that were removed. In cyclic photophosphorylation the electrons that are returned to the chlorophyll are the same ones that were removed.

C_3-type plants
Here the carbon dioxide combines with a five carbon (5C) acceptor substance to form a six-carbon compound which then breaks down into a three-carbon compound.

Examples are the broad bean, tomato, cucumber, and wheat.

C_4-type plants
Here the carbon dioxide combines with a three-carbon acceptor substance to form a four-carbon compound.

Examples are maize and sugar cane.

At high light intensities four-carbon plants have a photosynthesis rate two to three times greater than three-carbon plants. The leaves have a large internal surface area, short carbon dioxide diffusion pathways, and steep diffusion gradients,[368] all of which increase the rate of carbon dioxide uptake. The mesophyll cells feed their photosynthetic products into specialized **bundle sheath cells** which contain chloroplastids which are larger than normal and lack grana. These bundle sheath cells lie close to the phloem of the vascular bundles in the leaf and their products can be rapidly translocated away.

C_3 plants have a light-activated **photorespiration** distinct from normal respiration; thus at high light intensities some of the products of photosynthesis are broken down to carbon dioxide in special organelles, the peroxisomes. Some of this carbon dioxide is refixed by photosynthesis, but most of it escapes from the leaf. This is possibly an 'overflow' device for getting rid of excess photosynthetic products.

C_4 plants apparently have no photorespiration or, if they do, all of the carbon dioxide is refixed since no carbon dioxide is released at high light intensities. This adds further to their greater productivity.

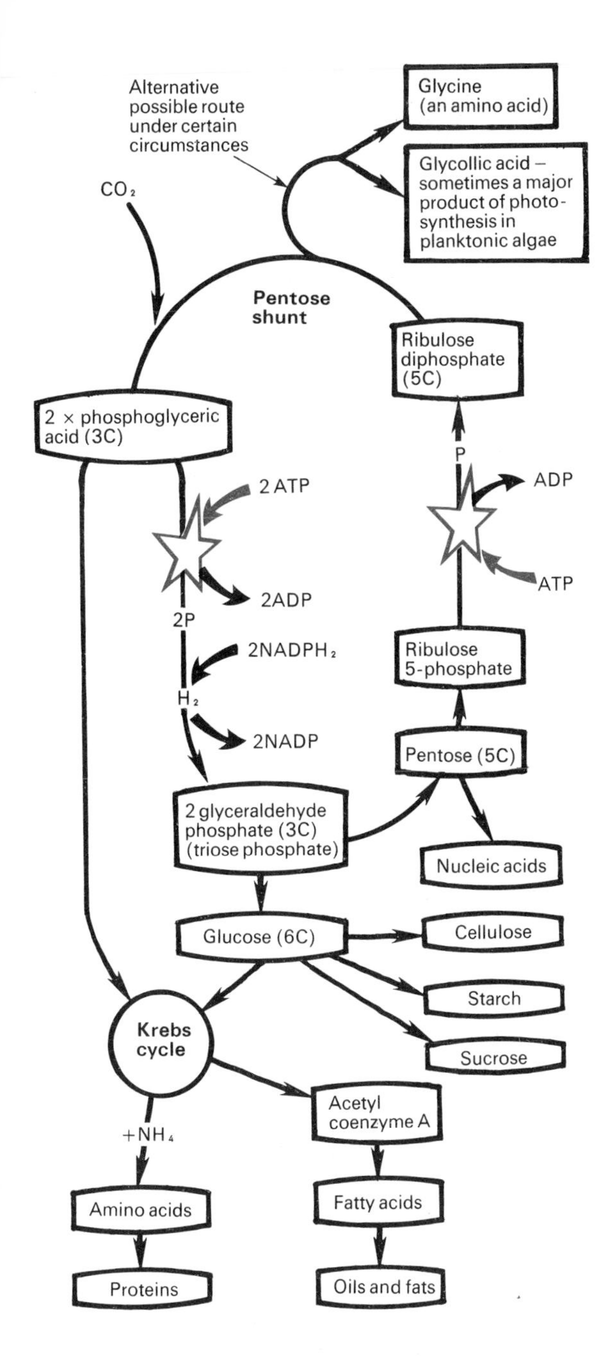

Figure 6.5 *Outline of dark reaction of photosynthesis (C_3 process)*
The complex series of steps in the pentose shunt are not shown, therefore the carbon atoms appear not to 'balance' in the flow chart. The glycollic acid produced by planktonic algae often acts as an overflow mechanism for photosynthesis as it can be released to the outside.

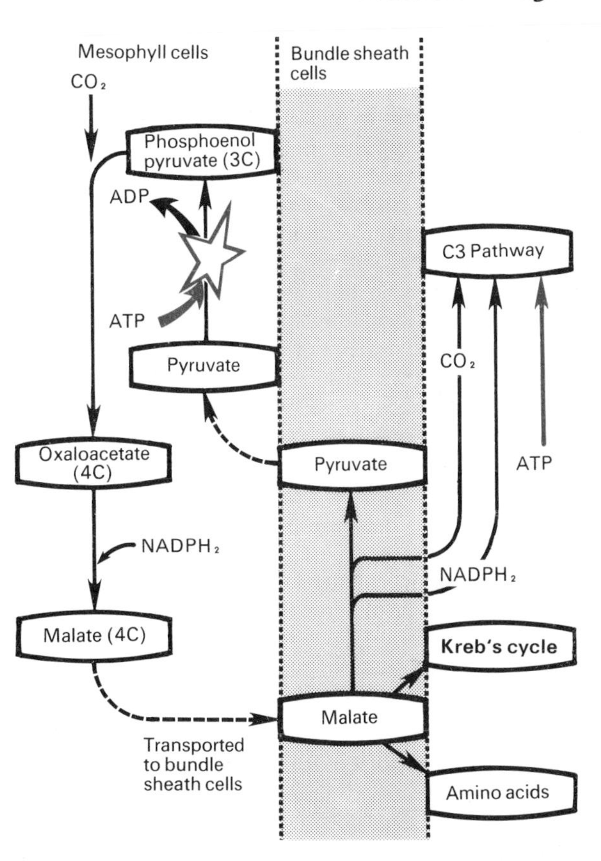

Figure 6.6 *Outline of dark reaction of photosynthesis (C_4 process—the Hatch and Slack pathway)*

Photosynthetic bacteria

These bacteria have photosynthetic pigments known as **bacteriochlorophylls** which are slightly different from the chlorophylls of green plants. They also differ in obtaining the hydrogen for the reduction of carbon dioxide from hydrogen sulphide or simple organic compounds like alcohols.

Chemosynthetic bacteria

Chemosynthesis differs from photosynthesis in that the energy for the synthesis of organic materials from inorganic materials comes from inorganic chemical reactions and not light. Important members of this group are the nitrifying bacteria,[111] e.g. *Nitrosomonas*, *Nitrococcus*, and *Nitrobacter*.

Fungi

Fungi and some bacteria lack photosynthetic pigments and any powers of chemosynthesis, and are therefore unable to synthesize organic compounds from inorganic materials. They are therefore **heterotrophic**, depending on a supply of ready-synthesized organic material, as do the animals.

Fungi may be **saprophytic**, when they feed on dead and decaying organic matter by the secretion of enzymes on to the organic material which digest it and render it soluble; the end products are absorbed by diffusion over the surface of the organism. (Animals that feed in this way are known as **saprozoic**.)

They may also be **parasitic** on a living host; although this is a mode of life rather than of feeding because, like the saprophytes, they still digest organic material extracellularly and then absorb the soluble end products.

Inorganic nutrients

Various chemical elements[361] and inorganic ions[362] are required for all aspects of a plant's physiology, including their incorporation into the organic products of photosynthesis. These elements and ions which are sometimes called minerals are obtained from the soil. However, parasitic plants (e.g. mistletoe) gain all of their necessary minerals, in both an organic and an inorganic form, from the xylem sap of the host plant.

Many minerals are needed by both animals and plants, but sodium and chlorine, which are needed in large amounts by higher animals, are not important to plants, and some which are essential to plants (e.g. boron) have no known function in animals.

Uptake

The mineral nutrients are either present in the soil particles or in the soil solution. Although the mineral nutrients may be present in the soil, their availability depends on a complex of factors including interactions between the different minerals. For example, the uptake of positive cations is accompanied by the uptake of negatively charged anions and vice versa, so that electrical neutrality is maintained inside the root. The rate of uptake is also affected by the ratio of the concentrations of different minerals in the soil as a result of antagonism and competition between them.

The nature of the soil also affects the uptake of these substances by the plant root hairs. Generally clay particles in the soil have a net surface negative charge which holds positively charged cations (e.g. hydrogen, H^+; potassium, K^+; sodium, Na^+; ammonium, NH_4^+; magnesium, Mg^{2+}; and calcium, Ca^{2+}). These ions are freely exchangeable for other similarly charged ions and represent a **cation store** that can be utilized by the plant, either directly or via the soil solution. One suggested exchange mechanism is known as the carbon dioxide hypothesis.

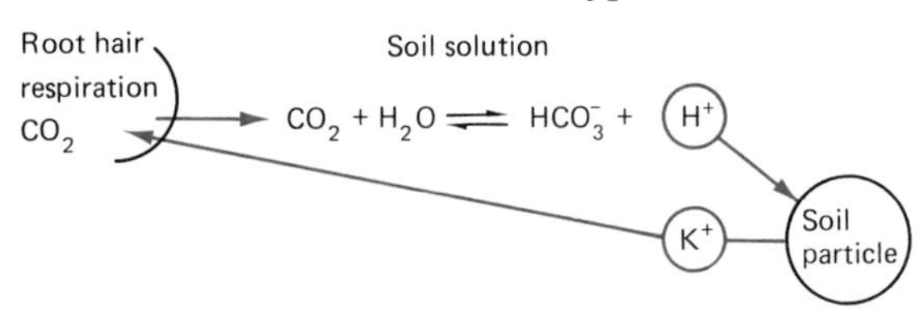

In general, negatively charged anions are not held by the soil particles and most are free in the soil solution. They are thus easily **leached** or washed from the soil and must be continuously replaced from decomposing organic matter via natural cycles, for example the nitrogen and phosphate cycles.[111]

There is no single pathway of mineral uptake. Some are drawn up to the leaves in the transpiration stream through the cellulose cell walls, only passing through cell membranes and cytoplasm at the endodermal passage cells.[167] Others are taken up into the root hair cytoplasm and then from cell to cell across the cortex to the vascular tissue by a complex of passive and active mechanisms.

Passive uptake mechanisms include mass flow mechanisms due to solvent drag in the transpiration stream,[168] whereby dissolved substances are swept along by the movement of the solvent. It also includes some ion exchange mechanisms which carry ions into the spaces of the cellulose cell walls, known collectively as the **cell free space**. Passive mechanisms, such as diffusion, may then result in the uptake of the minerals into the cytoplasm, but many will be taken up by **active mechanism** against prevailing diffusion gradients.

These active mechanisms usually involve carriers which transport the ions across the cell membrane from the cell free space into the cytoplasm using the energy from respiration. Active uptake mechanisms requiring respiratory energy seem to be most important in young seedlings, or in older plants under conditions of low transpiration.

Transport inside the plant

Once the minerals have reached the xylem[165] they are distributed in the transpiration stream to the various **sinks**, or points of utilization. The xylem sap contains about 1 per cent substances in solution, of which 75 per cent are organic compounds (such as amino acids synthesized in the roots using inorganic nitrogen) and 25 per cent are unassimilated mineral salts.

Since xylem tissue is dead, it cannot actively accumulate minerals or retain concentrations higher than those of the surrounding medium. The endodermal passage cells may be responsible for the active transport of minerals into the xylem tissue and the impermeable endodermal cells prevent the xylem from losing these minerals once they have accumulated.

Table 6.1 *Mineral nutrients of plants: macronutrients and micronutrients*

Element	*Form in which absorbed*	*Role in plant*	*Deficiency symptoms*	*Notes*
Macronutrients				
Potassium	K^+	Catalyst in protein synthesis, chlorophyll formation, and carbon assimilation. Role in cell membranes and disease resistance.	Leaves have pale yellow edges, necrosis or breakdown of leaf margins and tips. Increased amount of soluble carbohydrate and 'leaf scorch' due to low water content.	Most available in acid soils. Balance of nitrogen, phosphate, and potassium (N.P.K.) very important.
Calcium	Ca^{2+}	Component of middle lamella of cell walls, thus important in meristems.	Necrosis of terminal growing points, distortion of young leaves, chromosome abnormalities. Its absence may result in toxic amounts of magnesium entering.	Deficient in acid soils. Balance of sodium, potassium, and calcium important.
Nitrogen	NO_3^- NH_4^+	Part of amino acids, proteins vitamins, nucleic acids, ATP, and chlorophyll.	Very marked chlorosis, stunted growth, loss of lower leaves in extreme deficiency due to removal of amino acids from older leaves to growing regions.	Often deficient. Nitrogen from organic manures less easily leached than fertilizers, e.g. ammonium sulphate, sodium nitrate, gaseous ammonia injection into soil. Root nodules of legumes are a good source.
Phosphorus	H_2PO_4	Part of cytoplasm, nucleic acids, ATP. High concentration in fruits and seeds.	Dark green dead patches on leaves. Stunted growth.	Often deficient, not readily available in alkaline soils.
Magnesium	Mg^{2+}	Part of chlorophyll molecule and enzyme activater in phosphate metabolism.	Chlorosis of older leaves, i.e. from base up; veins remain green.	Often deficient in acid soils.
Sulphur	SO_4^{2-}	Part of some proteins and cofactors.	Chlorosis of young leaves, roots often enlarged.	Sulphuric acid in rain from SO_2 pollution prevents widespread deficiency.
Micronutrients				
Iron	Fe^{2+}	Involved in chlorophyll synthesis. Part of cytochrome pigments and some prosthetic groups.	Chlorosis of young leaves, as iron cannot be withdrawn from older leaves.	Less available on chalky soils as it is bound as insoluble ferric hydroxide.
Boron Chlorine Copper Cobalt Manganese Molybdenum Zinc	BO_3^{3-} B_4O_7 Cl^- Cu^{2+} Co^{2+} Mn^{2+} Mo^{3+} Zn^{2+}	Complex interactions, the role of many is not known. Mo is important in nitrogen fixation.	Difficult to identify direct and indirect effects.	These are the trace elements required in very small amounts.

The minerals mainly go to young growing regions which actively accumulate them. Some minerals do not reach the leaves in the concentrations expected from calculations concerned only with the concentration and rate of flow of the xylem sap. Moreover, a passive 'sweeping along' of nutrients in the sap would undersupply some regions and over supply others. Therefore some sort of control must be exerted by the plant.

In dicotyledons it is possible that the vascular cambium has a role in controlling the distribution of the mineral salts. Certainly, the main stream of sap and nutrients is next to the cambium in the youngest xylem elements, and any transfer to the phloem must pass through the cambium. This process, however, could not operate in the monocotyledons which lack vascular cambium.

Once the minerals have reached their destination they may be removed by competition to another, subsequently more active, region. The mobility of minerals within plants varies; thus phosphorus, sulphur, and nitrogen are easy to remove from the tissues, but calcium and iron are only recycled with difficulty and must be supplied from the root.

The roles of the various minerals are interconnected in many complex ways, but the main functions of the essential plant nutrients (those without which a green plant cannot complete its life cycle) are summarized in Table 6.1.

Insectivorous plants

These are an interesting case of specialization in green plants. They trap insects by a variety of ingenious mechanisms, and then digest them by secreting proteolytic enzymes and by harbouring decay bacteria. They all grow in regions where the soil is deficient in nitrogen, e.g. where either waterlogging or high acidity prevents bacterial decomposition of organic matter in the soil.

All known insectivorous plants can complete their life cycle without insects, but they supplement their nitrogen uptake from the soil by absorbing amino acids and nitrates from the digested insects. The insects are not used as an energy source since the plants photosynthesize normally.

ANIMAL NUTRITION

All animals are heterotrophic, that is they are unable to synthesize the organic materials they need from inorganic materials alone. Therefore their diet must include organic matter that has been synthesized at some time by plants, which will provide both materials and energy for use in metabolism. There is a wide range of nutritional requirements throughout the animal kingdom, but all animals require energy-yielding compounds (usually carbohydrates and fats), amino acids, vitamins, mineral salts, and water.

Energy-yielding compounds

Many invertebrates use a wide range of materials from which they can liberate energy by respiration, thus organic acids, lower fatty acids,[367] and alcohols are used. Vertebrates use carbohydrates, fats, and proteins; intermediate breakdown products of all these enter common enzyme pathways, and their metabolism is thus linked in complex ways.

Carbohydrates

These provide energy and organic carbon for animals. Some simple carbohydrates are described below.

MONOSACCHARIDES (e.g. glucose and fructose)

These require no digestion to render them diffusible; they are also the end products of the digestion of other carbohydrates.

DISACCHARIDES (e.g. sucrose, maltose, and lactose)

These are hydrolysed to monosaccharides during digestion; the reactions are catalysed by enzymes.

$$\text{sucrose} + \text{water} \xrightarrow{\text{sucrase}} \text{glucose} + \text{fructose}$$
$$\text{maltose} + \text{water} \xrightarrow{\text{maltase}} \text{glucose} + \text{glucose}$$
$$\text{lactose} + \text{water} \xrightarrow{\text{lactase}} \text{glucose} + \text{galactose}$$

POLYSACCHARIDES (e.g. starch, glycogen, cellulose, agar, inulin, and pectin)

Starch and glycogen are hydrolysed to glucose during digestion, but cellulose, although built up from glucose sub-units, is not digestible by most animals due to the absence of a cellulase enzyme. Wood-eating insects have symbiotic flagellates which produce enzymes that digest cellulose; mammals, particularly herbivores, have symbiotic bacteria in the gut which produce cellulase enzymes.

Fats

Most animals survive with little or no dietary fats, as these are readily synthesized from carbohydrates or the intermediates of protein metabolism. However, in man some polyunsaturated fatty acids are essential, and since they cannot be synthesized by the body, they must be supplied in the diet. One of these fatty acids is also essential for many insects, and no doubt for many other animals too.

Fats are also a source of fat-soluble vitamins, e.g. vitamin A.

Amino acids

Amino acids provide the organic nitrogen required by animals. Apart from some colourless flagellates which can obtain their nitrogen from nitrates or ammonia, all animals require ready-synthesized amino acids as part of their diet. From these **essential amino acids** the full range of necessary amino acids can be synthesized by means of various interconversions. Some protozoa can synthesize all the amino acids they need as long as they are supplied with any one type of amino acid, so they require only the amino group $-NH_2$ in the diet. All other animals require up to twelve essential amino acids in their diet, and about seven of these are a common requirement of most animals. Some of the non-essential amino acids stimulate growth and may be required in the diet only at certain stages of development.

Amino acids are taken up in the diet of most animals as proteins which are hydrolysed to their constituent amino acids. These are then absorbed and resynthesized into the protein of the animal for growth, repair, and general metabolic replacement. For example, the protein 'half-life' (the time in which about half the protein is replaced) in man for the liver, heart, and kidneys is only about ten days; for muscle, skin, and bone it is about 160 days; for the whole body it is about eighty days.

Proteins are also energy-yielding compounds, but this is secondary to their role of supplying amino acids.

Minerals

The term 'minerals' refers to inorganic chemical elements and ions.[362] They are required only in small amounts, but are essential for the healthy functioning of the body.

Although minerals are important in all functions of the body, they are particularly important in osmoregulation; the regulation of body acid–base balance; as structural components; as constituents of enzymes, hormones, and pigments; and as metabolic activators.

The effects of minerals on body function
A case of acrodermatitis enteropathica in a 20 year old man was successfully treated with oral doses of zinc sulphate. The disease is a rare hereditary condition which was only recognized as a disturbance of zinc metabolism in 1973.

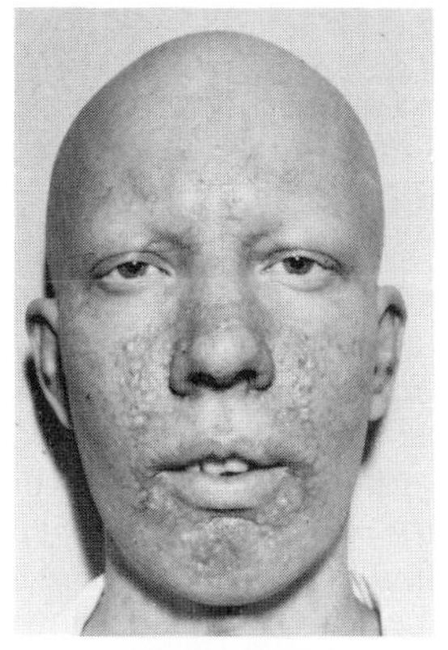
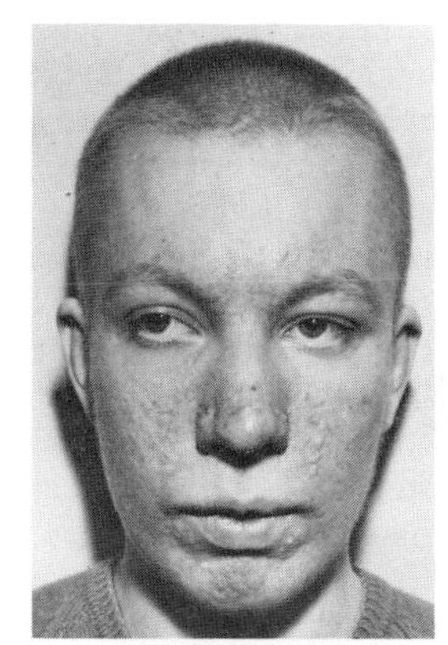
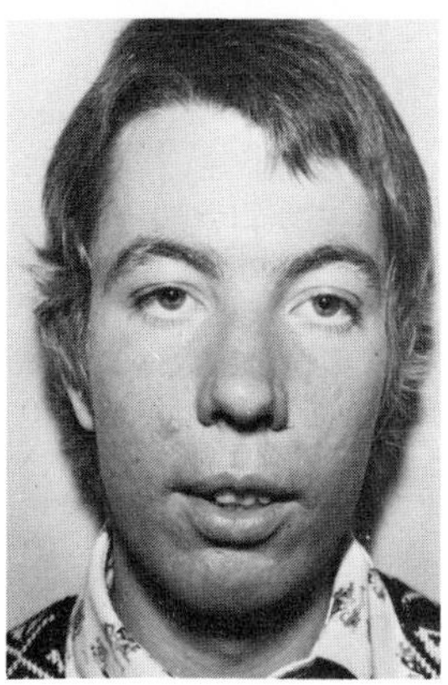
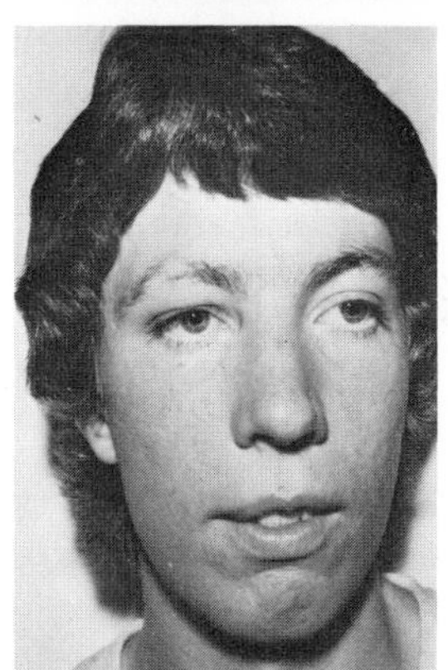

Vitamins

Vitamins are organic compounds which animals must obtain from their environment. They are needed only in very small amounts, but are extremely important in promoting and maintaining normal growth, development, and reproduction.

As different animals have different requirements the term vitamin can only be used with reference to specific animals, for example, vitamin C is a dietary requirement only for humans, monkeys, guinea pigs, fruit-eating bats, and the red-vented bul-bul bird, and is thus not a vitamin for the rest of the animal kingdom (as far as is known). As a group, vitamins are not related chemically.

In 1912 Funk coined the term **'vitamine'** from the idea that these accessory food factors were amines vital for life. With the subsequent discovery that not all were amines the word was modified to **vitamin**. The naming of individual vitamins began systematically, with vitamins A, B, and C named in sequence of their discovery. However, vitamin B was then recognized as being a group of substances which were therefore given distinguishing subscripts (e.g. B_1 and B_2). Indeed, vitamins A, D, K, E, B_{12}, and folic acid each consist of more than one substance. Other substances that were given a letter were subsequently shown not to be vitamins in the accepted sense (for example 'vitamin F' is now known to be two essential polyunsaturated fatty acids) and this led to gaps in the lettering sequence. To further add to the confusion vitamin K was named from the Danish word Koagulation! Because of this confusion the chemical names are to be preferred, as they also give information as to the nature and properties of the vitamins. However, the term 'B group' vitamins is still useful as its members are usually found together in food sources and have similar roles.

The vitamin requirements of most animals are unknown. Even in humans controversy still exists as to the exact requirements, and there is considerable variation in the requirement of vitamin C. However, the B group vitamins generally act as co-enzymes in fundamental chemical processes and are therefore needed by most animals, including the invertebrates.

Balanced diet in man

Carbohydrates

Carbohydrates are not essential, but they normally provide most of the energy of the diet and are necessary for the correct utilization of a limited protein supply (see Table 6.2).

In a natural diet, only small amounts of monosaccharides and disaccharides would occur, e.g. as glucose in fruits, fructose in honey, sucrose in onions, etc., and lactose in milk. The addition of large amounts of sucrose to the diet is a relatively recent development.

Table 6.2 *Average carbohydrate contents of some foods in g per 100 g*

Food	*Sugars*	*Starch*
Milk	4.8	0
White bread	0.3	54.0
Potatoes	0.4	17.6
Baked beans	5.2	5.1
Meat	0	0
Oranges	8.5	0

Starch grains are completely insoluble in water and need to be cooked in order to release the amylose for digestion by amylase enzymes. Glycogen, or animal starch, does not usually form part of the diet as it is hydrolysed to glucose in the muscles and liver after the death of the animal food source.

The indigestible polysaccharides, such as cellulose, provide the dietary fibre or roughage which satisfies the appetite and is necessary to provide resistance to the muscles of the gut wall during peristalsis.[145] It also retains water in the contents of the colon, and thus the transit time through the gut is shortened, which is thought to prevent the accumulation of putrefactive bacteria and consequent gut disorders.

Fats

Fats are a concentrated source of both energy and the fat-soluble vitamins A, D, E, and K. The polyunsaturated fatty acids linoleic and linolenic acids cannot be synthesized by the body and must be supplied in the diet, and are known as 'essential fatty acids'. (It has been claimed that only linolenic acid is required.)

Table 6.3 *Percentage of fat in foods*

Food	*Fat* (g per 100 g)	*Fatty acids (% of fat by weight)* *Saturated*	*Mono-unsaturated*	*Poly-unsaturated*
Cow's milk	3.8	62	30	3
Human milk	3.8	41	42	12
Cheese	34.5	62	30	3
Lamb	30.2	54	37	4
Herring	18.5	19	10	66
Butter	81.0	62	30	3
Margarine (soft)	81.5	32	40	23
Maize corn oil	99.9	15	29	51
Olive oil	99.9	11	74	10

Proteins

Proteins from animal sources yield the full range of essential amino acids, and are therefore known as proteins of **high biological value**. Proteins from plant sources do not yield the full range of essential amino acids for growth. Therefore they are known as proteins of **low biological value**, although a balanced mixture of plant proteins will yield a full range of essential amino acids. Wheat is a unique food due to the properties of the protein gluten, which enables dough to be made.

If carbohydrates and fats are in short supply, proteins are utilized as the major energy source. However, even when carbohydrates and fats are available, it is recommended that proteins should provide about 10 per cent of the daily energy requirements.

A daily protein intake of about 1 g per kg body weight per day is recommended, but there is considerable variation in individual requirements.

Table 6.4 *Average protein content of some foods, in g per 100 g*

Cow's milk	3.3	Contain the full range of amino acids required
Cheese	25.4	
Beef	18.1	
Soya beans	40.3	
Wholewheat flour	11.6	
Potatoes	2.1	
Peas	5.8	
Baked beans	5.1	
Roasted peanuts	28.1	

Minerals

As with all nutrients, a complex of factors affects the uptake of minerals so that it is difficult to determine the exact daily requirements. Thus in man the presence of cellulose fibres and phytic acid (found in wholemeal bread) decreases the uptake of available calcium, zinc, and iron; members of the cabbage family contain pregoitren which decreases iodine utilization; and calcium uptake is increased by vitamins D and C. In addition, individual requirements can vary widely depending on a complex of physiological factors.

Vitamins

It is difficult to assess the exact amount of vitamins required in a diet as, not only do individual requirements vary, but there are a complex of other factors involved. Provitamins are converted to vitamins at varying rates by the body; antivitamins oppose vitamin activity; some vitamins are in a bound form and are thus not available to the body; and, although synthesis by symbiotic bacteria of vitamins K, B_{12}, nicotinic acid, riboflavin, and folic acid occurs in the large intestine, the extent to which these are absorbable is not clear. Moreover, some bacteria actually remove vitamins from the diet.

A balanced diet

As is now clear, there can be no exact, recommended daily intake of nutrients. However, recommendations which have been produced give general guidance as to the nutrients required and more than satisfy the needs of the majority of the population.

Food additives

In industrial societies with large-scale food industries there is a relatively new component in the diet: artificial food additives. These are added for a variety of reasons, such as for preservation, processing, flavouring, colouring, and as a nutritional supplement. There is great confusion as to what levels of additive should be allowed and whether, in fact, they serve any useful purpose at all. Indeed, many are suspected of being positively harmful. There will never be clear cut conclusions to this debate as not only do individuals vary in their reaction to certain substances, but the effectiveness or toxicity of additives varies with changing feeding habits.

This confusion is clearly illustrated by the different standards in different countries as to allowable additives; thus in GB at one time there were thirty-three food colours allowed whilst ten were allowed in the USA. That they are an important component of man's diet is illustrated by the estimate that we absorb additives at a rate equivalent to twelve aspirin-sized tablets daily, or about 1.5 kg per year; in the USA the figure is closer to 4 kg per year per person. There are some 3000 additives in use, of which about half are for preservation.

Table 6.5 *Minerals: source, function, and daily requirements*

Mineral	*Main sources*	*Function*	*Estimated average adult daily requirement*	*Total body content (in g)*
			Macronutrient (daily requirement in g)	
Calcium	Dairy products Bread	Formation of bones and teeth; blood clotting; muscle contraction	1.1	1000
Phosphorus	Cheese Yeast extract	Formation of bones and teeth; part of DNA, RNA, ATP; acid–base balance	1.4	780
Sulphur	Dairy products Legumes	Part of thiamin; part of keratin	0.85	140
Potassium	Meats Potatoes	Osmoregulation; acid–base balance; nerve transmission	3.3	140
Sodium	Cheese Salt	Osmoregulation; acid–base balance; nerve transmission	4.4	100
Chloride	Meats Salt	Osmoregulation; acid–base balance; gastric acid	5.2	95
Magnesium	Cheese Greens	Energy metabolism; calcium metabolism	0.34	19
			Micronutrient (daily requirement in mg)	
Iron	Liver Cocoa	Haemoglobin and myoglobin; cytochromes; intracellular bacteriocide	16.0	4.2
Fluoride	Sea-food Water Tea	Bone and teeth formation; prevention of decay	1.8	2.6
Zinc	Meats Legumes	Many enzymes; protein metabolism	13.0	2.3
Copper	Liver Legumes	Formation of haemoglobin; energy release; formation of melanin	3.5	0.07
Iodine	Fish Iodized salt	Thyroxine	0.2	0.01
Manganese	Tea Cereals	Bone development; amino acid metabolism	3.7	0.01
Chromium	Meats Cereals	Glucose uptake by cells	0.15	0.001
Cobalt	Meats Yeast Comfrey	In vitamin B_{12}	0.3	0.001

Table 6.6 *Vitamins: their source, functions, and deficiency symptoms*

Vitamin *Name*	*Letter*	*Main sources*	*Function*	*Deficiency symptoms*
Retinol	A	Liver, milk, greens, carrots	(a) Formation of visual purple (b) Healthy membranes (c) Resistance to disease (d) Normal growth	(a) Night blindness (b) Xerophthalmia (c) Infection (d) Poor growth
Calciferol	D	Fish-liver oil, butter, sunlight on skin	(a) Uptake of calcium and phosphorous (b) Calcification of bone	Rickets in children Osteomalacia in adults
Tocopherol	E	Milk, egg yolk, wheatgerm, greens	(a) As an antioxidant (it is easily oxidized and conserves vitamins A, C, D, K and polyunsaturated fatty acids.)	Never demonstrated in man Sterility in rats
Phylloquinone	K	Liver, egg yolk, greens	Involved in blood clotting	Impaired clotting, and haemorrhage
Thiamine	B_1	Meat, milk, cereals, yeast	Coenzyme in release of energy	1. Beri-beri
Riboflavine	B_2 (USA old G)	Milk, fish, greens	Coenzyme in release of energy	1. Araboflavinosis
Niacin	(Part of old B_2)	Meat, fish, wheatgerm, greens	Coenzyme in release of energy	1. Pellagra
Pantothenic acid	B_5	Eggs, cereals	Coenzyme in release of energy	Fatigue, poor muscle co-ordination 'Burning feet' syndrome
Pyridoxine	B_6	Meats, potatoes, cabbage	(a) Coenzyme for amino acid metabolism (b) Catalyses conversion of amino acid tryptophan to niacin (c) Aids in release of energy from glycogen	Anaemia Hyper-irritability Loss of weight Convulsions
Biotin	(old H)	Milk, egg yolk, cereals, legumes	(a) Coenzyme in release of energy (b) Metabolism of fatty acids	Lassitude Dermatitis
Folic acid	Bc	Liver, kidney, greens	(a) Synthesis of adenine, guanine and thymine (b) Amino acid synthesis	Macrocytic anaemia
Cobalamin	B_{12}	Meat, liver, yeast, comfrey	(a) Maturation of red blood corpuscles (b) Carbohydrate metabolism (c) Growth factor	Pernicious anaemia
Ascorbic acid	C	Citrus fruits, greens	(a) Forms collagen and inter-cellular materials (b) Promotes use of calcium in bones and teeth	Scurvy

Vitamins are classified according to their solubility, thus A, D, E, and K are fat-soluble, while the B complex (8) and C are water soluble.

Table 6.7 *Vitamins: estimated average adult daily requirement*

Vitamin		*Estimated average adult daily requirement*	
Name	*Letter*	*Units* (mg)	*Food containing this amount*
Retinol	A	0.75	83 g margarine
Calciferol	D	0.0025	50 g sardines
Tocopherol	E	not known	—
Phylloquinone	K	not known	—
Thiamine	B_1	1.2	175 g wheatgerm
Riboflavine	B_2 (USA old G)	1.7	170 g wheatgerm
Niacin	(Part of old B_2)	18	150 g liver
Pantothenic acid	B_5	10	not known
Pyridoxine	B_6	2.0	190 g wheatgerm
Biotin	(old H)	not known	—
Folic acid	Bc	0.4	480 g spinach
Cobalamin	B_{12}	0.003	200 g cheese
Ascorbic acid	C	30	200 g lettuce

Table 6.8 *Recommended daily intake of nutrients (Department of Health and Social Security, 1969)*

Age ranges years	*Energy* MJ	*Energy* kcal	*Protein* *Recommended* (g)	*Protein* *Minimum requirement* (g)	*Calcium* (mg)	*Iron* (mg)	*Vitamin A (retinol equivalent)* (μg)	*Thiamin* (mg)	*Riboflavin* (mg)	*Nicotinic acid equivalent* (mg)	*Vitamin C* (mg)	*Vitamin D* (μg)
Infants												
Under 1	3.3	800	20	15	600	6	450	0.3	0.4	5	15	10
Children												
1	5.0	1200	30	19	500	7	300	0.5	0.6	7	20	10
2	5.9	1400	35	21	500	7	300	0.6	0.7	8	20	10
3–4	6.7	1600	40	25	500	8	300	0.6	0.8	9	20	10
5–6	7.5	1800	45	28	500	8	300	0.7	0.9	10	20	2.5
7–8	8.8	2100	53	30	500	10	400	0.8	1.0	11	20	2.5
Males												
9–11	10.5	2500	63	36	700	13	575	1.0	1.2	14	25	2.5
12–14	11.7	2800	70	46	700	14	725	1.1	1.4	16	25	2.5
15–17	12.6	3000	75	50	600	15	750	1.2	1.7	19	30	2.5
18–34 sedentary	11.3	2700	68	45	500	10	750	1.1	1.7	18	30	2.5
18–34 moderately active	12.6	3000	75	45	500	10	750	1.2	1.7	18	30	2.5
18–34 very active	15.1	3600	90	45	500	10	750	1.4	1.7	18	30	2.5
35–64 sedentary	10.9	2600	65	43	500	10	750	1.0	1.7	18	30	2.5
35–64 moderately active	12.1	2900	73	43	500	10	750	1.2	1.7	18	30	2.5
35–64 very active	15.1	3600	90	43	500	10	750	1.4	1.7	18	30	2.5
65–74	9.8	2350	59	39	500	10	750	0.9	1.7	18	30	2.5
75 and over	8.8	2100	53	38	500	10	750	0.8	1.7	18	30	2.5
Females												
9–11	9.6	2300	58	35	700	13	575	0.9	1.2	13	25	2.5
12–14	9.6	2300	58	44	700	14	725	0.9	1.4	16	25	2.5
15–17	9.6	2300	58	40	600	15	750	0.9	1.4	16	30	2.5
18–54 most occupations	9.2	2200	55	38	500	12	750	0.9	1.3	15	30	2.5
18–54 very active	10.5	2500	63	38	500	12	750	1.0	1.3	15	30	2.5
55–74	8.6	2050	51	36	500	10	750	0.8	1.3	15	30	2.5
75 and over	8.0	1900	48	34	500	10	750	0.7	1.3	15	30	2.5
Pregnant, 2nd and 3rd trimesters	10.0	2400	60	44	1200	15	750	1.0	1.6	18	60	10
Lactating	11.3	2700	68	55	1200	15	1200	1.1	1.8	21	60	10

Table 6.9 *Summary of daily dietary components*

	Number of main types	*Weight (in g)*	*% energy requirement*	*Energy content (total 3000 Kcals)*	*Notes*
Sugar	4	168	21	630	At present they make up 30% of total carbohydrate intake. Better if lower.
Starch	1	392	49	1470	Only readily available in cooked form.
Fats	Various	67	20	600	Have fat soluble vitamins A, D, E, and K associated.
Essential fatty acids	2	1	Insignificant	Insignificant	Sufficient even in poor diets.
Protein	Various	70	10	300	Must be balanced for amino acids (8 essential, 12 non-essential).
Minerals	15	17	—	—	7 macronutrients, 8 micronutrients.
Vitamins	13	0.064	—	—	Some can be stored in the body and are therefore not needed daily.
Fibre	1	10	—	—	Important in the movement of food through the gut.
Water	1	1000	—	—	As fluids. About additional 1000 g derived from food.

Table 6.10. *Food additives*

Additives	*Notes*
1 Nutritional	
Cuprous iodine Potassium iodide	Added to table salt at 0.01%, provides iodine for thyroxine production. Prevents goitre in regions with low iodine in food and water.
Fluoride	Added at 1 ppm to drinking water to prevent tooth decay.
Calcium	Added to bread flour as 0.0025% chalk.
Iron	Added to bread at 1.65 mg/100 g (0.00165%).
Vitamin B_1	Added to bread at 0.000 24%.
Vitamins A and D	Margarine must be nutritionally equal to butter, therefore vitamins A and D are added.
2 Preservation	
Sulphur dioxide	Fruit preservation, preserves colour; has an antiseptic value, slows browning. Present as sulphites in wine.
Sodium nitrate Sodium nitrite	Added to salt solutions in curing meats. Changes flavour, acts as an antibacterial agent (particularly *Clostridium botulinum*). Converts fresh myoglobin to nitrosomyoglobin the pigment of cured meat. About 227 g (½ lb) of cured meat eaten per week per head of population, all cured with nitrates and nitrites. Nitrites react with amines either in the meat or in the gut to produce nitrosamines, 80 out of 120 of which induce tumours in animals. The conversion is speeded up by heat and they tend to accumulate in the fat rather than the lean.

Table 6.10 *(continued)*

3 Flavouring	
Monosodium glutamate	Imparts a meaty taste, extremely large amounts shown to be toxic in animals. Banned from baby food.
Saccharin	No calorific value but 500 times sweeter than sugar.
Cyclamates	Sweetner, banned in 1970.
Chloroform	Added to cough medicines and toothpastes. Banned in USA.
Vanillin	Vanilla taste.
Piperonal Ethyl acetate Cinnamaldehyde	'Fruity' flavouring.
4 Colouring	Most common are water-soluble coal tar products, 29 identified as positively dangerous. No colouring allowed in milk, tea, coffee, fresh fruit, and vegetables.
Yellow OB and AB	Once used to colour butter and margarine, it is one of the most carcinogenic substances known.
Red no. 4	Once widely used in jellies and jams etc. Banned in 1964.
Citrus red	Colours oranges.
Brown F.K.	'F.K.' means 'for kippers'! Also in sausages, boiled sweets, ginger biscuits, soups. Possible reaction with nitrites to form nitrosamines. Banned in USA and Common Market.
Blue VRS	Once used to colour processed peas, all blues now severely restricted due to toxicity.
5 Indirect	
Animal feed additives	Arsenical additives to poultry feed. Antibiotics, tranquillizers, and female hormones for increased growth.
Pesticides	Remain as residues on and in fresh and preserved food, e.g. DDT, arsenic, lead, mercury, etc.
Growth inhibitors	For storage of vegetables, e.g. maleic hydrazide, as a sprout inhibitor on potatoes and onions, produces liver tumours in mice.
Packaging	Components of packaging materials which migrate into food, e.g. acrylonitrile from plastic bottles.
Sucrose and salt	Preserve by the osmotic withdrawal of water from micro-organisms, also flavour and nutritional.
Creosote	Found in smoked meats from the burning wood, or by direct injection.
Benzoic acid	Preserves vegetables. Banned in yoghourts, sauces, and preserved fruit, but it is still found.
Boric acid	Once a widely used antiseptic, less so now as known to be toxic in large amounts.
Formaldehyde	Once added to milk and cream. Now banned.

Feeding mechanisms

The ways in which animals obtain their food can be placed into three main groups, although there are many which are hard to place in a particular category.

Microphagous (or filter) feeders

These are aquatic animals that feed on relatively small food particles. Due to the minute nature of their food they usually tend to feed continuously. The commonest methods involve cilia which are found in all phyla except the Nematoda[333] and the Arthropods.[335] Examples of microphagous animals are *Paramecium*[225] and *Mytilus*[334] (the marine mussel).

Macrophagous feeders

These feed on relatively large food particles, and thus only feed occasionally, usually after a specific chemical mechanical stimulus. Also, these larger particles need to be broken up by mechanical and chemical action (see *Amoeba*,[224] *Hydra*,[332] *Periplaneta*[308] (cockroach), and most vertebrates (including man)).

Fluid feeders

Most animals which feed on fluids are insects with specialized tubular mouthparts, e.g. *Cimex lectularius*[308] (bed-bug), *Pediculus capitis*[309] (head louse), *Anopheles gambiae* (malarial mosquito), Lepidoptera[337] (butterflies and moths), and aphids[172] (greenfly).

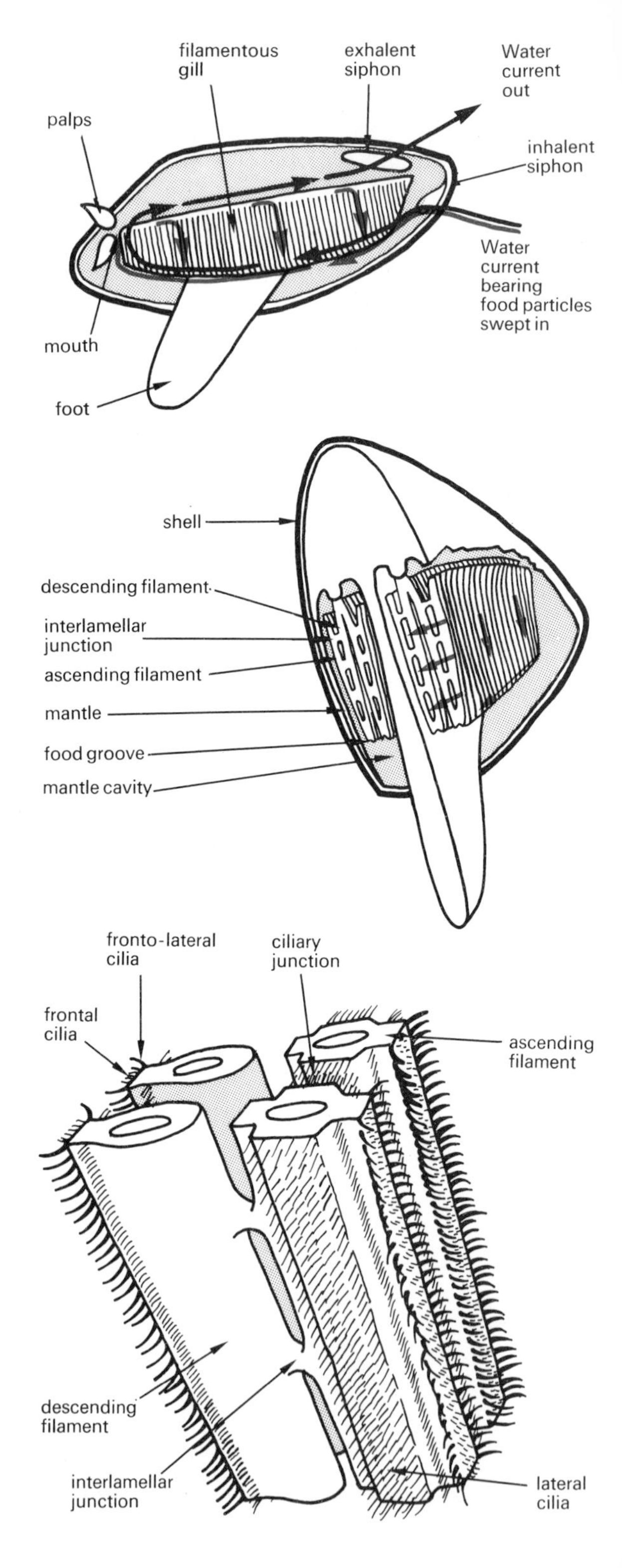

Figure 6.7 *Filter feeding in Mytilus*

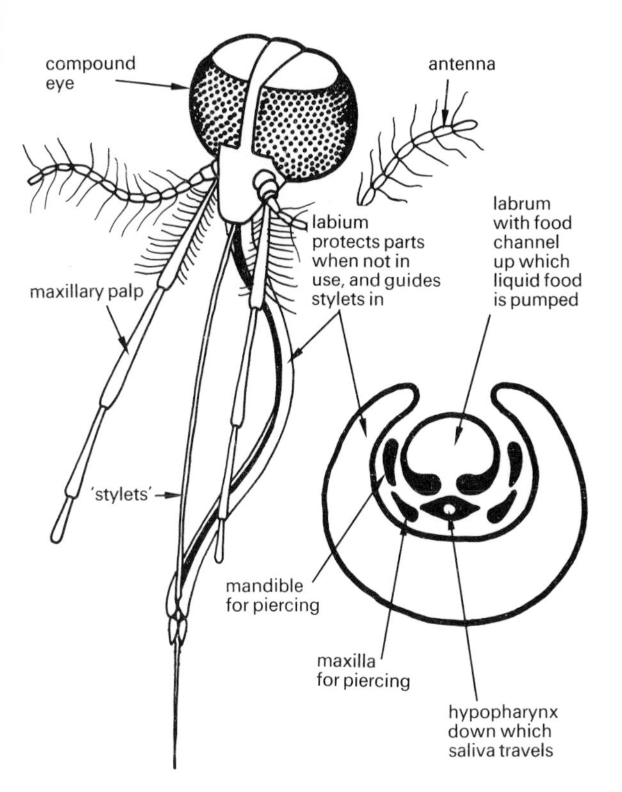

Figure 6.8 *Mouthparts and feeding in the female mosquito*
In the male the mandibular and maxillary stylets are absent and feeding from fluid at exposed surfaces is the rule.

Digestion

Digestion can be both extra-cellular and intra-cellular.

Extra-cellular digestion occurs outside cells and usually involves both mechanical and chemical processes. These result in the breakdown of complex (often insoluble) substances into the simple, soluble end-products of digestion which can then be absorbed through the cell membranes. Extra-cellular digestion occurs in the body cavity or **enteron** of Coelenterates, and the gut of those animals that possess one.

Intra-cellular digestion occurs within cells, sometimes in food vacuoles, as in phagocytic[224] cells such as *Amoeba*; in some types of white blood cells; and in the cytoplasm itself of all cells, especially in organelles such as the lysosomes.[4]

Digestion in mammals

GUT STRUCTURE AND FUNCTION

The gut is a muscular secretory structure with a large surface area well supplied with blood vessels for the absorption of the soluble end-products of digestion. It is exposed to much mechanical wear, and the lining epithelium undergoes active mitotic division, being completely replaced in man about every two days. The rate of cell replacement in the stomach wall alone is estimated at about 500 000 cells per minute. The epithelium secretes enzymes in acid and alkaline fluids, and it receives secretions from the salivary glands, the liver, and the pancreas.

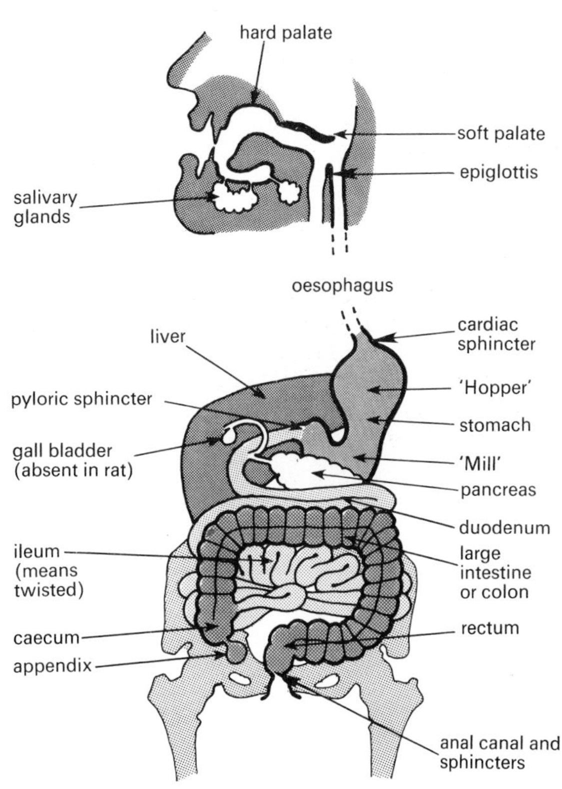

Figure 6.9 *Mammalian (human) gut structure*

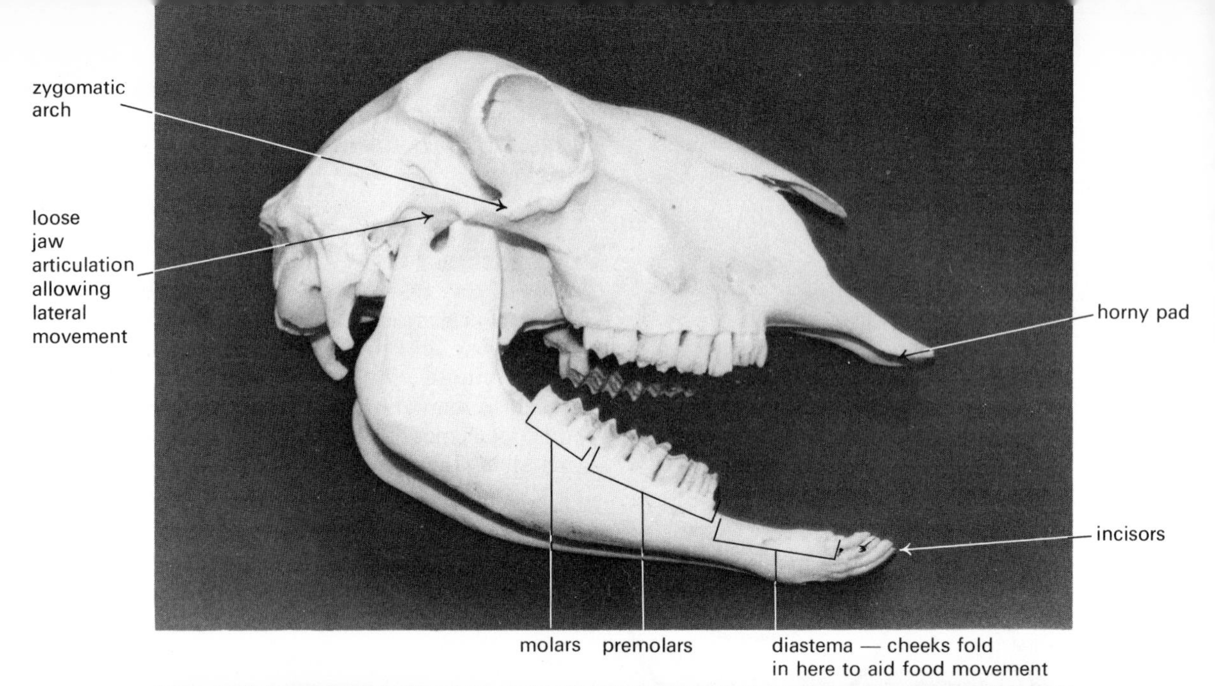

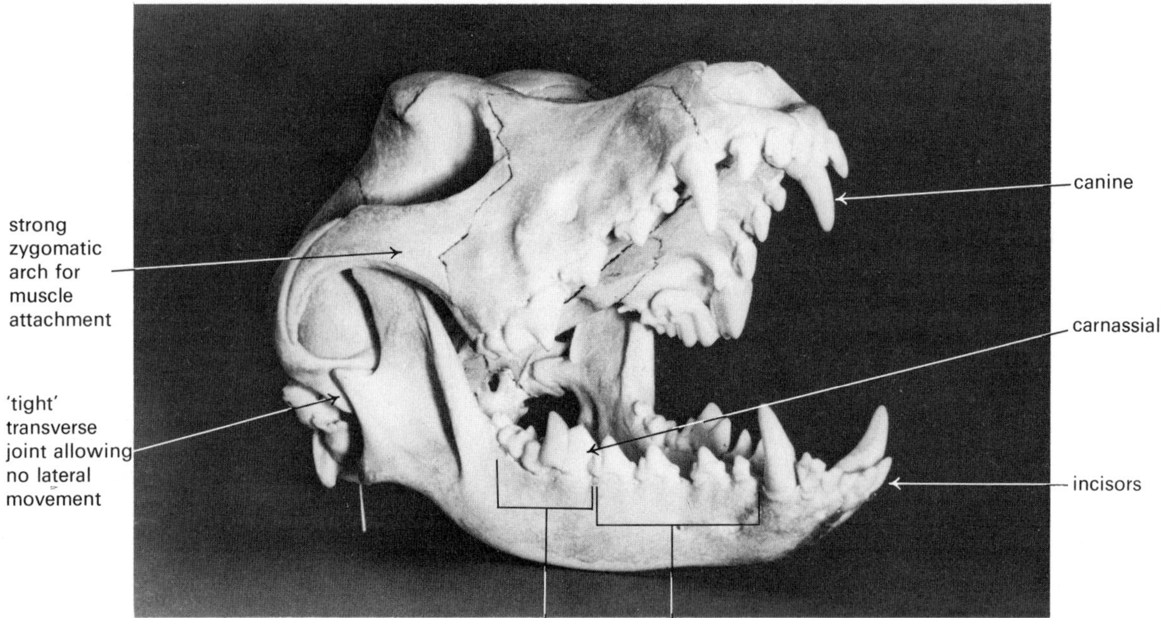

The dentition of a herbivore (above) and a carnivore (below)

THE MOUTH

Mastication involves the teeth, tongue, cheeks, and lower jaw. It increases the surface area of the food for chemical digestion and forms the food into a bolus for swallowing. The food is mixed with saliva, a slightly acid solution of salts, mucin, and salivary amylase enzyme which hydrolyses the amylose of cooked starch to maltose. It has an optimum pH of 6–7 and thus functions only until inhibited by the stomach acid. Amylase is absent in the saliva of carnivores. Saliva also moistens and lubricates the food for swallowing.

Mammals are **diphyodont** so they have two sets of teeth: the milk dentition and the permanent dentition. They are also **heterodont**, which means they have different types of teeth for particular functions. The dentition, jaw structure, jaw muscles, jaw articulation, and skull structure are closely related to the particular diet and thus differ in carnivores, herbivores, and omnivores.

THE OESOPHAGUS

After mastication and swallowing, the bolus of food passes down the oesophagus by waves of muscular contraction known as **peristalsis**, and enters the stomach when the cardiac sphincter muscle relaxes.

THE STOMACH

This is a muscular food reservoir which stores the food for a short time before passing it on, in small amounts as chyme, to the small intestine. Contractions of the stomach mix the gastric juice with the food and also aid in its mechanical breakdown. There is some digestion and absorption here.

Gastric glands secrete gastric juice which contains the following:

Mucus to protect the stomach wall from self-digestion.
Hydrochloric acid at pH 1 which kills bacteria and makes calcium and iron salts suitable for absorption in the intestine.
Rennin (in young mammals) which converts the soluble milk protein caseinogen into an insoluble form which can be attacked by pepsin.
Pepsinogen, which is activated to pepsin by hydrochloric acid (pepsin digests proteins to peptides).
Lipase, which has a weak, fat-splitting action, and an **intrinsic factor** for the later absorption of vitamin B_{12}.

Little absorption occurs in the stomach, but some glucose, minerals, water, vitamins, alcohol, and drugs are taken up into the blood stream.

THE SMALL INTESTINE

The small intestine is the main region of digestion and absorption, as it has a large surface area with folds, villi,[146] and microvilli,[10] which are well-supplied with blood capillaries.

Digestion in the small intestine
The wall secretes alkaline **succus entericus** or intestinal juice containing the following:

Amylase, which converts amylose in starch to maltose.
Maltase, which converts maltose to glucose.
Lactase, which converts lactose to glucose and galactose. (Lactase is absent in some adults, who then show lactose intolerance. Lactase production is encouraged by regular milk drinking.)
Sucrase, which converts sucrose to glucose and fructose.
Lipase, which converts fat, emulsified by alkaline bile from the liver, to glycerides, fatty acids, and glycerol.
Erepsin, a mixture of peptidases, which converts peptides to amino acids.
Enterokinase, which activates pancreatic trypsinogen to trypsin, which converts proteins to amino acids. The **trypsin** in turn activates pancreatic **chymotrypsinogen** to **chymotrypsin**, which converts proteins to amino acids.
Mucus, which protects the lining from the digestive enzymes.

The small intestine also receives secretions from the pancreas and the liver.

Table 6.11 *Summary of digestive enzymes and their action*

Secretion	*Enzymes*	*Substrate*	*Products*
Saliva	Amlase (ptyalin)	Amylose	Maltose
Gastric juices	Rennin (only in young)	Caseinogen	Casein
	Pepsin	Proteins	Peptides
	Lipase	Fats	Fatty acids and glycerol
Intestinal juice	Amylase	Amylose	Maltose
	Maltase	Maltose	Glucose
	Lactase (induced by regular milk drinking)	Lactose	Glucose and galactose
	Sucrase (invertase)	Sucrose	Glucose and fructose
	Lipase	Fats	Fatty acids and glycerol
	Erepsin (mixture of peptidases)	Peptides	Amino acids
	Enterokinase	Trypsinogen	Trypsin
Pancreatic juice	Amylase	Amylose	Maltose
	Maltase	Maltose	Glucose
	Trypsin (ogen)	Proteins	Peptides
	Peptidases	Peptides	Amino acids
	Lipase	Fats	Fatty acids and glycerol
	Chymotryp-sin (ogen)	Proteins	Amino acids
	Nucleases	Nucleic acid	Nucleotides
	Elastase	Elastin	Amino acids

The pancreas

This consists of two distinct parts: the endocrine Islets of Langerhans,[215] which secrete the hormones insulin and glucagon, and the exocrine part which secretes the pancreatic juice into the small intestine via the pancreatic duct.

About 1.5 dm^3 pancreatic juice is secreted per day in man. It contains water, alkaline salts to neutralize the acid from the stomach, lipase, amylase, maltase, trypsinogen, peptidases, and chymotrypsinogen. The secretion of the alkaline fluid is stimulated by the hormone secretin,[215] and the secretion of the enzymes is stimulated by the hormone pancreozymin.[215]

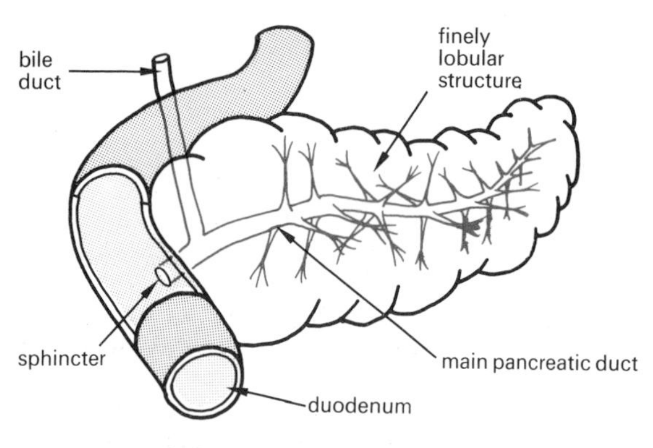

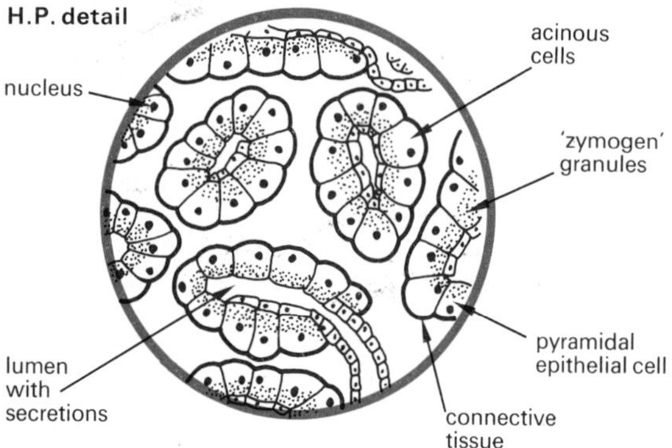

Figure 6.10 *Pancreas structure*

Absorption in the small intestine

The absorption of the end-products of digestion takes place over the large surface area of folds, villi, and microvilli. The villi are well-supplied with blood capillaries and lacteals. Absorption is aided by the '**villus pump**' which is a rhythmic movement of the villi caused by contraction of the smooth muscle of the muscularis mucosa. This movement also helps to squeeze the end-products of fat digestion, which are absorbed by the lacteals, into the main lymph vessels.

Substances absorbed in the small intestine include monosaccharides, amino acids, some larger protein fragments, and, in some cases, whole proteins (for example, when the suckling young absorb antibodies from the mother's milk). In adults the absorption of whole protein can cause an allergic reaction due to the presence of foreign protein in the blood. Fatty acids, and glycerol, vitamins, and minerals salts, are also all absorbed here.

Large intestine

The **caecum** is well-developed in herbivores and contains symbiotic bacteria which secrete cellulases to digest the cellulose cell walls of plant material. This breaking down of the cellulose yields monosaccharides and releases the proteinaceous contents of the cell for digestion. The bacteria also synthesize vitamins, but in man it is doubtful whether these are absorbed.

The **colon** is the main region of absorption of water and minerals.

THE LIVER

Most of the absorbed end-products of digestion are carried in the hepatic portal vein to the liver. Exceptions are most of the fatty acids and glycerol, which are absorbed as such and then resynthesized into fat which travels as an emulsion in the lymphatic system to the venous system and thence around the body.

Digestive functions of liver

The liver secretes bile, which is stored in man in the gall bladder. It is an alkaline fluid and about 1 dm^3 per day passes into the duodenum. It contains bile salts which emulsify fats, activate pancreatic lipase, increase the efficiency of carbohydrate and protein digesting enzymes, and increase the uptake of vitamins A, D, and K. The bile also stimulates the release of the digestive hormone secretin.[215]

Summary of non-digestive functions

(a) Destruction of old red blood corpuscles. The iron is stored as ferritin for reuse, the rest of the haemoglobin is broken down into the bile pigments (bilirubin and biliverdin) which pass to the kidney to colour the urine as urobilin or to the gut to colour the faeces as stercobilin.

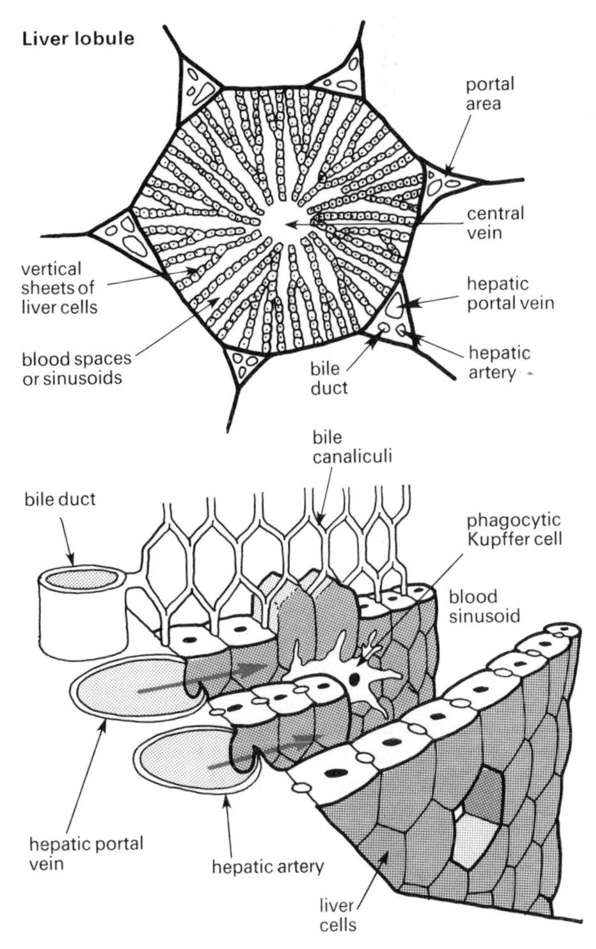

Figure 6.11 *Liver structure*
The so-called 'classic' liver lobule is only seen in the pig liver, in other mammals the structure of the liver is far from clear.

(b) Storage of excess carbohydrates and other carbon-containing compounds as glycogen.[366]
(c) Metabolism of fat.
(d) Amino acid metabolism, including deamination and urea formation.[193]
(e) Synthesis of vitamin A.
(f) Storage of vitamins A, D, B_{12}.
(g) Synthesis of blood proteins.
(h) Contributes part of the reticulo-endothelial defence system.[293]
(i) Detoxyfication, e.g. the degradation of fat-soluble drugs such as barbiturates, the breakdown of alcohol formed naturally by fermentation in the gut, inactivation in the male of female hormones from the adrenal cortex, and of male hormones in the female.
(j) The manufacture of red and white blood cells in the embryo.

Experimental and applied aspects

The complex series of steps in photosynthesis were unravelled by the use of two techniques that are used widely in research into metabolic pathways : the use of radioactive tracers and chromatography.

Plants are exposed to **radioactive isotopes**[361] of elements involved in the processes under investigation, and the radioactivity of these isotopes, which are chemically the same as the natural form of the element, allows their passage through the various metabolic steps to be traced.

For example, in order to determine whether the oxygen released in photosynthesis is derived from the carbon dioxide or from the water, plants are exposed in one case to carbon dioxide labelled with the radioactive tracer of oxygen ^{18}O, and in another case to water labelled with ^{18}O. If this is done, the ^{18}O only appears in the evolved oxygen in the latter case, thus demonstrating that the evolved oxygen in photosynthesis is derived from water and not from the carbon dioxide.

In order to trace the path of carbon (from carbon dioxide) through the complex series of steps in the dark reaction of photosynthesis, plants are exposed to carbon dioxide labelled with the radioactive isotope ^{14}C (a weak β emitter), for varying amounts of time. The plants are then killed as rapidly as possible to fix the reactants, which are then extracted and analysed. It is possible to use large plants in this procedure but there are difficulties, due to the size, in killing the tissues and fixing all the reactants at the same time throughout the plant. The most accurate work has therefore been done (originally by Calvin *et al.*) using suspensions of unicellular algae (*Chlorella pyrenoidosa* and *Scenedesmus obliquus*) which can be rapidly killed by the addition of a fixative to the suspending solution. (The ^{14}C is incorporated into bicarbonate ions $-H^{14}CO_3$ in the solution.) The experiment is run for a very short time at first, and for a slightly longer period on each subsequent occasion, so that the sequence of reactions and reactants can be discovered. Once the cells have been killed, the organic matter is extracted and analysed by **chromatography**.

There are many different chromatography techniques, but the original one (which is still widely used) is **paper chromatography**. A concentrated spot of the mixed extract is placed near one end of some absorbent paper which is resting in an organic solvent, and the whole is enclosed in a water-saturated atmos-

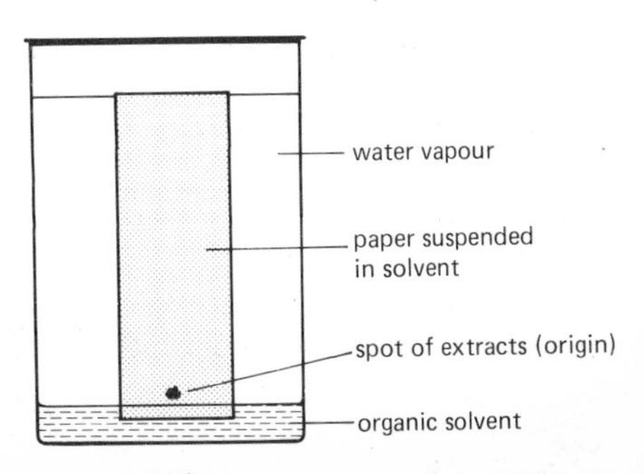

phere. The cellulose fibres in the paper absorb water from the atmosphere. This forms the stationary aqueous phase.

The organic solvent moves by capillary action along the length of the paper. This is known as the mobile phase. The different compounds in the mixture to be analysed are carried along the paper according to their relative solubility in the organic solvent and in water. As the solvent front moves along the paper the different compounds distribute themselves between the solvent and the water absorbed by the fibres of the paper. Those that are more soluble in water are carried the shorter distances while those that are more soluble in the organic solvent are carried further within a given time. The distance travelled is given as a so-called 'R_f value:

$$R_f = \frac{\text{distance compound has moved from its origin}}{\text{distance of the solvent front from the origin}}$$

The R_f value is characteristic and descriptive of a compound.

Unfortunately, chromatography in one direction only can not always separate out every compound. Therefore **2D paper chromatography** is employed to further separate closely located substances. In 2D chromatography the process is repeated on the same piece of paper, but using a different solvent at right angles to the first. This results in the separation of any overlapping compounds.

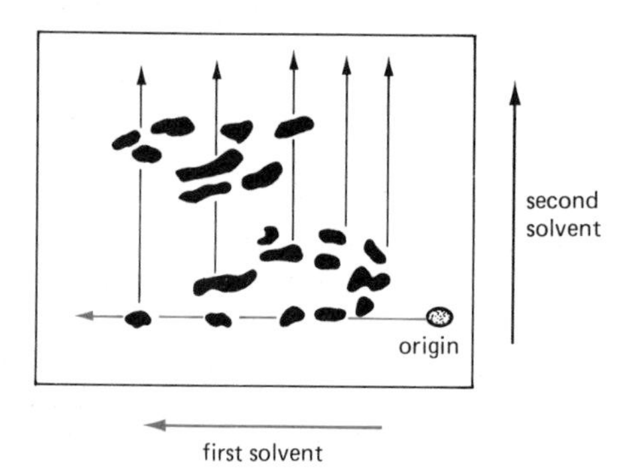

Since the compounds are usually colourless, the problem remains of their location and identification. When not using radioactive tracers, special sprays are available which locate the different compounds as coloured areas. If a radioactive tracer is used the paper is exposed to an X-ray sensitive film for a few days, during which time the radioactivity of the tracer is recorded. The film is then developed to give an **autoradiograph**.

(**Thin layer chromatography** utilizes a thin layer of material such as silica gel, alumina, etc., supported on a flat plate, instead of paper. It has several advantages, for example it takes a shorter time, the resolution is good, it is better for lipid separation, and the substances are more easily extracted.)

The substances can be identified in a variety of ways. If their R_f value is known, and it is clearly distinct from any other, this will identify the compound. The position of the spots can be compared to the position of those of known compounds run through the same process, either on a different piece of paper, or on the same piece of paper. Another method is to cut the actual area out from the original chromatography paper, extract the substance by a solvent, and analyse it chemically.

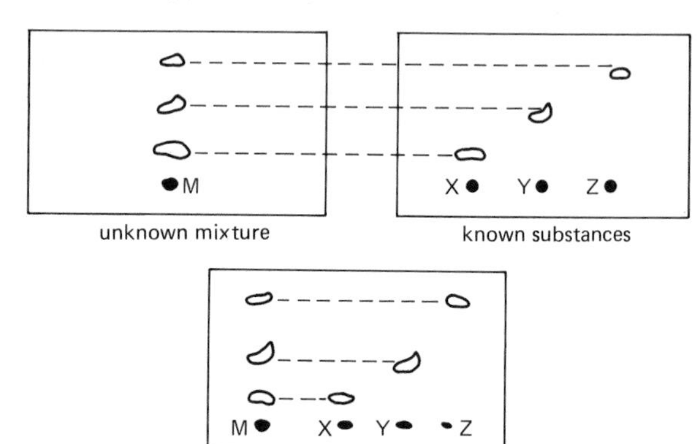

Using these methods, complex metabolic pathways may be elucidated, and in this way the first stable intermediate formed in photosynthesis was identified as being 3-phosphoglyceric acid (PGA).

The rapidity of photosynthetic reactions is indicated by the fact that 20–30 radioactive compounds can be extracted from algae that have been exposed to ^{14}C for only about 30 seconds. In fact, all stable intermediates of the Calvin cycle become saturated with ^{14}C within 3–5 minutes. It was also discovered that about 30 per cent of the carbon uptake in photosynthesis is incorporated directly into amino acids. This shows that amino acids are formed directly in photosynthesis, and not only via carbohydrates, as was once believed.

Guided example

In the seventeenth century Johann Baptista Van Helmont, one of the first physiological chemists, performed an experiment which he considered demonstrated that plants could change water into wood. He planted a willow tree in a large pot containing oven-dried soil. The soil was covered with a perforated iron plate to keep out dust; distilled water or rain was added when needed.

When planted, the young willow tree weighed five pounds, five years later it weighed about 169 pounds. (Van Helmont did not weigh the leaves which fell in the four autumns that passed.) The oven-dried soil weighed 200 pounds at the start of the experiment, and when oven-dried and reweighed at the end, had apparently lost only 2 ounces in weight. Van Helmont concluded in his *Ortus Medicine* that 'therefore 164 pounds of Wood, Barks, and Roots, arose out of water onely.'

Taking this account as accurate, answer the following questions.

1 By what processes did the willow tree increase its weight over the period of the experiment?
By photosynthesis producing organic compounds from carbon dioxide, and by the absorption of water. Carbon makes up at least 50 per cent of the dry weight of wood, and the water content is about 50 per cent of its fresh weight.

2 Can plants in fact change water into wood?
No, not as such. During photosynthesis water dissociates and the hydrogen ions are incorporated into a reduced carrier molecule which is subsequently used in the synthesis of carbohydrates, some of which can eventually be incorporated into the structure of wood.

3 What substances from the environment, other than water, contribute to the formation of 'Wood, Barks, and Roots'?
Mainly carbon dioxide from the atmosphere, but also mineral nutrients from the soil.

4 Why did the dried soil weigh less at the end of the experiment?
Due to the removal of mineral nutrients from the soil by the growing willow tree. The minerals are utilized in the process of photosynthesis for incorporation into complex organic molecules. They also play a direct role in the metabolism of cells. They are only required in small amounts.

5 Van Helmont's results indicate that the dried soil had only lost 2 ounces over the 5-year period. Do you think this figure is an accurate representation of the amount of inorganic minerals removed from the soil by the growing tree?
No. Although the exact amount of mineral nutrients removed from the soil by a plant is hard to judge, estimates of the mineral content of plants vary from about 8 per cent of the dry weight of stems up to about 15 per cent of the dry weight of leaves. A reasonable estimate for the whole tree would not be less than about 10 per cent of the dry weight.

Leaves can have a water content of up to 90 per cent of their fresh weight, and the water content of wood is about 50 per cent. Therefore the dry weight of the 5-year old willow tree could be estimated very roughly as being not less than about 50 per cent of the fresh weight, that is about 80 lb. Therefore a very low estimate of the inorganic mineral content of the willow tree would be 10 per cent of 80 lb—8 lb—all of which would have been obtained from the soil.

Questions

1 In the figure below, a diagram of the apical region of a root is shown in relation to mineral salt accumulation and respiration.

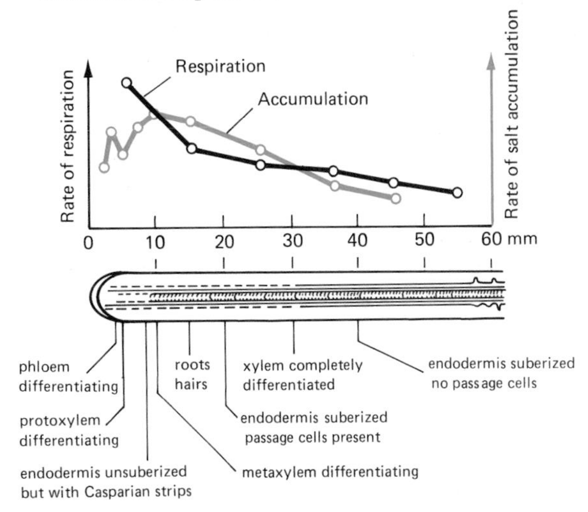

(a) (i) At what distance from the root tip are salts accumulating most rapidly?
(ii) Why is it necessary for the plant to concentrate the salts taken up from the soil?

(b) List the features of root structure shown in the diagram that are important in relation to salt accumulation.

(c) Salt accumulation and respiration appear to be correlated. From your knowledge of salt uptake in roots, explain a possible physiological basis of the apparent correlation. (C)

2 Give an account of the fixation and utilization of carbon in plants which is sufficiently detailed to explain the following statements.

(a) (i) The first product of carbon dioxide fixation is 3-phosphoglyceric acid (PGA). (ii) Light energy is required only to provide the ATP and electron donors (reduced co-enzymes) needed to reduce PGA and maintain a supply of ribulose diphosphate. (iii) There are alternative uses for ATP and the electron donors in addition to those of carbon dioxide fixation.

(b) (i) Starch grains begin to grow close to grana. (ii) The formation of cellulose and pectins occurs near to the plasmalemma. (JMB)

3 A filamentous green alga was placed in a liquid containing a large number of mobile aerobic bacteria and then illuminated along its length by light which had passed through a prism which had split it into its spectral components as shown by the scale below. The illustration shows the distribution of the bacteria 10 minutes after illuminating by this method.

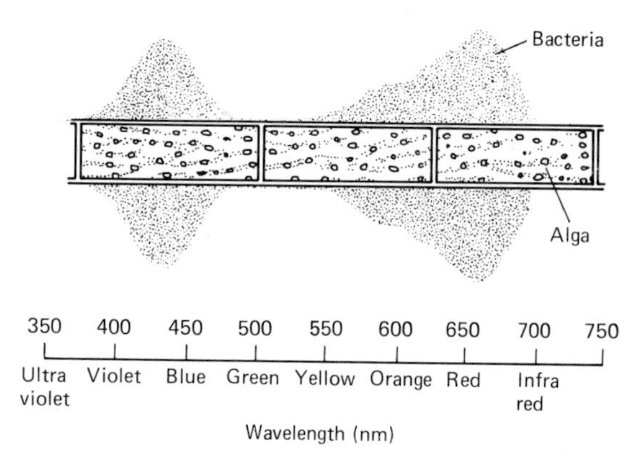

(a) Name a *genus* of filamentous green alga which could have been used in the experiment.
(b) How do you explain the distribution of the bacteria in this experiment?
(c) From the above experiment which colours of light do you consider are the most effective for photosynthesis?
(d) Most of the light falling on the algal cells is not absorbed. What happens to the light which is not absorbed?
(e) State *one* effect which ultra-violet light has on living organisms.
(f) Which gas in the upper atmosphere is responsible for absorbing most of the ultra-violet light emitted by the Sun? (O)

4 In a study of the productivity of the River Cam the radioactive isotope ^{14}C was used to detect the rate of carbon fixation. At the same time counts were made of the numbers of plant cells present.

The ^{14}C techniques involved releasing a certain amount of the isotope as $Na_2{}^{14}CO_3$ into a known volume of river water and suspending this at the same depth at which the sample had been taken. The samples were left to photosynthesize for one hour, and the particular set of results given below came from an experiment done around midday in midsummer.

Depth	*Light intensity (arbitrary values)*	*Diatoms per litre* ($\times 10^6$)	*Rate of photosynthesis (as mg C fixed per* 10^6 *diatoms)*
surface	100	3.5	6
1 metre	40	7.4	20
2 metres	35	10.3	14
bottom (3 metres)	10	15.2	10

(a) On a single piece of graph paper plot curves which illustrate (i) the relation between light intensity and numbers of diatoms per litre and (ii) the relation between light intensity and the rate of photosynthesis.
(b) From the curves comment on and account for the relationship between light intensity and (i) the numbers of diatoms, (ii) the rate of photosynthesis.
(c) How might you test in the laboratory the hypotheses you have put forward above?
(d) Cultures of diatoms under laboratory conditions can be made to photosynthesize much more rapidly than those in natural environments such as the river studied. What limiting factors to photosynthesis might be operating in the river but not in the laboratory?
(e) In all the above experiments no account was taken of the rate of respiration of the plant cells during the experiment so that the figures for photosynthesis only represent apparent values. Using the ^{14}C techniques (or any other with which you may be more familiar) explain how the true rate of photosynthesis might have been measured. (AEB)

5 Discuss the importance in the nutrition of man of (a) vitamins, (b) mineral salts, and (c) fats. (L)

6 Describe the general layout of the mammalian alimentary canal. Using *named* examples, show how this basic arrangement is modified to deal with the ingestion and digestion of different kinds of food. (L)

7 What kinds of food are taken in as small particles?
Using *named* examples, describe the mechanisms by which animals collect and selectively accept small particles of food. (L)

8 Give an account of ciliary feeding in a *named* animal. Compare ciliary feeding with one other specified type of feeding mechanism in another *named* animal, relating any differences to the way of life of the animals concerned. (L)

Further reading

Fogg G. E., *Photosynthesis* (London: English Universities Press, 1972).
Ministry of Agriculture, Fisheries, and Food, *Manual of Nutrition* (London: HMSO, 1976).
Sutcliffe J. F. and Baker D. A., *Plants and Mineral Salts*, Institute of Biology Studies in Biology No. 48 (London: Arnold, 1974).

7 Respiration

Gaseous exchange

The uptake of oxygen and the giving off of carbon dioxide during respiration occur as a result of the respiratory activities of the cytoplasm and the mitochondria within the cells of organisms. In this way energy is obtained from a variety of organic compounds.

Plants

Small plants with large surface area to volume ratios carry out gaseous exchange by diffusion over their entire surface. Larger plants, unlike larger animals, still tend to have a large surface area to volume ratio, due to the structure of the leaves and their pattern of branching growth. They therefore still carry out gaseous exchange by diffusion over most of their surface, via the stomata.[169]

Plants do not require as much energy per unit mass as animals do, since they have lower metabolic rates and, in woody plants especially, much of the tissue consists of dead xylem which does not respire. Thus plants do not carry out as much respiratory gaseous exchange per unit mass as animals. Furthermore, the consideration of respiratory gaseous exchange in higher plants is complicated, in the light, by the process of photosynthesis: respiration uses the oxygen released by photosynthesis and photosynthesis uses the carbon dioxide released by respiration. Indeed, respiration and photosynthesis are closely linked; both produce ATP[367] and both have many common intermediates and enzymes which interlink with all other biochemical pathways in the cell. However, exactly how closely they are interconnected in the living cell is hard to judge.

The two processes are *spatially* separated; photosynthesis in the chloroplastids and respiration in the mitochondria, but these organelles are often closely associated with each other. Some studies indicate that the two processes are completely separate, even to the extent that the oxygen produced in photosynthesis is not used in respiration. In this case, conversion of ADP + P to ATP in photosynthesis will not reduce the rate of respiration, which therefore continues at the same rate in the light as in the dark. In other types there is a 70 per cent reduction in the respiration rate in the light, indicating that in these cases the two processes are clearly linked. To complicate the picture further, C_3 plants have a light-activated photorespiration.[130]

Animals

Aquatic animals with a permeable surface and a sufficiently large surface area to volume ratio can use their whole body surface for gaseous exchange.

With increasing size, and thus decreasing surface area to volume ratio, transport systems are found and, although the entire surface can still act in gaseous exchange as in the earthworm, it is more usual to find special respiratory surfaces in both water- and air-breathing types. Respiratory surfaces must have a large surface area for exposure to the external medium, be thin and permeable for the rapid diffusion of gases, and be well supplied with blood capillaries.

AQUATIC RESPIRATION

Fish

The relatively low concentration of oxygen in water, when compared to air, means that gills must be much more efficient than the respiratory surfaces of air-breathing animals.

Gills have a very large surface area of delicate **gill lamellae** which are supported by the water.

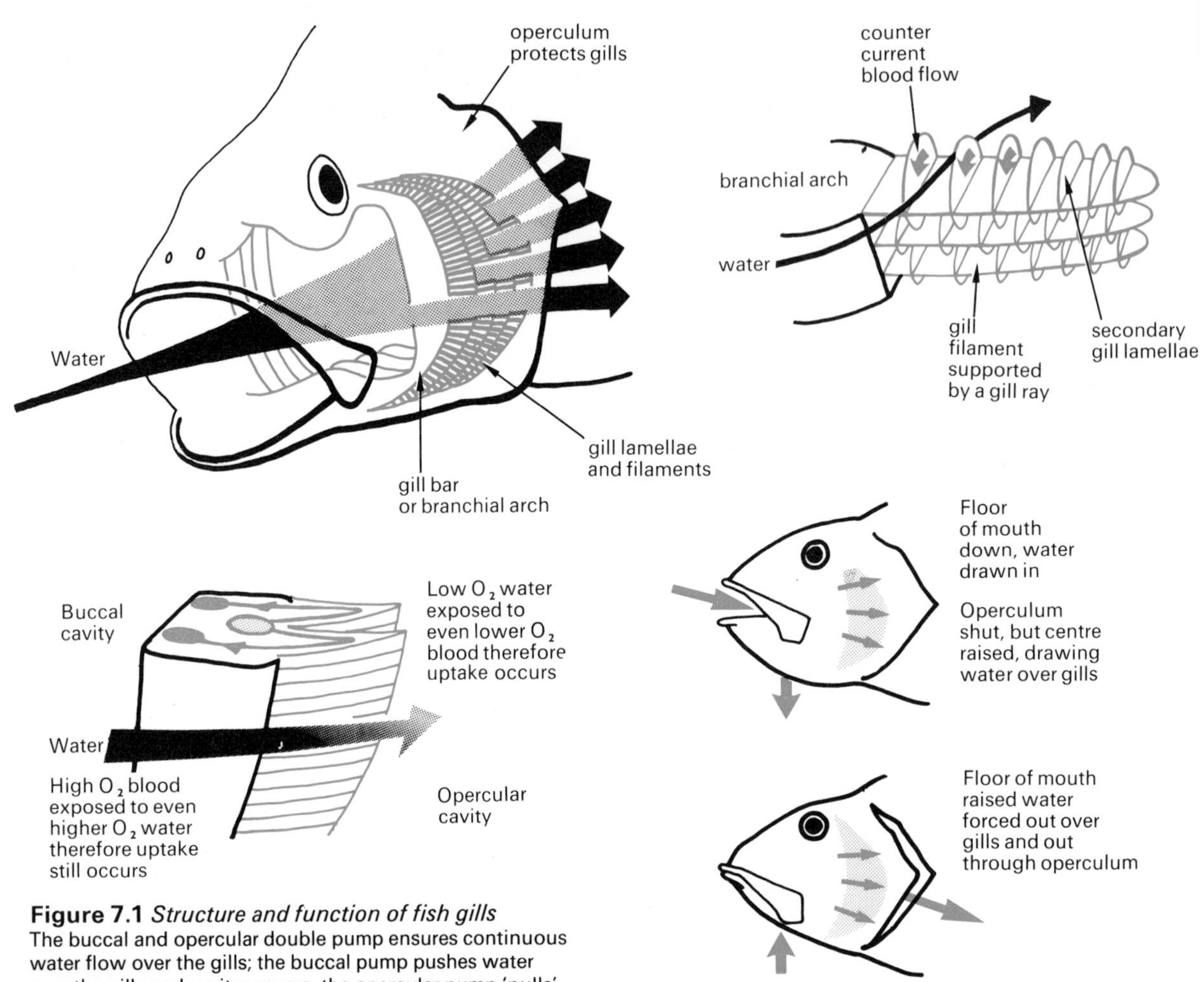

Figure 7.1 *Structure and function of fish gills*
The buccal and opercular double pump ensures continuous water flow over the gills; the buccal pump pushes water over the gills and, as it recovers, the opercular pump 'pulls' water through the gills by 'suction'. The through-flow of water provides sufficient 'ventilation' of the gills, and the counter-current exchange ensures high loading of the blood with oxygen.

There is obviously no risk of desiccation by evaporation, therefore there is no mucous layer produced. This shortens the diffusion distance between the oxygen in the water and the blood in the gill capillaries.

Gills have a counter-current blood supply which results in the efficient uptake of oxygen and removal of carbon dioxide. Carbon dioxide is also rapidly removed due to its relatively great solubility and fast diffusion in water. Powerful muscles are needed to pump large volumes of water over the gills; as much as 25 per cent of the absorbed oxygen can be used by these ventilating muscles. The through flow ventilation ensures that the whole of the gill surfaces are efficiently ventilated. Bony fish gills can extract up to 80 per cent of the oxygen present in the water.

Gills are also sites of **osmotic**[370] exchanges.

RESPIRATION IN AIR

Air is much easier than water to move over the respiratory surface, and it has a much higher oxygen content. Oxygen diffuses about three million times faster in air than in water. However, any large, thin, permeable, well vascularized surface exposed to the air will lose water by diffusion and evaporation. To prevent this loss, mucus is usually secreted over the respiratory surface. This static layer of water and mucus increases the diffusion distance between the air and the blood, and therefore decreases the efficiency of the surface. Furthermore, air does not support such delicate surfaces as does water. Hence air-breathing animals tend not to have such large permeable surfaces as are found in the gills of aquatic animals of comparable size.

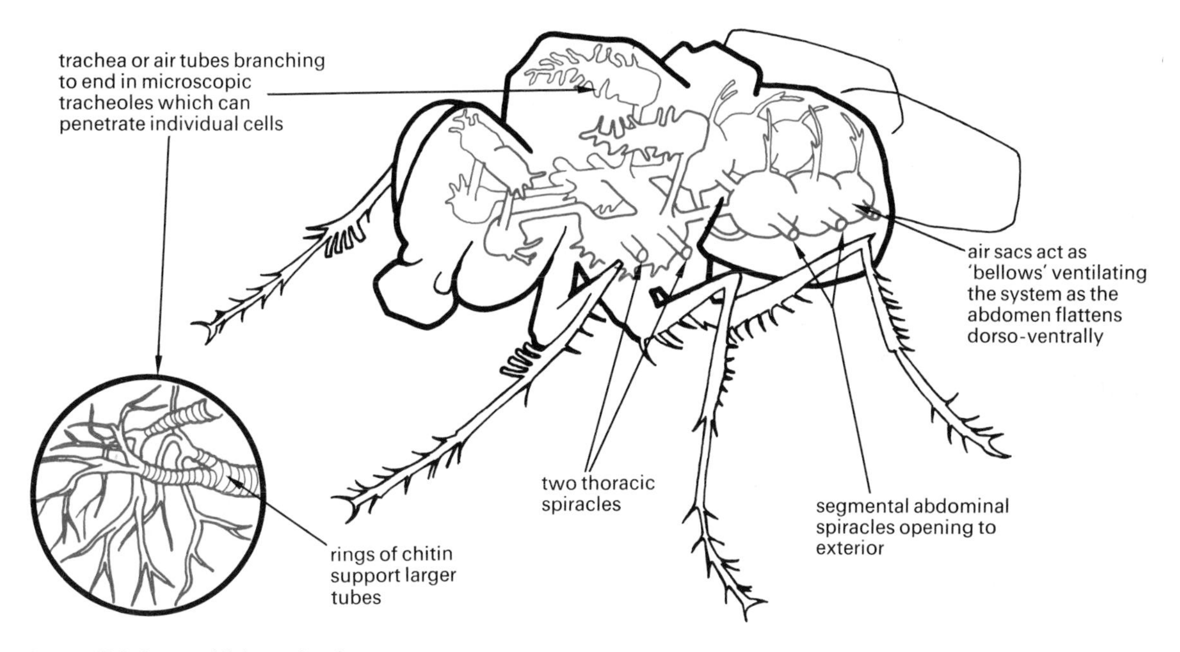

Figure 7.2 *Insect (fly) tracheal system*
The air sacs act as 'bellows' ventilating the system as the abdomen flattens dorso-ventrally. The controlled opening and closing of the spiracles in conjunction with the ventilating movements leads to through-flow of air from the anterior to the posterior.

Insects
The insects are unique in having a gas exchange and transport system separate from the blood system. This system of air tubes, or **trachea**, gives direct and rapid transport of oxygen in the gaseous state to the respiring tissues. This efficient transport of oxygen allows a high metabolic rate, and helps to explain why the insects are the only group of invertebrates to achieve flight.

The trachea open to the exterior by controlled vents called **spiracles**. When the insect is at rest, the movement of air within the system is mainly by simple diffusion. During exercise, muscular contraction telescopes the abdomen, causing dorso-ventral changes in volume. These volume changes result in internal pressure changes, which act on thin-walled air sacs in the trachea to move air through the system. By opening and closing the anterior spiracles in co-ordination with the more posterior ones, through flow of air is achieved. This is much more efficient than in and out tidal flow.

The trachea branch into **tracheoles** which can eventually penetrate individual cells. At rest, the tracheoles are filled with fluid which acts as a diffusion barrier between the cells and the oxygen in the tracheoles. During exercise, when the oxygen demands are greater, the fluid is withdrawn from the tracheoles by osmosis due to the rapidly metabolizing cells accumulating osmotically active waste products. These draw the water into the cells via their semi-permeable membranes and enable oxygen to reach the cells directly in gaseous form.

Carbon dioxide is removed mainly by the blood and is diffused out through the exoskeleton, which is permeable to gases.

Amphibia
The frog has two main respiratory surfaces: the lungs and the skin. The lungs are simple, with a relatively small surface area (for each millilitre of air in the lungs there is about 20 cm^2 of lung surface, compared to about 300 cm^2 to every millilitre of air in man). There is no thorax, and the lungs hang in the abdominal cavity. They are ventilated by the **buccal pump**, a pumping movement of the floor of the mouth.

The mouth and oesophagus remain closed throughout the following sequence of ventilating movements, which will, for convenience sake, be considered as starting when the lungs are empty.

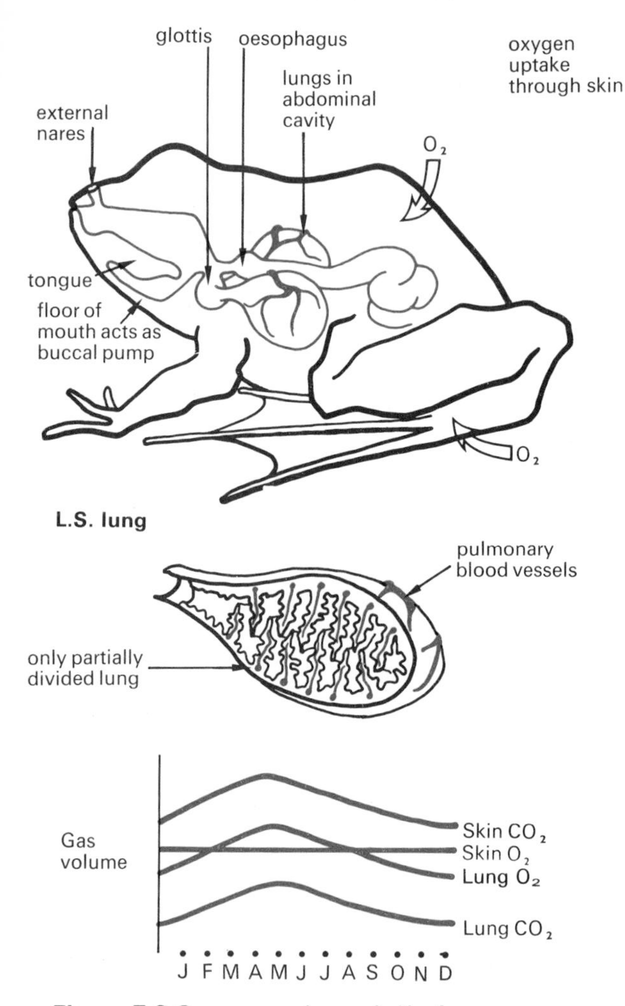

Figure 7.3 *Gaseous exchange in the frog*

(a) The nostrils (**nares**) open, the floor of the buccal cavity is lowered, and air is drawn into the buccal cavity.
(b) The nostrils shut, the floor of the buccal cavity is raised, and air is forced into the lungs.
(c) The floor of the buccal cavity is lowered again and air is drawn out of the lungs.
(d) The nostrils open, the glottis shuts off the lungs, and the floor of the buccal cavity is raised, forcing air out of the nostrils.

The skin acts as a respiratory surface, and has an extremely well-developed blood supply via the large cutaneous arteries and the extensive capillary beds. In addition, the heart has a single ventricle which allows most of the oxygenated blood returning from the skin to be pumped directly around the body without first going to the lungs. Due to the much higher concentration of oxygen in air than in water, the skin can absorb more oxygen when the frog is exposed to air than when it is submerged.

Some amphibia are completely aquatic, and some are almost completely terrestrial. Surprisingly, the lungs tend to be more important in completely aquatic types, where they are used to supplement the rather poor cutaneous uptake.

Reptiles

Reptilian lungs have an increased internal surface area when compared to amphibian lungs, due to the presence of **alveoli**. There are ribs, but no true muscular diaphragm. Some cutaneous respiration takes place in secondarily aquatic types, and in some lizards 85 per cent of the carbon dioxide is lost via the skin.

Birds

Birds have small, compact lungs consisting basically of a system of air tubes. Extending from these are large, non-respiratory air sacs which penetrate the wing bones and the abdominal cavity. By allowing air to pass right through the lungs, the ventilation is efficient, and by occupying space that would otherwise be fluid-filled, the air sacs decrease the weight of the body for flight. They also dissipate body heat into the expired air during flight.

Birds possess ribs, but no true diaphragm; however, ventilation of the lungs is aided by contraction of the flight muscles during flight.

Mammals

Mammalian lungs have a larger internal surface area than those of other vertebrates, and the possession of a true muscular diaphragm, forming a complete thoracic cavity, aids the efficiency of ventilation. However, the problems of tidal flow decrease their efficiency when compared to those of the birds.

Table 7.1 *Composition of inspired and expired air*

Inspired air	*Expired air*
21% oxygen	17% oxygen
79% nitrogen	79% nitrogen
0.03% carbon dioxide	4% carbon dioxide
Variable water vapour	Saturated water vapour
Atmospheric temperature	Body temperature
Bacteria	Bacteria

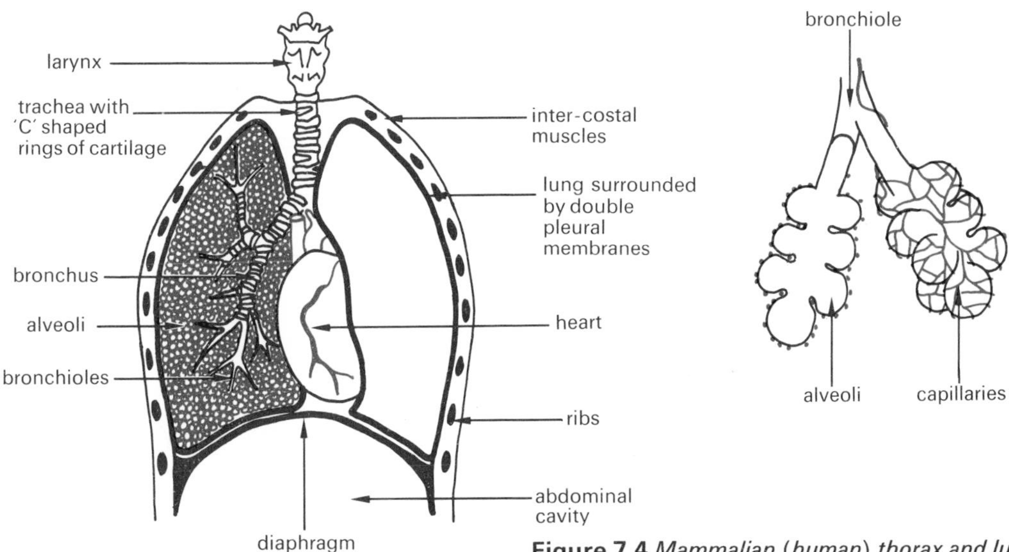

Figure 7.4 *Mammalian (human) thorax and lungs*
The lung surface is coated with a surfactant which prevents the collapse of the alveoli on expiration. The pleural membranes and pleural fluid hold the lungs on to the thoracic cavity wall and prevent friction during breathing movements.

One such problem is the so-called **dead space** of the respiratory system, which consists of two components. The **anatomical dead space** is that part of the system not vascularized for oxygen uptake, namely the mouth, pharynx, and bronchial tree as far as the terminal bronchioles. The **physiological dead space** is composed of those areas of lung which are undersupplied with blood in relation to their ventilation.

When standing at rest, about 30–50 per cent of the lungs may be undersupplied during some parts of the cardiac cycle, but during exercise few, if any, parts of the lungs are undersupplied and the physiological dead space is negligible.

Figure 7.5 *Mammalian (human) ventilation and lung capacities*
Inspiration: intercostal muscles contract, ribs and sternum move up and out, diaphragm contracts and moves down, the volume of the thorax is increased. Therefore the internal thoracic pressure is decreased and the external atmospheric pressure forces air into lungs.
Expiration: intercostal muscles relax, ribs and sternum move down and in, diaphragm relaxes and moves up, the volume of the thorax is decreased. Therefore the internal thoracic pressure is increased and the air is forced out of lungs.

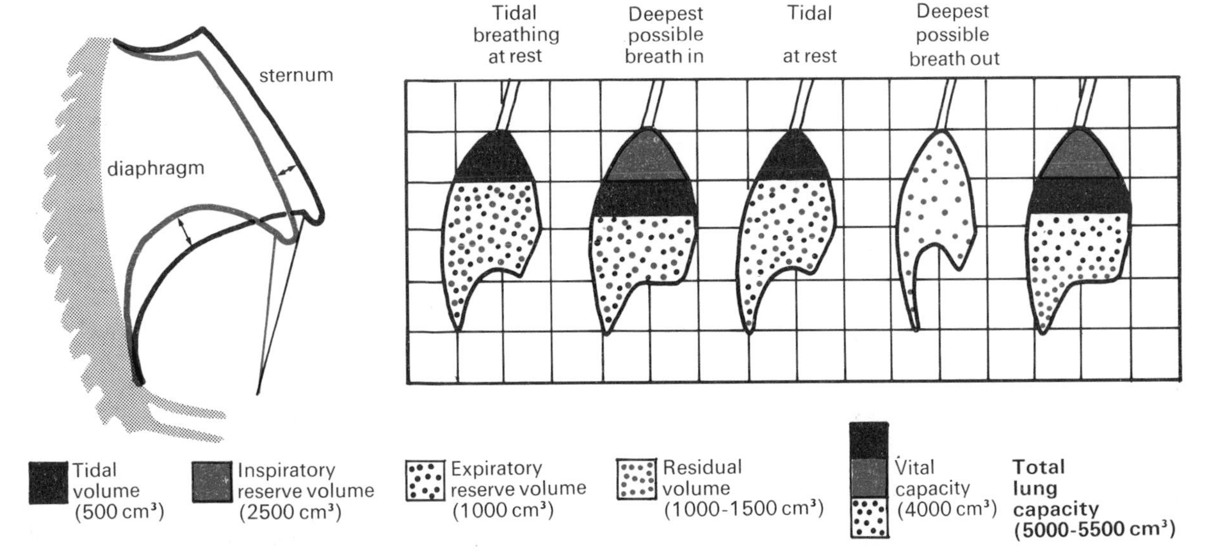

Control Normally respiratory movements are involuntary, being controlled by rhythmic discharges of nerve impulses from the respiratory control centres in the pons of the hind brain[199] and in the medulla oblongata.[199] These impulses travel down the phrenic nerves to the diaphragm, and down the intercostal nerves to the intercostal muscles.

The rate of respiration is dependent mainly upon the carbon dioxide concentration of the blood, which has a direct effect upon the respiratory centres in the medulla oblongata. The chemoreceptors in the carotid[181] and aortic[181] bodies are only affected by high levels of blood carbon dioxide when the level of oxygen in the blood is low. Both an increase in body temperature and a decrease in the oxygen concentration of the blood increase the sensitivity of the respiratory centres to the carbon dioxide.

When the alveoli are stretched at the end of inspiration, stretch receptors in the bronchioles send impulses to the inspiratory control centre, which cause the respiratory muscles to relax. This leads to expiration and is known as the Hering–Breuer reflex. In normal breathing expiration is passive, i.e. it requires no muscular effort, but during exercise expiration becomes active.

Various reflexes, such as swallowing, coughing, hiccuping, and yawning, can over-ride the basic rhythmic reflexes. The voluntary control of breathing is an essential part of speech in man.

Transport of respiratory gases

OXYGEN TRANSPORT

Oxygen transport invariably involves a blood pigment. All such pigments contain a metal in an organic complex. Most pigments operate continuously, but some store oxygen which is released only under conditions of low oxygen concentration, thus acting as an emergency system for oxygen transport. Blood pigments also serve as buffers[363] controlling the pH of the blood, and in the transport of carbon dioxide.

There is a wide variety of pigments found throughout the animal kingdom.

Haemocyanin, with copper as its metal, occurs in the plasma of many molluscs and arthropods.

Haemerythrin, with iron as its metal, is found in the corpuscles of some annelida.

Chlorocruorin, with iron as its metal, is found in the plasma of some annelida.

Haemoglobins, containing iron, are the most widespread of all pigments and are found throughout the animal kingdom. It would seem that the haemoglobin type of molecule has arisen many times with similar haem- or iron-containing groups and different protein portions. Invertebrate haemoglobin is thought, by some, to be sufficiently different in structure to merit the separate name of **erythrocruorin**.

Thus the haemoglobins are a group of pigments which are not necessarily similar in structure.

Each molecule of mammalian haemoglobin consists of four sub-units, and each sub-unit consists of an iron-containing haem group combined with a protein globin part. Each sub-unit combines reversibly with one molecule of oxygen. Thus one molecule of haemoglobin can carry four molecules of oxygen.

The formation of oxyhaemoglobin from deoxyhaemoglobin is thought to involve a physical holding of the oxygen by the haemoglobin sub-units, as a result of their special 3-dimensional structure. The oxygen is easily displaced by carbon monoxide, which combines irreversibly with haemoglobin to form **carboxyhaemoglobin**. In man each red blood corpuscle contains about 300 000 000 molecules of haemoglobin.

Haemoglobin must associate with oxygen at the respiratory surface, and dissociate or release it at the respiring tissues. The amount of oxygen uptake and release depends on a complex of factors, including the concentration or partial pressure of oxygen, the concentration of carbon dioxide, and the type of haemoglobin. The behaviour of haemoglobins under differing conditions can be illustrated by oxygen equilibrium curves.

Myoglobin is a special type of haemoglobin molecule found in some muscles, and is similar in structure to just one sub-unit of haemoglobin and thus only combines with one molecule of oxygen. However, it has a greater affinity for oxygen than haemoglobin and will therefore take oxygen from it, thus acting as a 'middle man' for the transfer of oxygen between the muscle tissue and the haemoglobin of the blood.

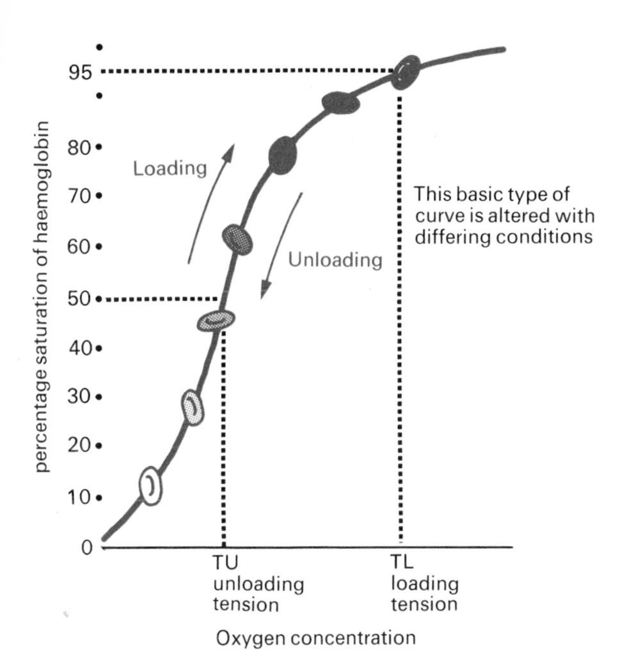

Figure 7.6 *Association/dissociation curves of mammalian haemoglobin*
The most important difference between the haemoglobins of different animals is in the oxygen concentrations at which they load and unload. This feature is related to the environmental conditions in which the different animals live.

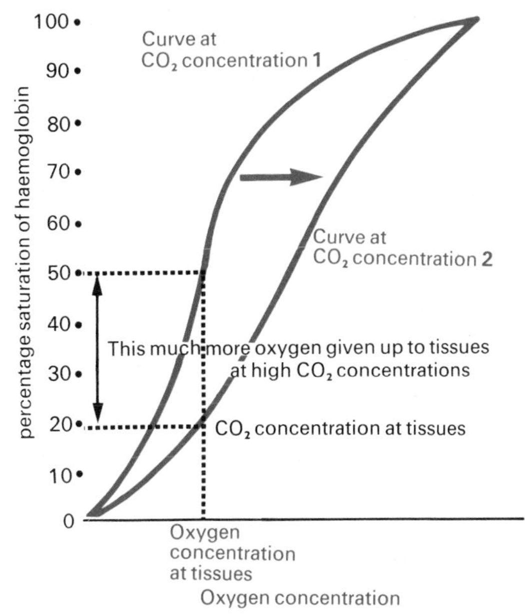

Figure 7.7 *The Bohr shift of mammalian haemoglobin association/dissociation curves*
The curve at CO_2 concentration 2 is 'shifted to the right' by an increased CO_2 concentration. This is known as the Bohr shift and results in the blood releasing oxygen more easily to the tissues at low oxygen concentration.

Figure 7.8 *Association and dissociation curves of mammalian maternal and foetal haemoglobin*
Foetal haemoglobin loads up higher (A) from lower oxygenated haemoglobin in maternal placental capillaries (B). This is a significant factor in viviparity.

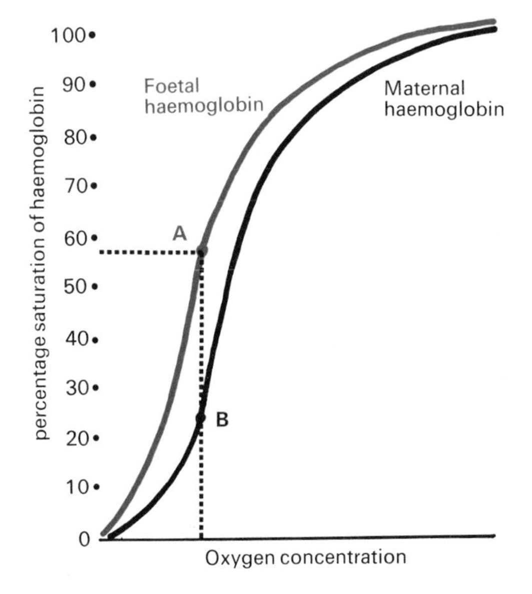

Figure 7.9 *Association and dissociation curves of mammalian myoglobin*
Myoglobin shows no Bohr shift; if it did its curve would move towards that of haemoglobin, and the advantageous effect of the higher affinity for oxygen would be lost. Myoglobin loads up high with oxygen from low oxygen haemoglobin and acts as an oxygen store for exercising muscle.

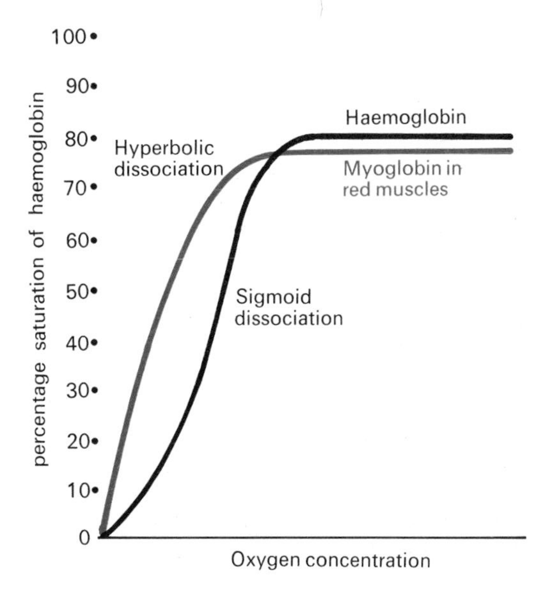

CARBON DIOXIDE TRANSPORT

Carbon dioxide is carried in mammalian blood in three ways. About 10 per cent is carried in solution in the plasma as carbon dioxide and as carbonic acid (H_2CO_3), about 80 per cent is carried as sodium bicarbonate ($NaHCO_3$) and potassium bicarbonate ($KHCO_3$) in the red corpuscles, and about 10 per cent is carried in combination with the protein part of haemoglobin as **carbamino-haemoglobin**, and to a lesser extent in combination with the proteins of the plasma. It is released from the blood in the lung capillaries and diffuses out into the alveoli, from where it is exhaled.

Outline biochemistry of cellular respiration

Cellular respiration is a biochemical activity carried out in the cytoplasm of all cells, and results in the transfer of energy from complex organic compounds to the energy-carrier molecules of adenosine triphosphate (ATP). This 'trapped' energy is subsequently used in the metabolism of the cell. The main site of ATP formation in animal cells are the **mitochondria.**[5]

Green plants also produce ATP in their chloroplasts during photosynthesis for the production of complex organic compounds; plant cells therefore have fewer mitochondria than animal cells.

The process of energy transfer from organic compounds to ATP involves many complex enzyme-controlled chemical reactions in which the organic compounds are oxidized to CO_2 and H_2O along a variety of pathways. However, the main pathway for the oxidation of starch in plants, glycogen in animals, and glucose in both, is the **Embden–Meyerhof–Parnas (EMP) pathway** leading into the **Kreb's or citric acid cycle.**

The EMP pathway is also known as **anaerobic glycolysis** as it can proceed in the absence of oxygen. This anaerobic part of respiration occurs in the general cytoplasm of the cells outside of the mitochondria, and yields relatively little energy.

The pyruvic acid formed as an intermediate enters the mitochondria and, if sufficient oxygen is present, is completely oxidized to CO_2 and H_2O, with a corresponding transfer of large amounts of energy to ATP via the Kreb's cycle. The oxidations involved are all due to the removal of hydrogen from the substrate. This hydrogen is passed down a series of carriers and eventually combines with oxygen to form water. It is during this process that energy is transferred into ATP.

Other respiratory substrates, for example fats and proteins, can also be oxidized via interconnections with these pathways, indeed there are complex links between all aspects of carbohydrate, fat, and protein metabolism, including synthetic reactions.

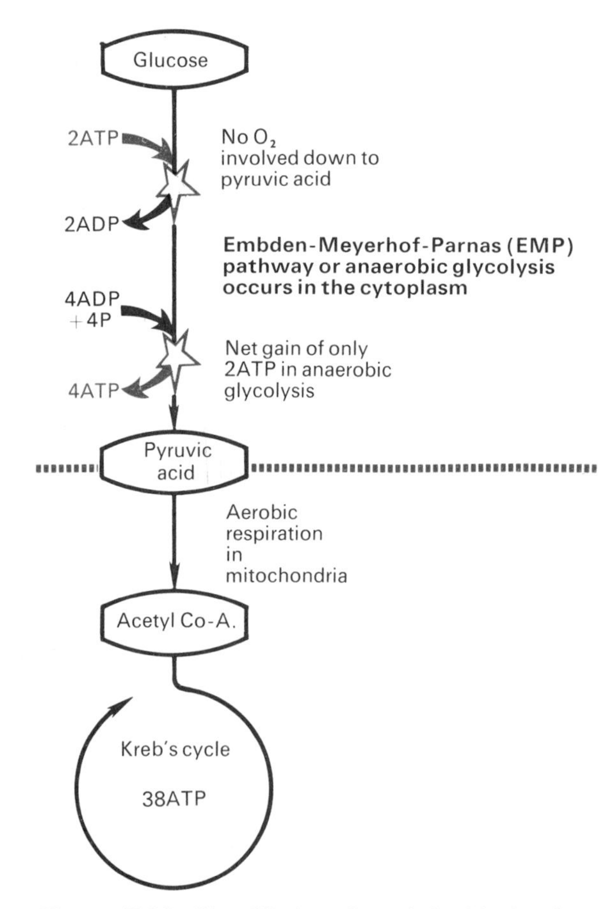

Figure 7.10. *Simplified outline of the biochemistry of respiration*

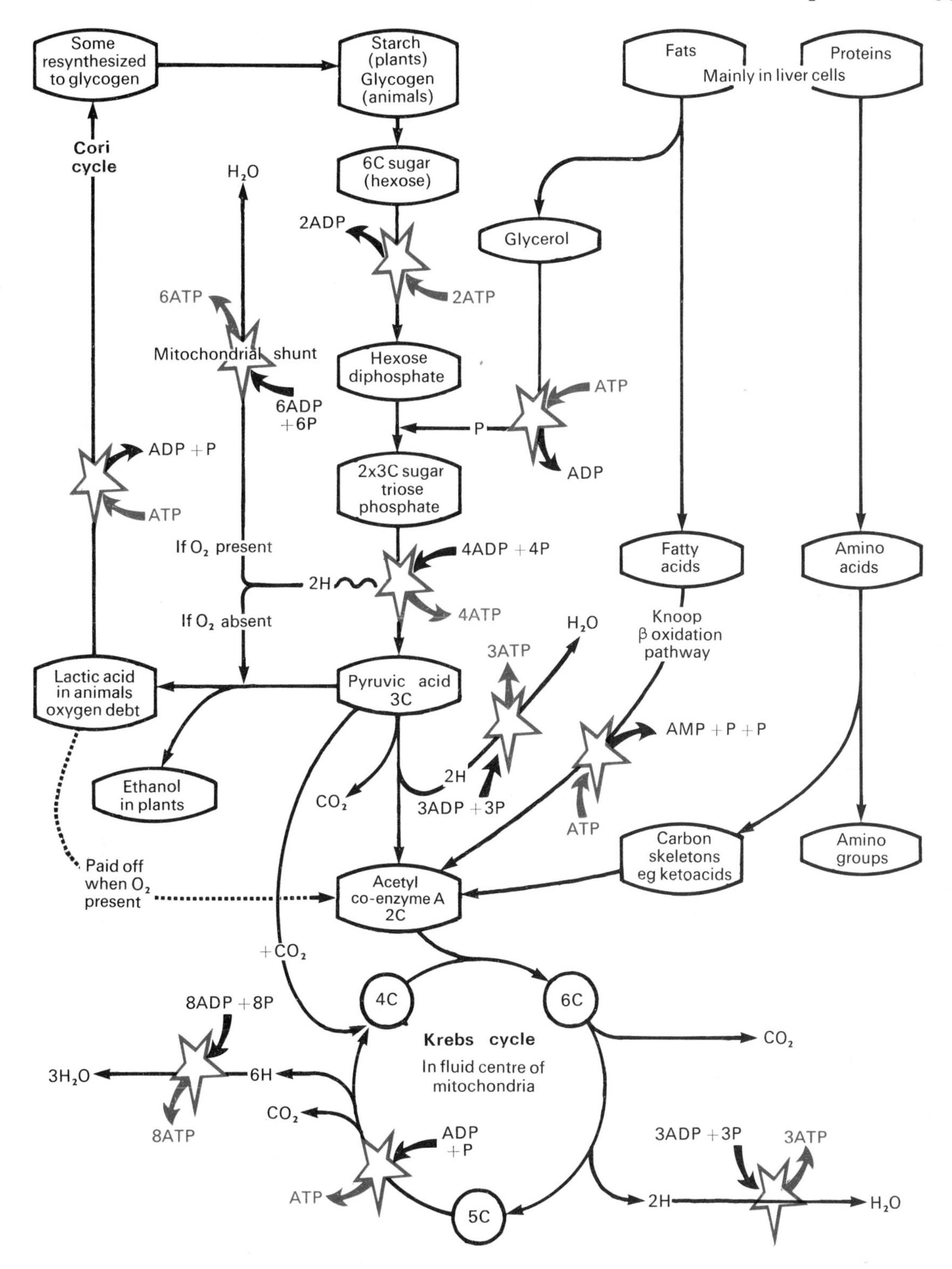

Figure 7.11 *Biochemistry of respiration*
All the reactions are enzyme-controlled. Each molecule of glucose gives 2 turns of the cycle.

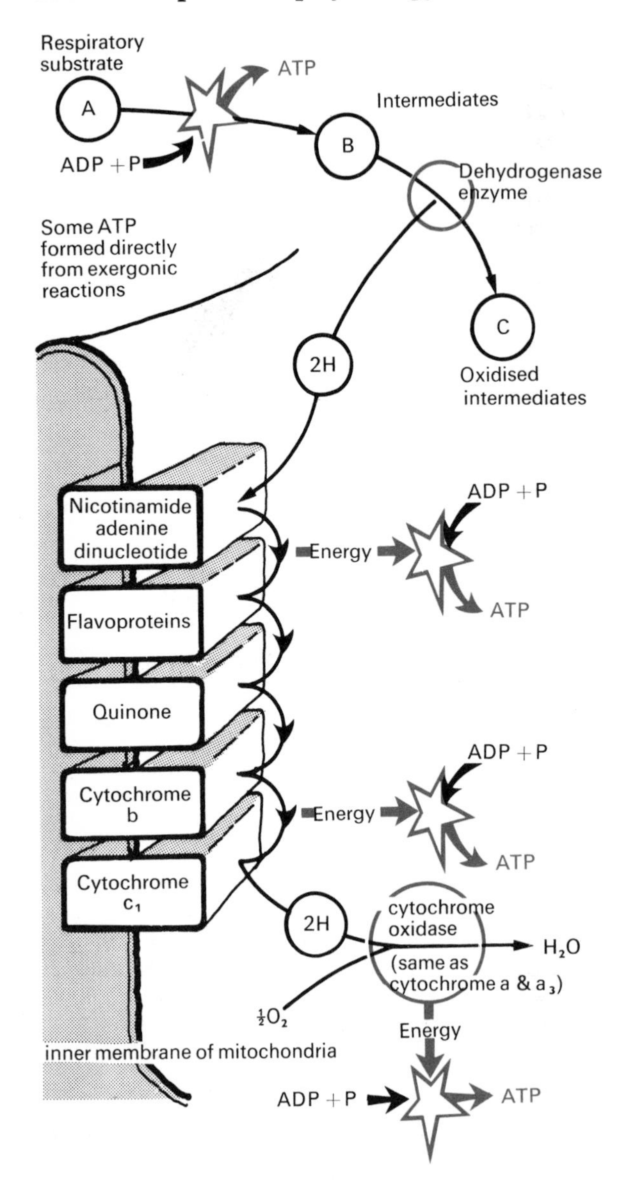

Figure 7.12 *The cytochrome chain and ATP formation*
The cytochrome chain is located in the inner membranous cristae of the mitochondria. The 'links' are arranged in an ordered sequence. For oxidation to take place via the cytochrome chain, ADP + P must be present. When all the ADP + P is converted to ATP, oxidation stops, and in this way ATP is made at the same rate as it is required. Little ATP is stored as such.

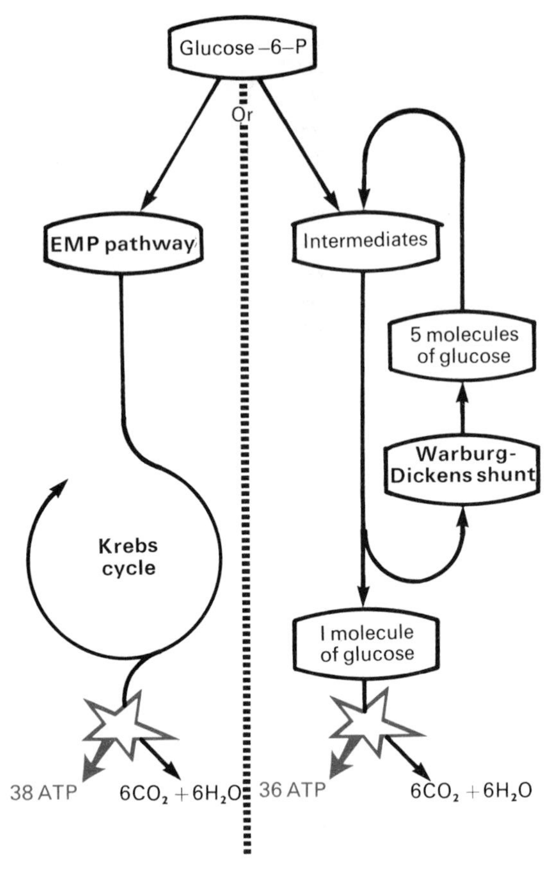

Figure 7.13 *Alternative respiratory pathway*
The pathway involving the Warburg–Dickens shunt is an additional pathway found in many organisms. There are fewer reactions and fewer enzymes. It shows certain similarities with the pentose shunt in photosynthesis.

Efficiency of respiration

The efficiency of respiration is calculated by expressing the amount of energy which is successfully transferred into ATP as a percentage of the amount which can be liberated by the complete oxidation of that substance. That energy which is not trapped in ATP is lost as heat, and although in this context this is considered to be inefficient, this heat aids metabolism by increasing the rate of chemical reaction, especially in homoiotherms.[244] The efficiency of respiration is difficult to calculate in the living cell as a whole since there are so many interconnecting pathways, but it is possible to calculate figures for the oxidation of a mole of substrate, for example:

Glucose contains about 686 kcals mol^{-1}.
The oxidation of one mole via the EMP and Kreb's pathway yields 38 mole of ATP.
Each mole of ATP traps 8 kcals.

$$\therefore \text{percentage efficiency} = \frac{38 \times 8}{686} \times 100 = 41\%$$

Respiratory quotient (RQ)

The respiratory quotient is the ratio of the amount of carbon dioxide produced to that of oxygen taken up during the oxidation of a substrate in unit time. It provides some measure of the main substrate being respired at any given time, for example:

$$\underset{\text{glucose}}{C_6H_{12}O_6} + \underset{\text{oxygen}}{6O_2} \longrightarrow \underset{\text{carbon dioxide}}{6CO_2} + 6H_2O$$

$$\therefore \quad RQ = \frac{6CO_2}{6O_2} = 1$$

Fats have a RQ of 0.7, but proteins vary so much in composition that estimates for the RQ range between 0.5 and 0.8. Anaerobic respiration will have a RQ of infinity as there is carbon dioxide output but no oxygen uptake, for example:

$$C_6H_{12}O_6 \longrightarrow 2C_2H_5OH + 2CO_2$$

$$\therefore \quad RQ = \frac{2CO_2}{0} = \infty$$

However, mixtures of substrates and of aerobic and anaerobic respiration will prevent any simple conclusion being drawn as to the nature of the substrate being used.

Metabolism

The term metabolism refers to the sum total of all the chemical reactions necessary for the maintenance of life. It can be divided into two major sub-divisions, catabolism and anabolism.

Catabolism refers to those reactions involved in the breakdown of complex substances into simpler end products, and includes digestion and respiration. Catabolic reactions *release* energy.

Anabolism refers to those reactions involved in the synthesis of complex substances from simpler ones, and includes the building up of storage, structural, and functional materials. Anabolic reactions *require* energy.

Metabolic rate

Such are the complexities of the reactions involved in metabolism that it is impossible to measure the metabolic rate itself. It is therefore usually considered in terms of the energy released by the body, which is assumed to be proportional to the total metabolic rate, and which is determined indirectly from knowledge of the respiratory gaseous exchange.

It is assumed that all the energy produced comes from aerobic respiration of carbohydrates and fats, and the RQ is assumed to be 0.75, which corresponds to an energy output of 20.14 kJ/l O_2 (4.82 kcal/l O_2) consumed. The oxygen consumption per minute can be measured, and consequently the heat production per minute can be calculated, for example:

RQ = 0.75
Energy output = 20.14 kJ/l O_2 consumed
Oxygen uptake per minute = 0.25 dm^3
$\therefore$ heat production per minute

$$= 0.25 \times 20.14 \text{ kJ}$$
$$= 5.03 \text{ kJ min}^{-1}$$

For comparison purposes between animals of different sizes the metabolic rate is expressed as kJ m^{-2} of surface per hour.

BASAL METABOLISM

Basal metabolism is the metabolism required to maintain the basic life processes, which include the functioning of the nervous system, the heart and circulation, the glands, the respiratory system, and the production of heat to maintain the body temperature.

To determine the basal metabolic rate (BMR) the factors that affect metabolism must be eliminated or standardized as much as possible. The BMR is therefore measured under **standard resting conditions**; that is whilst asleep, twelve to eighteen hours after eating, in an equable temperature at which there is no heat loss or gain. It is assumed that the energy produced under these conditions is that required to maintain the basic life processes.

FACTORS AFFECTING THE METABOLIC RATE

Many factors affect the rate of metabolism.

(a) The surface area to volume ratio is important since it is the volume of the body that produces the energy, most of which is lost as heat over the surface area. The larger the surface area in relation to volume, the faster must be the metabolic rate to replace the losses.

(b) Metabolism is faster in the young and decreases with age.

(c) The hormones adrenalin and thyroxine have a powerful stimulating effect on the metabolism.

(d) Eating increases the metabolic rate. This is partially due to the stimulation of digestion and absorption, but mainly due to the increase in the activity of the liver in dealing with the absorbed end products of digestion. (This stimulation of metabolism by eating is known as the **specific dynamic action** (SDA) of food.)

(e) Muscular activity increases metabolism.

(f) If the external temperature fluctuates beyond the efficiency range[246] then metabolism is increased.

Many other factors such as sleep, emotions, and infection, and, in the female, pregnancy, menstruation, and lactation, also affect the rate.

Experimental and applied aspects

The elucidation of the biochemical pathways of cellular respiration involves the use of the same experimental techniques as the analysis of the biochemical pathways of photosynthesis, including, for example, radio-active tracers,[147] chromatography,[147] and enzyme inhibitors.

Enzyme inhibitors block specific enzyme-catalysed reactions and lead to the accumulation of intermediates which can be analysed and identified. Thus, if an inhibitor blocks a particular step in the metabolic pathway, the intermediate compound immediately preceeding that step will accumulate first, the one before that slightly later, and so on down the chain.

A —enzyme 1→ B —enzyme 2→ C —enzyme 3 (inhibited)→ D —enzyme 4→ E

A → B → C C accumulates first

A → B → C B accumulates slightly later

A → B → C A accumulates slightly later

Other biochemical clues, such as similarity between compounds (for example in the above diagram A and B will be closer in chemical structure than A and C), further help the build-up of the correct interpretation of a sequence of reactions.

With respect to gaseous exchange during respiration, some simple experimental techniques can be used which are based on one or other of two procedures, namely the detection of pH changes, and the measurement of volume (or pressure changes). In both cases, when working with plant material, there are problems in relation to the gaseous exchanges of photosynthesis.

Those experiments involved with the detection of **pH changes** are based on the fact that carbon dioxide, which is evolved during respiration, is a so-called 'acid-gas', giving rise to carbonic acid in solution. Hence, by the use of indicator solutions and titration methods, the amount of CO_2 evolved under various circumstances can be measured. This in turn gives a measure of the rate of respiration.

Those experiments involved with the detection of **volume changes** simply measure changes in the volume (or pressure) within a closed system which occur as a result of an organism's respiration. The **respirometer** or spirometer is a piece of apparatus which measures those changes. It consists basically of a chamber of six to seven litres capacity, which is sealed with a water seal so that the chamber can move up and down without at any time being open to the atmosphere. The chamber is counterbalanced so that it moves only in response to changes in volume in the apparatus, and not in response to its own weight.

The apparatus is filled with oxygen, and the subject breathes in and out through a valve which ensures unidirectional flow through the apparatus. A cylinder of CO_2 absorbent material removes CO_2 from the exhaled air and prevents it re-entering the chamber. The pen records the up-and-down movement of the chamber as a trace on calibrated paper on the drum of the kymograph, moving down when the subject breathes in, and up when the subject breathes out. The calibrations are in 250 ml squares, and allow the actual volume changes in the chamber to be monitored. The kymograph drum is set to rotate at a convenient speed (e.g. 20 mm per min), which allows the timing to be monitored.

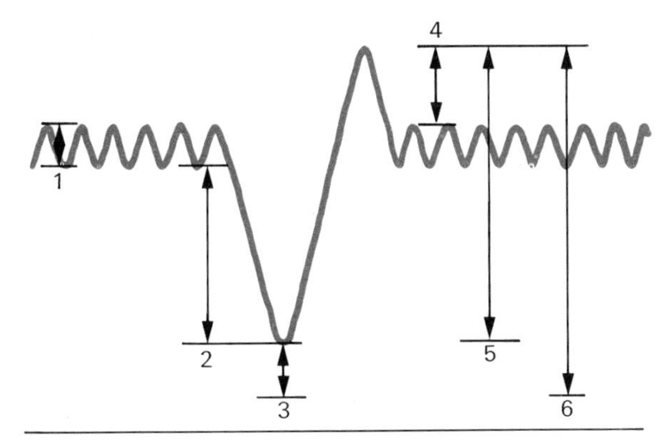

Figure 7.14 *Kymograph trace showing respiratory volumes*
(Pen lever goes down on breathing in and up on breathing out.)
1 Tidal volume. 2 Inspiratory reserve volume. 3 Residual volume. 4 Expiratory reserve volume. 5 Vital capacity. 6 Total lung capacity (estimated by multiplying the expiratory reserve volume by 6). The 'dead space' is estimated by finding the weight in pounds, and this figure (in cm^3) is taken as the volume of the 'dead space'.

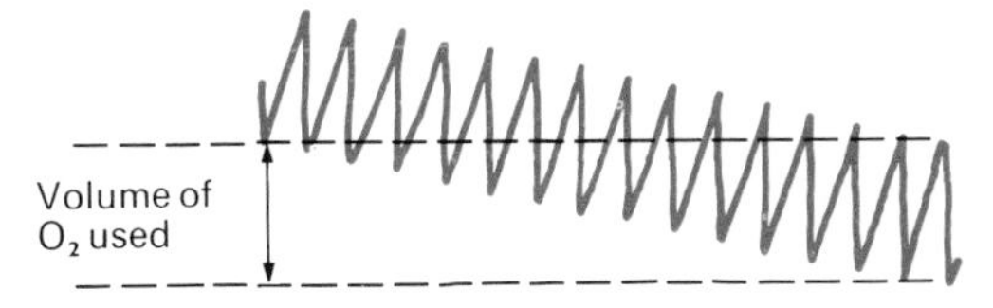

Figure 7.15 *Kymograph trace showing oxygen uptake*
Subject breathes tidally for a period of one or more minutes; the CO_2 is absorbed by the absorbent. The volume of CO_2 evolved is the same as the volume of O_2 uptake, therefore the amount by which the chamber drops is equal to the volume of oxygen uptake.

Guided example

There are many problems for air-breathing animals diving to any depth, and due to their more rapid metabolism these problems are greater for diving birds and mammals than for reptiles.

Maximum diving duration times

Alligator	2 hours
Sperm whale	90 minutes
Blue whale	15 minutes
Seal	15 minutes
Razor bills	52 seconds
Man breathing air	3 minutes
Man saturated with oxygen	13 minutes

1 Being air-breathing, all diving tetrapods must take a supply of oxygen down with them for use during the dive. In what form, and in what places within the body, will this oxygen be carried?
All types take down a certain volume of oxygen in a gaseous form in the lungs, and in a combined state in the blood, tissue fluids, and muscles. The various components of this oxygen store vary in importance in different animals, thus the blood store is most important in the seal, and the muscle store important for the whale.

2 Many diving tetrapods have large blood volumes, high red blood cell counts, and large stores of myoglobin, but the oxygen carried is not enough for aerobic respiration during a long dive. A seal can carry enough oxygen for about 5 minutes and yet can remain submerged for 15 minutes. How can this be achieved?
The extra time is achieved by anaerobic respiration, which leads to the accumulation of lactic acid in the tissues. The peripheral circulation to the skin and muscles is shut down and blood is shunted to the brain. The muscles continue to respire anaerobically until their oxygen store in the myoglobin is depleted; they then begin to accumulate lactic acid. This lactic acid must eventually be oxidized using oxygen, and thus represents what is known as an oxygen debt.

3 When is it possible for the oxygen debt to be paid off?
The oxygen debt is paid off on surfacing, when the peripheral circulation is reopened and the lactic acid released as a surge into the blood. Divers are capable of very rapid oxygen uptake, thus seals can absorb about 50 per cent of the available oxygen in the air, about three times the average figure for mammals.

4 What other substance will also accumulate during the dive, as a result of respiration?
During submergence carbon dioxide also accumulates, and as this usually stimulates breathing, divers are less sensitive than other vertebrates to this rise in blood carbon dioxide concentration.

5 Submergence also causes **bradycardia**, the reflex slowing of the heart, e.g. in the seal the rate drops from 80 beats per min to 10 beats per min. Why is this rather unexpected?
The dive is a very active period, and under normal circumstances increased activity increases the heart rate. It is also a period when the CO_2 concentration increases, which normally acts as a stimulus for an increased heart rate in other vertebrates.

6 Pressure increases by one atmosphere (14.7 pounds per square inch) for every 10 m (33 ft) descended and this causes the nitrogen in the air in the lungs to be forced into solution in the blood. If man surfaces rapidly this dissolved nitrogen comes out of solution as small bubbles in the blood. These bubbles interfere with the circulation, giving rise to the condition known as the 'bends'. How might this problem be overcome in diving mammals, all of which surface rapidly?
Diving mammals overcome this problem by avoiding having large volumes of air in contact with the vascular respiratory surface of the lungs. For example, some whales regularly descend to great depths (greater than 1000 m) and thus face considerable pressure problems. Their lungs are not large in proportion to their body and are not completely filled with air on diving. As the pressure rises and the lungs become compressed, the air is displaced into the bronchi and trachea, and the alveolar surface becomes reduced and thickened. Hence the amount of nitrogen diffusing into the blood is small.

Questions

1 What are the properties of a respiratory surface? Show how these properties are exemplified by the gas-exchange structures of (a) a fish, (b) an earthworm, and (c) a mammal. (L)

2 Survey briefly the characteristics of a respiratory surface.
By reference to a *named* aquatic animal and a *named* terrestrial animal, compare (a) the nature of the respiratory surfaces and (b) their modes of functioning. (L)

3 (a) What are the principal properties of an efficient blood pigment?
(b) Name *two* blood pigments and state a group of animals in which each pigment may be found.
(c) The graph below shows the relationship between the percentage oxygen saturation for two blood pigments and the oxygen concentration in the surrounding air.

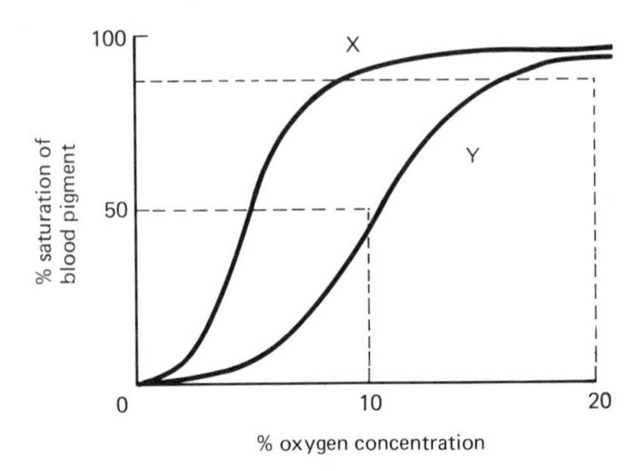

(i) What will be the effect on the saturation of the blood pigments X and Y, of reducing the external oxygen concentration from 20 per cent to 10 per cent?
(ii) Which of the pigments would be more suitable for a mud-dwelling animal? Briefly give reasons for your answer. (L)

4 (a) Describe the methods and mechanisms of gas exchange in (i) a green plant, (ii) an insect, (iii) a fish, and (iv) a mammal.
(b) Compare the properties of air and water as respiratory media. (L)

5 (a) Describe how oxygen reaches the respiring cells in (i) a tree, (ii) an insect.
(b) How are the processes of gas exchange and transport related to overall size in a tree and an insect? (L)

Further reading

Bryant C., *The Biology of Respiration*, Institute of Biology Studies in Biology No. 28 (London: Arnold, 1977).

Hempleman H. V. and Lockwood A. P. M., *The Physiology of Diving in Man and Other Animals*, Institute of Biology Studies in Biology No. 99 (London: Arnold, 1978).

Hughes G. M., *Comparative Physiology of Vertebrate Respiration* (London: Heinemann, 1962).

8 Transport

TRANSPORT IN PLANTS

Surface area to volume ratios are important in determining whether plants have specialized transport systems or not. Generally the lower plants—the algae, fungi, and Bryophyta—do not have specialized transport systems; although in some of the Bryophyta there are specialized vascular tissues for the transport of water and organic materials in the gametophyte[30] stem.

With the aerial sporophyte[31] generation and its dominance in the life cycle, transport systems are necessary. In terrestrial plants the roots must be in the soil to obtain water and the photosynthetic leaves must be exposed in the atmosphere to sunlight and carbon dioxide. Thus transport systems are necessary between the two, and are found in the Pteridophytes, Gymnosperms, and Angiosperms (sometimes known collectively as the Tracheophyta as they possess specialized water-conducting elements known as tracheids).

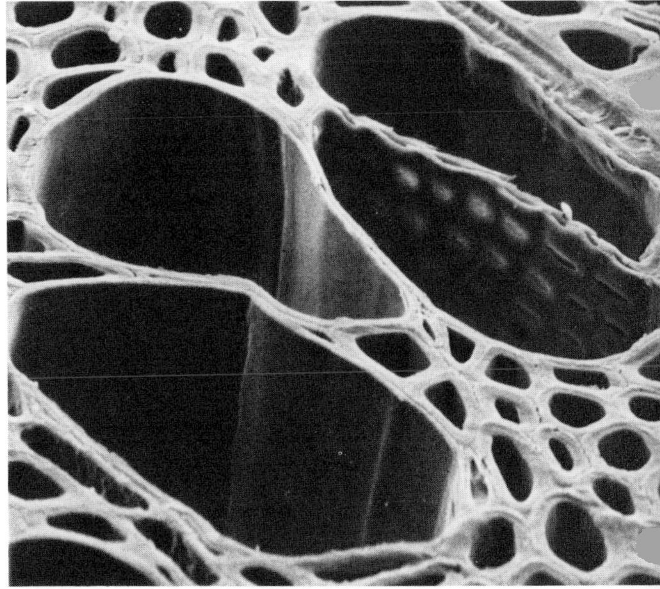

A concentric band of xylem in a woody stem (*Knightia excelsa*) transverse section (above) and tangential section (below) (×1250)

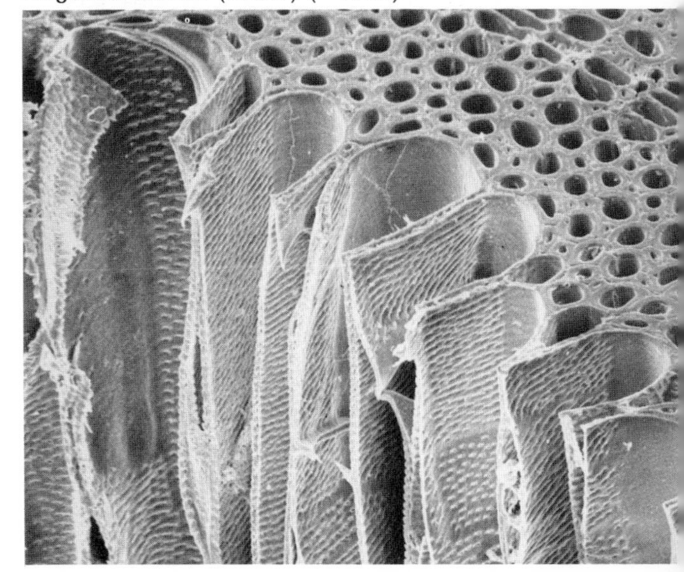

Water transport

The mass flow water transport system in Angiosperms is the xylem tissue. This consists of elongated tracheids and vessels with strong lignified walls to withstand the negative pressures inside the system. Lignin is also waterproof, and this ensures that the water is restricted to the large clear lumen which is produced as a result of the death of the living contents. The walls are pitted to allow for some lateral transport between elements.

Radial transport in plants

The path of the radial transport of water in the roots, from the soil solution to the xylem, is still not fully understood. There are two main possible pathways, either through the cytoplasm and vacuoles of the root hairs and parenchyma cells of the root cortex—a route of high resistance; or

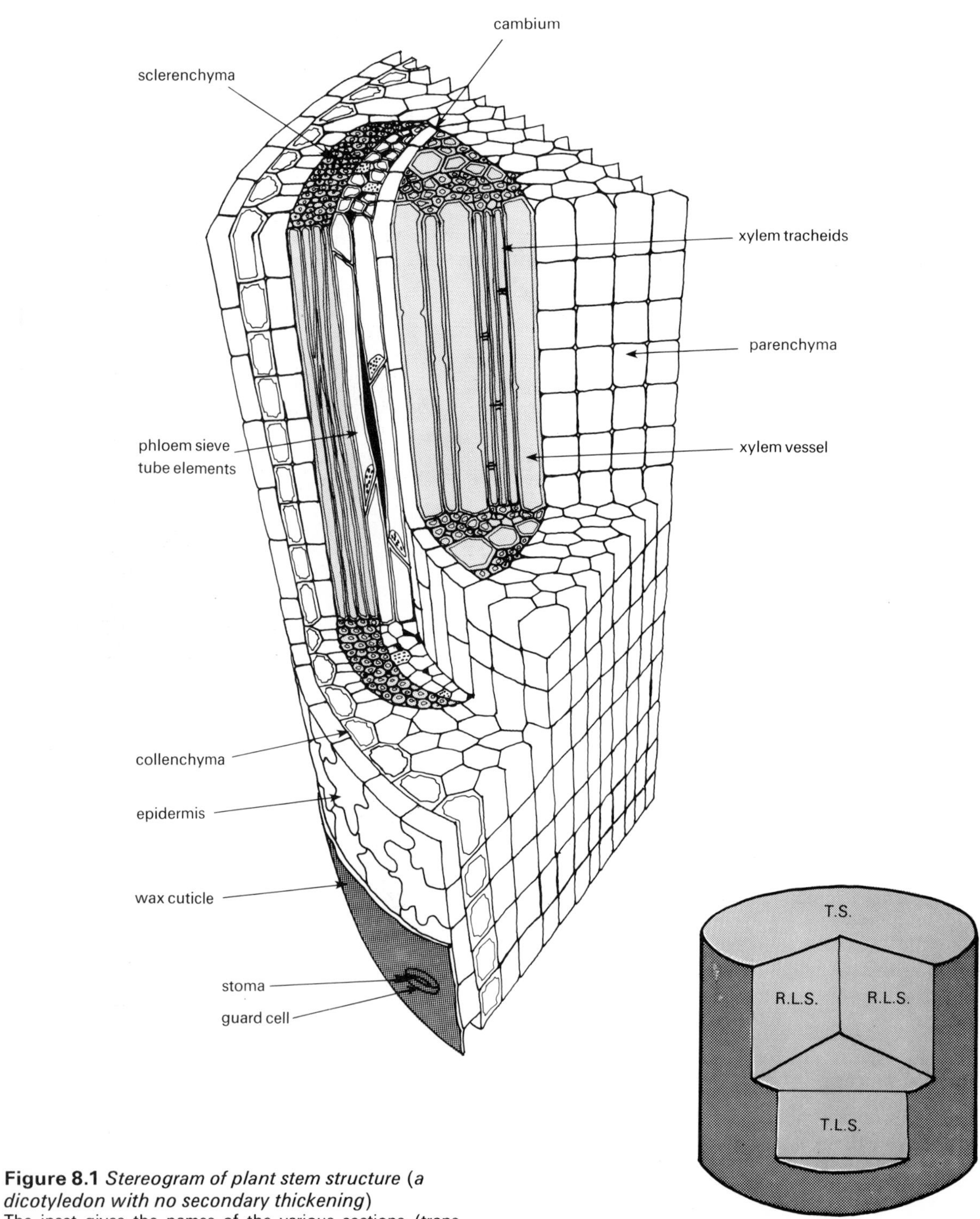

Figure 8.1 *Stereogram of plant stem structure (a dicotyledon with no secondary thickening)*
The inset gives the names of the various sections (transverse, radial longitudinal, and tangential longitudinal).

through the porous cellulose cells and intercellular spaces (except through the endodermis)—the route of least resistance. In *Pelargonium* water is found to travel across the root sixty times faster than the rate expected via the vacuolar route, indicating that the cell wall route is being used.

The vacuoles are separated from this main pathway by cytoplasmic barriers of high resistance, but they can 'tap-off' water from the main stream if they develop a requirement for water.

Between the parenchyma cells of the root cortex and the xylem tissue is the **endodermis**. These cells have their radial and transverse walls impregnated with waterproof lignosuberin—in the Casparian strip. Later in development the strip spreads over the whole of the radial wall and inner tangential walls. At intervals unthickened thin-walled passage cells remain, usually opposite protoxylem elements of the vascular bundle. Therefore any water and dissolved solutes must enter and pass through the cytoplasm of, first of all, the endodermal cells, and later, when their tangential walls are sealed, through the cytoplasm of the passage cells. This ensures that the water and solutes come 'under the control' of living cytoplasm at least once before entering the xylem. The endodermis also prevents any backflow of water from the xylem to the cortex if the vascular bundle develops a positive pressure, and may play a role in the active accumulation of salts by the root.[149]

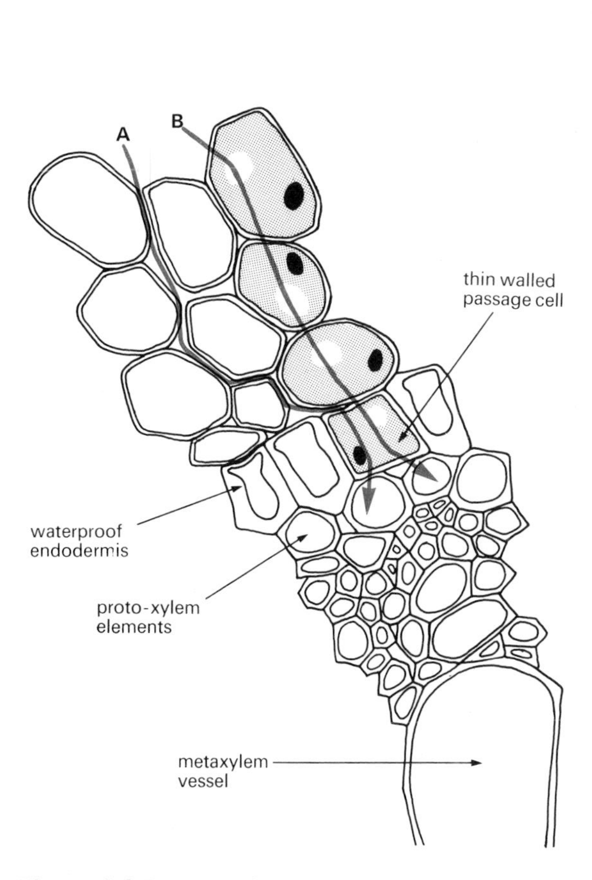

Figure 8.2 *Passage of water across a root*
A Water is drawn by a cohesive transpiration stream through the intercellular spaces and cell free space of the cell walls. This is the path of least resistance.
B Water is drawn in by an osmotic gradient from cell to cell. This is the path of highest resistance.

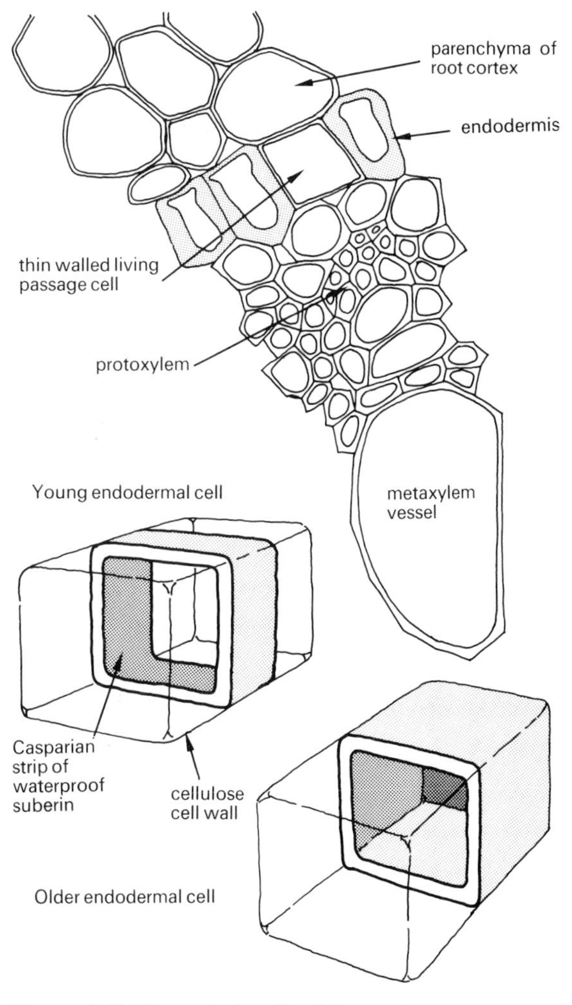

Figure 8.3 *The endodermis of the root*
The Casparian strip is added to by extra suberin until the entire cell wall is suberized, thus preventing the passage of any substance.

Movement of water into the xylem

It is still not clear exactly how water enters the xylem from the cortex, but there are two main possibilities. The cortical cells and endodermal cells may actively secrete salts into the xylem. This would produce a local region of high osmotic concentration, and the water in the cortex would therefore move into the xylem by osmosis. Another possibility is that the water is 'pulled' into the xylem by the cohesive forces of the transpiration stream.

It is also possible that a combination of these processes results in the movement of water.

Path of upward movement

The water and dissolved solutes move up the stem, through the leaves, to the atmosphere, in the xylem tissue. The pattern of transport depends on the distribution of the functioning xylem, e.g. it may be confined to scattered vascular bundles or, as in tree stems, it may be restricted to the outer few annual rings—the sapwood.

Mechanism of upward movement or transpiration

Although there is still some controversy, the most widely accepted explanation of the mechanism for the upward transport of water is that known as the Dixon–Joly Cohesion–Tension Theory, first formulated in 1894. It is based on purely physical forces. Evaporation from the leaf pores or stomata causes a gradient of water need across the mesophyll cells of the leaf, which causes water to be withdrawn from the xylem.

Cohesive[364] forces between water molecules result in a column of water being drawn up the plant as the water evaporates from the leaves in the process known as **transpiration**. This 'pull' generates a tension which results in the sap being under a negative pressure. It can produce a measurable decrease in stem diameter (and even trunk diameter in pine) at the height of transpiration. Calculations of the magnitude of the cohesive forces required for the tallest trees show that the cohesive forces of water and xylem sap are more than sufficient both to support the necessary water column and to overcome the resistance to flow in the xylem elements.

The column of water is held at the top by imbibition and adhesive forces between the water and the cellulose cell walls, which again are of sufficient magnitude to support the highest column.

The process requires a continuous column of water, and theoretically any air bubbles would break the cohesive 'pull'. The xylem vessels of many trees do contain air at certain times, especially in winter, when wind movements can break water columns and freezing can force air out of solution. One function of the high positive root pressure in the spring could be to force such gases back into solution. Also the formation of new elements allows the re-establishment of continuous water columns. However, trees can maintain rapid flow, even if many vessels are blocked by air, because air breaks do not threaten the total cohesive system.

Although the Dixon–Joly Theory is generally accepted, there are two main objections. First, there is a lack of proven hydrostatic gradient with height. According to the theory, the sap pressure should decrease one atmosphere for every ten metres increased in height. In fact, work with vines shows that under conditions of rapid transpiration the sap pressure is often higher at the top.

Secondly, the theory does not consider the problems of growth in relation to water transport. Newly differentiated xylem fills with water by osmotic withdrawal from previous years' xylem, thus initially water rises to the top of a tall tree in a series of annual steps. This implies that the pattern of growth is more important in the establishment of the water column than the cohesion theory suggests.

Factors affecting the rate of transpiration

EXTERNAL FACTORS

When considering the effect of external factors it is assumed that internal factors are constant, and the stomata are fully open.

Air movements

Diffusion of water vapour from the stomata results in layers of humidity (known as diffusion shells) building up between the stomatal opening and the air. The thickness of these layers varies according to the velocity of the air; they become thinner as air movements increase. The thinner the shells the more rapid is the rate of transpiration. By making the leaves bend, the air movements may also cause the mass flow of air into and out of the leaves, thus increasing water loss.

Relative humidity of air
With decreasing relative humidity, transpiration increases because the water from the leaf can evaporate more easily. The relative humidity is dependent mainly on temperature—it decreases as temperature increases—and a rise of 10 °C doubles the steepness of the humidity gradients from leaf to air, thus increasing transpiration.

Light intensity
The main effect of light is on the stomata, and results in their daytime opening and a consequent increased transpiration rate. In addition it has an effect on leaf structure,[128] which can in turn affect transpiration.

Availability of soil water
Any factor that decreases the availability of soil water will decrease transpiration by limiting the water supply. In extreme cases this will lead to wilting.

INTERNAL FACTORS

Water retention
With decreasing water availability, the water in the plant is held more tenaciously by an increase in the imbibition of water by cellulose; transpiration is thus decreased.

Internal surface area
The greater the area of cells exposed to internal air spaces, the faster will be transpiration under given conditions; for example, in general trees grown under good light conditions have a higher surface area than shade-grown trees.

Species characteristics
Any factors typical of a particular species may affect transpiration; factors such as rooting characteristics which affect water uptake therefore also affect transpiration.

Stomatal number and behaviour
There are usually more stomata on the lower side of dorsi-ventral leaves, although some (e.g. laurel) have stomata only on the lower side; bifacial leaves (e.g. grasses) have equal numbers on both surfaces. Generally the number decreases from the tip to the base and from the margin to the midrib of a leaf. The number can vary with the environmental conditions under which the plant develops, for example in moist conditions there are less per unit area, whereas in dry conditions there are, rather surprisingly, more per unit area, although fewer in total number as the total leaf area is usually reduced. On average there are about 300 per mm^2 of leaf surface. They are also found in a reduced form on floral organs, and on submerged plants.

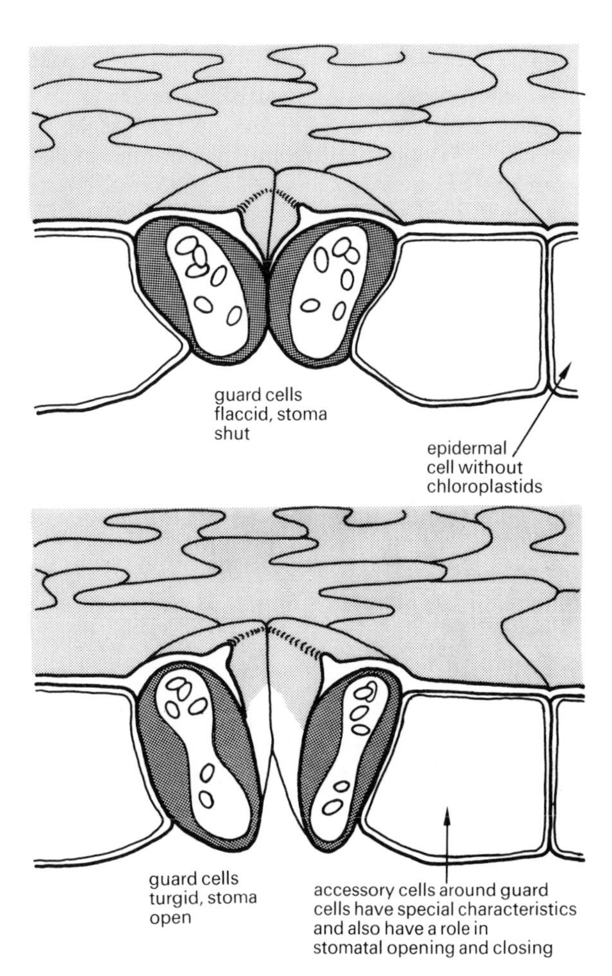

Figure 8.4 *Stomata*
In dorsi/ventral leaves there are more stomata on the lower surface. Bifacial leaves (e.g. daffodil) have equal numbers on both surfaces. Stomata are also found on herbaceous stems and in a reduced form on floral organs. Some submerged aquatic plants have a few vestigial stomata.

Mechanism of stomatal opening The mechanism by which stomata open and close is still not fully understood, but all theories depend on explanations of the changes in the shape of the guard cells due to changes in their water content or turgidity. When the guard cells are distended with water, or turgid, the stomata are open; when they are not so distended and are flaccid they are closed.

The presence of chloroplasts in most guard cells could indicate that photosynthesis is involved in some way. The simple explanation is that the guard cells produce sugars in the light by photosynthesis, and that these sugars increase the concentration of the cell contents so that water is drawn in by osmosis. This increases their turgidity and results in the opening of the stomata.

However, this cannot explain the opening and closing of all stomata. For example, changes in the rate of sugar production by photosynthesis would be too slow to account for the rapid opening sometimes seen. Also some guard cells lack chloroplasts; some stomata open in the light in carbon dioxide-free air; and in xerophytes[184] the stomata can open at night to allow the temporary dark fixation[129] of carbon dioxide needed in photosynthesis during the day when the stomata are closed to conserve water.

Effect of starch–sugar balance The balance between sugar and starch interconversion may be more important than the direct formation of sugars by photosynthesis. Thus when the sugars are converted to starch the osmotic concentration decreases, resulting in flaccidity and closure, and vice versa. In guard cells, starch content increases at night and decreases in the day, the reverse of occurrences in other cells. Also, guard cells retain starch in prolonged darkness, even when the rest of the plant is starved of carbohydrate, indicating that this starch may have a special function. This is seen elsewhere only in the geosensitive starch grains of the root whch detect the direction of gravity.[221] Several influences may be operating:

(a) It is possible that light activates enzymes which hydrolyse starch, so that sugars increase in the day.

(b) Changes in acidity may be important; for example, under acid conditions starch production is encouraged. Acid conditions of about pH 5 would arise from the accumulation of respiratory carbon dioxide at night, and the conversion of sugars to starch with subsequent stomatal closure would result. During the day photosynthesis would remove the carbon dioxide, lower the acidity, and result in the hydrolysis of starch to sugars with subsequent opening of the stomata. In plants lacking starch (e.g. the onion and most other monocotyledons) there could be a similar interconversion of fructosans and fructose.

(c) If opening depended entirely on photosynthesis, either directly on the formation of sugars, or indirectly by the removal of respiratory carbon dioxide resulting in a drop of acidity and a consequent increased hydrolysis of starch, the action spectrum[128] for the opening should coincide with that for photosynthesis—but it does not, and there is quite a large difference. Opening is greater in light from the blue end of the spectrum, and it is suggested that in some way blue light stimulates the hydrolysis of starch.

(d) At normal light intensities the ratio of photosynthesis to respiration in mesophyll cells is 10, but in guard cells it is only 1 due to the small amount of photosynthetic activity. As the balance is so delicate in guard cells, changes in the relative rates of the two processes would be more effective in controlling guard cell movements.

(e) There could be a non-photosynthetic method of carbon dioxide assimilation in which carbon dioxide is directly incorporated with an acceptor molecule to form organic acids. This would be more effective in adjusting the pH than the photosynthetic mechanism.

Function of stomata There are two main functions: to allow for the uptake of carbon dioxide and to control the rate of water loss. Under conditions for photosynthesis, the loss of water via the transpiration stream can be seen as an unavoidable consequence of having pores for carbon dioxide absorption. However, changes in stomatal size affect water loss more critically than carbon dioxide gain; for example if the pore area is reduced to 6 per cent of its maximum, the transpiration rate is reduced to 25 per cent of its maximum under given conditions, but the assimilation rate of carbon dioxide is reduced to only 50 per cent of its maximum. Also stomatal control of water loss is greatest at high wind velocities, when it is needed most.

The function of transpiration

There are three main functions of transpiration. It provides a **transport stream** bringing water and dissolved minerals to aerial photosynthetic parts from the roots. It exerts a **cooling effect**; although the maximum effect rarely exceeds 5 °C and, under very hot conditions when the cooling effect is most required, the stomata tend to close, thus stopping transpiration. It aids **water and salt uptake** by the roots; but even when transpiration ceases salt uptake can still be adequate.

Cuticular transpiration

Cuticular transpiration is only significant when the stomata are closed, as for most plants it represents less than 10 per cent of total water losses. It is controlled mainly by the thickness of the cutinized cuticle.

Transport of organic material

Specialized vascular tissue is not found in the lower plants and it is only in the higher vascular plants that such tissue is seen for the transport of organic material. Such transport is known as **translocation** and occurs from the point of origin or **'source'** of the organic material to the point of destination or **'sink'**. If the 'source' is in the roots and the 'sink' is in the aerial parts, then the organic material will move up the transpiration stream in the xylem. Otherwise, the specialized transport system is the **phloem**.

Table 8.1 *Transport in plants*

Material	*Source*	*Sink*
Soluble carbohydrates (e.g. glucose, sucrose)	(a) Photosynthesizing cells (b) Any cell converting insoluble starch to soluble sugars (c) Deciduous leaves prior to leaf fall (d) Seed storage tissues when germinating (e) Fruit storage tissues when germination occurs	(a) Any respiring cells with inadequate photosynthesis (e.g. root cells, stem cortex, medullary ray cells, flowers, meristems, etc.) (b) Any storage tissue (e.g. corms, bulbs, rhizomes, fruits, seeds, medullary rays, etc.)
Amino acids	(a) Mainly young leaves (b) Mature rootlets in some cases (c) Deciduous leaves prior to leaf fall (d) Seed storage tissues when germinating (e) Fruit storage tissues when germination occurs	(a) Any cell requiring amino acids for growth, repair or maintenance (e.g. apical and lateral meristems)
Hormones	(a) Apical meristems (b) Developing seeds (c) Leaves	(a) Zones of elongation (b) Fruits (c) Flower buds (d) Lateral buds

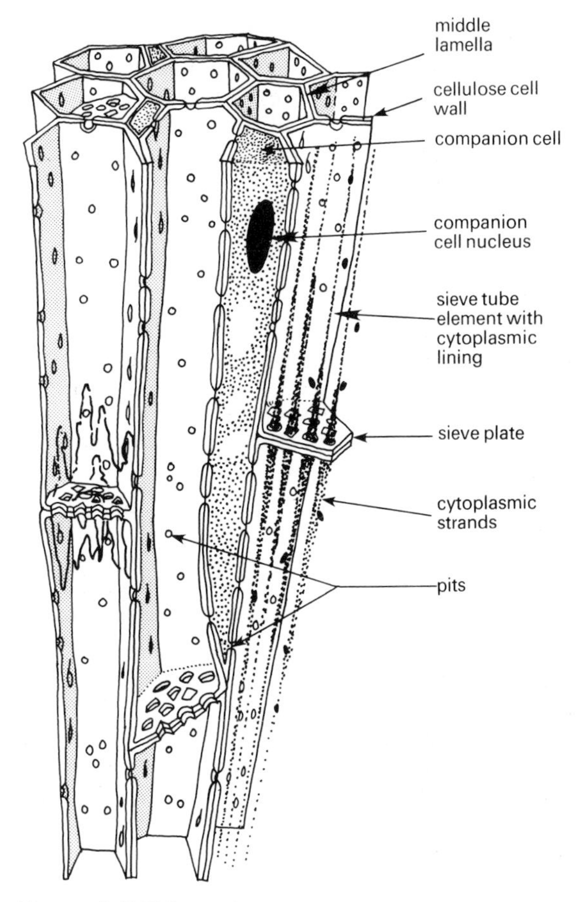

Figure 8.5 *Phloem tissue*

The mechanism of flow

Diffusion is not sufficient to account for the observed rates of flow in the phloem, since they are up to 60 000 times faster than could be explained by diffusion alone. There are two main theories, neither of which is entirely accepted. However, any explanation must depend on the vital activities of the cytoplasm, since sieve tubes must be living in order to carry out translocation.

The **pressure flow theory** depends on the 'source' regions having a higher osmotic pressure than the 'sinks', due to the presence of greater concentrations of osmotically active materials. As a result of this and the fact that the two regions are in direct cytoplasmic contact it is suggested that the organic materials are forced along the resultant pressure gradient.

The **cytoplasmic streaming theory** suggests that the actively streaming cytoplasm carries the organic materials and diffusion plays a role in moving materials across the sieve plates between the sieve-tube elements. However, streaming is only active in young elements, and even here the fastest measured rate appears to be too slow to account for the observed rates of translocation.

Both these theories imply that all materials travel at the same speed and in the same direction. There is evidence to suggest that this is not so, for example various injected materials can be seen to move at different rates from the carbohydrates in the translocation stream. There are also some specialized cells associated with phloem tissue which are somehow involved in the translocation mechanism, called the **transfer cells**. These are found next to sieve tubes, especially in small leaf veins and cotyledons of some species of Leguminosae. They have a characteristic dense cytoplasm and many organelles. Where present, they appear to be invloved in transferring solutes to and from the sieve tubes of the phloem.

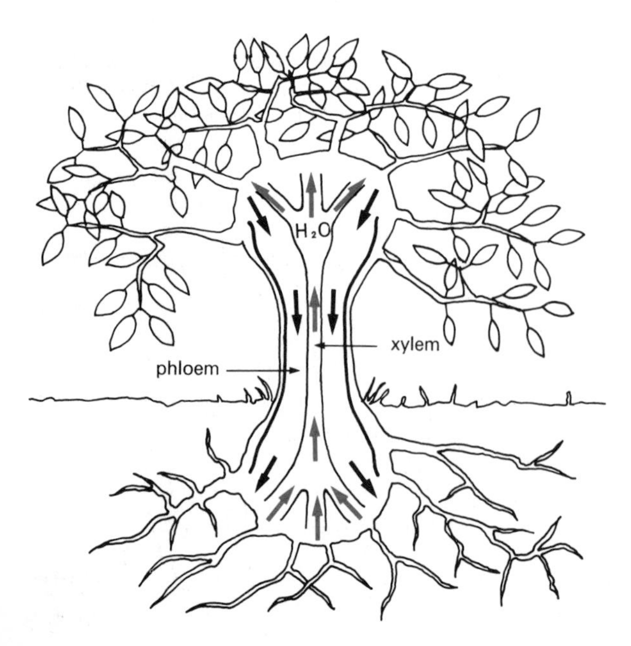

Figure 8.6 *Mass flow transport in phloem*
Water passes from xylem to phloem by osmosis. High pressure builds up in the phloem, forcing organic material in solution through the phloem tissue. The explanation requires the phloem to show a high internal positive pressure. This is the case, as can be seen when feeding aphids are knocked off a plant stem and the phloem contents exude from the broken feeding stylets which remain protruding from the phloem sieve-tube elements.

TRANSPORT IN ANIMALS

In the smaller invertebrates such as the protozoa and Coelenterata, and in some larger ones, such as the Platyhelminthes with a flattened ribbon-like body, there is a large surface area in relation to the body volume; in other words there is a large surface area to volume ratio. Therefore diffusion is sufficient for the internal transport of materials throughout the body, since no part of the body volume is far away from the surface at which the exchanges of materials with the environment occur.

In those animals whose surface area to volume ratio is below a certain critical size (all those other than the smaller invertebrates) diffusion is no longer sufficient for the internal transport of materials, and mechanical pumping of blood and body fluids, which are specialized as transporting media, is found.

Blood is moved by the contraction of specialized regions of the tubular blood vascular system in which it circulates. Such regions range in complexity from contractile vessels to true chambered hearts, but in all cases the contraction of the muscle-fibres are co-ordinated to produce flow in one direction, with the aid of one-way valves.

Pumping mechanisms

Contractile vessels

These vessels occur at all levels of the animal kingdom, but in some organisms they are the only propulsive mechanism; for example the Annelida, with their contractile dorsal and ventral vessels and their lateral pseudohearts which are also specialized contractile vessels.

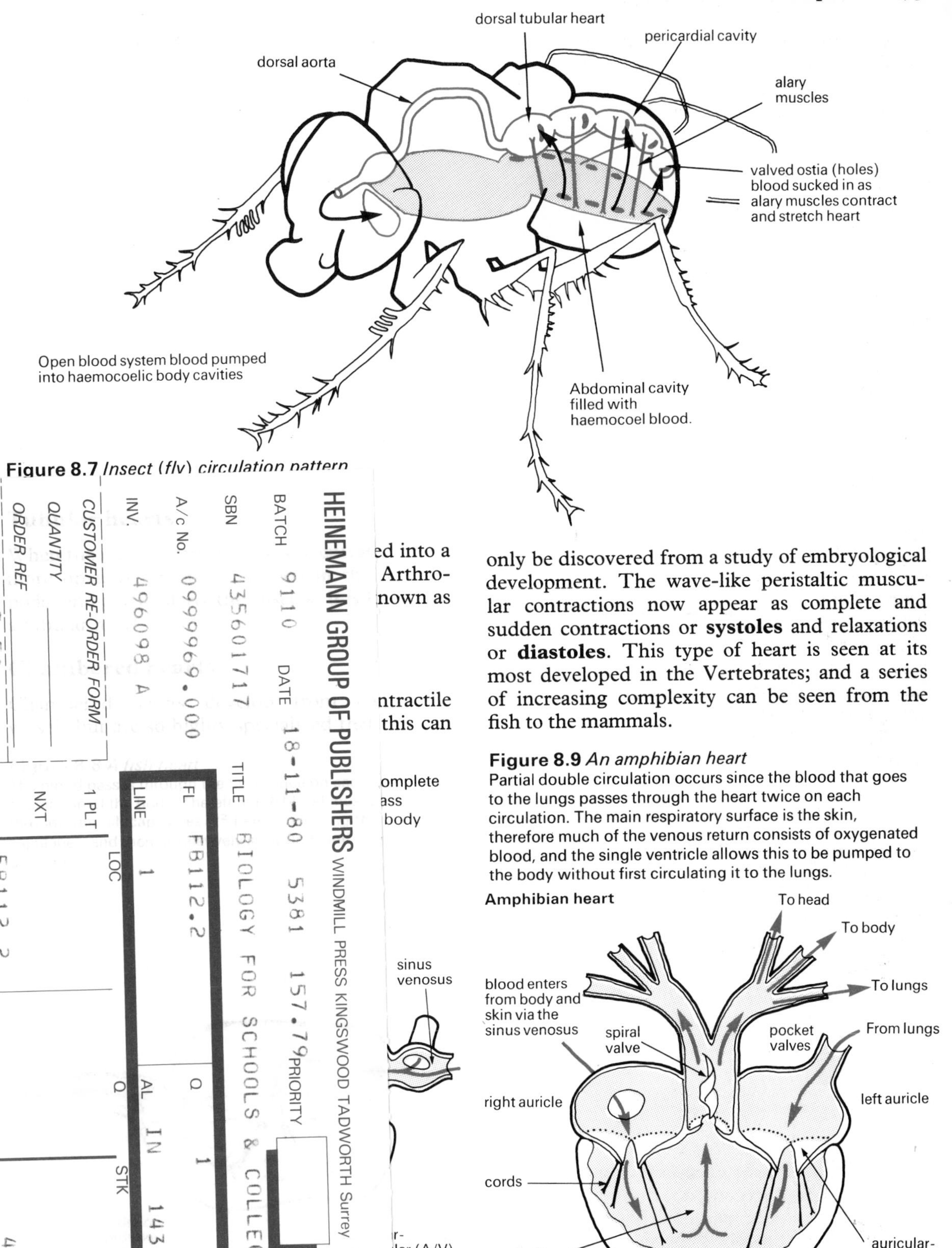

Figure 8.7 *Insect (fly) circulation pattern*

...ed into a ... Arthro- ... nown as ... ntractile ... this can ... omplete ... ass ... body

only be discovered from a study of embryological development. The wave-like peristaltic muscular contractions now appear as complete and sudden contractions or **systoles** and relaxations or **diastoles**. This type of heart is seen at its most developed in the Vertebrates; and a series of increasing complexity can be seen from the fish to the mammals.

Figure 8.9 *An amphibian heart*
Partial double circulation occurs since the blood that goes to the lungs passes through the heart twice on each circulation. The main respiratory surface is the skin, therefore much of the venous return consists of oxygenated blood, and the single ventricle allows this to be pumped to the body without first circulating it to the lungs.

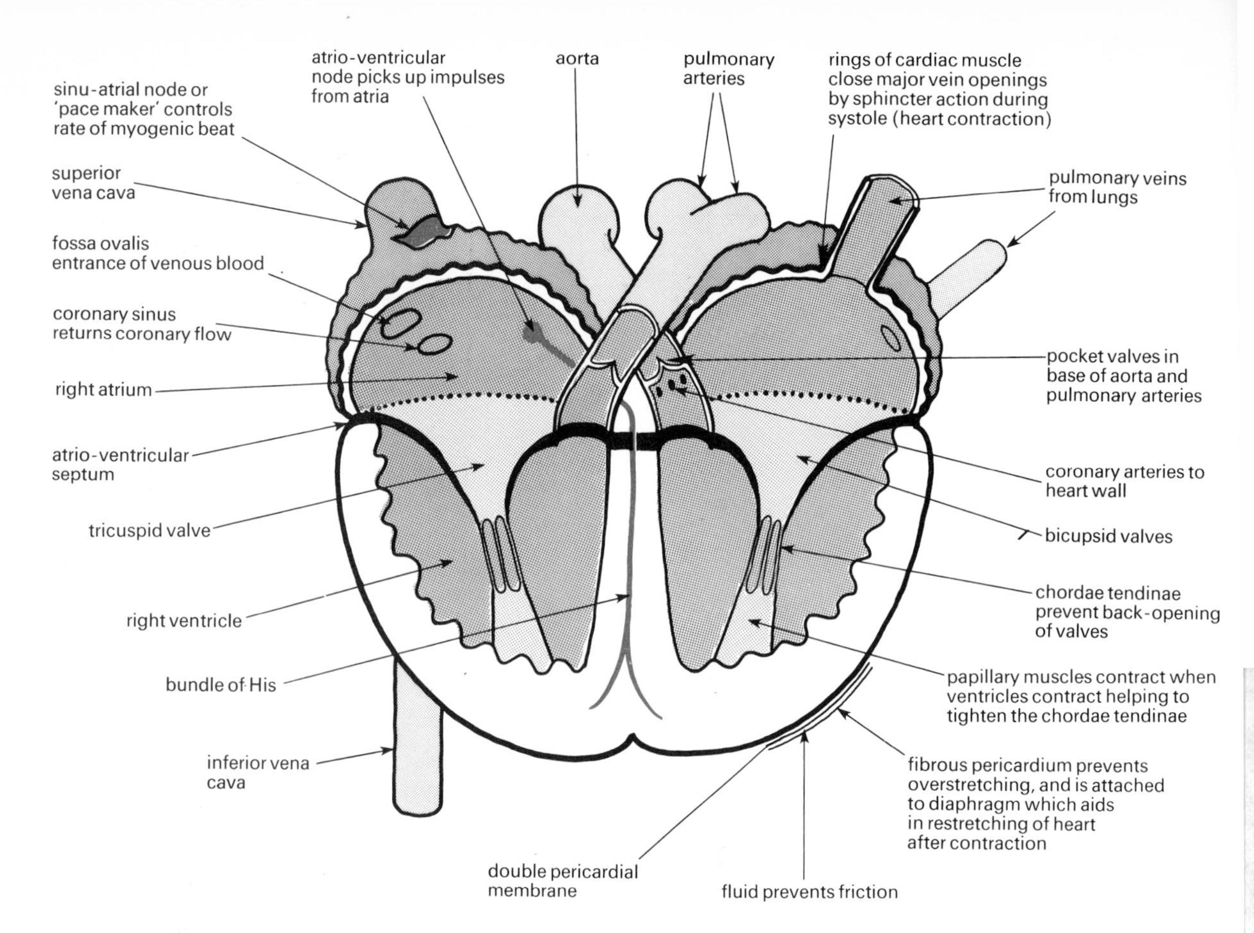

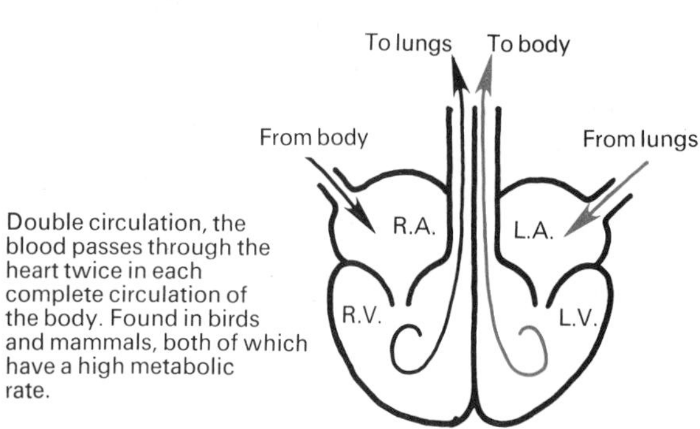

Double circulation, the blood passes through the heart twice in each complete circulation of the body. Found in birds and mammals, both of which have a high metabolic rate.

Figure 8.10 *A mammalian heart*

PROPERTIES OF CARDIAC MUSCLE

Cardiac muscle has certain special features not seen in any other type of muscle. For example, it is **myogenic** and so contracts rhythmically without any nervous stimulation although the rate of contraction is controlled by the sinu-auricular node (SAN) or pacemaker.

During contraction it is absolutely **refractory**, and so does not respond at all to further stimuli. Summation effects are therefore impossible, and it cannot develop a tetanus or continuous contraction; nor can it incur an oxygen debt.

CARDIAC CYCLE

The result of these co-ordinating impulses is that the heart beats in a co-ordinated rhythmic sequence known as the **cardiac cycle**.

SPREAD OF CARDIAC IMPULSE

The fibrous band of tissue, the atrio-ventricular (AV) septum, which runs in the AV groove, prevents the co-ordinating impulses from the sinu-auricular node spreading from the auricles to the ventricles. The AV node picks these impulses up and they are transmitted to the base of the ventricles by the Bundle of His (a tract of nervous tissue), where they initiate ventricular systole or contraction. The effect of this system is to ensure that the waves of contraction are in the direction of the intended blood flow.

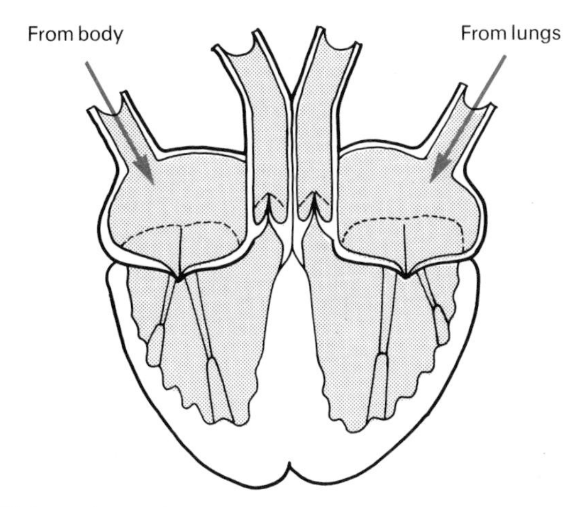

(1) Any point may be taken as the 'starting point' of a cycle. In this case it is taken as when the heart is empty, all valves are closed, and the atria are filling.

(2) As the atria fill, the pressure within them rises and the AV valves are pushed open. The ventricles fill, and the flow of blood into the heart slows and stops.

(3) The atria contract (atrial systole) and the ventricles become overfilled and stretched.

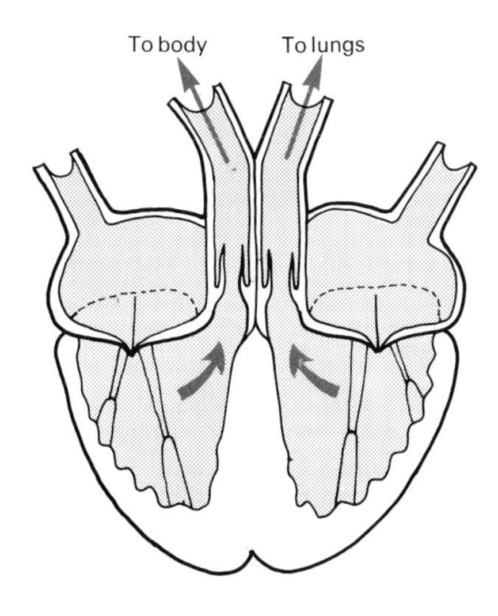

(4) The ventricles contract (ventricular systole); the ventricular pressure rises steeply, shutting the AV valves (first heart sound) and opening the pocket valves.

(5) The ventricles relax (ventricular diastole); any backflow of blood in the vessels closes the pocket valves (second heart sound).

(6) The cycle is then repeated.

Figure 8.11 *Cardiac cycle*

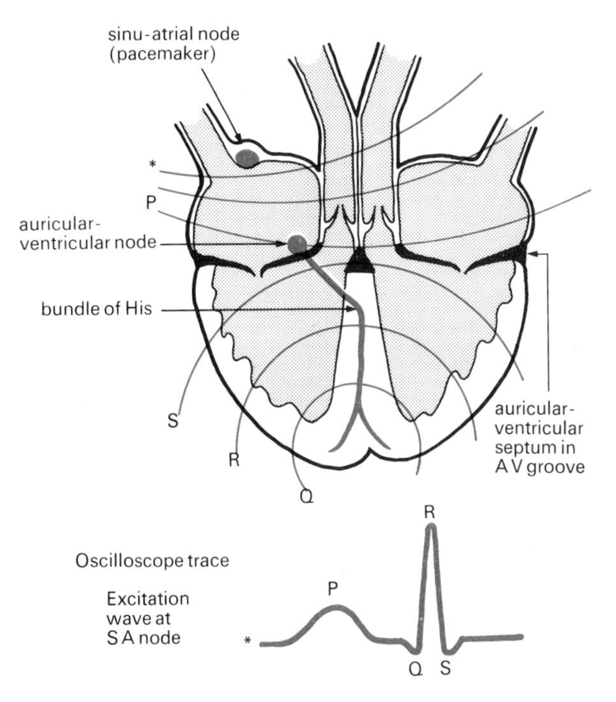

Figure 8.12 *Spread of cardiac impulses*

ALTERATION OF THE HEART RATE

The basic rhythm of the cardiac cycle, or heart beat, can be increased or decreased in response to various demands made on the body.[251] This alteration in the rate can be produced by nervous and hormonal action. The parasympathetic[202] part of the autonomic nervous system decreases the rate, and the sympathetic[202] part increases the rate. The hormones adrenaline[214] and thyroxine[214] also increase the rate.

Patterns of blood supply

From the heart, the blood travels in a system of blood vessels which supply the organs in different patterns. The **series pattern** of vessels, which supplies organs one after the other in a sequence, is most efficient when events in one capillary bed are directly dependent on the function of the preceding bed, for example the gut and the liver, where the serial arrangement is necessary for the maintenance of a constant blood composition.

The **parallel pattern** of vessels, which supplies several organs at the same time, ensures that blood reaches a region undepleted of essential substances, and not loaded with waste products. If one organ is damaged, the parallel pattern also ensures that the blood supply to other organs is not interrupted; and control of blood distribution can be achieved by closing off certain sections while leaving other sections unaffected.

Both patterns are found in the distribution of blood vessels in most animals.

Circulatory systems can be classified as open or closed. In **open systems** the blood is only enclosed in vessels for part of the circulation and is free in blood spaces in other regions; in **closed systems** the blood flows entirely within a closed system of vessels.

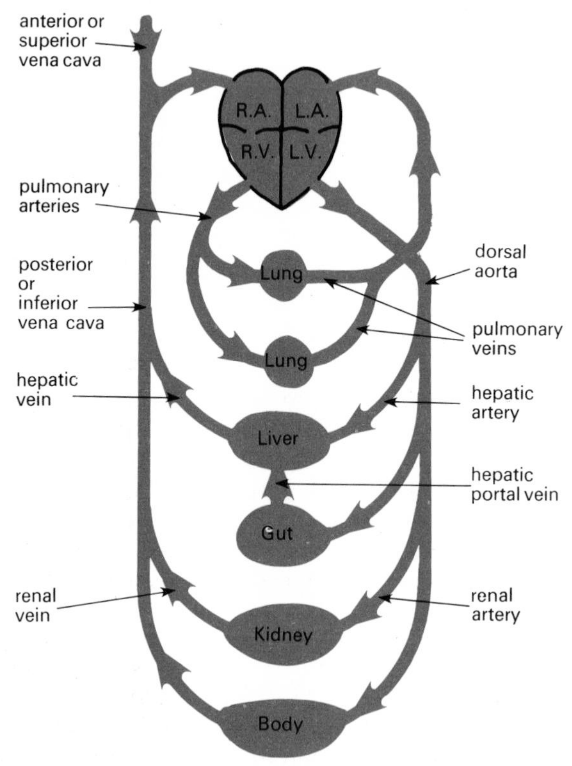

Figure 8.13 *Simplified mammalian circulatory system*

Open systems

Open systems are typical of arthropods, molluscs, and cartilaginous fish. In these the blood pressure is low due to low peripheral resistance as the blood flows freely in the blood spaces, and the pressure often varies widely throughout the system. The regulation of pressure in open systems is not understood, but it is mainly dependent on body muscle activity rather than the rate of heart contraction.

Closed systems

Closed systems are typical of most vertebrates. They have low blood volume, rapid blood flow, and high blood pressure due to the high peripheral resistance presented by the smaller vessels. Fluid exudation occurs at the capillary beds as the result of the high blood pressure, and this fluid acts as a carrier system between the blood in the vessels and the tissues.

VESSELS

The vessels carrying blood away from the heart are known as **arteries**, and are of two main types. The **large elastic arteries** are those close to the heart, with thick elastic walls capable of extending to accommodate the cardiac output, for example the aorta, common carotids, and subclavian arteries.

The **muscular arteries** continue peripherally from the large elastic ones. They have thick, smooth muscle walls, the tone of which is important in maintaining blood pressure and controlling distribution of the blood to various parts of the body. (The larger arteries have a blood supply to their own walls, by means of small vessels.) The muscular arteries lead into the **arterioles**, which lead into the **capillaries**.

The capillaries are the smallest branches of the blood vessels, and branch finely throughout all the tissues. They lead into the **venules**, which are thin-walled and are important in exchanges with the tissues. The venules lead into the **veins**, which carry the blood back to the heart.

The veins are larger in cross section and more numerous than the arteries, therefore they have a larger capacity than the arterial system. The walls are thinner and less elastic than those of the arteries, but they are more supple. Where the veins enter the heart there is a layer of cardiac muscle in the vein walls. At intervals there are pocket valves, in man usually in pairs. They consist of thin connective tissue membranes containing a network of elastic fibres and covered with endothelium.

The role of the vessels in maintaining the circulation of the blood

Together with the pumping action of the heart and its dynamic response to various demands, the blood vessels also assist in the circulation of the blood around the body, according to need. The elastic arteries are stretched by the blood which a contraction of the ventricles pumps out, and they recoil when the ventricles relax, thus assisting in pumping the blood and producing a smoother flow of blood through the capillaries (although the intermittent surge of blood through the arteries can still be felt as the **pulse**).

The walls of the muscular arteries and arterioles are in a state of variable tone or of contraction which is controlled from the vasomotor centre in the medulla oblongata,[201] and, in addition to adjusting the blood pressure, this control of the degree of contraction of the arterioles can be used to change the distribution of blood to various tissues and organs.

In addition the arterioles have pre-capillary sphincter muscles which by contracting and relaxing can control the flow of blood through the capillaries. When the pre-capillary sphincter is contracted the blood flows through the **arterio-venous shunts**, which directly connect the arterioles and venules, thus bypassing the capillaries.

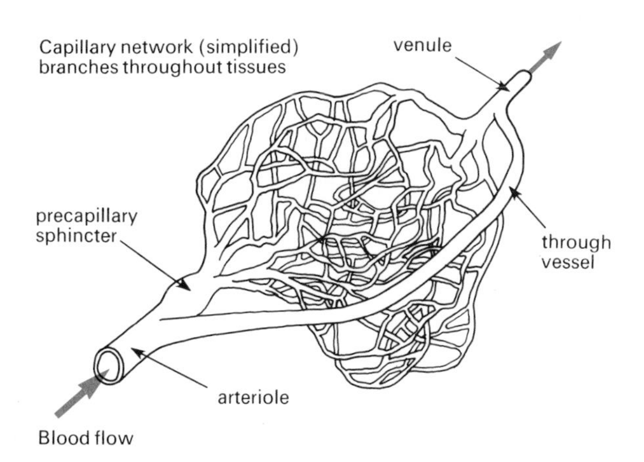

Figure 8.14 *Capillary shunt*

If all the capillary beds were open at the same time they would contain the entire blood volume of a mammal. It follows that many must be shut off at any given time. This gives rise to a type of competition between organs for blood. A good example is seen between the capillaries of the gut wall and those of the skeletal muscles; during exercise the blood is withdrawn from the gut wall and diverted to the skeletal muscles. Thus there is a good physiological reason why one should rest after eating.

The veins receive blood at very low pressure. Their thin, non-muscular walls and large cross section present little resistance to the flow of blood back to the heart, and the residual blood pressure moves the venous blood along. In addition, the movement of the skeletal muscles massages the veins and, in conjunction with the unidirectional pocket valves, this promotes venous return to the heart. The valves also prevent any backflow under gravity in certain regions, although in other regions gravity aids venous return—as from the head region in man.

The respiratory pump also aids venous return. Since the pressure in the thorax is always lower than atmospheric pressure, venous blood is drawn into the veins of the thorax. Also the descent of the diaphragm increases the abdominal pressure and so helps to 'squeeze' the blood into the veins of the thorax.

BLOOD PRESSURE IN CLOSED SYSTEMS

When the blood is enclosed in a continuous system of tubes, for a given blood volume the arterial blood pressure depends on two main factors: the cardiac output from the left ventricle and the peripheral resistance due to the smaller vessels resisting the flow of blood.

$$\text{arterial blood pressure} = \text{cardiac output} \times \text{peripheral resistance}$$

The muscular arteries and arterioles are in a state of variable tone or contraction which maintains the blood pressure. The arterioles are also very sensitive to adrenaline, and the small amount of this hormone normally present in the blood helps to maintain their tone. This contributes to the maintenance of the blood pressure.

As vessels branch, the cross-section of each branch decreases but the total cross-section increases. The rate of flow thus *decreases*, and the blood moves through the capillaries relatively slowly. This aids the exchange of substances between the blood and the tissues.

In mammals the **pulmonary** or lung circulation has a lower pressure than the **systemic** or body circulation. This is because the right side of the heart is less muscular than the left; the peripheral resistance is lowered by the thin elastic walls of pulmonary arteries; and the arterioles have less smooth muscle in their walls. This prevents the delicate pulmonary system from being damaged, and there is little tissue fluid formation.

The flow in the pulmonary system is faster since the total area of cross-section of the vessels in the lungs is smaller than that in the systemic system. In the systemic system the flow is slower and the pressure is higher, allowing effective tissue fluid exchange and kidney function.

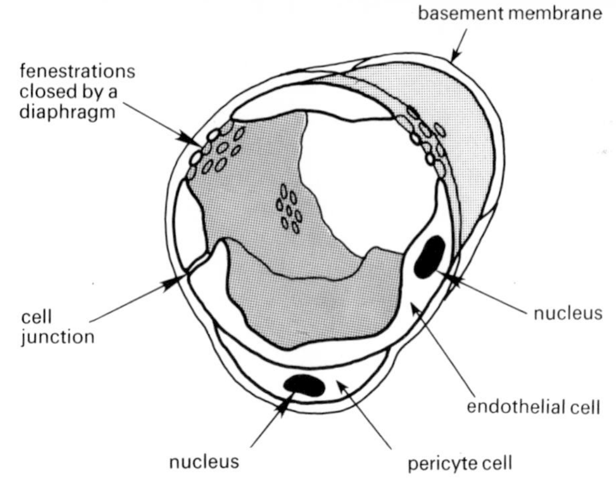

Figure 8.15 *Capillary structure—fenestrated type*
These are found supplying gut villi, endocrine glands, and renal glomeruli; all are regions of very active exchanges. In the capillaries of the glomeruli the fenestrations are not closed by a diaphragm and fluid exudation is very rapid.

EXCHANGE ACROSS CAPILLARY WALLS

Capillaries present a large surface area for exchange between the blood and the tissues. In the closed system of mammals, blood only comes into direct contact with the tissues in the sinusoids or blood spaces of the liver, spleen, and some endocrine glands. All other tissues are bathed in tissue fluid from the plasma exuded or pushed out under pressure between the capillary endothelial cells and through the pores in the fenestrated capillary walls.

In fenestrated capillaries the rate of exchange is increased, in spite of the diaphragm and basement membrane which cover the pores or fenestrations. When the diaphragm is absent, as in the glomeruli of the kidneys,[190] exchanges are even faster.

There are special modifications for fluid exudation in the kidney which relate to its special function of urine formation. The glomeruli are capillary beds within arterioles, therefore the blood pressure remains high along their length and there is no reabsorption of fluid at the venular ends (as seen in ordinary capillaries); the

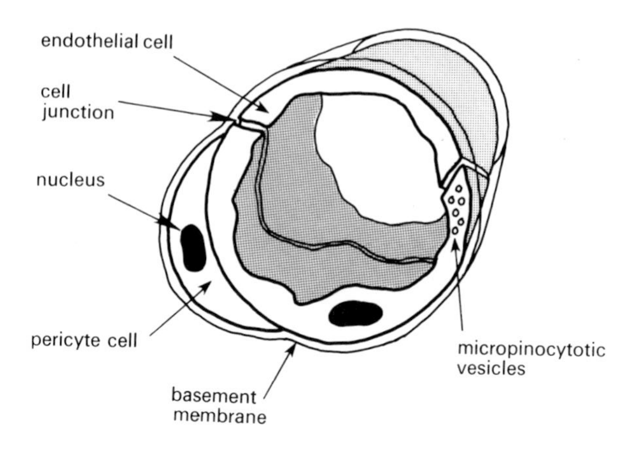

Figure 8.16 *Capillary structure—continuous type*
These are found supplying smooth, striped, and cardiac muscle and have no pores in their walls. In this type the micropinocytotic vesicles are important in exchanges.

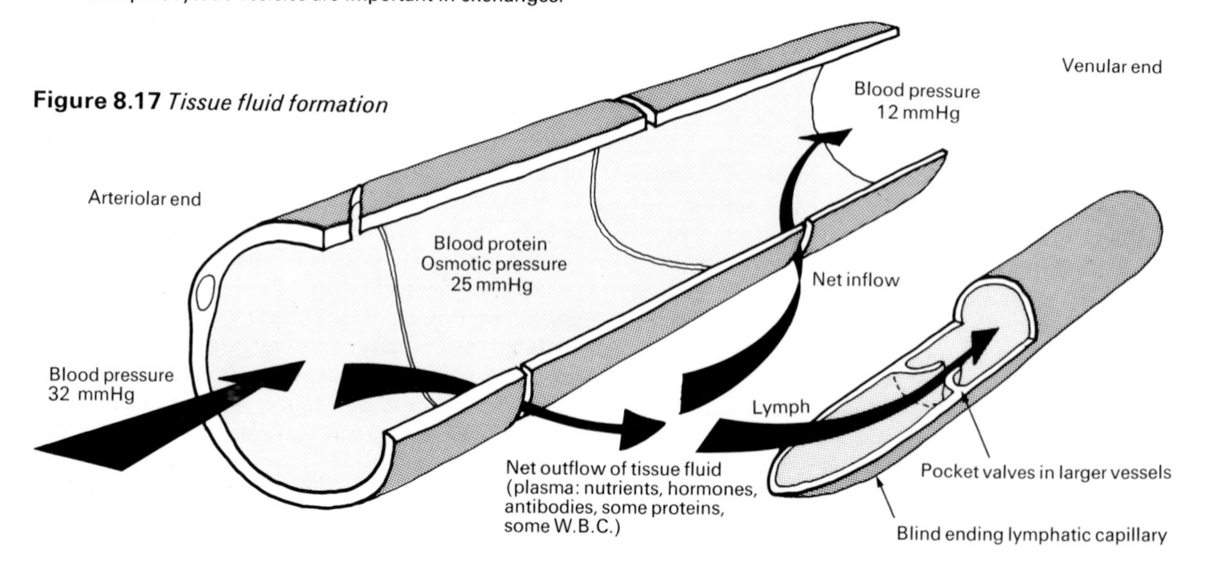

Figure 8.17 *Tissue fluid formation*

endothelium of the capillaries has numerous open pores; and the tissue fluid drains into tubes leading to the exterior.

In the capillaries of the brain there are very tight junctions between the cells which limit fluid exchanges. In addition the brain tissues actively exclude many substances. This limited exchange between the blood and the brain tissues is a form of protection, and gives rise to the concept of the **blood-brain barrier**.

Lymph and the lymphatic system

That part of the tissue fluid which is not reabsorbed by the blood capillaries enters the lymphatic capillaries through their endothelial walls, and becomes **lymph**.

Lymph has a higher protein content than the tissue fluid and contains lymphocytes, a particular type of white blood cell. The endothelium of the blood capillaries normally allows a little blood protein to escape into the tissues. This escaped protein cannot pass back into the blood capillaries, but it can enter the lymph capillaries. If the lymphatic capillaries did not drain the tissues, the blood protein would accumulate in the tissue fluid and, due to its osmotic concentration, would hold increasing amounts of water in the tissues.

The blind-ended lymphatic capillaries lead into wider lymph vessels which are similar in structure to the veins, having thin walls with pocket valves. Contraction of skeletal muscles massages the vessels and the valves ensure unidirectional flow. This, together with the respiratory pump (as in the venous return of blood to the heart), aids the draining of tissue fluids. Fish, amphibia, and reptiles, have lymph hearts with striped muscle and valves which pump lymph in the system.

During exercise the blood pressure may increase, more tissue fluid is exuded, the lymph vessels are massaged more vigorously, and therefore the tissue fluid is drained more quickly. All these factors increase the speed of supply of oxygen and nutrients to, and removal of waste products from, the tissues.

At intervals in the system are swellings of reticular tissue, called **lymph nodes**, which play an important role in the defence of the body against infection.[291]

The lymph is returned to the circulation in the anterior venae cavae.

Structure and function of mammalian blood

Blood is the specialized fluid tissue of the transport system. It consists of fluid plasma and suspended cells.

Plasma

Plasma consists of 90 per cent water, and 10 per cent materials in suspension and solution. Plasma proteins, formed mainly in the liver, are responsible for maintaining the blood's viscosity. (This is another factor in the maintenance of peripheral resistance and therefore arterial blood pressure.) They are also important in the buffering[363] capacity of the blood. The proteins include **serum albumin**, to which the plasma calcium is bound; **serum globulins**, of which some are antibodies and some are important in the carriage of hormones and vitamins; and **fibrinogen**, which is important in blood clotting.[293]

There are also many dissolved inorganic and organic materials, some of which are part of the plasma and others transported materials, e.g. amino acids, glucose, fats, urea, creatine, bicarbonate, some oxygen, etc.

Erythrocytes (red blood corpuscles or RBCs)

Red blood corpuscles are small, non-nucleate, biconcave discs in man. The lack of nucleus permits the biconcave shape (which increases the surface area for gaseous exchange) without decreasing the volume of haemoglobin carried. The total surface area of the RBCs in man is about 3500 m^2. Their average diameter in man is 7–8 μ, which is the same as the diameter of the capillaries. They therefore move through the capillaries in single file. This increases the number of RBCs coming close to the tissues and aids the diffusion of oxygen to the cells. The RBCs have pliable membranes which also aid their passage through the capillaries. They contain haemoglobin which combines reversibly with oxygen to form oxyhaemoglobin. If the haemoglobin were not contained in corpuscles the viscosity of the blood would be too high and, with a molecular weight of 68 000, the haemoglobin would be filtered by the kidneys.

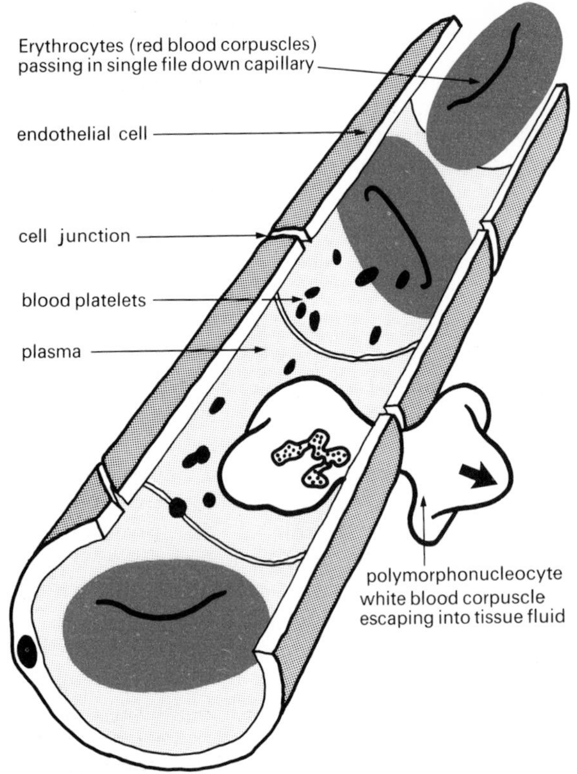

Figure 8.18 *Composition of blood*

Leucocytes

Leucocytes are larger, nucleated cells important in the body's defence against disease.[292] There are several different types, of which two main ones are **granulocytes**, which engulf bacteria by phagocytosis, and **lymphocytes**, which form antibodies.[291]

Platelets

Platelets are fragments of megakaryocytes, or large nucleated cells, from the bone marrow, and are important in blood clotting.

Experimental and applied aspects

There are many different experimental techniques for determining the route taken by transported substances in plants. For example, the fact that the transport of water in the transpiration stream occurs in the xylem and is not dependent on the activities of living cells can be demonstrated by immersing the cut end of a leafy stem in a poisonous solution coloured with a suitable dye. The poisonous solution is found to travel to the leaves successfully (eventually killing them), and sections of the stem can be cut to demonstrate the presence of the dye mainly in the xylem (although some lateral spread can occur).

However, the use of metabolic poisons or inhibitors can prevent the translocation of organic compounds within the plant, thus showing that translocation does require the activities of a living tissue.

Ringing experiments can also be employed; these entail removing a ring of all the living tissues (which includes the phloem) external to the xylem of a woody plant, and observing the effect on the transport of different substances. For example, ringing is seen not to interfere with transpiration, thus demonstrating that water is moved up the plant in the xylem. However, it leads to the swelling of the outer tissues (phloem) above the ring due to the accumulation of organic materials, and to the eventual death of the roots, thus demonstrating that organic materials are translocated down to the roots from the photosynthetic leaves in the living tissues external to the xylem (confirmed to be in the phloem by other experiments).

Such studies of transport in plants can be further augmented by the use of **radioactive tracers**. If a plant has its roots immersed in water containing radioactively labelled $^{32}PO_4$ (phosphate ion), and is subsequently sectioned and the sections used to make autoradiographs, then the $^{32}PO_4$ is found in both the xylem and the phloem. However, if the xylem is lagged from the phloem by some barrier material, such as grease-proof paper, which prevents the lateral transport of materials from one to the other, the $^{32}PO_4$ is found only in the xylem. Similarly, if a ringed plant is placed in water containing $^{32}PO_4$, then the $^{32}PO_4$ can soon be detected in the leaves, showing that its transport in the xylem was not prevented by the removal of the phloem.

Interestingly, studies on the rate of movement of radioactive tracers within the xylem and phloem can yield some surprising results. For example, the rate of transport of $^{32}PO_4$ in the xylem has no simple relationship to the rate of transpiration; and compounds containing ^{14}C and compounds containing ^{15}N have been detected moving in different directions in the phloem. Neither of these results would be expected according to the classic theories of transport in the xylem and phloem.

Patterns of distribution throughout the plant can be detected by making autoradiographs of entire pressed plants which have been exposed to some radioactive tracer. For example, if a leaf of a photosynthesizing plant is exposed to $^{14}CO_2$ for varying periods of time, then pressed, oven dried, and autoradiographed, the position of the translocated products of photosynthesis will be shown. In another experiment a leaf of a plant may be exposed to $^{14}CO_2$, and the plant then cut into its different parts. Each part may then be completely oxidized, and the evolved CO_2 absorbed into a solution. The radioactivity of the CO_2 can thus be measured accurately (using a scintillating counter). In this way the percentage distribution of the radioactive products may be measured in the different parts of the plant after a leaf of the plant has been exposed to $^{14}CO_2$ for different lengths of time.

Guided example

The mammalian foetal circulation and the changes it undergoes at birth clearly show the way in which the structure of the heart and the pattern of circulation are adapted to the uptake and transport of the respiratory gases.

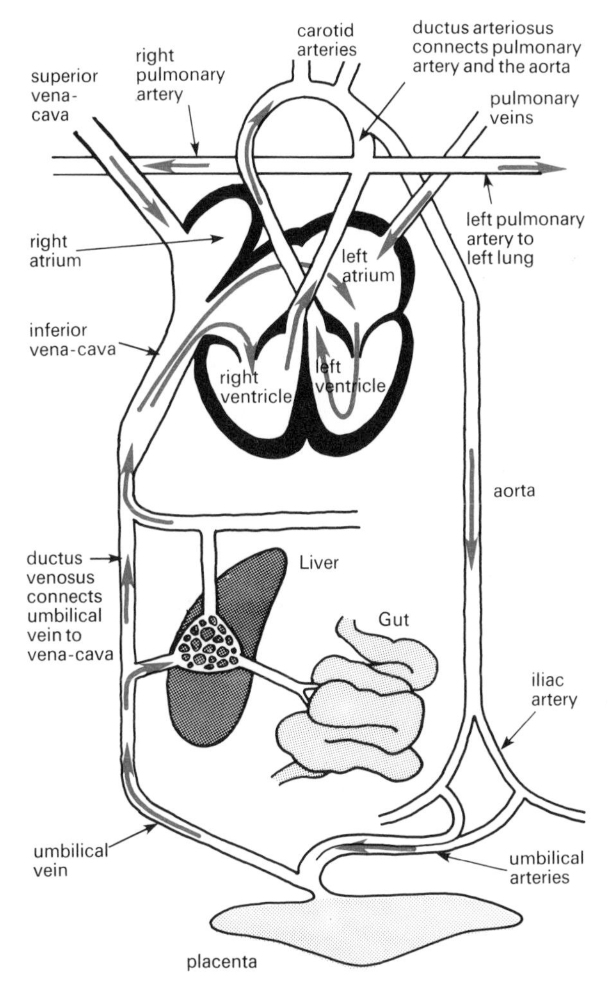

Figure 8.19 *Mammalian foetal circulation*

Study the annotated diagram and answer the questions provided.

1 In the foetus about 50 per cent of the blood entering the right atrium passes through an opening, the foramen ovale, into the left atrium. What reasons can you think of to explain this?
The blood entering the right atrium is partially oxygenated, since much of it is coming from the placenta, and thus needs to enter the general circulation as soon as possible without first circulating around the non-functional lungs.

2 Of the 50 per cent of the blood entering the right atrium that does not pass through the foramen ovale, about four-fifths passes through the ductus arteriosus into the main aorta. What reasons can you think of to explain this?
As in the previous answer, the ductus arteriosus allows most of the blood to by-pass the lungs.

3 The unexpanded lungs offer a high resistance to the flow of blood through the pulmonary circulation, and this further reduces the amount of blood passing to them. Why is it necessary, however, for some blood always to take this route?
The lung tissues still require a supply of blood to transport oxygen and nutrients to their living cells, and to remove their waste products.

4 The flap valve of the foramen ovale opens into the left atrium, and is kept open during foetal life by the pressure of blood from the inferior vena cava. With the severance of the umbilical cord and the opening of the pulmonary circulation, the valve is shut and anatomical fusion of the valve with the atrial wall occurs within a few days of birth. What changes in the pressure cause the valve to shut?
An increase in the pressure in the left atrium (due to the increase in blood returning from the lungs) and a decrease in blood pressure in the inferior vena cava (due to the interruption of the umbilical blood flow).

5 Within minutes of birth the rise in oxygen concentration in the blood coming from the lungs causes the ductus arteriosus to constrict. What effect will this have on the circulation?
All of the blood from the right ventricle will pass to the lungs.

6 If the foramen ovale is not sealed, and the ductus arteriosus is not completely constricted and blocked after birth, what do you think would be the main physiological effects?
As the lung by-passes would still be functioning, the pulmonary circulation would not be fully perfused with blood and therefore gaseous exchange would be impaired. A mixture of oxygenated and deoxygenated blood would be pumped around the body.

Questions

1 Compare the structure of a xylem vessel with that of a sieve tube. Show how differences between them may be related to their different functions in the plant. (L)

2 (a) What are the specifications for a successful closed circulatory system?
(b) Describe how the circulatory system of a mammal meets the varying needs of the tissues. (JMB)

3 By means of large, fully annotated diagrams describe the pathways along which water moves from the soil solution into the xylem vessels of the root. Explain concisely the differences in the way root tissue absorbs and transports a substance such as calcium nitrate in comparison with water. (L)

4 By means of labelled drawings only describe the structure of the tissue in which sugars and amino acids are transported in a plant stem.

Outline one possible mechanism of transport of these solutes in this tissue. What evidence supports this hypothesis? (L)

5 Young seedlings were allowed to absorb radioactive isotopes of calcium and phosphorus (incorporated in a soluble ion) for one hour. They were then removed to a non-radioactive solution and were sampled 5 minutes, 6 hours, and 48 hours after the radioactive treatment.

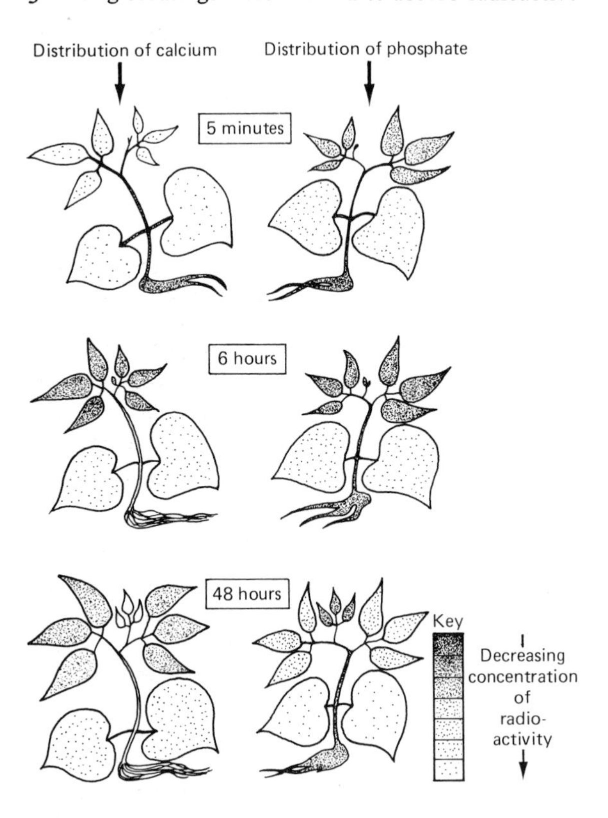

The shading on the diagrams is directly proportional to the amount of radioactivity found to be present in any one part of the plant.

(a) Compare the distribution of calcium and phosphate after 5 minutes.
(b) Compare the distribution of calcium and phosphate after 6 hours.
(c) Compare the distribution of calcium and phosphate after 48 hours.
(d) Make a general comparison between the mobility of the calcium and phosphate in the seedlings.
(e) Correlate one universal function of phosphate in living tissue with the distribution in this experiment. Clearly specify this function.
(f) The results with these seedlings are also characteristic of the behaviour of calcium and phosphate in a deciduous tree. Draw a calcium cycle and a phosphate cycle, indicating the similar and dissimilar features of the recycling of these two ions which would take place in a deciduous wood.
(g) Describe an experiment using these isotopes and woody privet shoots to determine the extent to which these elements are translocated in the phloem or the xylem. (AEB)

Further reading

Muir A. R., *The Mammalian Heart*, Oxford/Carolina Biology Reader (Oxford: Oxford University Press, 1971).

Neil E., *The Mammalian Circulation*, Oxford/Carolina Biology Reader (Oxford: Oxford University Press, 1975).

Richardson M., *Translocation in Plants*, Institute of Biology Studies in Biology No. 10 (London: Arnold, 1968).

Sutcliffe J., *Plants and Water*, Institute of Biology Studies in Biology No. 14 (London: Arnold, 1969).

9 Osmoregulation and excretion

Osmoregulation

Osmosis is the passage or diffusion of water, from a dilute or **hypotonic** solution to a more concentrated or **hypertonic** solution, through a semi-permeable membrane which is more permeable to water molecules than to dissolved solute molecules, in an 'attempt' to equalize the concentrations, that is to make them **isotonic**.

Osmoregulation is the control of osmotic concentrations within an organism; thus it is the control of the amount of water and dissolved substances (mostly mineral salts) in the cell and in the organism as a whole.

Although the control of water and of salts are considered here as if they were separate processes, it must be remembered that in fact they are in many ways inseparable. It must also be remembered that the uptake of water and salts does not depend solely on the principles of osmosis, which are the relative strength of solutions and the nature of membranes. It must also depend upon the overall metabolism of the cell, as a large component of salt movement is active and dependent upon respiratory energy, and much water movement depends in turn upon this primary active salt exchange. Thus salts may be moved actively and water may then follow passively by osmosis.

Different environments

SEA WATER

The concentration of sea water varies throughout the oceans, e.g. the Mediterranean is more concentrated than the Atlantic, and there are local areas of dilution around estuaries and in intertidal zones. However, an average figure for the salinity of sea water is 34.5 parts per thousand ($^{0}/_{00}$). In this environment the problems are **water loss** by osmosis and **salt gain** by diffusion.

FRESH WATER

This also varies with local conditions, but any water with a salinity of less than 0.5 $^{0}/_{00}$ may be considered as fresh. In this environment the problems are of **water gain** by osmosis and **salt loss** by diffusion.

BRACKISH WATER

Water that is between sea water and fresh water in concentration is termed **brackish**, and it is arbitrarily defined as water with a salinity between 0.5 and 30 parts per thousand. This includes estuarine waters, where variations in tides and river flow produce highly variable osmotic conditions, and some intertidal zones. Here the problems are extremely **variable**.

TERRESTRIAL ENVIRONMENT

Here the problem is mainly one of water loss by **evaporation**. Usually salts are readily available in the diet, and osmoregulation in terrestrial organisms is mainly concerned with water conservation.

Water balance in terrestrial animals is achieved by balancing the processes of water loss with those of water gain. Water is lost by evaporation from the skin and respiratory surfaces; in the urine and the faeces; and by glandular secretion. Water is gained by drinking; as free water in food; and as a by-product of respiration.

The plant cell as an osmotic system

Osmoregulation is essentially a feature of animal cells, as plant cells have the strong cellulose cell wall surrounding the cell membrane which resists over-expansion due to excess water entry from dilute or hypotonic solutions. Animal cells, which do not have this cell wall, are liable to disruption of the delicate cell membrane when exposed to hypotonic solutions and must therefore osmoregulate more actively. Also, plant cells can tolerate a wider range of water content than most animal cells and still function normally.

The plant cell wall is permeable to water; the vacuolar contents represent an internal solution of a certain concentration; and the cell membrane, the vacuolar membrane, and the cytoplasm all act as semi-permeable membranes. In the plant cell these three are treated theoretically as a single membrane of negligible thickness. Thus, in a plant cell, the passage of water by osmosis is considered as occurring between the external solution and the vacuolar sap.

Although the cell wall is fully permeable to water, it resists the entry of water into the cell by its elastic recoil when stretched, which gives rise to the **wall pressure**. It can also resist loss of water from the cell. The cytoplasm is closely applied to it, even penetrating it via the plasmodesmata, and thus as the cell contents shrink away from the wall with water loss, a negative wall pressure or 'suction effect' is developed which tends to pull water back into the cell. The basic equation describing the osmotic behaviour of a plant cell (where O.P. is osmotic pressure) is:

(external O.P. − internal O.P.) − wall pressure
= the force that is left to draw water into the cell

So the difference between the concentration of the vacuolar sap and the external solution, minus the opposing wall pressure, gives the force that is left to draw water into the cell. This force is variously called suction pressure, diffusion pressure deficit, or more recently, **water potential**. The equation can be used to illustrate the relationship between these components under differing conditions.

When in a hypertonic solution, water is withdrawn from the cell. As the cytoplasm withdraws from the cell wall, **plasmolysis** is said to occur. Generally, the process is reversible. (Plasmolysis does not normally occur in nature, for example even flaccid cells in a wilting leaf do not usually become plasmolysed.) The stage at which the cytoplasm just begins to pull away from the cell wall is known as **incipient plasmolysis**. At this point the wall pressure (W.P.) is zero and the suction pressure (S.P.) is taken as zero. It is therefore possible to obtain an approximate measure of the internal osmotic concentration of the plant cell:

$$\text{S.P.} = (\text{O.P.}_{\text{int}} - \text{O.P.}_{\text{ext}}) - \text{W.P.}$$
$$0 = (\text{O.P.}_{\text{int}} - \text{O.P.}_{\text{ext}}) - 0$$
$$\therefore \text{O.P.}_{\text{int}} = \text{O.P.}_{\text{ext}}$$

Thus the osmotic concentration of the plant cell is taken as being that of the external solution, which is known.

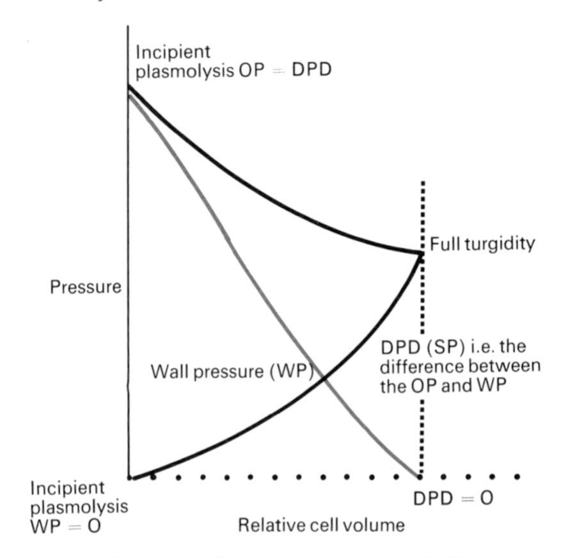

Figure 9.1 *Hoffler diagram showing variations in pressure within a plant cell from incipient plasmolysis to full turgor pressure*
There is a range of elasticity in plant cell walls. Old, thick-walled cells may hardly change their volume as water is absorbed and lost, but thin-walled cells may change in volume by up to 40%.

The whole plant as an osmotic system

Water and salts are absorbed by the roots from the soil and transported up the plant to the leaves, from which the water evaporates. The control of this system is a form of osmoregulation. In plants there are no specific osmoregulatory organs, but in extreme conditions of water shortage certain structural modifications decrease the rate of water loss. Plants with these modifications are known as **xeromorphic** and can be divided into xerophytes and halophytes.

Xerophytes are found in regions of restricted water supply and conditions of potentially high transpiration. When well-supplied with water they tend to transpire faster than other plants, but when water is absent they have the ability to reduce their water loss drastically, and thus survive long periods of drought.

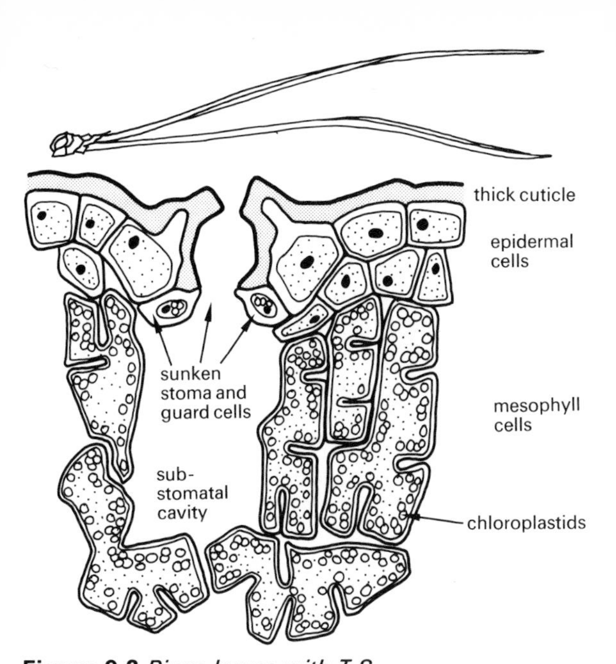

Figure 9.2 *Pinus leaves with T.S.*
The leaves or 'needles' have a reduced surface area and are grooved with sunken stomata in the grooves; both features reduce evaporative water losses in transpiration.

The cytoplasm can survive almost complete desiccation; leaves are usually reduced, leaving the stem as the main site for photosynthesis; stomata are reduced in number and usually protected in pits; and there is a thickened cuticle. In some, to prevent water loss in the heat of the day, the stomata only open at night. Leaf fall also reduces transpiration losses and occurs in temperate regions with the onset of winter, when the soil water is not readily available due to the low temperatures. In arid regions leaf fall in some plants occurs in dry periods, sometimes several times a year.

Some xerophytes (e.g. *Ammophila*) show leaf folding or rolling which reduces water loss through the stomata. Extensive root systems, for the extraction of greater amounts of water from the soil, are also found. Succulent xerophytic plants (e.g. the cacti) store water in special tissues on which they can survive dry periods.

Halophytes are found where the water supply appears to be abundant, but where in fact conditions of physiological drought exist due to the high salt content (e.g. salt marshes and estuaries). Some of these are capable of controlling their salt content by an active osmoregulatory mechanism, for example the Australian saltbush (*Atriplex spongiosa*) has special epidermal bladder cells in which excess salt is actively deposited. Eventually these bladder cells either burst or drop from the leaf, removing the excess salt to the exterior.

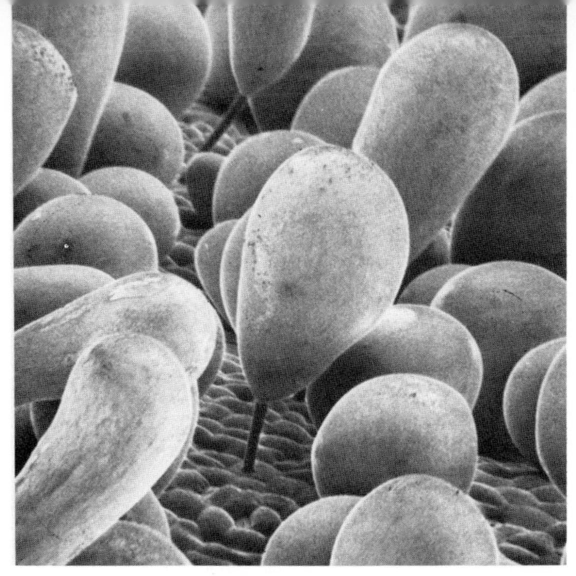

Salt bladders on the leaf of the saltbush *Atriplex spongiosa* (× 460)

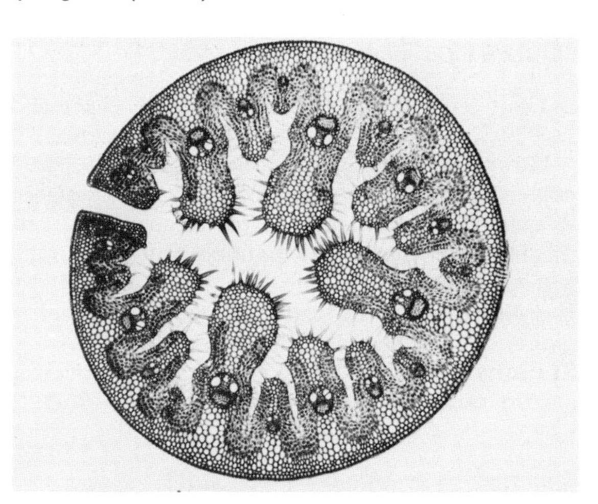

T.S. marram grass (*Ammophila*)

Osmoregulation in animals

Animals fall into two main groups with regard to their osmoregulation: the poikiliosmotic osmotic conformers and the homoiosmotic osmotic regulators.

Poikiliosmotic osmotic conformers do not actually regulate as such, but 'adjust' to various external solutions by allowing their body fluids to equilibriate with the external concentration.
Euryhaline species tolerate wide external, and therefore internal, fluctuations and are also known as tissue-tolerant species.
Stenohaline species tolerate only limited external, and therefore internal, fluctuations and are thus only found in environments of constant concentration.

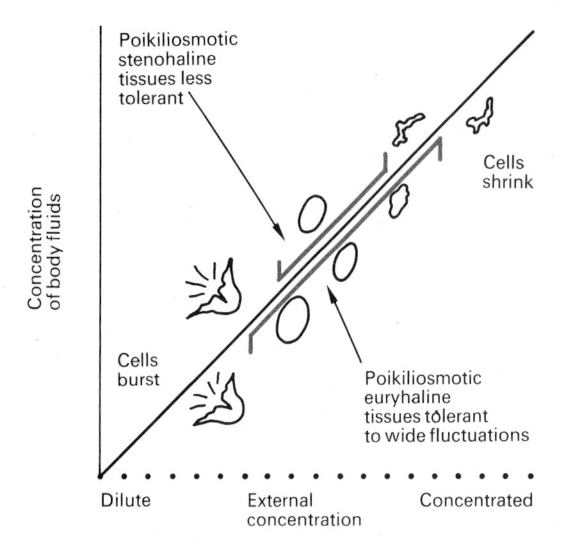

Figure 9.3 *Osmotic conformers*

Homoiosmotic osmotic regulators do have powers of osmotic regulation. They are thus able to maintain a constant internal concentration in the face of external fluctuations, but they vary widely in this ability.
Euryhaline species can regulate over a wide range of external concentrations.
Stenohaline species can only regulate over a narrow range of external concentrations.

Osmoregulation in sea water

INVERTEBRATES

Generally, the cytoplasm of sea-dwelling invertebrates is isotonic[370] to sea water, their surfaces are permeable, they are unable to osmoregulate, and their tissues are not tolerant to wide fluctuations in concentration; they are thus poikiliosmotic and stenohaline. However, the proportions of salts in the cytoplasm are not the same as in sea water and, as the surfaces of the animals are generally permeable, salts tend to enter and leave by diffusion. There must be some powers of active salt regulation to prevent this, and consequently there is some control of water movement.

Protozoa
Generally, the cytoplasm of the Rhizopoda[331] (e.g. marine *Amoeba*) is isotonic to sea water, although, if this is so, it is difficult to understand how they obtain their supply of pure water. When placed in dilutions of sea water they are able to develop contractile vacuoles. However, most Ciliophora[331] (e.g. marine ciliates) do have contractile vacuoles in sea water. This indicates that their cytoplasm is hypertonic[370] and that water does enter by osmosis.

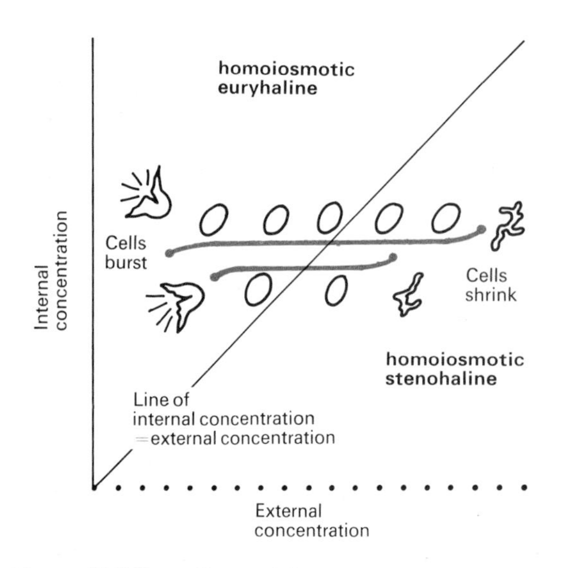

Figure 9.4 *Osmotic regulators*

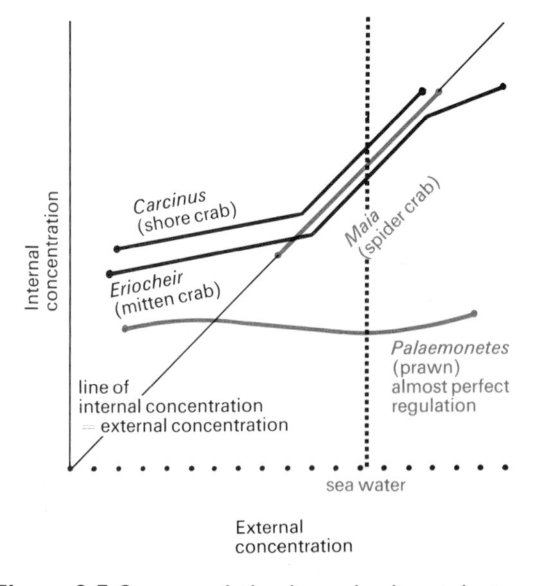

Figure 9.5 *Osmoregulation in marine invertebrates*

VERTEBRATES

The concentrations of inorganic salts in the body fluids are maintained at levels much below those in sea water, and all show powers of osmoregulation.

Elasmobranchs (cartilaginous fish)
The concentration of the body fluids due to inorganic salts is hypotonic to sea water, and water loss and salt gain over the membranes of the body would therefore be expected to occur. However, the blood plasma has an overall concentration slightly hypertonic to sea water, due to the retention of the nitrogenous wastes, urea and trimethylamine oxide.

The kidneys have a high threshold to the filtration of nitrogenous wastes and any that are filtered are reabsorbed by specialized regions of the kidney tubules. The gills and pharynx are also relatively impermeable to nitrogenous wastes, which prevents their loss by diffusion.

All tissues of the elasmobranchs, except the brain and the blood, synthesize urea, and certainly elasmobranch tissues tolerate levels of nitrogenous waste that would be lethal to other vertebrates; some studies even maintain that the heart cannot beat in the absence of urea. As a result of this retention of nitrogenous waste the body fluids are slightly hypertonic to sea water, water enters by osmosis, and a constant flow of hypotonic urine is produced. Excess salts are excreted by the rectal gland.

The keratinized waterproof skin and scales of the elasmobranchs helps cut down the overall water and salt exchanges.

Teleosts (bony fish)
The body fluids of the teleosts have a concentration equivalent to about 50 per cent sea water; they are thus very hypotonic and tend to lose water by osmosis and gain salts by diffusion over the large surface area of the gills.

Teleosts drink sea water and actively take up salts in the gut wall. This gives rise to local areas of hypertonicity, and water then enters passively by osmosis. The absorbed excess salts are then actively excreted out via special cells on the gills.

The kidneys are generally reduced or even aglomerular[190] to reduce the potential water loss via filtration, so only small volumes of urine are produced. The kidney tubules excrete some urea and also reabsorb some salts as required. Urea is also lost over the gills. Again, the generally waterproof skin and scales reduces the overall salt and water exchanges.

Osmoregulation in brackish waters

ESTUARINE WATERS

The point reached by the highest tide represents the lower limit of truly fresh-water animals, while the point reached by the maximum fresh water flow marks the upper limit of the truly marine animals. Mobile animals can migrate up and down with the tide to maintain themselves in their optimum osmotic conditions, but fixed animals are limited in their distribution according to the maximum range of concentrations in which they can survive.

Nearly all estuarine animals are euryhaline, that is they can live in wide ranges of concentrations of external medium. Some are osmoconformers with tissue tolerance, others are osmoregulators maintaining a homeostatic regulation of the body fluids. Those with small surface area to volume ratios can survive fluctuations simply on the basis of there not being sufficient surface for rapid exchanges to occur.

INTERTIDAL WATERS

The fluctuations in salt concentration are less here than in estuarine waters, and are due mainly to the effect of rainfall on shallow sea water. However, the problems faced by organisms in this region are complicated by water loss due to evaporation from permeable surfaces which may be exposed to a drying atmosphere between tides. Thus the common shore crab (*Carcinus maenus*) can maintain its body concentration in dilutions of sea water, and can also survive for up to eight days in air by tolerating a rise in blood concentration of about 33 per cent.

Osmoregulation in fresh water

Successful adaptation to fresh water requires a low permeability to water, efficient powers of water elimination and salt conservation, and the ability actively to absorb salts from the surroundings. In addition, salts are absorbed via the diet.

INVERTEBRATES

Protozoa

There is a very large surface area to volume ratio in protozoa, the plasmalemma is permeable to water, and the cytoplasm is hypertonic to fresh water. A comparatively large volume of water therefore continually enters the organism by osmosis. This tends to increase the cell volume and dilute the cytoplasm, and must therefore be eliminated to prevent disruption of the cell. It is actively eliminated via the contractile vacuole against the prevailing osmotic gradient using energy from respiration. The rate of contraction is proportional both to the external concentration and to the temperature, in addition, any factor that affects respiration will affect the rate.

The continual expulsion of water also flushes away valuable salts, which must be replaced by active uptake from the surrounding water and from the diet.

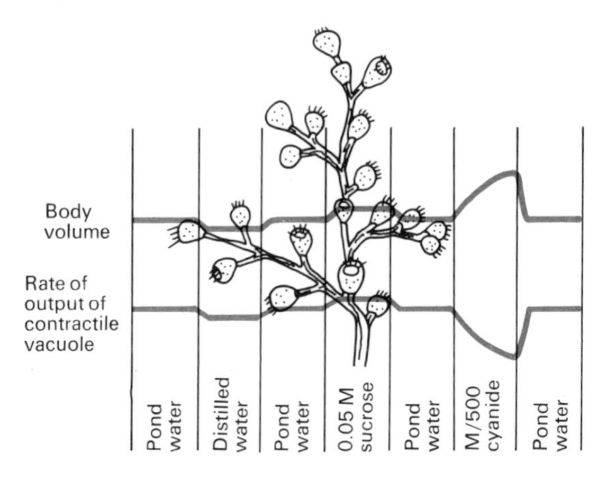

Figure 9.6 *Osmoregulation in the fresh-water protozoan Zoothamnium*

Pond water is hypotonic. Distilled water is even more hypotonic. 0.05 sucrose is less hypotonic than pond water, but still hypotonic to the cytoplasm. M/500 cyanide inhibits respiration and therefore inhibits the action of the contractile vacuole; it is also hypotonic, therefore the body volume rises. At these concentrations the effect is reversible.

Other invertebrates

The larger fresh-water invertebrates have a smaller surface area to volume ratio than the protozoa, and their body surfaces are less permeable to water. Thus, although the body fluids are more hypertonic, proportionally less water enters by osmosis.

VERTEBRATES

These consist of only one major group, the teleosts; although the amphibia can also be considered as a fresh water group with regard to their osmoregulation.

Teleosts

These are very hypertonic to fresh water, which enters over the large, well-vascularized surface of the gill, and as much as 30 per cent of their body weight of water is taken up each day. The kidneys are well-developed, producing a large volume of dilute urine which, although some salts are reabsorbed, represents a drain on the salt content of the body. To counteract this, special cells on the gills actively take up salts from the fresh water against the prevailing diffusion gradient. Teleosts do not generally drink water.

Amphibia

Amphibia are very similar to the teleosts in their osmoregulation. They are hypertonic to fresh water and large amounts of water enter across the skin when they are submerged. The well-developed kidneys produce large volumes of dilute urine which are a drain on the body salt reserves. The skin actively absorbs salts against the prevailing diffusion gradients to make up this loss.

It is interesting to note that many amphibia, when faced with water shortage such as under conditions of **aestivation** (which is a dormancy during the dry season), produce urea as their nitrogenous waste instead of ammonia. The urea accumulates in the body until water is available again. This switch is necessary because of the toxicity of ammonia.

The trend to urea accumulation under conditions of physiological drought is reminiscent of the elasmobranchs. Indeed *Rana cancrivora*, the crab-eating frog of the Philippines, can survive in salinities up to 80 per cent of that of sea water by maintaining a blood level of urea of about 3 per cent. This is comparable to the level found in the elasmobranchs, and serves the same purpose in that it raises the internal body fluid concentration so that it is hypertonic to the surrounding salt water, and can therefore absorb water by osmosis.

Amphibia never drink, therefore all water uptake must be either through the skin, or from the food.

Osmoregulation in the terrestrial environment

The main problem of terrestrial animals is essentially that of water loss by evaporation, given that sufficient water and salts are available in the diet.

INVERTEBRATES

Due to the large surface area to volume ratio of invertebrates, there is a great danger of evaporative water loss over their permeable surfaces.

Insecta
The general exocuticle is permeable to water, but the thin, outer epicuticle of wax or grease forms an effective barrier to the passage of water. In fact, it is much more impermeable than would be expected from its thickness, and this is due to the special alignment of the molecules. Such epicuticles have a **transition temperature** above which this molecular arrangement is disrupted, with a resulting rapid rise in the rate of water loss. This disruption is irreversible for wax cuticles, but with grease it is reversible. In aquatic insects the epicuticle prevents osmotic exchanges over the body surfaces.

About 30 per cent of the total water loss is through the articular membranes.[229] Although the tracheal system[153] drastically reduces water loss by evaporation from the respiratory surfaces, the greater proportion of water loss still occurs through the spiracles. The main organs of osmoregulation are the Malpighian tubules which conserve water by producing a very hypertonic urine with insoluble crystalline uric acid.[194]

VERTEBRATES

The larger size of vertebrates reduces the surface area to volume ratio when compared to the invertebrates, but the problem of evaporative loss over their permeable surfaces remains.

Reptilia
The **cleidoic egg**, with waterproof embryonic membranes and supporting shell, is found in this truly terrestrial group of vertebrates, and is an adaptation to breeding on land.

Reptiles have a keratinized skin and scales which cut down water loss. The kidneys have reduced glomeruli, and produce insoluble uric acid[194] which conserves water in excretion. In addition, the terminal part of the cloaca reabsorbs water from the faeces and the nitrogenous waste products.

Birds
Birds have a cleidoic egg, the kidneys produce uric acid, the urine is hypertonic, and the cloaca reabsorbs water (as in the Reptilia); but they have a large water loss via the respiratory surfaces due to their high metabolic rate and temperature.

Mammals
When compared to the reptiles, mammals are generally more permeable to water, and their greater metabolic rate increases evaporative losses. The problem of water conservation is complicated by the process of temperature control in which water is evaporated to lower the body temperature.[244] The skin is keratinized, which renders it almost waterproof, and the body hair traps a still layer of air which reduces evaporative losses. As in the reptiles and birds, the permeable lung surfaces only connect with the exterior through restricted openings, but substantial losses still occur via the respiratory ventilation movements (up to 0.5 l per day in man).

It is interesting to note the situation in **desert-living mammals** which are under continual water stress. An example is provided by the camel which can survive up to sixty days on green food only and seventeen days on dry food only. The hump is a central depot of subcutaneous fat, which leaves the rest of the body relatively free to act as an efficient radiator of heat, especially at night. During the day the body temperature can rise to 41 °C before the onset of sweating, and this accumulated heat is radiated off at night so that in the morning the body temperature can be as low as 34 °C. Camels can tolerate a loss of 40 per cent of their body water (20 per cent would be fatal to other mammals) and recent work on their red blood corpuscles indicates that these can withstand wide fluctuations in plasma concentration without being damaged.

Thus the camel mainly survives by 'putting up with it' rather than by 'avoiding it'. When water is available, they are capable of rapid water uptake, to an extent that would cause water intoxication in other mammals.

Kidney function

The kidneys are the mammalian organ of osmoregulation. They maintain the normal composition, volume, and reaction of the body fluids—a process which includes nitrogenous excretion. In man they filter and adjust the entire body fluids about fifteen times per day.

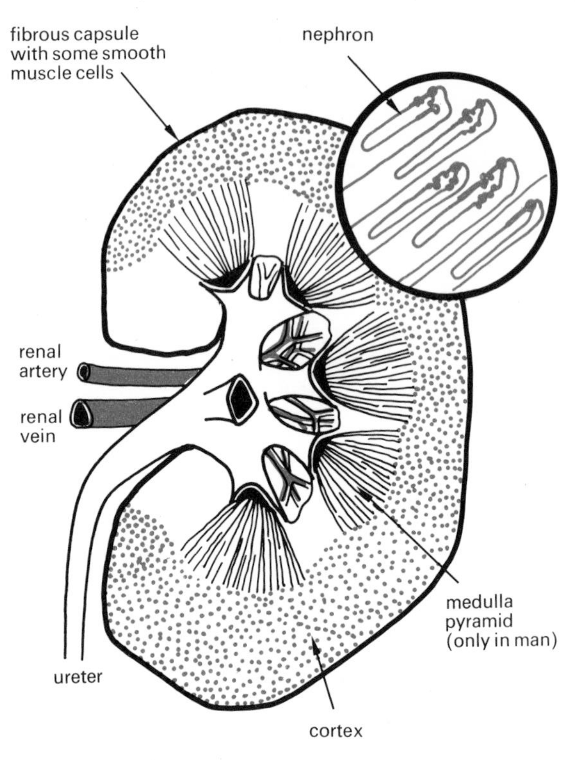

Figure 9.8 *L.S. mammalian (human) kidney*

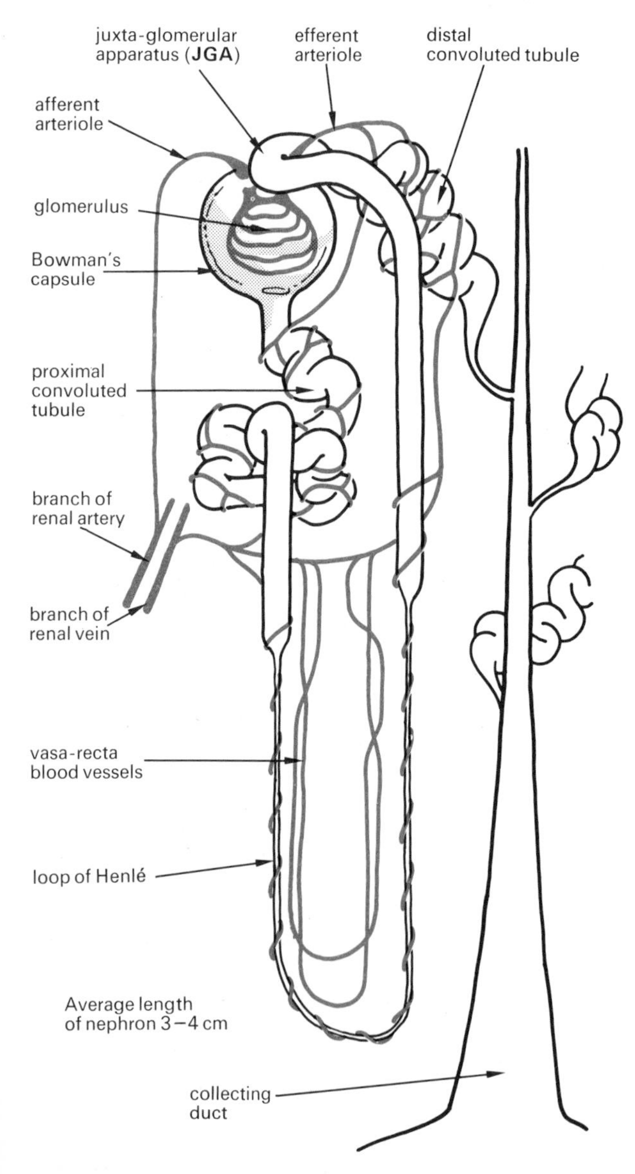

Figure 9.7 *Nephron or uriniferous tubule*

Glomerular filtration

This process is known as **ultra-filtration under pressure**. The renal blood pressure is high due to the normal hydrostatic pressure of the blood and the fact that the glomeruli are capillary beds between arterioles and not between an arteriole and a venule.[178] The blood pressure forces the plasma through the walls into the tubules. The threshold of filtration is at the size of molecules (such as haemoglobin) with a molecular weight of 68 000. One reason why haemoglobin must be enclosed in the red blood corpuscle is to prevent its loss by filtration. The filtration pressure is opposed by the osmotic pressure of the blood so that the **effective filtration pressure** is the difference between the two.

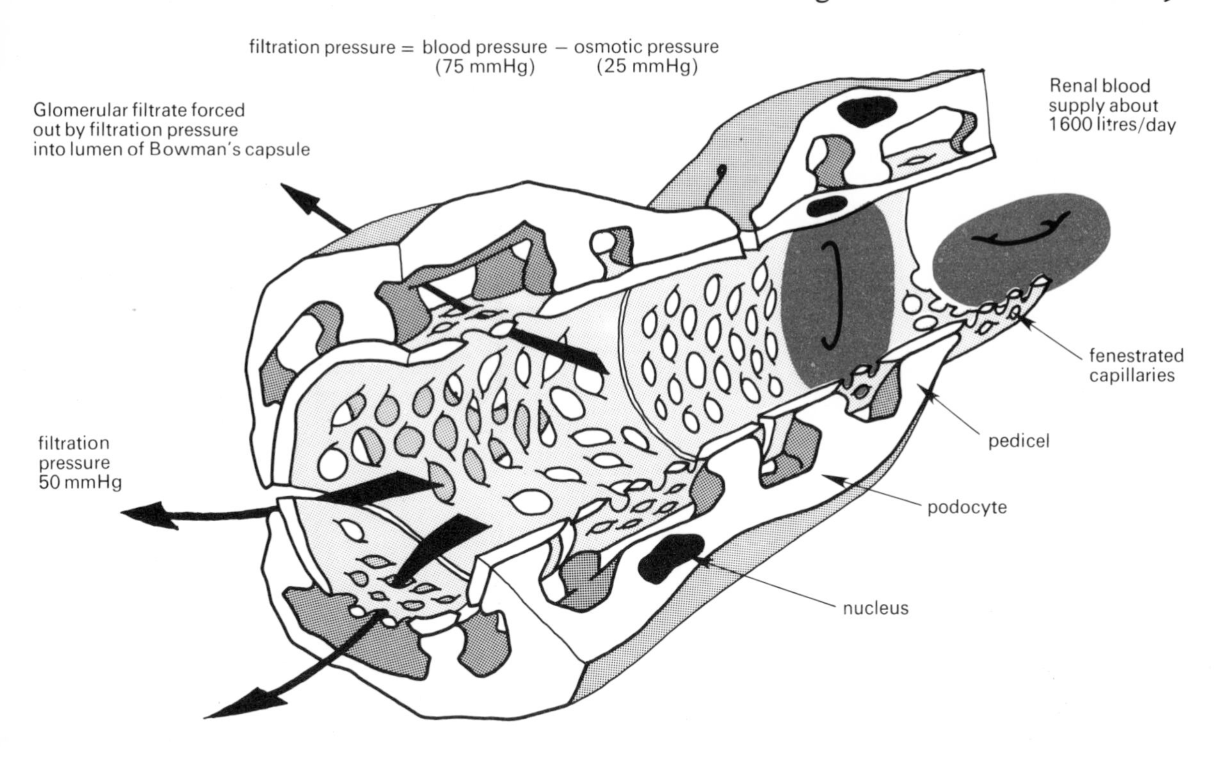

Figure 9.9 *L.S. glomerulus capillary to show glomerular filtration*
The podocytes make up the wall of the Bowman's capsule which completely surrounds all the capillaries. The arrangement of the pedicels allows free passage of the filtrate into the lumen of the Bowman's capsule and the rest of the nephron. In man about 170–200 litres of filtrate are produced each day, 99% of which is reabsorbed by the kidney tubule.

The juxtaglomerular apparatus (JGA) monitors the blood pressure, the plasma volume, and the plasma sodium ion concentration. If any of these drop the JGA secretes the hormone **renin** into the bloodstream. This acts on angiotensinogen (a protein in the bloodstream) to produce angiotensin which raises the blood pressure by causing the constriction of the efferent arterioles and other blood vessels. Angiotensin also stimulates aldosterone secretion from the adrenal cortex which results in sodium retention by the kidneys, and a subsequent retention of water and an increase in blood volume.

Proximal convoluted tubules

These tubules are the major region of reabsorption of useful substances from the filtrate, and 88 per cent of the water, all of the glucose, and about 80 per cent of the salts are reabsorbed here. The proximal tube also adds substances to the filtrate by active secretion. At the end of the region in which all these changes occur, the Loop of Henlé receives a fluid similar in concentration to that of the filtrate, but with only 12 per cent of its volume.

The Loop of Henlé, distal convoluted tubule, and collecting duct

Further reabsorption of useful substances and tubular secretion occur, but the loop is particularly important in the regulation of the pH of the body fluids which are kept alkaline despite an overall production of excess acid by body metabolism. Excess acid is eliminated and useful bicarbonates are reabsorbed.

The loop is impermeable to water; but salts, mainly sodium chloride and urea, are reabsorbed by the loop and continually recycled in the vasa recta blood vessels and build up locally high concentrations in the tissues surrounding the loop and the collecting duct. This leads to the ability to produce a hypertonic urine via the **hairpin countercurrent multiplier system.**

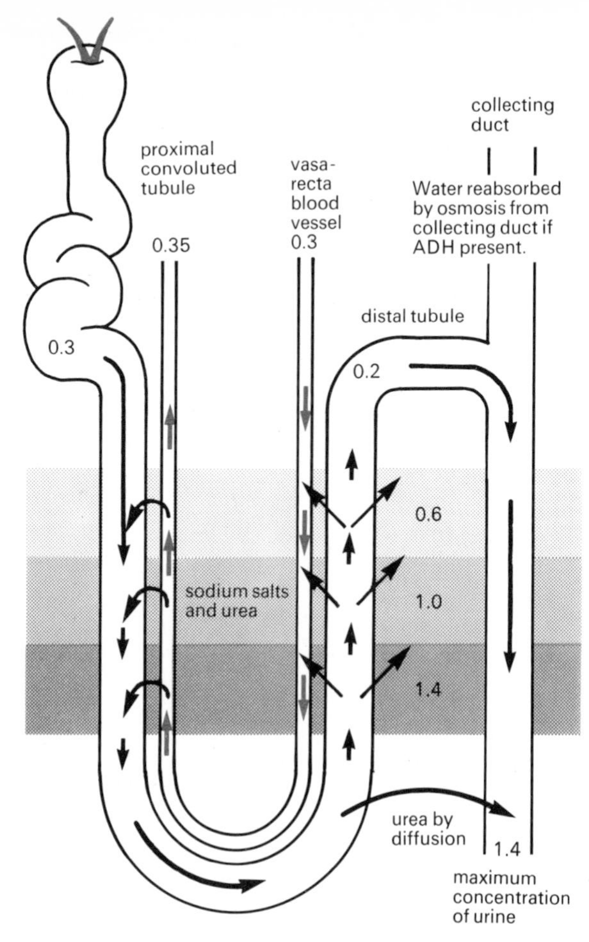

Figure 9.10 *Hairpin countercurrent multiplier system*
The Loop of Henlé is only found in birds and mammals. (Figures are of concentration in osmoles per litre.) If ADH is present, water is reabsorbed by osmosis from the collecting duct.

Control of water reabsorption

Osmoreceptors in the carotid bodies in the neck monitor the osmotic concentration of the blood and send information to the hypothalamus[200] of the brain which, in turn, sends information to the neurohypophysis of the pituitary.[213] The pituitary then secretes an anti-diuretic hormone (ADH) which travels in the bloodstream to the distal convoluted tubule and the collecting ducts where it increases the permeability of their walls to water. In the absence of ADH these walls are virtually impermeable to water, so that all the remaining filtrate is lost. Thus the sequence of events is as follows.

If the **blood concentration increases**, ADH secretion is increased; the permeability of the collecting duct walls to water is increased; more water is reabsorbed; therefore the blood concentration is reduced and the urine becomes hypertonic and decreases in volume. In addition the thirst centres in the hypothalamus of the brain are stimulated by increased blood O.P.

If the **blood concentration decreases**, ADH secretion is decreased; less water is reabsorbed; the urine becomes hypotonic and increased in volume; therefore the blood concentration is increased.

Absence of the ADH due to pituitary malfunction leads to continual and rapid loss of large volumes of dilute urine (up to 24 litres per day in man), a condition known as diabetes insipidus. Alcohol inhibits ADH activity and thus produces similar effects.

FLUID VOLUME CONTROL

Increased blood concentration can be due to either excess salt and/or decreased water; and decreased blood concentration can be due to either decreased salt and/or increased water. Blood volume control is therefore also important, and volume and pressure receptors in the walls of the major blood vessels are also involved in fluid homeostasis[243] in that they relay information to the hypothalamus.

The end product of all these processes is **urine**, which has a variable composition depending upon the state of the body.

Table 9.1 *Average composition of human urine*

	g per 100 cm^3	*Increase over plasma concentration*
Water	96.00	Variable up to × 4.7
Urea	2.00	× 60
Sodium	0.30	× 1
Potassium*	0.15 (This is added by tubular secretion)	× 25
Chloride	0.60	× 2
Ammonia	0.04	× 400
pH	4.7–8.0	—
Proteins	0.00	—
Urochrome pigments (residues from the metabolism of tissue protein)	Variable	—
Total volume	1.50 l per day	—

This composition varies with the metabolism and the particular conditions of salt and water balance of the body.

The minimum obligatory volume in man is 300 ml per day, which is necessary for the elimination of poisonous wastes.

Nitrogenous excretion

This is essentially a feature of animals rather than plants. Nitrogenous wastes arise mainly from amino acid metabolism. Plants synthesize amino acids according to their needs and thus do not produce an excess which needs eliminating. Also their low metabolic rate, when compared to that of animals, reduces the turnover of amino acids in general protein metabolism. As a result there is little production of poisonous nitrogenous waste products; what little there is may be deposited in insoluble form in inter- or intra-cellular spaces, or removed from the plant by leaf fall.

Animals must take in many of their amino acids in their diet and any excess to their needs must be eliminated. Moreover, their more rapid metabolic rate and powers of locomotion and movement lead to a larger turnover of amino acids in protein metabolism and a larger production of nitrogenous waste products. Nitrogenous wastes also come from the metabolism of a variety of nitrogen-containing compounds, for example about 5 per cent of nitrogenous waste in animals comes from nucleic acid metabolism.

Products of protein degradation

AMINO ACIDS

Excretion of amino acids occurs in some aquatic invertebrates, but since it is wasteful of carbon it is not widely found. In man, only traces of amino acids are found in the urine. In order to conserve the valuable 'carbon skeletons' of the amino acid molecules, the amino group is removed in a process of **deamination** and the remainder of the molecule is incorporated into the general carbohydrate metabolic pathways. The amino group is removed as ammonia, which is either excreted as such, or converted to urea or uric acid prior to excretion.

AMMONIA

Ammonia is very toxic, due to its alkalinity, and must be removed as rapidly as possible. It is very soluble and easily diffusible, and is only excreted as an end product of nitrogenous excretion in aquatic animals where plenty of water is available for its rapid removal, as in many aquatic invertebrates and the fresh-water, bony fish.

Animals with ammonia as the end-product of their nitrogenous waste production are known as **ammonotelic**.

UREA

Urea is formed from the ammonia of protein breakdown which occurs in the ornithine cycle in which the amino group is converted to urea via a series of intermediate compounds. The enzymes of this cycle are found mainly in the liver of mammals, and in the kidneys of birds.

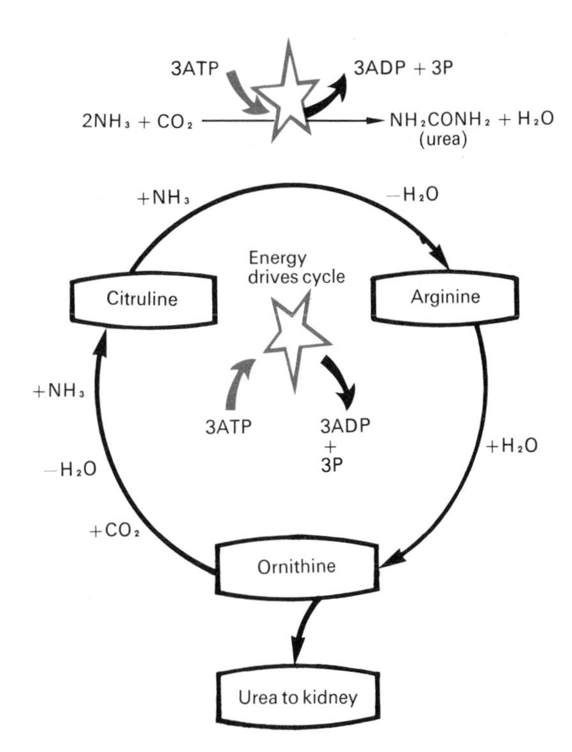

Figure 9.11 *Outline of the ornithine cycle*
This is not necessarily the only system for urea production.

Urea has a carbon:nitrogen ratio of 1:2, thus its excretion conserves carbon relative to the nitrogen lost. It has a neutral pH, a relatively low toxicity, and the tissues can tolerate it better than ammonia; it can therefore be stored and concentrated in a bladder.

Animals with urea as the end-product (e.g. mammals) are known as **ureotelic**.

URIC ACID

This has a C:N ratio of 5:4 and thus represents a proportionally greater loss of carbon to the body than urea. It is almost totally insoluble and therefore non-toxic. It precipitates from solution before the concentration rises to a level that might cause osmoregulatory problems. It is stored and excreted in crystalline form. Animals that have uric acid as their end-product are known as **uricotelic** (e.g. birds, reptiles, insects, and gastropod molluscs).

In man, nucleic acid metabolism gives rise to small amounts of uric acid (0.6–2.0 g per day). Man is therefore **uricogenic**; uric acid is made but is not the main nitrogenous waste end-product. The only other mammal known to produce uric acid is the Dalmatian dog!

Summary

Nitrogenous excretion is never limited to a single end-product in any animal group: all excrete a mixture of compounds, but one usually dominates. The relationship between the main excretory product and the availability of water is an important adaptive feature of the animal in its environment.

Experimental and applied aspects

The simplest type of experimental technique in the investigation of the osmotic concentrations of plant cells and the osmoregulatory powers of aquatic and semi-aquatic amimals is to immerse them in solutions of varying concentrations and to measure any subsequent changes in size, weight, or turgidity; whichever is the more convenient.

With plant material, observations can be made (under the high power of the microscope) of the behaviour of single cells immersed in solutions of different concentrations. That concentration of external solution which just causes incipient plasmolysis is taken as being isotonic to the vacuolar contents at that particular time. Alternatively, that concentration of external solution which causes plasmolysis of 50 per cent of the cells in a block of tissue can be considered as being isotonic to the average cell in that tissue. If the results are plotted on a graph it is possible to derive that particular point even if that particular concentration of solution was not one of those actually used.

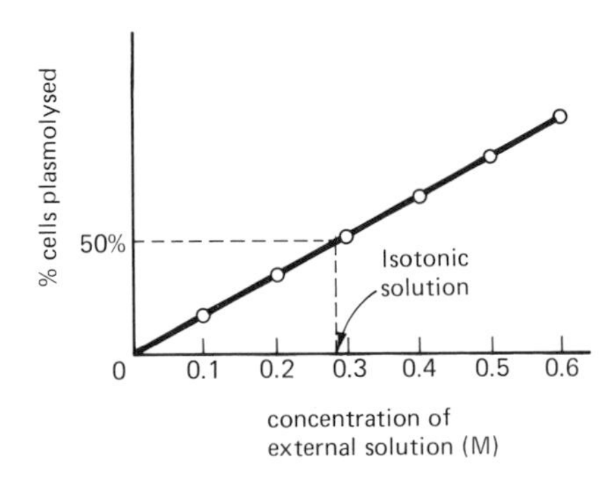

Similarly, changes in weight or length of blocks of tissue in different external concentrations can be plotted.

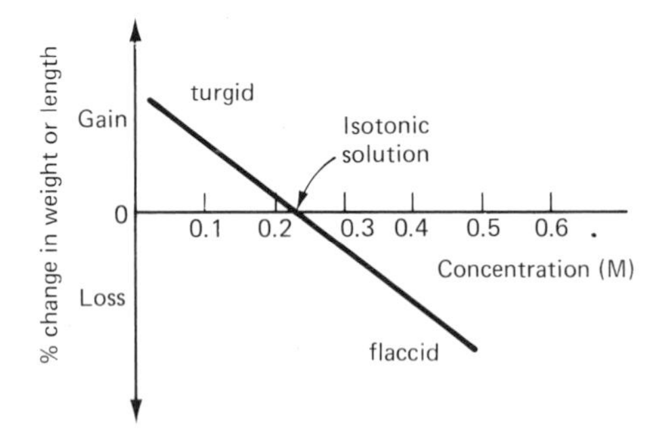

With more advanced techniques, such as micropipetting, in which samples of the actual vacuolar fluid can be taken and analysed, direct measurements of the internal concentrations of plant cells are possible.

Similar studies with animals are complicated by the fact that they usually osmoregulate. Thus an animal that shows no weight or volume change in a particular concentration of solution is not necessarily isotonic to it—it may well be carrying out active osmoregulation to maintain the constant weight or volume. Therefore the volume and concentration of any urine produced should also be measured, a difficult procedure with small aquatic animals.

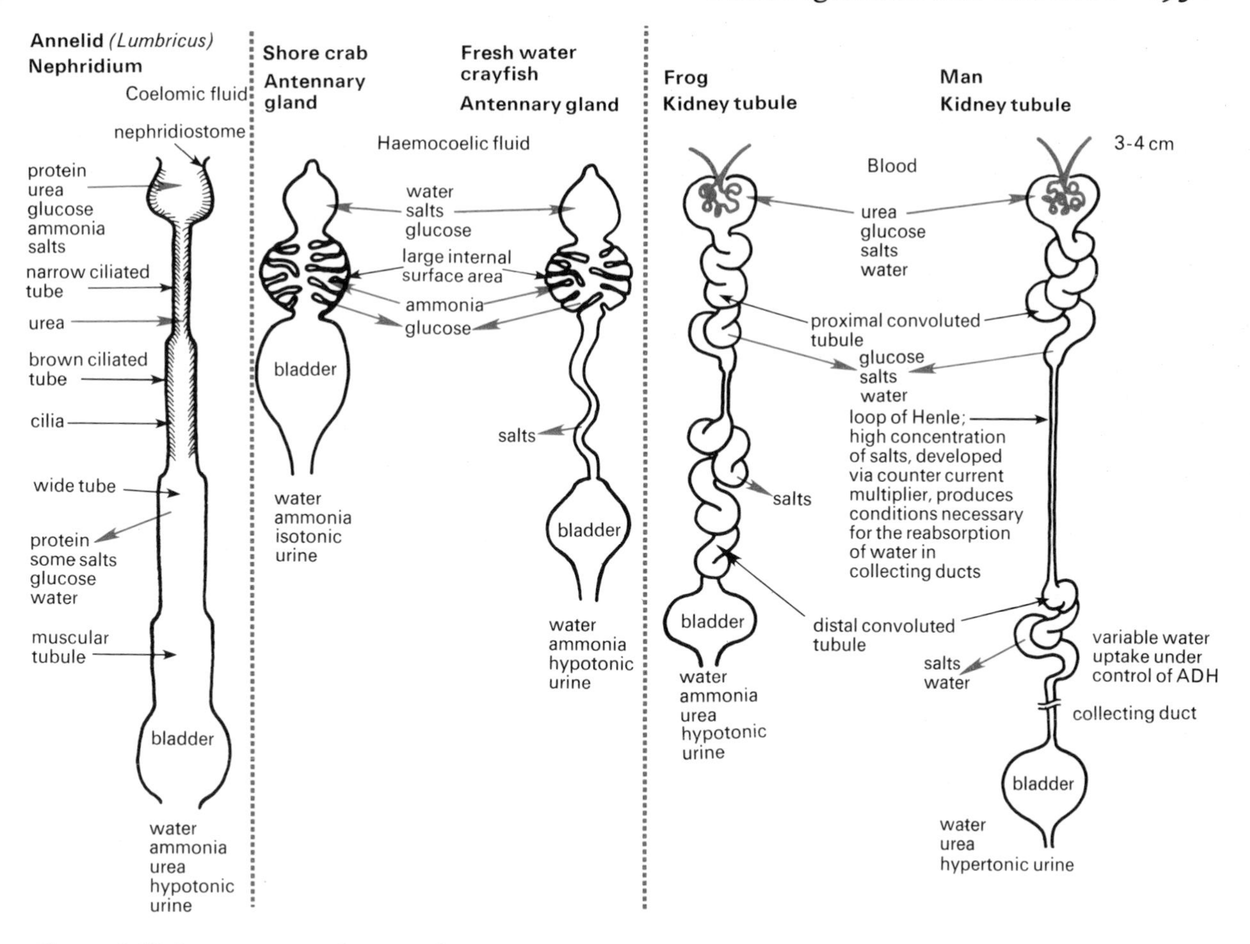

Figure 9.12 *Some osmoregulatory and excretory tubules*

Urine strengths are in relation to body fluid concentration. Urine strength can vary, but these given are typical.

Guided example

1 What similarities in structure are there between the different osmoregulatory structures in Figure 9.12?
Some of the main ones are that they are all tubular; they all have an enlarged region in close contact with the blood or body fluids; and they all lead into enlarged regions known as bladders leading to the exterior.

2 What similarities in function are there between them?
They all remove water and dissolved materials from the blood or body fluids, and reabsorb some before draining urine to the exterior. They are thus all involved in osmoregulation and excretion.

3 Which of the animals that have these structures are capable of conserving water?
Only man can conserve water by excreting a hypertonic urine. An isototonic urine does not conserve water as it is the same strength as the blood and its elimination simply reduces the fluid volume of the animal but does not affect its concentration. A hypotonic urine obviously results in the loss of water.

4 What structural difference is there between the two Crustacean tubules, which is also seen between the two vertebrate tubules?
Both the crayfish antennary gland and the human kidney tubule have extra lengths of tubules not seen in the others.

5 What is the functional significance of the difference between the Crustacean tubules?
The extra length of the tubule in the crayfish reabsorbs salts which are scarce in fresh water. The shore crab does not have this problem as salts are plentiful.

6 What is the functional significance of the difference between the vertebrate tubules?
The Loop of Henlé and the collecting duct in the human kidney tubule are both involved in the production of a hypertonic urine, which conserves water. The amphibian tubule is unable to produce a hypertonic urine and, to this extent, amphibia are poorly adapted to life on land (or in the sea).

7 Why do all such structures have a high metabolic rate?
Because of the active secretion and reabsorption of various substances in response to the continuing needs of the body.

Questions

1 Insects carry out gaseous exchange by means of air tubes, which communicate with the atmosphere by means of several spiracles (openings) on the surface of the body. Some insects are able to carry out ventilating movements of the abdomen.

Three groups of ten cockroaches were placed in desiccators over fresh anhydrous calcium chloride. A fourth group of ten cockroaches was placed in a desiccator lacking calcium chloride. In the first desiccator the air was enriched with 4 per cent carbon dioxide. In the second desiccator the insects were dusted with a fine abrasive powder. In the third desiccator the air was normal, apart from the effect of the calcium chloride, and the insects were untreated. The graph shows the loss of water in each case over a period of three hours.

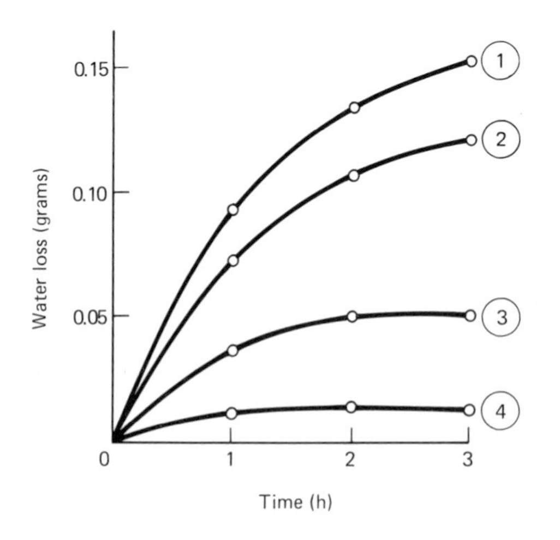

Give possible reasons for the results obtained and suggest one follow-up experiment. (Anhydrous calcium chloride is a dehydrating agent.) (SUJB)

2 What do you understand by the terms osmotic potential and membrane permeability as applied to the water relations of animals?

By reference to *named* examples, specify the osmoregulatory problems faced by a fresh water, a marine, and an estuarine species. In each case briefly outline the means by which the animal overcomes these problems. (L)

3 (a) What is meant by (i) osmotic pressure, (ii) turgor pressure, (iii) suction pressure (or diffusion pressure deficit or water potential) and (iv) plasmolysis?
(b) Discuss whether a plant membrane should be considered to be semi-permeable or differentially permeable.
(c) Explain what you would expect to happen when a plant cell having an osmotic potential equivalent to 5 atmospheres and a suction pressure of 2 atmospheres is (i) immersed in pure water; (ii) immersed in a hypertonic sucrose solution; (iii) immersed in strong acid. (L)

4 An experiment was carried out to estimate the water potential of the cells of a potato tuber.

Six dishes were filled with sucrose solution, molarity ranging from 0.1 to 0.6. A seventh dish contained distilled water. A large number of similar discs was cut from a potato tuber and stored in a covered dish until all were ready. They were then washed quickly, dried carefully, and weighed. The same number of discs was then placed in each of the prepared dishes.

After two hours the discs were dried and reweighed. The percentage change in mass was calculated for each set of discs. The results are shown in Fig. 1.

Fig. 2 shows osmotic pressure plotted against the molarities of the sucrose solutions used.

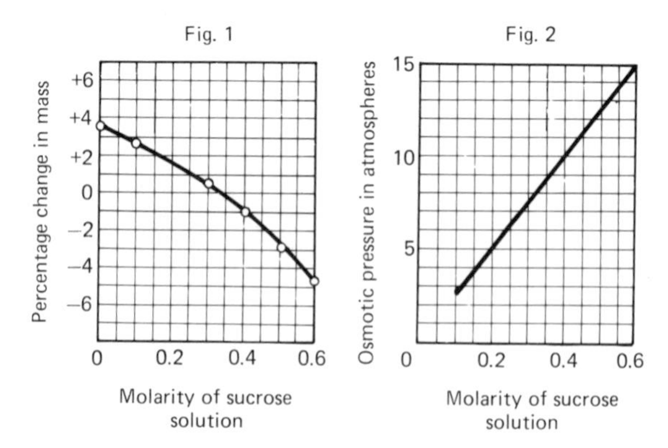

(a) Explain how you would prepare a 1.0 M solution of sucrose. (Relative molecular mass of sucrose = 342.)
(b) What volumes of 1.0 M sucrose solution and distilled water would you use to make up 100 cm of (i) a 0.2 M solution of sucrose, (ii) a 0.6 M solution of sucrose?
(c) Why were the potato discs (i) kept in a covered dish until weighed, (ii) dried carefully before weighing?
(d) Using the graphs, determine the water potential of the potato tuber cells, in atmospheres.
(e) If the investigation had been performed using solutions of potassium nitrate, the solutions would have exerted higher osmotic pressures than equivalent molarities of sucrose. Explain why this would occur. (JMB)

5 In an experiment, two species of protozoan were exposed to different dilutions of sea water for one hour and the number of vacuolar contractions counted. The following results were obtained:

Concentration of sea water per cent	*Number of vacuolar contractions per hour* Species 'A'	Species 'B'
100	0	0
80	0	0
65	0	0
50	6	0
35	20	25
25	42	45
20	56	58
15	64	65
10	63	74
5	20	82
0 (distilled water)	22	90

(a) Explain precisely how you would carry out the above experiment.
(b) Plot the results of the experiment as a graph.
(c) Interpret the results and explain them as fully as you can.
(d) In another experiment small quantities of a mercury compound were added to the 15 per cent concentration of sea water containing the specimens. What results would you expect? Explain your answer. (L)

6 The graphs, A and B, show the changes in weight of two species of shore crab during immersion in sea water diluted by distilled water in the ratio 1:4. Comment on, and suggest reasons for, these results. Describe a suitable control. (SUJB)

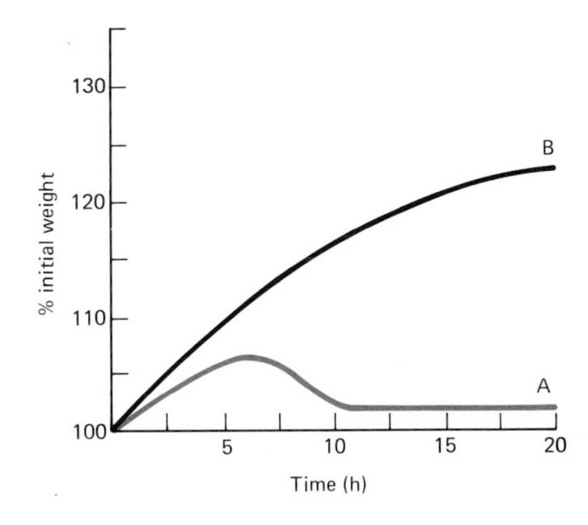

Further reading

Moffat D. B., *The Control of Water Balance by the Kidney*, Carolina/Oxford Biology Reader (Oxford: Oxford University Press, 1971).

10 Sensitivity and co-ordination

NERVOUS SYSTEMS

Central nervous system

In unicellular organisms the entire cell functions as a sensitive and reactive system. In multicellular organisms, all cells retain a degree of sensitivity, but there is a differentiation of cells specialized for this function: the nerve cells. These are essentially the same in all animals, usually being cells with long, fine extensions or fibres. Further nervous complexity is achieved by the increased complexity of the interconnections between them.

Coelenterates have developed specialized nerve cells which are interlinked to form nerve nets; in many cases there are specialized tracts of nervous tissue where the nerve cells are associated closely together.

Platyhelminthes have concentrations of nerve cells, in which the nerve cell bodies form masses or **ganglia** and the fibres form nerve tracts.

The Annelids have their well-developed ganglia and nerves arranged into a true central nervous system, with lateral branches forming the peripheral nervous system.

Arthropods, particularly the insects, have nerve masses, which are sufficiently complex to be called brains, together with a system of ganglia and nerves.

In general molluscs do not have particularly complex nervous systems, but the cephalopod molluscs (octopus and squid) have many ganglia fused to form a brain which is comparable in complexity to that of some vertebrates.

The vertebrates have well-developed central and peripheral nervous systems. The central nervous system consists of the brain and spinal cord, the peripheral system is formed by the lateral nerves. The vertebrate brain develops as an anterior enlargement of the dorsal hollow spinal cord, and is closely associated with the major sense organs. The basic pattern of fore, mid, and hind brains is found in all vertebrates, but is obscured in the mammals, and especially in man, by the tremendous development of special areas, in particular the cerebral hemispheres.

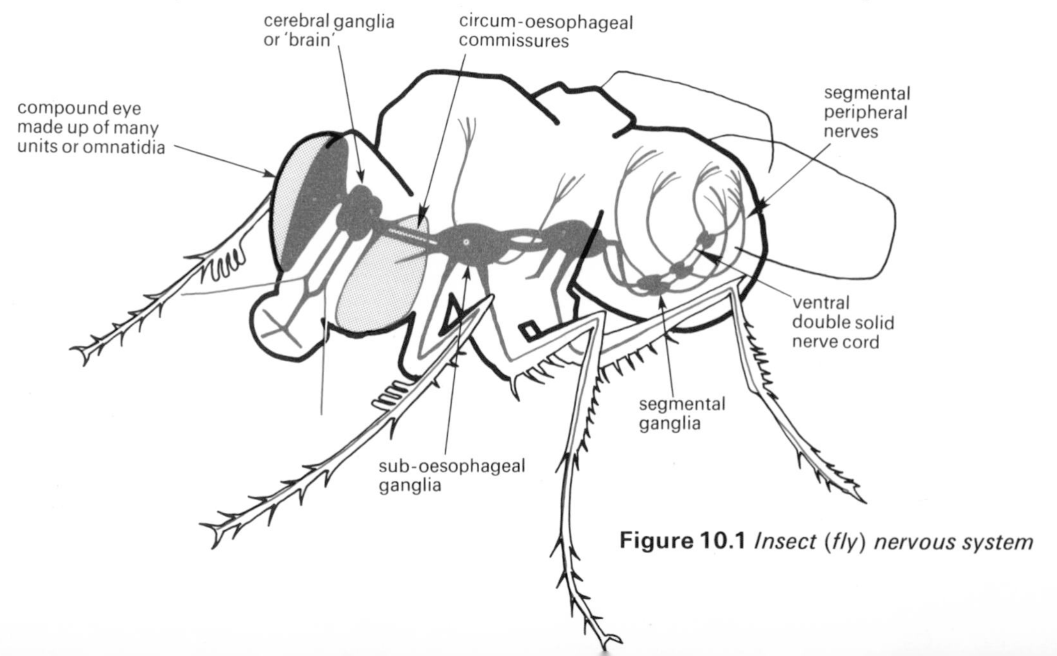

Figure 10.1 *Insect (fly) nervous system*

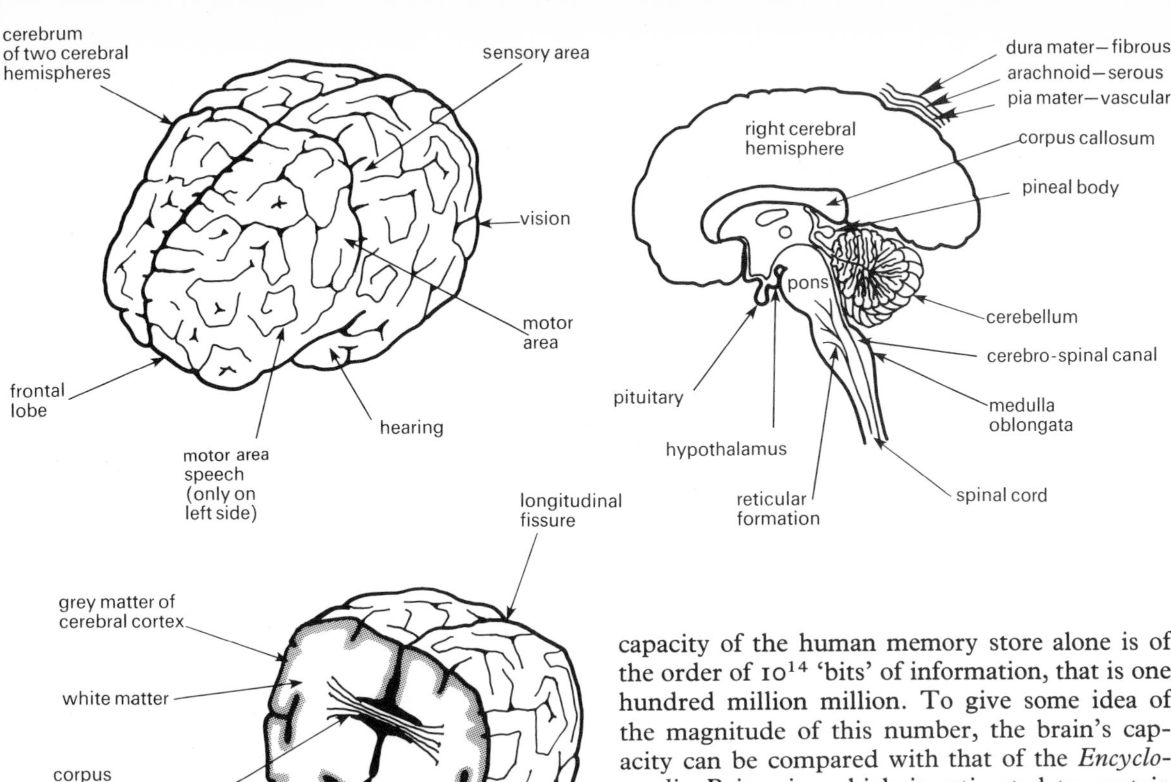

Figure 10.2 *Structure of the human brain*
The meninges (dura mater, arachnoid, and pia mater) cover the brain and spinal cord.

Mammalian brain

CEREBRAL HEMISPHERES

The cerebral hemispheres are well-developed in all mammals, and the greater part is formed by the roof or **neopallium** which is particularly well-developed. The complexity of the functions of the brain depends on the number and complexity of the interconnections between the neurone cell bodies. These are located in the outer cortex of grey matter of the brain, which is about 3 mm thick in man, and which contains 90 per cent of all the nerve cells in the body. In higher mammals, and the primates and man in particular, the cerebral hemispheres and cerebellum have a much folded surface which increases their surface area and underlying grey matter, and hence increases the complexity of the neuronal structure of the brain.

Some idea of the complexity of the brain can be obtained by considering the amount of information it must handle. It is estimated that the capacity of the human memory store alone is of the order of 10^{14} 'bits' of information, that is one hundred million million. To give some idea of the magnitude of this number, the brain's capacity can be compared with that of the *Encyclopaedia Brittanica* which is estimated to contain 2×10^8 'bits' of information, that is two hundred million.

The neurones of the cerebral cortex are continually active, even in the absence of stimulation. In the drowsy state with little sensory input there is a basic rhythm of activity of about 10 cycles/s. This is termed the α rhythm.

The term **committed cortex** is used to describe those areas of the brain where a clear function has been identified, for example hearing, sight, smell, and movement.

The term **uncommitted cortex** refers to those regions of the hemispheres, especially in the frontal and prefrontal lobes, where no clear function has yet been identified. These are assumed to be the areas of man's 'higher' mental activities, such as consciousness, learning, and memory.

Although each hemisphere is an independent entity in itself, there is a slight structural and functional asymmetry between them. At birth the left hemisphere is generally larger than the right and usually develops into the dominant partner which, in man, in effect does the talking, reading, and writing. The minor side (usually the right) recognizes shapes and works in a more 'global' intuitive fashion. Integration of the two hemispheres is achieved by the exchange of information across the **corpus callosum**, a large tract of interconnecting nerve fibres.

CORPUS CALLOSUM

The corpus callosum is a broad band of nerve fibres linking the two cerebral hemispheres which is particularly well-developed in primates and especially so in man.

Although one hemisphere is usually dominant in man, in most respects the two hemispheres are wholly functional independent units, each dealing with one half of the body. A constant interchange of information across the corpus callosum integrates the activities of the two halves of the brain, allowing them to share learning and memory, and to integrate bilateral sensations.

The function of the corpus callosum is dramatically illustrated in people in which it has been severed. Such people function quite normally but the two cerebral hemispheres do not share their experiences. The speech centre is in the left hemisphere, as is that part of the visual cortex dealing with impulses from the right eye. Thus if the right eye is covered and the person is shown an object, they can describe it but they cannot name it. However, if they then touch it, they can name it immediately, since the touch sensations reach the speech centre without having to cross the corpus callosum!

ASSOCIATION CENTRES

Groups of neurones known as the ganglia or **nuclei** are primary correlation centres linking sensory and motor nerves into complex reflex systems. Around these primary areas, and linking them together, are association centres which are involved in a more complex analysis and treatment of information. These reach their highest development in man, particularly in the cerebral hemispheres and the cerebellum. Among those centres unique to man is that which integrates several senses, for example sight, sound, and touch, so that they may be associated in relation to particular objects.

RETICULAR FORMATION

The reticular formation is an ill-defined nerve network in the central part of the brain stem. In some ways it acts as a filter for the stream of impulses from the body to the higher centres of the cerebral cortex. However, it also arouses the cerebral cortex to a state of awareness, with the conscious state depending on an interaction between the cortex and the reticular formation. It is thus also involved in the loss of consciousness that occurs with sleep. Serotonin, produced in the brain, acts on the reticular formation to induce light sleep, and noradrenaline likewise induces deep, dreaming, or so-called paradoxical, sleep.

The reticular formation is also involved in all voluntary and reflex motor activity. The cerebral cortex is only a potentially powerful machine without the reticular formation.

THALAMUS

The thalamus monitors all the sensory input into the brain and is the centre of sensations of pain and pleasure.

HYPOTHALAMUS

This contains groups of cell bodies or nuclei which integrate all the metabolic homeostatic[243] reflexes. All impulses in the autonomic nervous system[202] originate in this region as a result of information received from all parts of the nervous system and blood stream. The hypothalamus is also the main nervous control centre of the endocrine system,[212] synthesizing hormones or 'releasing factors' which cause the release of hormones from the anterior pituitary. It also makes the hormones vasopressin[213] and oxytocin[213] which are neurosecretions transported by nerve fibres to the posterior pituitary. Here they are stored until their release is triggered by the hypothalamus.

Through its connection with the cerebral cortex, medulla oblongata, spinal cord, autonomic nervous system, and endocrines, the hypothalamus is involved in the regulation of heart action, vasoconstriction and vasodilation, respiration, gut function, temperature control, water balance, appetite, reproduction, and many other functions.

Since it also controls the major autonomic and hormonal systems, it can be considered to be the site of the 'biological clock', the existence of which is postulated to explain the control of the circadian, monthly, annual, and life-long physiological rhythms that exist.

CEREBELLUM

The cerebellum is well-developed in mammals and particularly so in man. The surface is convoluted, increasing the cortical grey matter as in the cerebral hemispheres. It is the centre of the co-ordination of voluntary muscle activity, but it does not actually initiate such activity. The initiation of voluntary muscle activity comes from the cerebral hemispheres.

Mammals have direct nervous connections between the cerebral hemispheres and the cerebellum which are not seen in other vertebrates.

MEDULLA OBLONGATA

Many major reflexes have their control centres located in the medulla oblongata, for example the cardiac centre, and vasomotor centre.

BRAIN BIOCHEMISTRY

Little is yet understood about the detailed chemical functioning of the brain. That 10 per cent by volume of the brain cortex is composed of synapses,[205] and that such complex brain functions as consciousness, learning, and intelligence are dependent on synapse functions, indicates the importance of neurohumoural transmitters in the brain. These transmitters in the brain include noradrenaline, acetylcholine, serotonin, and dopamine.

Other chemicals produced in the brain include various peptide substances, including enkephalin, which is a natural opiate, reducing pain; and the so-called 'factor S' which gradually builds up when the animal is awake and leads to sleep.

The continuous synthesis of transmitter substances requires a rich supply of materials and energy. Brain cells carry no respiratory substrate reserves and therefore must be kept continuously supplied with glucose, along with oxygen, from the blood. In man the brain uses 20 per cent of the body's glucose, 20 per cent of the basic oxygen intake, and receives 20 per cent of the blood volume, that is about 800 ml/min. The metabolic rate and blood supply of the cerebral grey matter is about the same as maximally contracting skeletal muscle. The blood supply to the grey matter is not related to the number of nerve cells but to the number of synapses.

BRAIN GROWTH

The tremendous growth of the brain is one of the main features of mammalian, and particularly human, development. The brain growth-spurt in man occurs from about six months after fertilization, at which time the adult number of neurones is present, to about two years of age. It is already 25 per cent of the adult weight at birth, and in the first two years of life it grows at the rate of about 1–2 mg/min, reaching about 75 per cent of the adult weight in this time. During this period the nerve cells grow and branch, forming the complex pattern of synapses which is the basis of human consciousness, learning, and intelligence.

This period is really the only time that the brain can develop properly and is thus a period where correct nutrition is essential. The cerebral cortex in particular is susceptible to malnutrition. There is evidence from mammals other than man that during this growth period the formation of certain interconnections will not take place in the absence of specific sensory inputs; an interesting example of interaction with the environment.

Spinal cord

The spinal cord continues from the medulla oblongata to almost the end of the vertebral column. It is a hollow tube with a fine central canal lined with ciliated epithelium which circulates the cerebro-spinal fluid. It is made up of nerve cells, fibres, and neuroglia cells. Arising from the cord are pairs of segmentally arranged spinal nerves. The cord relays impulses to and from the body and the brain, and is the centre of a large number of spinal reflexes.

Figure 10.3 *Spinal cord and reflex arc involving two neurones*

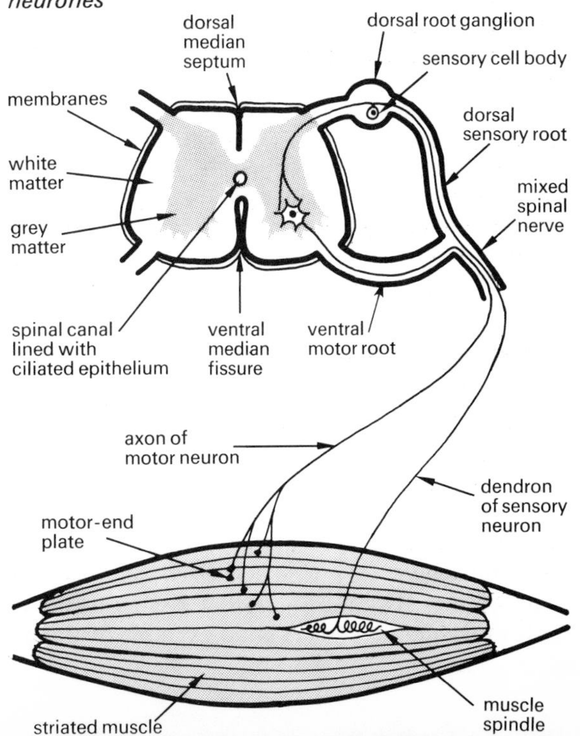

Peripheral nervous system

Cranial nerves

The cranial nerves are paired segmental nerves of a similar origin to the spinal nerves, but which have become concentrated anteriorly by the complex development of the cranium in vertebrates, and especially in mammals. They include the sensory nerves from the sense organs of the head; the motor nerves to the extrinsic eye muscles that move the eyes, and the facial muscles, which are particularly well-developed in man; and nerves which make up part of the autonomic nervous system.

Spinal nerves

The spinal nerves occur in segmentally arranged pairs and consist of sensory and motor nerve fibres running to and from the spinal cord. The nerve fibres are either **somatic**, supplying mainly the skin and voluntary muscles; or **visceral**, supplying the gut, involuntary muscle structures, and various glands. The somatic and visceral fibres are both sensory and motor, and it is the visceral motor fibres that make up the **autonomic nervous system**.

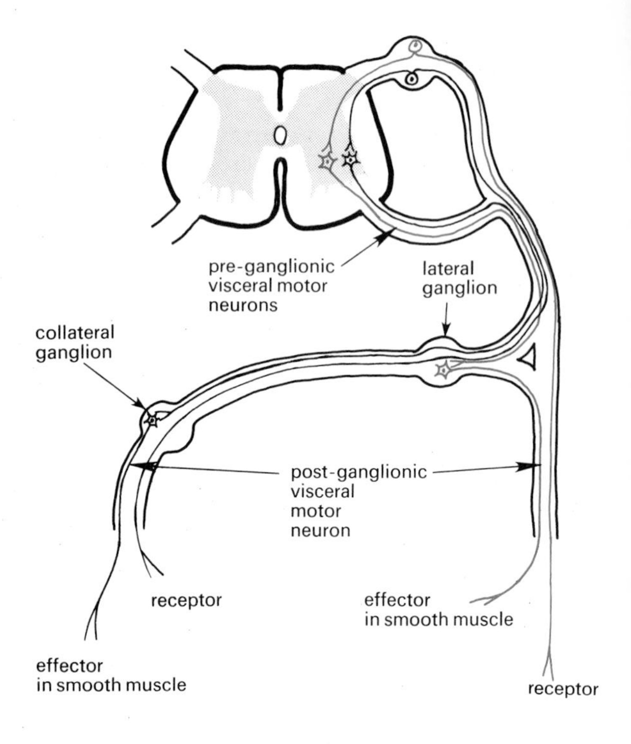

Figure 10.4 *Autonomic nerve pathways*

Autonomic nervous system

The autonomic nervous system is made up of the visceral motor fibres running to the involuntary smooth muscle structures and glands. In this system there are always two nerves on any pathway from the spinal cord to the effector, which synapse in a ganglion. The pre-ganglionic fibres are medullated[204] and the post-ganglionic are non-medullated.

The system operates below the level of consciousness, as the name autonomic suggests, and is involved in the maintenance of homeostasis in conjunction with the hypothalamus of the brain. It is separated into two parts: the sympathetic and the parasympathetic systems.

The **sympathetic system** consists of fibres which leave the spinal cord in the thoracic and lumbar regions and form chains of sympathetic ganglia on either side of the vertebral column. From here postganglionic fibres run to the effectors.

In the lower vertebrates there is basically one pair of sympathetic ganglia in each body segment, but in the mammals many ganglia are fused into larger centres and the clearly segmental arrangement is lost. A good example of this is the fusion of the collateral ganglia of the abdomen into the 'solar plexus'.

The **parasympathetic system** consists of fibres leaving the cranial nerves and sacral region of the spinal cord. The ganglia of this system are found in or near the effectors, and the nerves do not supply the limbs as do those of the sympathetic system.

Many glands and organs are supplied or innervated by both systems, which are often opposite or antagonistic in their action, but which in some cases also work together. However, many tissues have no parasympathetic supply and their activities are increased and decreased by the sympathetic system alone.

The sympathetic system generally tends to act as a unit and prepares the body for emergencies, whereas the parasympathetic system tends to act separately on individual organs and regulates the basic maintenance activities of the body. Amongst many other effects the sympathetic system increases the heart rate, relaxes the bronchial muscles, increases the secretion of adrenalin, and constricts gut sphincters. In contrast, the parasympathetic system decreases the heart rate, constricts the bronchial muscles, promotes gastric gland secretion, and relaxes gut sphincters.

Nervous tissue

Nervous tissue basically consists of actively conducting nerve cells, or neurones, with their satellite, non-conducting, packing and support cells, or **neuroglia**.

Neurones

There is a considerable variety of shape and size between the different types of neurone. The longest are the large motor cells with a relatively small cell body but with an extension or axon fibre up to 90 cm in length. As well as their special conducting properties, neurones have pronounced powers of protein synthesis, and the large nucleus and nucleolus, the many ribosomes and mitochondria are all associated with active synthetic and secretory functions. The cytoplasm also shows very active streaming which is thought to be important in the transport of materials down the long fibres for maintenance and repair, and for the production of neurosecretions.

In multi-polar neurones there are many dendrites receiving information but only a single axon carrying information out of the cell. This indicates that the cell body must have an **integrative function**, that is it can process the incoming information in some way.

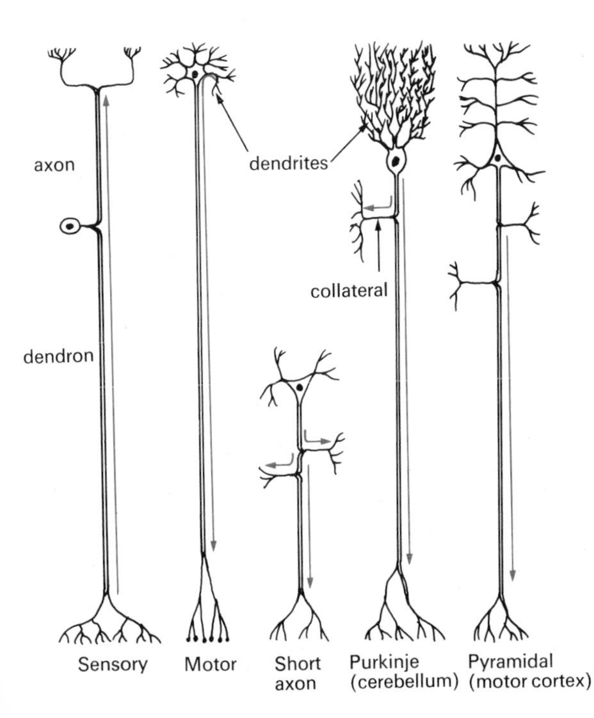

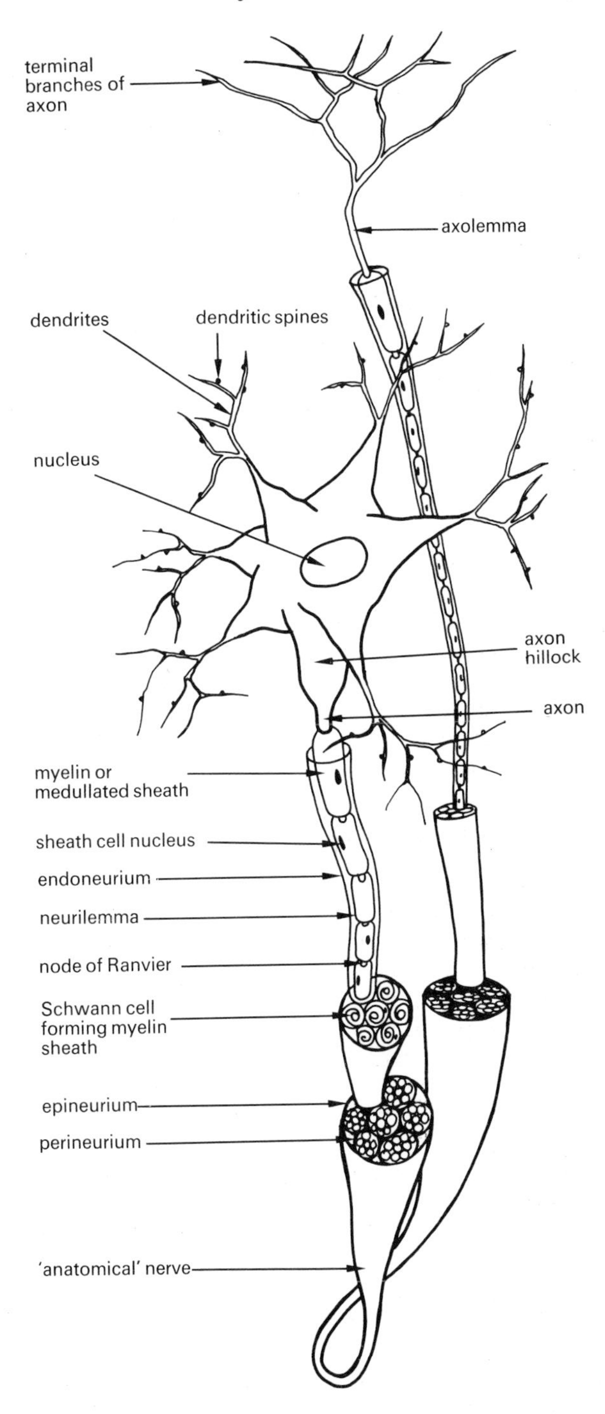

Figure 10.5 *Neurones and nerve*

Figure 10.6 *Several types of neurones for a variety of functions*

Satellite cells or glial cells

There are several types of satellite cells, including Schwann cells, oligodendrocytes, and astrocytes.

Schwann cells surround the axons and dendrons of all neurones in the peripheral nervous system. However, in the **medullated** or **myelinated nerve fibres** these Schwann cells form an expanded fatty sheath around the axon and the dendron fibres. There are several Schwann cells along the length of each fibre, and the points where they are separated from each other are known as the Nodes of Ranvier.

In the **non-medullated fibres** the Schwann cells are still present but do not expand to form a fatty sheath. Non-medullated fibres are found in the post-ganglionic visceral motor fibres of the autonomic nervous system and in somatic fibres of less than 1 μ in diameter.

Oligodendrocytes carry out a similar process of myelination, but within the central nervous system.

Astrocytes are usually found in the brain, close to blood capillaries, with their 'end feet' in contact with the endothelium of the capillary wall. A role in the transport of nutrients from the blood to the neurones has been suggested, but has not been proven.

Nerve impulse

The impulse is initiated by the stimulation of the neurone either by an impulse from another neurone or by a stimulus from a sensory terminal. An impulse is a physico-chemical change, involving alterations in the electrical and chemical state of the neurone.

RESTING POTENTIAL

All living cells show a difference of electrical charge or potential across their membranes. This is the result of a complex of passive and active mechanisms affecting the distribution of charged ions inside and outside the cell.

Nerve cells are adapted to emphasize this tendency and to utilize it in the formation and transmission of nerve impulses. When at rest the neurone has a resting potential of about 50–75 mV across its membrane, being more positively charged outside and more negatively charged inside.

ACTION POTENTIAL

When stimulated, the neurone membrane suddenly becomes permeable to sodium ions which rush in and make the inside of the neurone positive to the extent of 40 mV, in a process known as depolarization. It is the passage of this wave of depolarization along the neurone that constitutes the impulse.

An impulse is either transmitted in full or not at all, according to the 'All or Nothing Law'. It also travels without decrement; that is it is as strong at the end of the fibre as it was at the beginning. Since the action potential does not vary with the nature of the receptor, the type of sensation, such as heat, cold, touch, etc., does not depend on the character of the impulse.

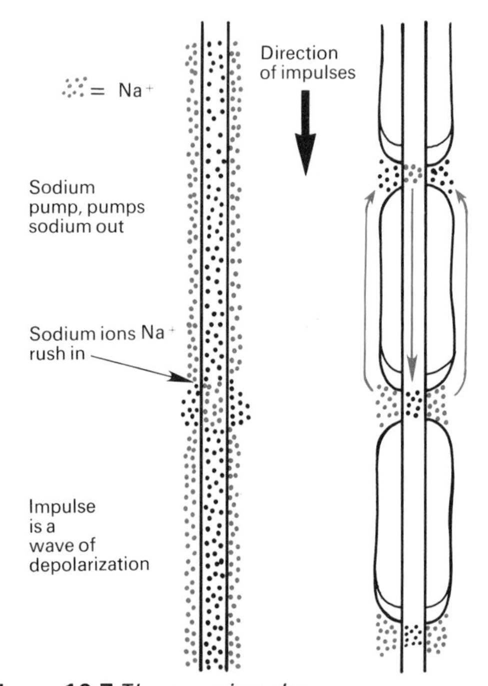

Figure 10.7 *The nerve impulse*
In a myelinated fibre the current flow is restricted to the nodes of Ranvier in what is known as saltatory conduction. This speeds the rate of transmission and conserves energy.

The speed with which the impulse travels is proportional to the diameter of the nerve fibre, reaching a maximum of 120 m s^{-1} in the largest medullated nerve fibres. The myelin sheath plays an important role in increasing the speed of transmission of impulses along the fibre. In myelinated fibres the current flow is restricted to the Nodes of Ranvier in a process known as **saltatory conduction**, which also serves to conserve energy.

REFRACTORY PERIOD

After the passage of the action potential or wave of depolarization through a region, the resting potential must be regenerated. Whilst this process is going on that region is **inexcitable** or **refractory**. The resting potential is regenerated by means of the active sodium pump transporting the sodium ions back to the outside of the membrane, and by the membrane regaining its original impermeability to their inward rush. The large fibres recover in about 1 ms, and therefore theoretically could transmit impulses at a rate of 1000 each second. In fact the fastest observed rate is about 100 a second.

The synapse

Neurones (with very few exceptions) do not come into direct physical contact with each other. At the points where they communicate with each other they are separated by small gaps or **synapses**. These can occur between all parts of the nerve cells. One neurone may receive up to several thousand synaptic contacts, particularly in the brain.

Synapses can be considered as the 'decision points' of the nervous system, as a variety of factors, which can affect the functioning of the nervous system as a whole, also influence them. This would not be so possible if an electrical impulse travelled along an uninterrupted fibre.

In all parts of the nervous system, except the autonomic nervous system, nearly all the synapses are in the brain or the spinal cord; the exceptions are found in the retina (which develops embryologically from an outgrowth of the brain).

SYNAPTIC TRANSMISSION

When an impulse arrives at a synapse it triggers the release of a chemical transmitter substance which diffuses across the gap and stimulates the post-synaptic neurone to initiate an impulse. There is no common transmitter substance known in the central nervous system, but in the peripheral nervous system there are two main transmitter substances: acetylcholine and noradrenaline.

Those nerves that use acetylcholine are known as **cholinergic** and include the somatic nerves, the pre-ganglionic fibres of the autonomic nervous system, and the post-ganglionic fibres of the parasympathetic branches of the autonomic nervous system. Those that use noradrenaline[214] are known as **adrenergic** and include the post-ganglionic sympathetic branches of the autonomic system.

Acetylcholine is synthesized in the presynaptic terminals and, when an impulse arrives, is released in quanta or 'packets' from the storage vesicles. It diffuses rapidly across the gap and combines with special receptor sites on the postsynaptic neurone terminals, somehow triggering a wave of depolarization in that neurone. However, some nerves are inhibitory, and the transmission of the neurosecretion inhibits the initiation of an impulse in the postsynaptic neurone. The acetylcholine is rapidly removed by diffusion away, and by breakdown by the enzyme cholinesterase. This 'clears' the synapse for the next transmission.

Synapses are found in some invertebrates where there is a very tight junction between neurones, and the synaptic transmission is electrical.

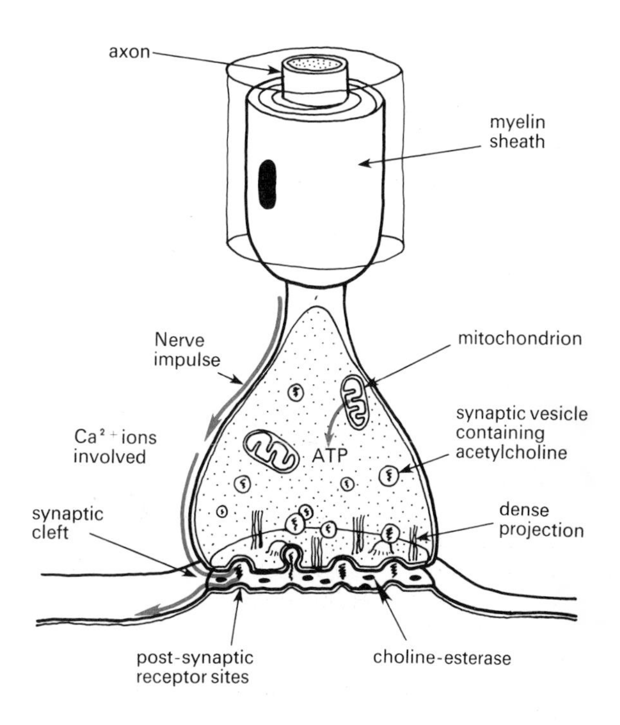

Figure 10.8 *The synapse and synaptic transmission*

The reflex arc

The neurone is the structural unit of the central nervous system (CNS), but the functional unit on which the CNS's activity is based is the reflex arc.

There is a hierarchy of reflex actions from the simplest, based on the spinal cord, to those of the higher centres of the cerebral cortex. All neurones within the system are interlinked and this enables conditioned reflexes, which blend into the higher behavioural patterns of animals, to be established.

Mammalian sense organs

Although there is a wide variety of sense organs, all share certain common characteristics.

(a) They are parts of the nervous system which respond to some stimulus or change in the environment.
(b) They **transduce** or change environmental energy into electrical energy in the form of a sequence of nerve impulses.
(c) They are generally specific, with morphological, histological, and physiological differentiation to concentrate on a particular type of stimulus. They are known as peripheral analysers because they analyse the complex of environmental stimuli at the 'edge' of the body.
(d) The varying intensity of stimuli is translated into varying frequency of transmission of impulses.
(e) With continuous or repeated stimulation most sense organs show **adaptation** in the form of a generally reduced sensitivity.
(f) Apart from some preliminary processing of the information in some sense organs before transmission of the impulses, the interpretation of these impulses as external events is a function of the brain.

Classification of the sense organs

Mammalian sense organs can be classified according to the type of stimulus to which they react.

Exteroceptors respond to external stimuli, including smell, taste, touch, sound, and light.
Interoceptors respond to internal stimuli, including any change in the interior homeostatic state.
Proprioceptors are stimulated by changes in the muscular and skeletal systems, and include the part of the ear concerned with balance.

The ear

BALANCE

The labyrinth or vestibular apparatus, consisting of the semicircular canals, the sacculus, and the utriculus, is a system of special proprioceptors which is stimulated by gravity and movements of the head. It is important in the maintenance of correct posture, expecially in man which has the erect bipedal habit. Its function depends mainly on movements of fluid affecting specialized collections of sensory cells: the cristae and maculae.

HEARING

Hearing involves the detection of sound waves or vibrations in the air. These have different frequencies (measured in cycles per second or herz (Hz)) which are detected by the ear and brain as differences in pitch. About 2000 gradations of pitch are detectable by the human ear. It is most sensitive to sound energy and pitch variations in the range used in speech (1000–3000 Hz). The ear is relatively insensitive to lower frequencies

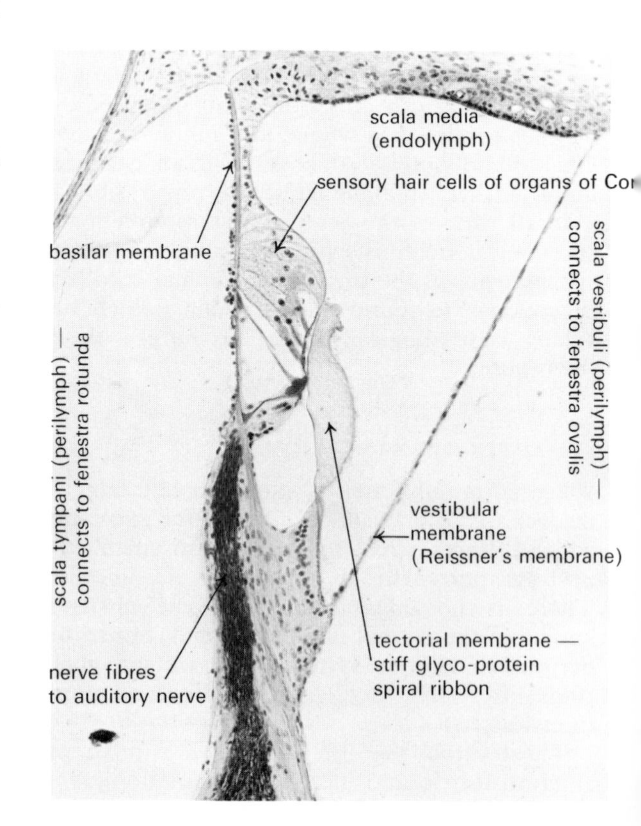

T.S. Organ of Corti

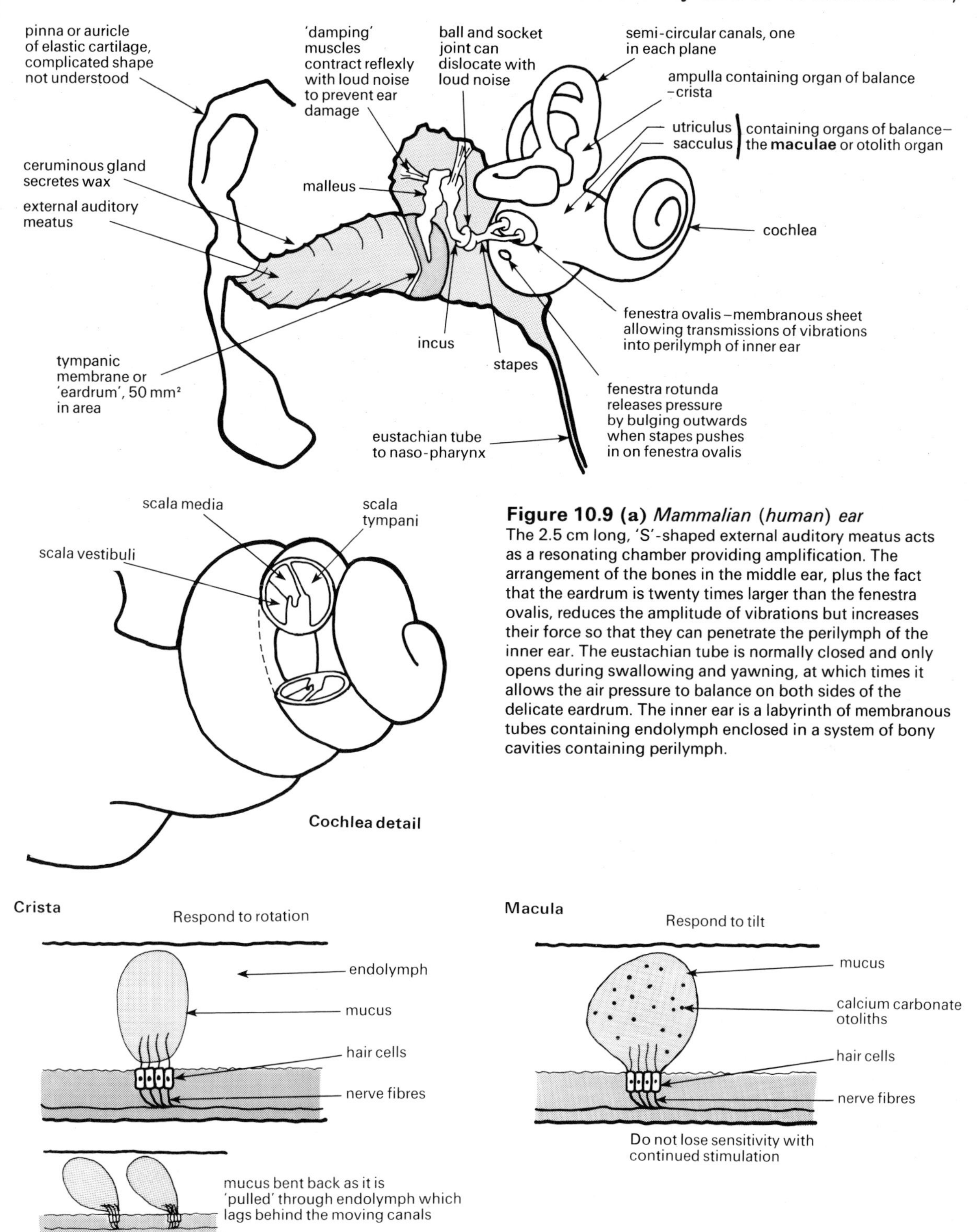

Figure 10.9 (a) *Mammalian (human) ear*
The 2.5 cm long, 'S'-shaped external auditory meatus acts as a resonating chamber providing amplification. The arrangement of the bones in the middle ear, plus the fact that the eardrum is twenty times larger than the fenestra ovalis, reduces the amplitude of vibrations but increases their force so that they can penetrate the perilymph of the inner ear. The eustachian tube is normally closed and only opens during swallowing and yawning, at which times it allows the air pressure to balance on both sides of the delicate eardrum. The inner ear is a labyrinth of membranous tubes containing endolymph enclosed in a system of bony cavities containing perilymph.

Figure 10.9 (b) *Function of cristae and maculae in balance*

and thus avoids hearing the vibrations generated by the muscles and skeleton during movement.

Other terms, such as loudness and noise, are more subjective and depend on a complex of psychological factors.

The ears can locate sound from the right or the left on the basis of the different time taken for the sound waves to reach each ear. The pinnae are thought to be involved in the location of sound from the front and the back, although exactly how is not clear.

In many mammals warning cries are in a range of frequencies where detection of source of sound is poor; cries of the young are in the range where detection of the source is good; both have survival value.

The sound of one's own voice is essential in the feed-back control of speech; in other words one must be able to hear oneself to be able to speak. When speaking one hears one's own voice by air and bone transmission, whereas the listener only receives air-transmitted sound which carries fewer of the lower frequencies; thus we never hear ourselves as others do.

Path of sound waves

The vibrations of the air which are to be interpreted as sound set up a sequence of vibrations from the tympanic membrane, through the malleus, incus, and stapes to the oval window membrane and then via the perilymph and endolymph of the inner ear to the basilar membrane and Organ of Corti. The basilar membrane vibrates, and the hair cells of the Organ of Corti, which are attached to the basilar membrane at one end and the stiff tectorial membrane at the other, are stimulated mechanically and transduce this information into nerve impulses. The basilar membrane is about 31 mm long and has about 25 000 receptors attached to nerve fibres. This description of the transduction of these vibrations into electrical vibrations by the mechanical movement of the sensory hair cells is a simplification of a very complex process involving a potential difference across the membranes between the endolymph and perilymph.

The impulses from each ear are transmitted via the auditory nerve and auditory pathways in the brain equally to the two cerebral hemispheres. The two interact in such a way that two ears are more than twice as good at detecting sound as one.

The eye

IRIS AND SIZE OF PUPIL

The iris is a muscular diaphragm with a central opening called the pupil. The size of the pupil controls the amount of light entering the eye. The iris contains smooth muscle fibres, laid down in a circular pattern, which, when they contract under the stimulation of parasympathetic nerves of the autonomic nervous system, cause the pupil to constrict. This happens in bright light to prevent too much light entering, and during near vision. The sympathetic nerves act on radially arranged smooth muscle which, when contracted, enlarges the pupil. This occurs in decreasing light and during distant vision.

LIGHT REFRACTION OR FOCUSING

Cornea

When light passes from one medium, such as air, to another, such as the transparent tissues of the eye, it is bent or refracted, and this leads to an image being focused on the retina. The greatest refraction in the eye occurs at the air–cornea surface. This refractive power is lost under water, so that the difference between the clarity of vision under water and in air is a measure of the contribution of the air–cornea surface to the focusing power of the eye.

Lens

The lens of the eye is biconvex and produces an inverted image on the retina. It is made of ribbon-like epithelia arranged in concentric layers in a highly elastic capsule. It has no direct blood supply and receives and eliminates materials via the aqueous humour.

The refractive power of the lens is about one-third that of the cornea but can be adjusted by changing the shape of the lens. This occurs during a process known as accommodation.

Accommodation

The eye at rest is focused on distant objects. The ciliary muscles are relaxed, the suspensory ligaments are taut and pull the lens thinner and flatter, and the image is focused on the retina. When focusing on objects less than 6 m away three things occur: the eyes converge on the object; the pupils constrict; and accommodation occurs when the ciliary muscles contract and release the tension on the suspensory ligaments.

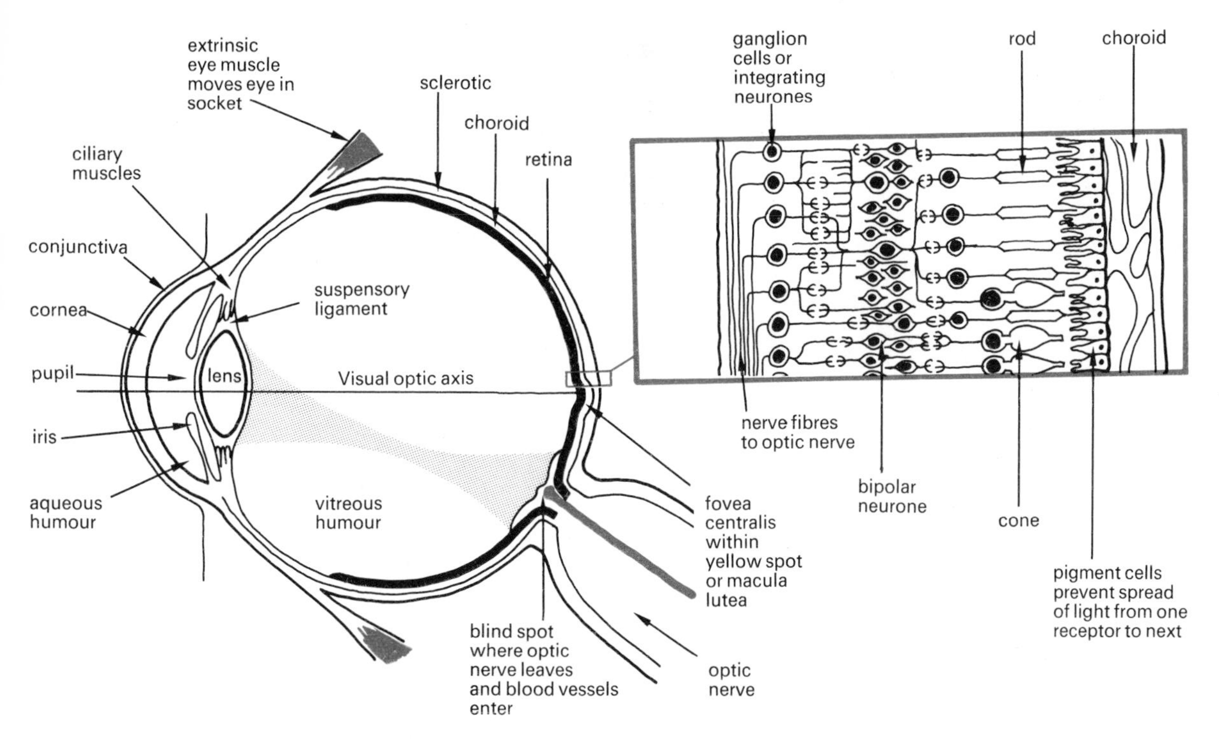

Figure 10.10 *Mammalian (human) eye (horizontal section)*

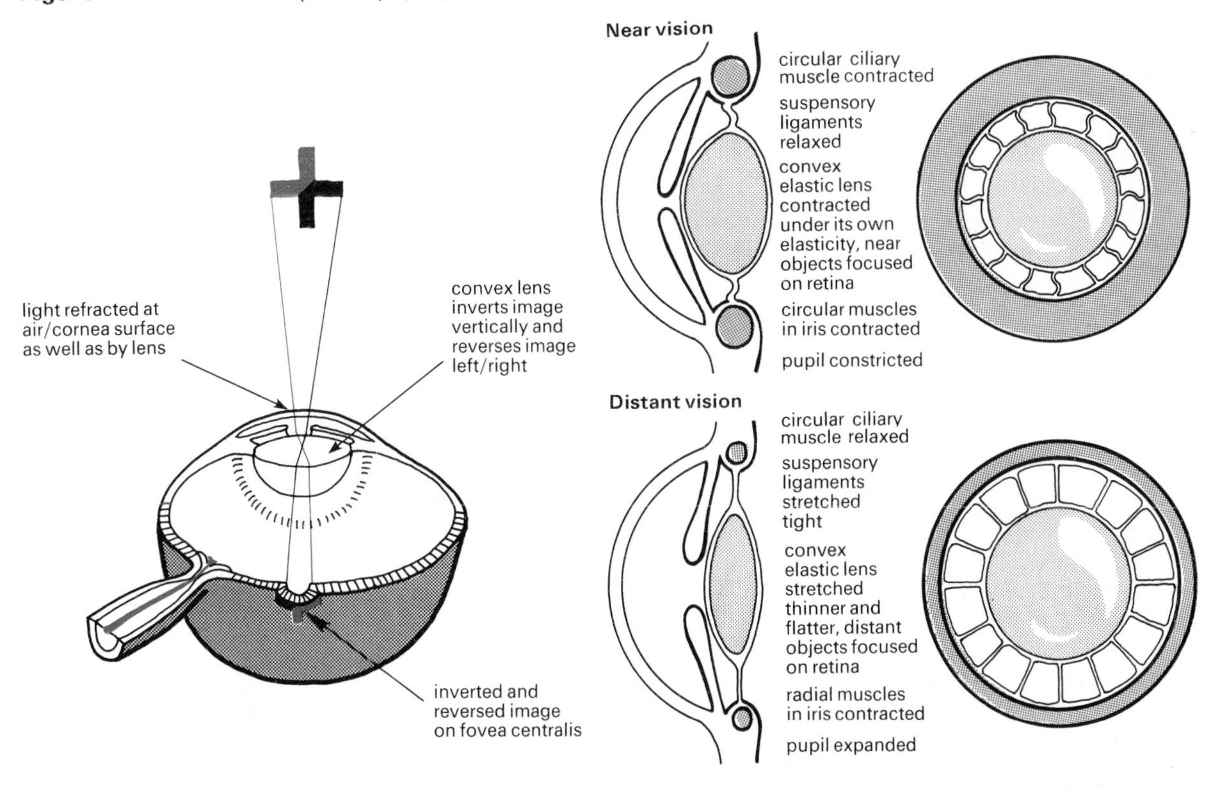

Figure 10.11 *Optical paths within the eye*

Figure 10.12 *Accommodation of the eye in distant and near vision*

This latter allows the elastic lens to become shorter and fatter. The change in the curvature of the lens brings the image of the near object to a focus on the retina. With age, the lens loses its elasticity which leads to a poorer ability to focus on near objects.

Chromatic aberration
Lenses tend to bend short wavelengths (the blue end of the spectrum) more than the red or long wavelengths; therefore the image of a point of white light is a blurred circle fringed with colour. This is known as chromatic aberration. The human eye lens has a large chromatic aberration which it overcomes by filtering out some of the blue end with yellow pigments.

The yellowish colour of the lens cuts out light near ultra-violet; as this yellowness increases with age, more and more blue is cut out. Lensless eyes give good vision in the ultra-violet range. The macula lutea or yellow spot contains the yellow xanthophyll pigment which will further absorb some of the blue end of the spectrum and help overcome chromatic aberration.

RETINA

The retina is the light-sensitive layer of the eye consisting of specialized sensory cells, called the rods and cones, and many integrating nerve cells. There about 125 million rods and 7 million cones

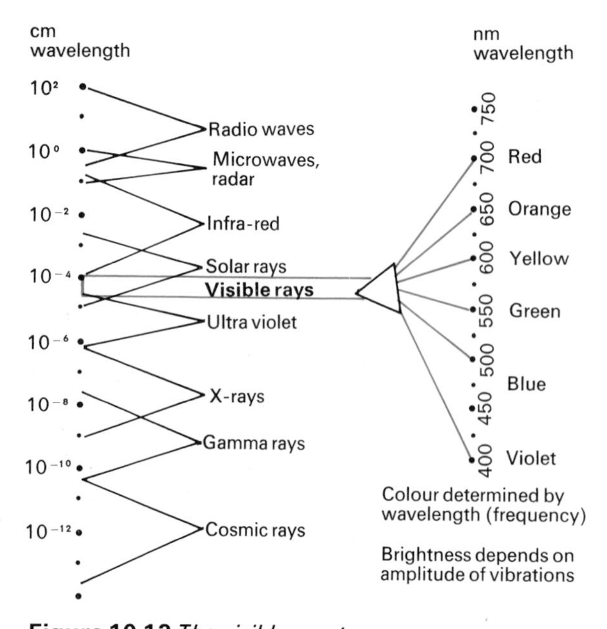

Figure 10.13 *The visible spectrum*

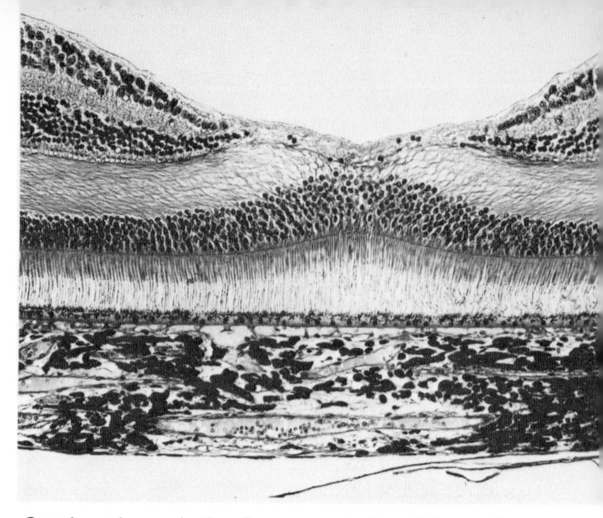

Section through the fovea centralis of the retina showing the deflection of the overlying nerve layers

in the retina, but only about 800 000 nerve fibres in the optic nerve transmit the impulses to the brain; thus considerable processing and integration of impulses must occur before transmission to the brain. Indeed, single nerve cells in the visual cortex of mammals are known to 'fire' only when the image of some specific shape falls on a specific retinal field. Retinal cells must, therefore, be integrated in some way to detect specific shapes and contours. The development of the retina in the embryo as an outgrowth of the developing brain supports the supposition that the retina is concerned not only with the detection of light but also with the partial processing of the information before transmission.

Cones
The cones are sensitive to bright light and colour. Exactly how they detect and transmit information about colour vision is still not known. The fovea centralis or central depression contains only cones, each of which is connected to its own nerve cell. The depression increases the number of cones exposed to the light at that point and the superficial nerve layers of the retina are deflected away from above the cones so that the light has an uninterrupted path to them. All of these features make the fovea the area of detailed daylight vision where the image is focused.

Rods
Rods are much more sensitive than cones and are involved in vision in dim light. They are insensitive to colour. There are no rods in the fovea but they increase in number towards the edge of the retina. Up to 300 rods may be con-

nected to a single nerve cell, giving poor visual acuity, but these regions are more sensitive in detecting movement.

Resolving power or visual acuity
The ability to distinguish fine detail depends on the optical system, the number of light-sensitive cells per unit area, and the number of light-sensitive cells connected to each optic nerve fibre. The finest limit is determined by the distance between three cones since, for the two outer cones to distinguish two points, the centre cone must not be stimulated.

Cones usually connect to individual optic nerve fibres and the fovea centralis depression increases the number of cones exposed to the focused image during daylight vision, thus there is high visual acuity in this region.

The rods, on the other hand, usually show convergence (many rods attach to each single optic nerve fibre) and, outside of the fovea centralis, the resolving power drops by about a third.

Transduction
This is achieved by the bleaching of the photopigments in the rods and cones. Most work has been done on the pigment, known as rhodopsin or visual purple, extracted from rods, which is bleached by light to produce a colourless protein, opsin, which initiates a nerve impulse in some way. The process is reversible and resynthesis of rhodopsin occurs, especially in low light intensities. Thus the rods become more sensitive in a process of dark adaptation.

$$\text{rhodopsin} \underset{\text{dark}}{\overset{\text{light}}{\rightleftharpoons}} \text{opsin} + \text{retinene} \quad (\nearrow \text{initiates nerve impulse})$$

$$\text{retinene} \underset{\text{NAD}}{\overset{\text{NADH}_2}{\rightleftharpoons}} \text{vitamin A}$$

DEPTH OF FIELD

The image focused on the retina is in two dimensions and the brain must build up a three-dimensional picture of the world from the information it receives. How this is done is still not completely understood, but the brain uses a series of visual clues including so-called 'one eye factors' and 'two eye factors'.

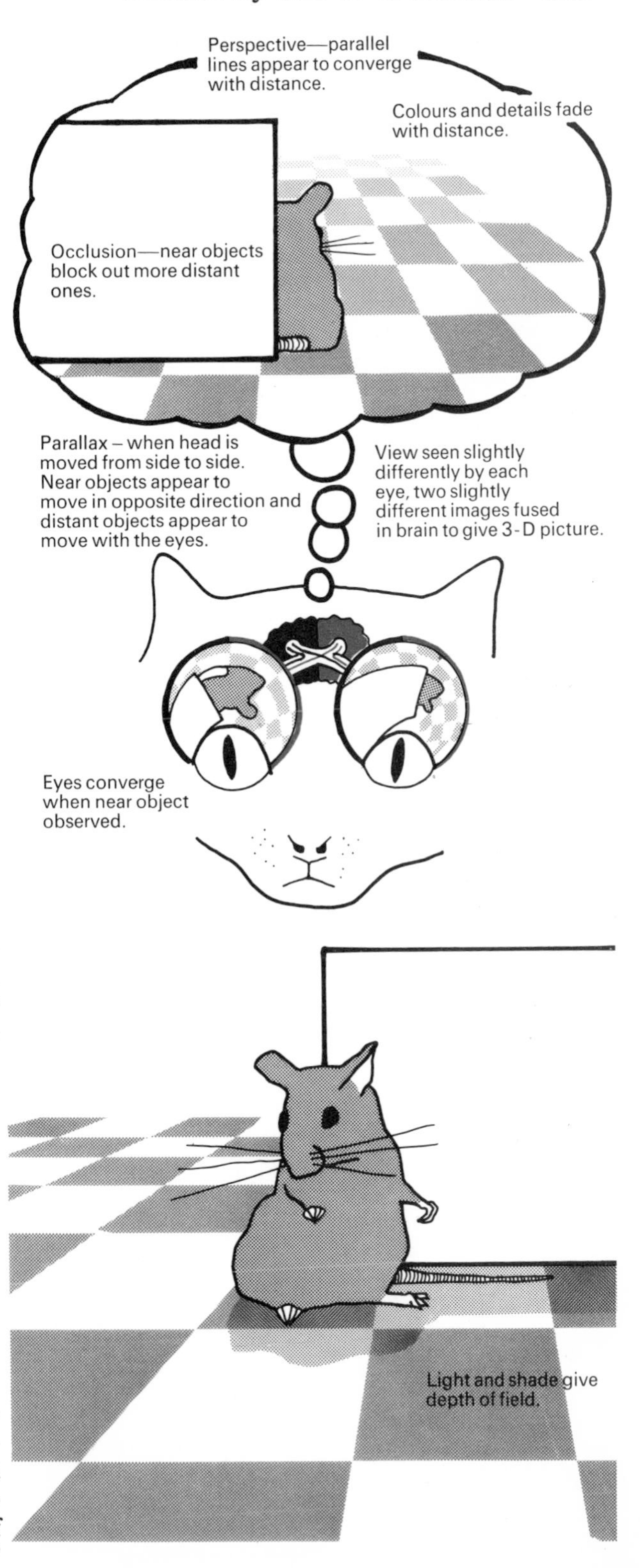

Figure 10.14 *Stereoscopic vision*

VISUAL CORTEX

The construction of the 'picture' of the outside world, which is built up on the basis of the sensory input from the retina, takes place in the visual cortex. Damage to this area produces corresponding blind patches or scotomas, but in a strange way people with such damage can still 'see' in these patches without 'realizing' it, since some nerve fibres from the eyes go to the subcortical optical lobe visual areas and allow 'sight' below the level of consciousness. Thus such people can tell when a light flashes in a blind patch caused by such damage, without 'knowing' how.

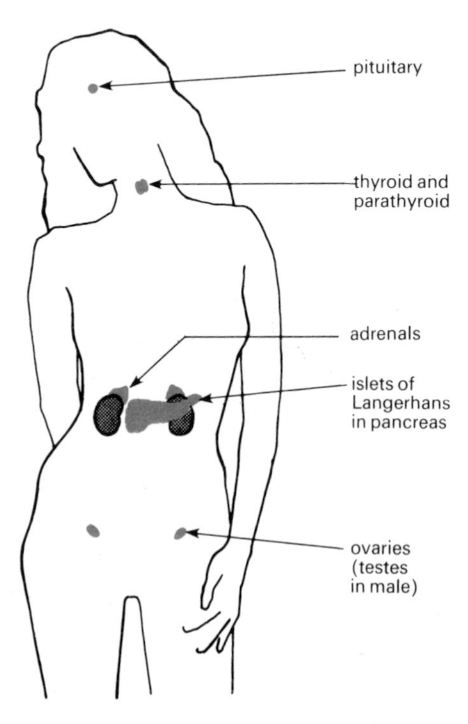

Figure 10.15 *Mammalian (human) endocrine system*

ENDOCRINE SYSTEMS

The other major co-ordinating system found in many invertebrates and all vertebrates, which is particularly well-developed in mammals, is the endocrine system. It consists of ductless glands which secrete hormones directly into the capillaries with which they are well supplied. Although widely scattered, the glands are linked by the blood stream and the autonomic nervous system, and act mostly as an integrated system.

The system plays a major part in homeostasis and is under the ultimate control of the hypothalamus through which it can be affected by states of the higher brain centres such as fear and emotion. Hormones, by controlling the balance of anabolism and catabolism, are also able to co-ordinate long term changes involved in growth and maturation which are less open to shorter-action nervous control.

Hormones

Hormones are organic secretions produced in small quantities and transported in the blood to exert their action at a distance from their site of production, usually on a specific target organ or system, but sometimes more generally.

They are capable of rapid diffusion, and are sometimes unstable, capable of being broken down by enzymes so that they do not accumulate. However, some do exert their effects over long periods. They are of varying chemical composition and include proteins, polypeptides, and steroids. They exert their effect by binding to receptor sites on cell membranes and triggering some internal cell mechanism involving AMP[367] (adenosine monophosphate).

The naming of hormones is still rather confused due to the difficulties of isolation and identification. Many hormones have a common name, such as growth hormone, as well as a more formal name, such as somatotrophic hormone; the names of such trophic hormones can be shortened, as in somatotrophin.

Pituitary

ADENOHYPOPHYSIS

The isolation and identification of the adenohypophysial hormones is very difficult, but there appear to be six.

Growth hormone (somatotrophic hormone (STH) or somatotrophin) acts directly on the tissues, promoting growth of the skeleton, muscles, and general body metabolism. It also causes the retention of calcium, potassium, and sodium in the kidneys. Some attribute a diabetogenic effect to this hormone, that is it is antagonistic to, and decreases the production of, insulin. However, others claim that there is a separate diabetogenic hormone. Somatotrophin is also essential for lactation in the female. This apparently unrelated collection of functions may well be a product of inaccurate analysis techniques.

Corticotrophin (adrenocorticotrophic hormone (ACTH)) regulates the growth and functioning of the adrenal cortex.

Thyrotrophin (thyrotrophic hormone, thyroid-stimulating hormone (TSH)) regulates the growth and functioning of the thyroid.

Follicle stimulating hormone (FSH) initiates the cyclic changes in the ovaries of the female, (the development of the Graafian follicles) and initiates sperm formation in the testes of the male.

Interstitial cell stimulating hormone (ICSH) or luteinizing hormone (LH) causes the release of the egg from the ovary, and thus causes the follicle to turn into a corpus luteum; it also stimulates the secretion of testosterone by the testis of the male.

Prolactin (lactogenic hormone) controls the activity of the corpus luteum of the ovary and has a role in milk production.

Luteotrophin (luteotrophic hormone (LTH)), may be a separate hormone with an effect on the corpus luteum stimulating it to secrete progesterone.

CONTROL OF ADENOHYPOPHYSIAL SECRETION

Neurosecretory cells in the hypothalamus secrete neurohormones, known as releasing factors, which are carried in a short portal vein to the adenohypophysis where they stimulate the release of the adenohypophysial hormones.

Homeostatic feedback mechanisms also operate, via the hormones of the glands stimulated by the adenohypophysial hormones; thus hormones such as thyroxine, oestrogens, and adrenal corticoids will inhibit the adenohypophysial secretion of their trophic hormones when present in high levels in the blood.

NEUROHYPOPHYSIS

The neurohypophysis consists of non-myelinated nerve fibres and non-secretory supporting cells. The nerve fibres originate in the hypothalamus, where the cell bodies of the neurones synthesize neurosecretions. These are carried as granules of colloidal material, which travel by cytoplasmic streaming down the axon to the neurohypophysis where they are released when an impulse travels down.

Oxytocin produces powerful contractions of the uterus, induces lactation, and facilitates the ejection of milk during suckling. The nervous stimulation of the mammary glands by the young during suckling causes the release of the hormone.

Anti-diuretic hormone (ADH) or vasopressin constricts arterioles, causing a rise in blood pressure which lasts longer than that induced by adrenalin; it also reduces the amount of water lost in the urine.[192]

Figure 10.16 *Neurosecretions of the mammalian hypothalamus*

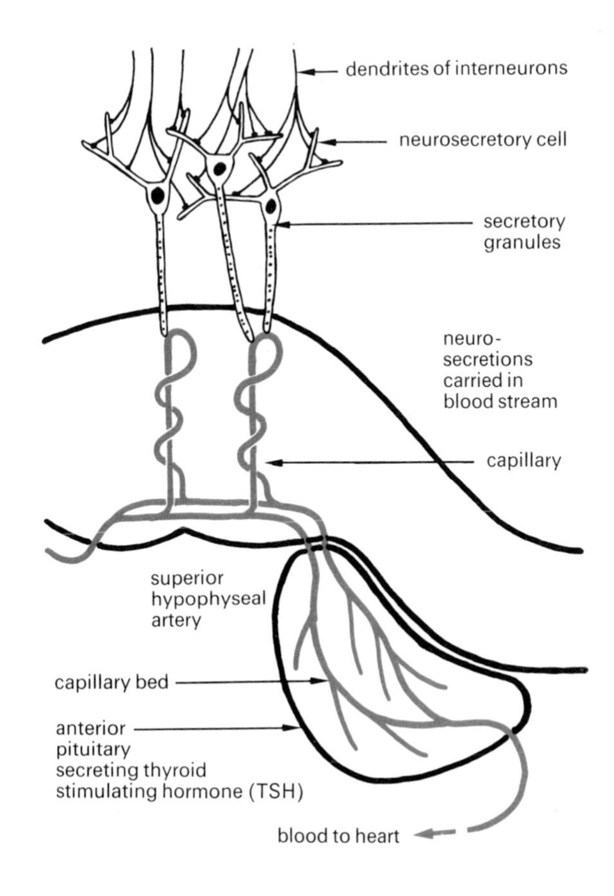

Thyroid

The thyroid consists of vesicles which absorb iodine from the circulation by the 'thyroid pump' and bind it to protein to form thyroglobulin. The thyroid is the only endocrine gland to store its product in vesicles rather than in cells. When required, the thyroglobulin is converted to **thyroxine**.

Thyroxine exerts a general stimulating effect on oxidation reactions in all cells, possibly by 'uncoupling' oxidation from ATP formation,[160] and is responsible for maintaining up to 50 per cent of the basal metabolic rate. It also controls normal brain development; is necessary for the correct functioning of the pituitary growth hormone; is involved in carbohydrate, protein, and fat metabolism; and has a role in most growth and development processes.

Parathyroids

The four parathyroids are small glands embedded in the thyroid. They secrete parathyroid hormone or **parathyrone** which regulates the calcium ion concentration of the plasma and body fluids within narrow limits. If the blood calcium drops, parathyrone causes the level to increase by the removal of calcium from the bones, by increasing the uptake in the gut, and by increasing reabsorption of filtered calcium in the kidney tubules. It also removes phosphate from the bones which is subsequently lost via the kidneys.

A second hormone, **calcitonin**, reverses these processes. Both hormones have interactions with vitamin D activity.

Adrenal cortex

The adrenal cortex has a completely different origin, development, structure, and function from the medulla. Its secretions, all of which are steroids, have a wide and varied effect.

Cortisol (hydrocortisone) controls the distribution of body fluids and ions, and maintains the blood pressure and kidney filtration rate. It also inhibits the release of histamine and thus prevents too strong an inflammatory reaction; and increases the use of protein for energy production.

Corticosterone has little known effect, but favours the conversion of carbohydrate to fat and has some affect on protein metabolism.

Aldosterone causes the retention of sodium ions by the kidney tubules and is thus important in osmoregulation.

Adrenocortical sex hormones are produced by the adrenal cortex, a secondary source of sex hormones after the gonads. They include androgens, oestrogens, and progesterone in both sexes. The androgens stimulate the growth of the body hair and increase the sex drive in the human female.

The adrenal cortex secretions, with the exception of aldosterone, are under the control of ACTH from the adenohypophysis of the pituitary. Aldosterone secretion is under the control of the juxtaglomerular apparatus[191] (JGA).

Adrenal medulla

The adrenal medulla is directly supplied by nerve fibres of the sympathetic nervous system but, unlike those of the autonomic nervous system, these nerves have no ganglia. In fact the adrenal medulla cells are derived from the post-ganglionic fibres themselves, and thus this region could be considered as part of the sympathetic nervous system.

Two main secretions are found, adrenaline and noradrenaline.

Adrenaline (epinephrine) affects all those structures supplied by the sympathetic nervous system and produces the same general effects. Thus it increases muscle tone, metabolism of blood sugar and serum fatty acids, metabolic rate, heart rate, and rate and depth of breathing. It also dilates the coronary and striped muscle blood vessels, and constricts those of the skin and of the gut (causing the gut to relax but the gut sphincters to contract). The sum total of these actions is to prepare the body for emergencies.

Noradrenaline is the precursor of adrenaline but is also secreted as a separate hormone. It has similar effects to adrenaline but, in contrast to adrenaline, tends to be secreted during activity, rather than in anticipation of it. It is also released in response to general stress. Noradrenaline maintains a high blood pressure, promotes the action of the heart, and increases vasodilation; but has little effect on the general metabolism.

Chromaffine tissue

The adrenal medulla tissue darkens when exposed to chromic acid or dichromate and for this reason is referred to as chromaffine tissue. Patches of chromaffine tissue are also found outside of the adrenal medulla, along the sympathetic nerve chain, and close to the blood vessels in the abdomen and pelvis. These patches of tissue also produce adrenalin and noradrenalin. However, unlike the adrenal medulla (which produces mainly adrenalin), they mainly produce noradrenalin.

The adrenal medulla is under the direct control of the sypathetic nervous system, which in turn is influenced by exercise, emotion, and stress. The response to **stress** over a long period passes through three stages together known as the general adaptation syndrome. The first stage is the alarm stage when the short term reactions of adrenaline and noradrenaline are activated. The second stage is the resistance stage when the adrenal glands enlarge (in mice provoked to fight for 5 min per day for five days, the adrenals enlarged by 38 per cent). Finally the exhaustion stage is reached which results in the collapse of the system, leading sometimes to sterility—as is often seen in animals at the bottom of dominance hierarchies.[237]

Islets of Langerhans

The Islets of Langerhans are patches of endocrine tissue scattered throughout the pancreas. Special staining techniques show them to be composed of two types of cell: α cells, which secrete **glucagon**, and β cells, which secrete **insulin**. Both of these hormones are involved in the complex regulation of the blood glucose level. For example, increased blood glucose affects the liver directly, causing it to convert glucose to glycogen, but it also stimulates the secretion of insulin from the β cells of the Islets of Langerhans, which further stimulates the liver and muscles to synthesize glycogen. Insulin also increases glucose utilization by the tissues and inhibits those mechanisms which tend to increase the blood glucose level.

Any drop in the level of blood glucose stimulates the liver directly to convert more glycogen to glucose (glycogenolysis). A drop in blood glucose also stimulates the hypothalamus to trigger the release of anterior pituitary diabetogenic factor (which is the principle antagonist to insulin), and of adrenalin from the adrenal medulla; both these increase the level of glucose.

Thyroxine secretion from the thyroid is also increased as the blood glucose level drops; this not only increases glycogenolysis but also promotes glucose formation from proteins and fats (neoglucogenesis or gluconeogenesis) and increases the breakdown of insulin.

The secretion of glucagon from the α cells in the Islets of Langerhans is increased and, although this also opposes the action of insulin, it acts only on the liver and not the muscles and is thought to have a relatively minor role.

Gastric mucosa

The presence of food and the products of digestion causes the release of the hormone **gastrin** into the blood. Gastrin stimulates the oxyntic cells of the stomach to produce hydrochloric acid after a lag period of 30–60 minutes.

Intestinal mucosa

Fat is the last type of food to leave the stomach and when it reaches the small intestine it stimulates the release of **enterogastrone** into the blood. **Enterogastrone** inhibits the release of gastrin which is now no longer needed since the stomach is empty of food.

The wall of the small intestine also secretes the hormones villikinin, secretin, pancreozymin, and cholecystokinin.

Villikinin stimulates the villus pump, which is the rhythmic contraction of the smooth muscle muscularis mucosa, which aids absorption and the emptying of the villus lacteals.[146]

Secretin travels in the blood to the pancreas and stimulates the secretion of watery pancreatic juice rich in bicarbonate but containing few, if any, enzymes. It also regulates the flow of intestinal juices or succus entericus.[145] The secretion of secretin is mainly stimulated by the presence of acid in the small intestine.

Pancreozymin stimulates the secretion of enzymes by the pancreas without increasing the volume of flow of the pancreatic juice. The secretion of pancreozymin is mainly stimulated by the presence of peptones in the small intestine.

Cholecystokinin stimulates the contraction of the gall bladder[146] and thus causes the expulsion of the bile into the small intestine. The secretion

Table 10.1 *Summary of mammalian hormones*

Area of function	*Hormone*	*Source*	*Main effects*	*Control stimulus*
General metabolism and growth	Growth hormone	Adenohypophysis	Overall growth	
	Thyroxin	Thyroid	Tissue respiration, basal metabolic rate	Thyrotrophin
	Parathyrone	Parathyroid	Calcium and phosphate metabolism of bones	Blood calcium
	Insulin	Pancreas	Increased glucose use and glycogen storage	Blood glucose
	Glucagon	Pancreas	Antagonistic to insulin	Blood glucose
	Hydrocortisone	Adrenal cortex	Carbohydrate, protein, and fat metabolism combats stress	ACTH
Circulation	Adrenaline	Adrenal medulla	Increased circulation to muscles	Nervous
	Noradrenaline	Adrenal medulla	Increased blood pressure	Nervous
	Angiotensin	Blood plasma	Increased blood pressure stimulates secretion of aldosterone	Renin
	Serotonin	Platelets	Blood pressure, gut movements	
	Histamine	Tissues	Inflammatory response	Injury
	Vasopressin (ADH)	Neurohypophysis	Increased blood pressure	
Osmoregulation	ADH (vasopressin)	Neurohypophysis	Increased water reabsorption by kidney	Blood concentration
	Aldosterone	Adrenal cortex	Sodium retention by kidney	Blood concentration
Digestion	Gastrin	Stomach wall	Stimulates secretion of gastric juice	Food in stomach
	Enterogastrone	Intestine wall	Inhibits secretion of gastrin	Fat in intestine
	Villikinin	Intestine wall	Stimulates 'villus pump'	Food in intestine
	Secretin	Intestine wall	Stimulates flow of watery pancreatic juice	Acid in intestine
	Pancreozymin	Intestine wall	Stimulates secretion of pancreatic enzymes	Peptones in gut
	Cholecystokinin	Intestine wall	Causes contraction of gall bladder	Fat in intestine
Reproduction	Follicle stimulating hormone (FSH)	Adenohypophysis	Causes ovary follicles to develop and testis to make spermatozoa	Hypothalamus releasing factor
	Luteinizing hormone (LH) (interstitial cell stimulating hormone (ICSH))	Adenohypophysis	Causes release of egg from ovary follicle to turn into corpus luteum and testis to make testosterone	Hypothalamus releasing factor
	Prolactin (luteotrophic hormone)	Adenohypophysis	Affects corpus luteum in pregnancy and milk secretion	Hypothalamus releasing factor
	Oxytocin	Neurohypophysis	Stimulates uterine contractions and ejection of milk	Hypothalamus and suckling
	Androgens	Adrenal cortex	Growth of body hair, anabolism	ACTH
	Oestrogens	Ovary	Secondary sexual characteristics. Uterine changes up to ovulation	Ovary development
	Progesterone	Ovary	Uterine changes after ovulation	Ovulation
	Placental hormones	Placenta	Maintain pregnancy	Placental development
	Testosterone	Interstitial cells of testes	Secondary sexual characteristics	LH (ICSH)
Endocrine glands	Thyrotrophic hormone (TH)	Adenohypophysis	Stimulates thyroid	Thyroxin levels
	Adrenocortico-trophic hormones (ACTH)	Adenohyphophysis	Stimulates adrenal cortex	Stress
	Diabetogenic hormone	Adenohypophysis	Decreases insulin production and action	Blood sugar
	Hydrocortisone	Adrenal cortex	Affects entire endocrine system	Stress

of cholecystokinin is mainly stimulated by the presence of fats and meat extracts in the small intestine.

Gonads

The gonads secrete the reproductive hormones, the functions of which are summarized in Table 10.1, but which are dealt with more fully in Chapter 3.

Serotonin

Serotonin is derived from broken down blood platelets. It increases blood pressure, influences gut movements, and is a transmitter and regulatory substance in the CNS.

Histamines

Histamines are released from damaged tissue. They cause extreme capillary dilation and permeability, allowing the escape of plasma proteins into the tissue fluid. They are also released in allergic reactions.

Pheromones or ectohormones

Pheromones are chemicals produced by one animal which, when received by other individuals of the same species, alter their behaviour or development in some way. Thus they function as external hormones. They include the general scents or smells used in marking territory, attracting the opposite sex, identifying members of the same society, and many others, which all alter the behaviour of other individuals in some way.

For example, if a pregnant female mouse smells a strange male soon after fertilization, implantation of the fertilized eggs in the uterus will not occur. This is referred to as the 'Bruce effect'. Another example is seen in the queen honey bee who produces unsaturated fatty acids in the mandibular glands. When these are licked by the workers and spread around the hive, the development of the ovaries of young workers is inhibited, and they are thus prevented from making more queen cells.

Experimental and applied aspects

Investigations of the activity of nervous systems involve the detection of the electrical activity of the nerve impulses. These can be detected and recorded using the **cathode ray oscilloscope**.

The oscilloscope has a cathode ray tube which contains a metal filament. When this filament is heated by the passage of an electric current it emits a stream of electrons which are focused to form a beam. The beam produces a spot of light on a fluorescent screen which can be deflected from side to side by applying voltages to the so-called X-deflection plates, and can be deflected in the vertical plane by the Y-deflection plates. The X-deflection plates can cause the spot to be swept very rapidly at constant speed from left to right, and this movement forms the time base for the observations. If the spot crosses the screen at a high enough frequency it appears as a horizontal line of light.

Platinum electrodes in contact with the nerve under study are connected to the oscilloscope input, and the greatly amplified action potentials are applied to the Y-plates. These cause the spot to be deflected vertically and the resultant of the horizontal and vertical movements forms a wave. The height of the wave depends on the amplification of the action potential, and its length depends on the frequency of the time base. Both can be adjusted so that the wave form best fills the screen.

Double beam oscilloscopes have two beams of electrons, the lower of which can be used to record any stimulations of the nerve employed in the experimental procedure. (Traces can be obtained in a similar way using a pen recorder which records an ink trace on a strip of calibrated paper.)

The diagram shows the trace from a double beam oscilloscope (A and B) during the compound action potential of a frog's sciatic nerve following electrical stimulation (S_1 and S_2). S_1 and S_2 are two shocks of equal strength delivered to the nerve separated by 2 ms. Since S_2 fell in the relative refactory period of the nerve, trace A shows only a single action potential.

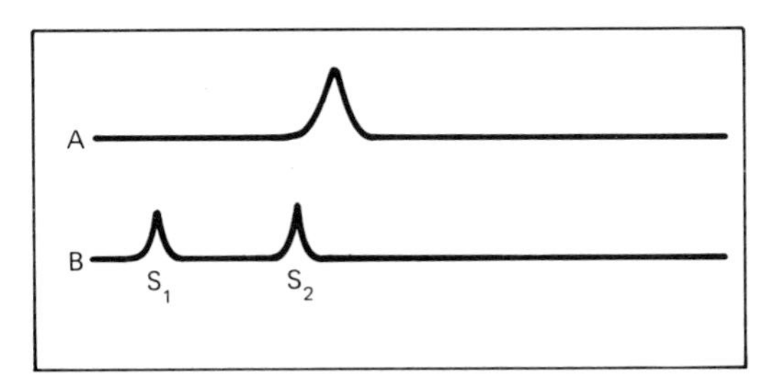

Guided example

1 The two co-ordinating systems of the body, the nervous and endocrine systems, are usually described separately. Do you think that they are in fact separ-

ate systems functioning independently of each other?
No. The two co-ordinating systems of the body are themselves co-ordinated to act in unison; in fact the relationships between the two are so close that they could in many ways be considered as components of a single system, particularly in the vertebrates. They are dealt with separately, more as a matter of convenience rather than as a reflection of their relative independence.

2 Does a description of nerve cells, in terms of their electrical activity only, give an accurate picture of their properties?
No. Nerves, as well as transmitting electrical impulses, are active secretory structures producing neurosecretions at synapses and at effector terminals. Special neurosecretory cells which look like neurones and which can transmit impulses are particularly adapted to the production of hormones. In these, the axons do not innervate effectors or synapse with other neurones. Such cells are usually gathered in groups and have swollen axon terminals where the secretions are stored before release. These ends are clearly associated with the circulatory system to form neurohaemal organs. The neurohypophysis of the pituitary is one such neurohaemal organ, with the cells originating in the hypothalamus.

3 Direct innervation by nerves of the autonomic nervous system can increase hormone secretion by the thyroid and by the Islets of Langerhans in the pancreas. How do you think a nervous stimulation could have this effect?
Probably by increasing the blood supply to these glands by vasodilation.

4 Why are the effects of adrenalin produced by the adrenal medulla so similar to those of the sympathetic nervous system?
The adrenal medulla is directly innervated by branches of the sympathetic nervous system and acts as a reservoir for its neurosecretions. These neurosecretions are subsequently released as the hormones adrenalin and noradrenalin which affect all those structures supplied by the sympathetic nervous system, and produce the same general effects.

5 In what ways can the endocrine system affect the activity of the nervous system?
In a complex of ways. For example, the correct balance of hormones is necessary to maintain optimum conditions for the functioning of the nervous system. This is clearly seen by the roles of insulin, glucagon, adrenalin, thyroxine, and growth hormone in maintaining the blood glucose level at its optimum; this is necessary for correct brain function. Also hormones can 'feedback' information to the brain from the body, which can result in an alteration of its functioning. This is particularly clearly seen in certain behaviour patterns which are only expressed when the correct hormonal balance is present.

Questions

1 Describe the essential features of hormones. Explain the role of hormones in the control of growth in (a) a plant and (b) a mammal. (L)

2 The pituitary is said to be a 'master gland' within the endocrine system of a vertebrate. Using specific examples of the hormones which it produces, describe instances of its activities which you feel reflect this description of the gland.
To what extent do you consider that the statement might mis-represent the role of the pituitary gland in relation to the other endocrine organs? (L)

3 (a) Make a fully labelled diagram to show the main parts of a mammalian brain.
(b) Briefly explain the functions of the main parts of the brain in a mammal.
(c) Give an illustrated account of the way in which nerve impulses are conducted in a mammalian nerve fibre. (L)

4 By reference to *either* a sensory neuron *or* a motor neuron explain how a nerve cell is structurally and physiologically adapted to the reception and transmission of a nerve impulse.
By reference to these neurons, and such others as you may care to specify, briefly describe and illustrate the system whereby the nerve cells in a vertebrate are linked together to pass information from a receptor to an effector organ. Give one example to show how the system works. (L)

5 (a) What are the biological advantages of (i) sight, and (ii) colour vision to animals?
(b) Make a large, clearly labelled diagram of the mammalian eye as seen in section.
(c) Briefly describe how accommodation for distance and light intensity is brought about. (L)

6 What are the biological advantages of a sense of hearing in mammals? Make fully labelled diagrams to illustrate the structure of the ear of a mammal. Give a concise account of the way in which sound vibrations are translated into nerve impulses. (L)

Further reading

Blakemore C., *Mechanics of the Mind* (Cambridge, Cambridge University Press, 1977).
Gregory R. L., *Eye and Brain* (London, Weidenfeld and Nicolson, 1977).
Lee J. and Knowles F. G. W., *Animal Hormones* (London: Hutchinson, 1965).
Lewis J. G., *The Endocrine System* (Harmondsworth: Penguin Books, 1973).
Nathan P., *The Nervous System* (Harmondsworth: Penguin Books, 1969).

11 Movement, response, and behaviour

MOVEMENT

Plant movement

There are several different types of plant movement, including non-living movement, automatic movement, turgor movement, and stimulus movement.

Non-living movements are not dependent on living cells and are usually associated with hygroscopic mechanisms, as in the annulus of the fern.[32]

Automatic movements are endogenous or spontaneous growth movements, such as in the circumnutation (circular movements) of the stem tips of climbers.

Turgor movements depend on changes in turgor of cells which, depending on their position, can lead to movement such as the leaf rolling seen in some xerophytes which is caused by the loss of turgor of thin-walled hinge cells.[185]

Stimulus movements occur in response to an external stimulus and include **tactic** movements where the whole plant moves, as in the motile algae; **nastic** movements which are non-directional, as in the opening and closing of flowers in response to changes in temperature and light; and **tropic** movements which are directional as in bending towards or away from the light.

General responses of plants to stimuli

Plants show many similarities to animals in their response to external stimuli. There is a **threshold** of stimulation below which there is no response, although a number of sub-threshold stimuli may accumulate and trigger a response. There also is a **latent period** or delay between the stimulus and the response; and the response is proportional to the size, that is the intensity and the duration, of the stimulus. However, the response is often restricted to a special zone, for example the zone of elongation in shoots and roots. Moreover, the sensitive zone is not highly differentiated, as in animals' sense organs, and it is in fact not clear how most stimuli are detected.

There is no nervous system comparable to that found in animals, and response and internal coordination is by means of hormones only. Due to the close relationship in plants of movement to growth, any external factor that affects growth will also affect the response to a stimulus.

Tropic movements

There are a wide variety of stimuli to which plants respond in a directional manner including photo (light), geo (gravity), hydro (water), rheo (water current), aero (air), thigmo (touch), chemo (chemical), and thermo (temperature).

The response may be positive or negative (directly towards or away from the stimulus) or at some angle to the stimulus. The bending of parts of the plant towards or away from a stimulus is caused by an unequal distribution of auxins in the zones of elongation. This leads to different cell extensions on either side.

PHOTOTROPISM

Stems are, up to a certain limit, positively phototropic, but at very high light intensities they may become negatively phototropic. They respond to unilateral light as long as there is a difference of at least 1 per cent in light intensity between the light and shade sides. Once the stem is pointing towards the light the response stops as the light is no longer unilateral. Maximum photosensitivity occurs towards blue light. This indicates that the receptor pigments involved are either carotenoids or riboflavine, which are known to absorb blue light.

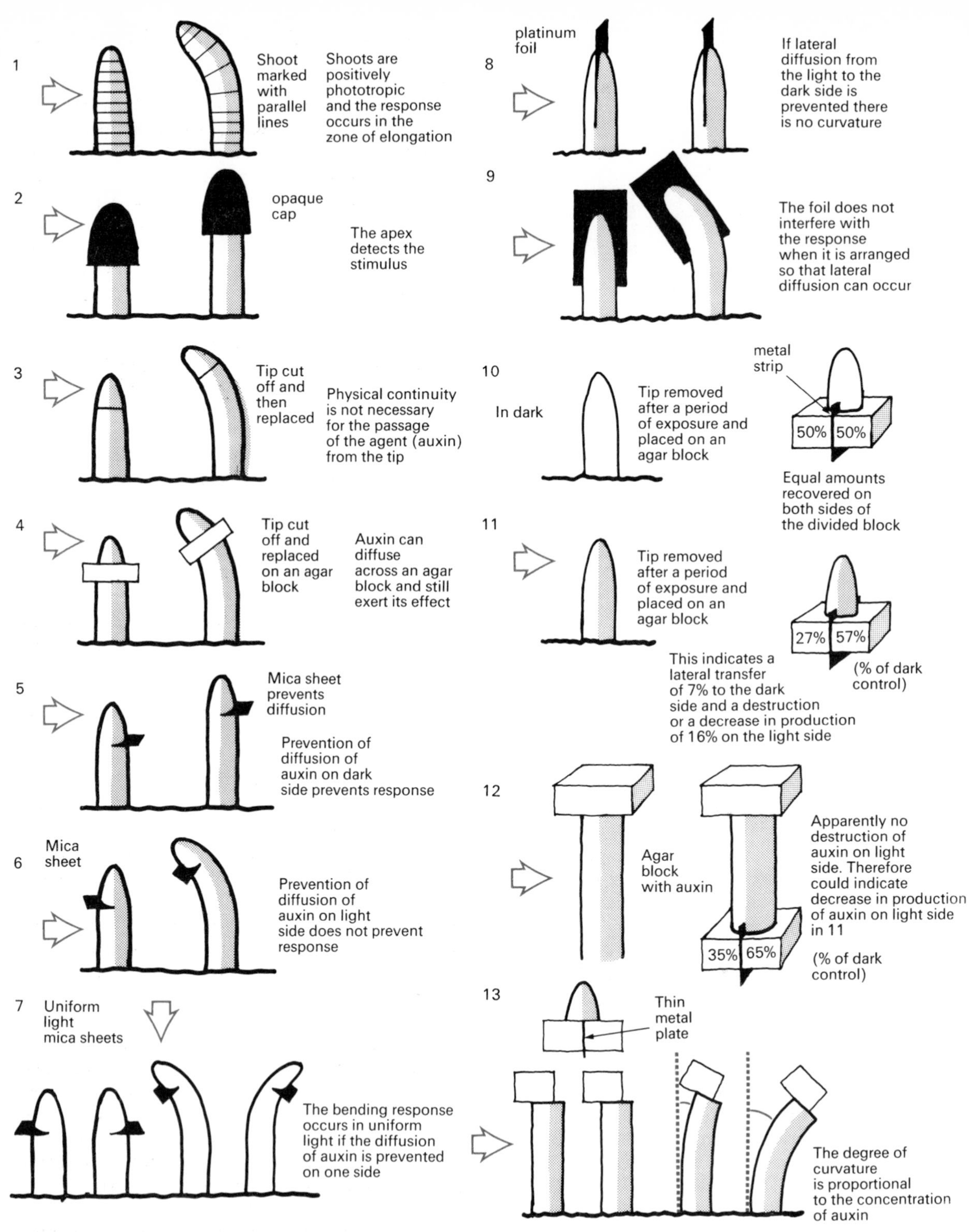

Figure 11.1 *Summary of some of the experimental evidence relating to the mechanism of phototrophism*
Radioactive labelling of auxin with ^{14}C has revealed little extra information. Whether the unequal distribution of auxin under unilateral light is the result of decreased auxin synthesis on the light side; increased auxin synthesis on the dark side, destruction of auxin on the light side, or lateral diffusion from the tip to the dark side is still not clear, although lateral diffusion is considered the main cause.

Side view of young tree

View of same plant from above to show the arrangement of the leaves for maximum exposure to light for photosynthesis

Leaves arrange their laminae at right angles to the incident light for maximum absorption of light for photosynthesis, but roots are aphototropic—they show no response to light. The bending of stems towards light is caused by an increased auxin concentration on the side away from the light, resulting in the cells on that side of the stem elongating more than the others. However, the cause of this unequal distribution of auxin in the zone of elongation is still not clear.

There is no clear evidence for the light destruction of auxins *in vivo*, and although there is more evidence for lateral translocation, there is no clear understanding of the cause or of the mechanism by which this can occur. Both destruction and lateral translocation may be involved, and in addition light may affect the synthesis of auxins in the tip.

GEOTROPISM

Primary roots are positively geotropic, lateral roots grow at right angles to gravity, and tertiary roots are insensitive to gravity. Main stems are negatively geotropic, while lateral stems and leaves grow at an angle to gravity. Modified stems such as rhizomes and runners grow at right angles to gravity.

Gravity acts on mass and is detected by cell inclusions which respond to gravity by redistribution. In geosensitive regions there are special mobile starch grains which accumulate on the lower side of cells. They persist even in extreme starvation and are seen in some monocotyledons, such as the onion, that do not normally produce starch. Low temperatures cause their disappearance and a consequent loss of the geotropic response.

All of these points indicate that these grains provide one way in which the stimulus of gravity is detected, although other cell inclusions can also act in a similar way since the geotropic response still occurs in many plants in the absence of starch grains. Gravity results in an unequal distribution of auxin so that the highest concentration is on the lower side in both roots and stems. Exactly how the movement of geosensitive particles can cause this redistribution is not known.

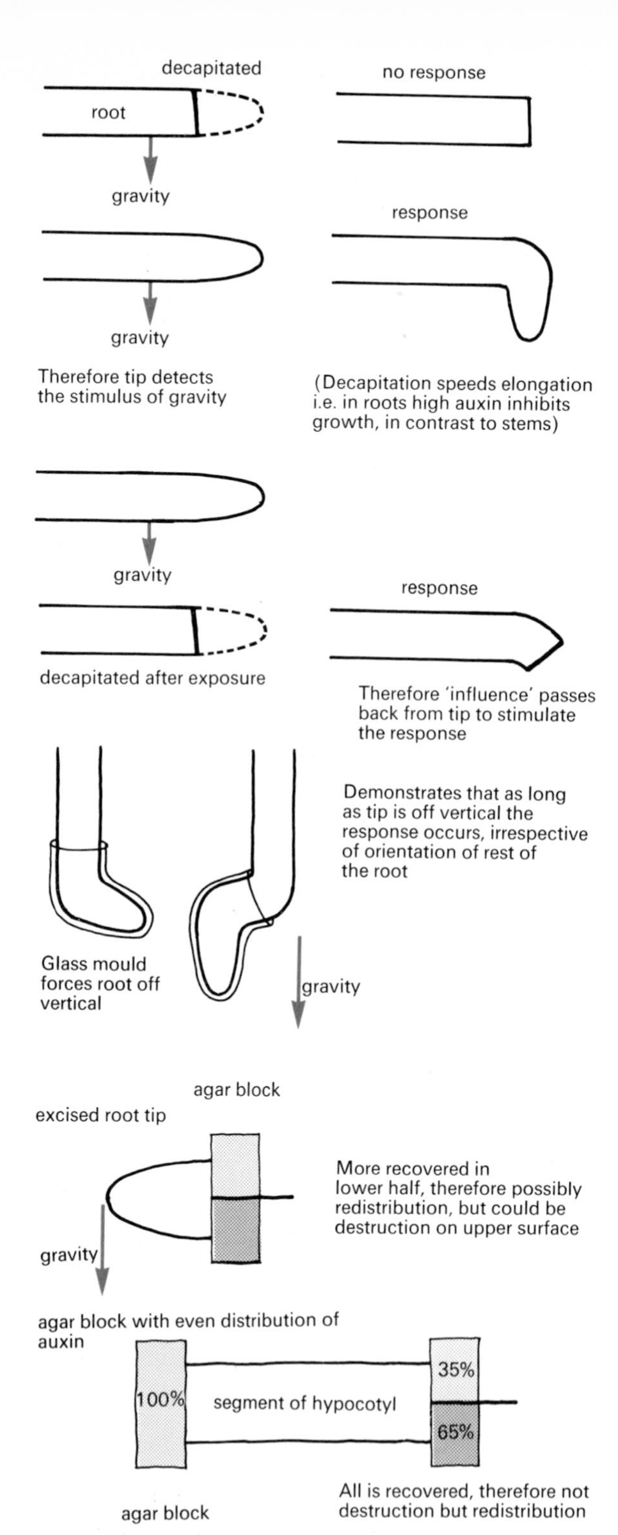

Figure 11.2 *Summary of some of the experimental evidence relating to geotrophism*
Auxin is also redistributed under gravity in stems, as is seen by the suppression of lateral bud development, and the stimulation of adventitious root formation, on the lower surface of horizontal stems.

This increased concentration on the lower side of roots inhibits cell extension, thus leading to a bending towards gravity. In stems, however, increased concentration increases cell extension and leads to a bending away from gravity. These opposite responses of root tissue and stem tissue to the same concentration of auxin is explained by their differing sensitivities to the hormone.

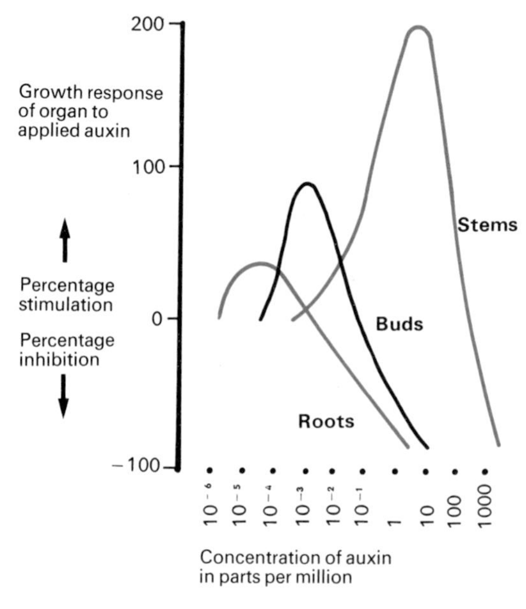

Figure 11.3 *Differential sensitivity of plant parts to auxin*

All the tropisms, especially photo, geo, and hydro, interact in a complex of ways to result in the overall orientation of the plant in its habitat.

Plant hormones

Plant hormones are only produced in extremely small amounts and their isolation, and consequently their identification, presents many problems. There are three main hormones known to occur naturally and to be essential for the normal development of plants. These are **auxin, gibberellin**, and **cytokinin**. Two other compounds occur naturally as plant growth hormones, these are **ethylene** and **abscissic acid** (abscissin). Others that have been suggested include **florigen**, the 'flowering hormone', and **traumatin**, the 'wound hormone', although clear evidence for their existence as separate hormones is lacking.

All plant hormones interact in complex ways so that it is sometimes difficult to identify specific functions for single hormones.

AUXIN

Auxin was isolated originally from human urine, 40 mg of so-called 'active material' being obtained from 150 l of urine. This 'active material' was subsequently identified as a substance known as indoleacetic acid (IAA). It reached the urine from plant material in the diet which was absorbed in the gut and subsequently excreted without damage; the human body thus acts as a natural concentrator of the hormone.

Auxin (IAA) is synthesized from the amino acid tryptophan in root and shoot tips, buds, expanding leaves, and seeds.

Auxin and other hormones can be transported in any tissue but, over longer distances, they travel mainly in the phloem and xylem. Transport in the parenchyma shows polarity, so that transport can only occur in one direction—from the site of synthesis to the site of action. However, this polarity decreases as the hormone travels further from its source.

Auxin has a wide variety of functions. It increases cell wall plasticity by loosening the bonding between the cellulose fibres. This decreases the wall pressure, allowing more water to enter, which thus stretches the cell wall and results in cell expansion. This is important in normal growth and in tropisms. Auxin also accelerates cell division and may be involved in the division of cambium cells and the formation of callous tissue over wounds. It accelerates the development of fruit, stimulates root initiation, inhibits lateral bud development, and affects flowering.

GIBBERELLIN

A fungus (*Gibberella fujikuroi*), parasitic on rice, causes rice plants to grow tall with long chlorotic or yellowish leaves, to flower early, and to produce a small crop.

Several compounds which cause these effects have been isolated from the fungus; one such is gibberellic acid. This has been found to occur naturally in all higher plants and is also known as gibberellin.

Gibberellin increases cell elongation. When applied externally to dwarf varieties it causes internode extension of the stem, resulting in the attainment of normal size; in plants like cabbages and lettuces it causes premature 'bolting'. (Auxin does not have such a strong effect when applied externally.)

It has a variety of other effects, including those on germination and flowering which are dealt with in Chapter 3.

Table 11.1 *Summary of plant hormones and their function*

Hormone	*Function*
Auxin	Increased cell elongation in growth and tropisms Increased rate of cell division in cambium Differentiation of vascular tissue Increased rate of cell division in wounds (traumatin?) Suppression of lateral bud development Initiates lateral and adventitious root formation Stimulates development of fruit Affects flowering
Gibberellin	Cell elongation Promotes germination of seeds Ends dormancy in buds Affects leaf expansion and shape Retards leaf abscission. Aids setting of fruit after fertilization Removes need for cold treatment in vernalization Affects flowering (florigen?)
Cytokinin	Increased rate of cell division Increased cell enlargement in leaves Stimulates lateral bud development Breaks dormancy in some species Promotes flowering in some species Stabilizes protein and chlorophyll
Abscissin	Generally antagonistic to gibberellins and cytokinins Retards growth Induces dormancy in buds Inhibits germination
Ethylene	Promotes ripening of fruit Hastens abscission
Traumatin	Heals wounds by callous formation
Florigen	Promotes flowering

CYTOKININ

Cytokinin was originally isolated from coconut milk (which is in fact the liquid endosperm of the coconut seed). It was found that the coconut milk contained some factor that stimulated cell division in isolated tissue fragments. This fraction was subsequently isolated as a natural hormone from other plant tissue and named cytokinin.

Its main effect is in increasing the rate of cell division, but it also has important interactions with auxin in the growth and development of plants (see Chapter 3).

ABSCISSIN

Abscissin is structurally related to vitamin A. It was originally isolated from cotton fruit, but is now recognized as being a plant hormone found in all higher plants. It is generally antagonistic to gibberellins and cytokinins, inhibiting stem elongation, germination of seeds, and the sprouting of buds. It is also involved in leaf abscission.

ETHYLENE

Ethylene is a product of plant metabolism and is classified as a plant hormone. Ethylene is synthesized mainly where there are high levels of auxin. It is particularly important in the ripening of fruit, and there is a sharp climacteric rise in its production in ripening fruits. It also triggers the synthesis of the enzymes involved in abscission and thus hastens this process.

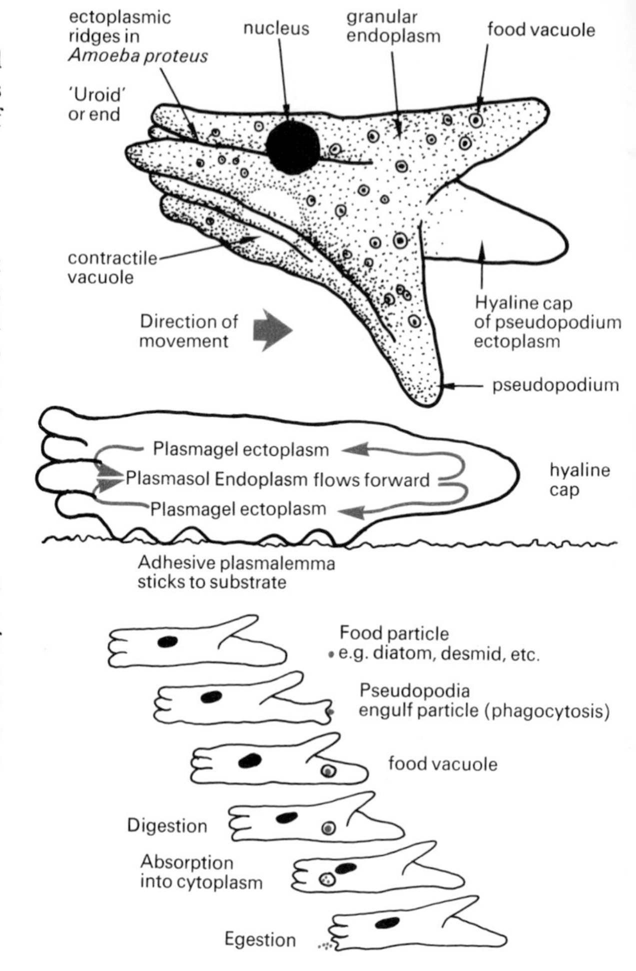

Figure 11.4 *Amoeboid movement*

Animal movement

Types of movement

AMOEBOID MOVEMENT

Amoeboid movement involves two fundamental characteristics of all living cytoplasm. Cytoplasm is **thixotropic**, which means it can change its state from semi-solid gel to liquid sol at constant temperature. Once in the sol form it exhibits streaming. Both of these processes are dependent on respiratory energy. Amoeboid movement makes use of both these phenomena during the formation of the pseudopodia by which the cell changes its shape and position. This type of movement is found in *Amoeba*, slime fungi, nematode spermatozoa and white blood corpuscles.

In many cases, such as *Amoeba*, pseudopodia are also used in feeding. They flow out around a food particle and enclose it in a food vacuole in a process known as phagocytosis.

CILIA

Cilia are fine extensions of the cell, have a complex structure, and move in a co-ordinated rhythmic way known as the **metachronal rhythm**. In small organisms, such as the ciliated protozoans and the ciliated larvae of many animals, the movement of the cilia results in locomotion. In these ciliated animals locomotion is closely related to feeding since the water that is moved by the cilia contains suspended food particles.

Cilia are also used for moving fluids and particles over surfaces, often again associated with feeding, as in the lamellibranch molluscs and some protochordates (e.g. *Amphioxus*). In addition they are found in annelid nephridia for the movement of wastes, in mammalian oviducts for the movement of eggs, and in the air passages of vertebrates for the removal of particles from the respiratory tract.

outer contractile filaments

T.S. cilium

central elastic filament

basal plate

firm limiting pellicle

basal granule acts as a pacemaker

rootlet ends near nucleus

Cilium stiff on effective stroke pushes water back.

Cilium relaxed on recovery stroke, therefore does not push water 'forward'.

Metachronal rhythm: each cilium is slightly out of phase with the next, resulting in smooth forward motion. If they were all to beat together progress would be 'jerky'.

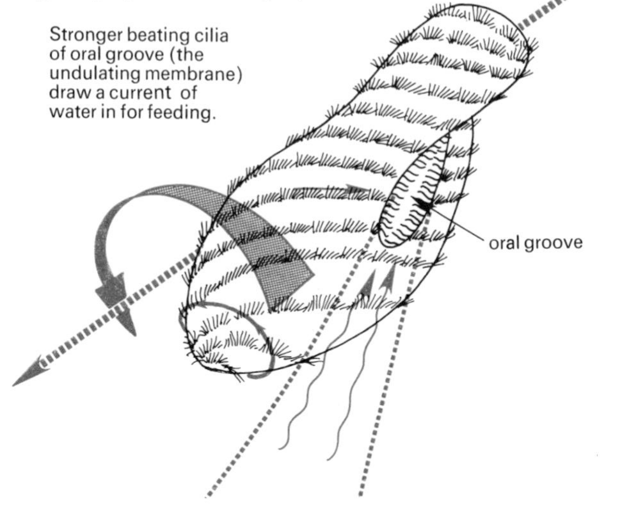

Figure 11.5 *Movement by cilia in Paramecium*
In *Paramecium* the cilia beat back to the right causing rotation; this prevents the stronger beating oral groove cilia from driving the animal in a circle. The rotation and the tendency to circle resolve into a spiral forward motion.

Cilia are not found in the arthropods, the Nematoda, or in the Plant kingdom (with very few exceptions).

FLAGELLA

Flagella are very similar in structure to cilia but are much longer and are usually only found singly or in pairs. Unlike cilia, they are found in plants as well as animals. They are used for locomotion in the flagellate protozoa (e.g. *Euglena* spp.), in some unicellular algae (e.g. *Chlamydomonas* spp.), and in many bacteria. They are also found widely in motile male gametes of both plants and animals.

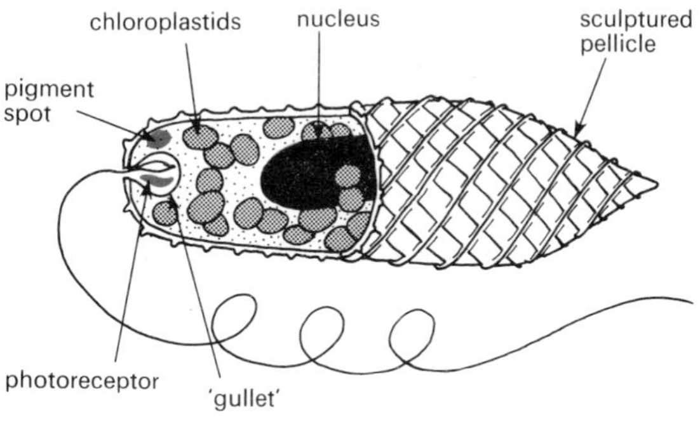

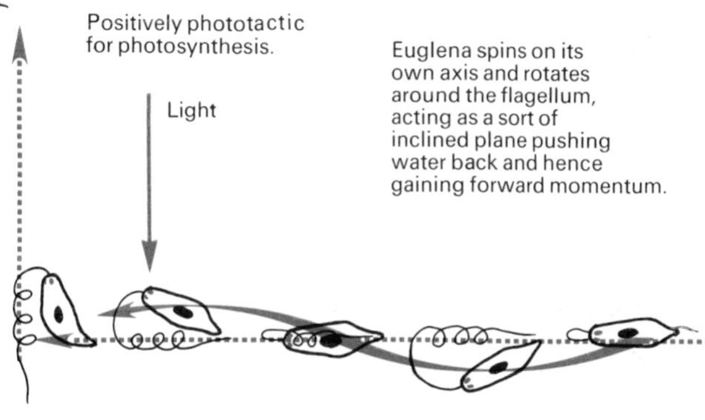

Figure 11.6 *Movement by flagella in Euglena*
The flagellum waves are actively generated along its length, they are not produced by a simple whiplash effect. Unilateral light is blocked from the photoreceptor by the pigment spot as *Euglena* spins and rotates. This is detected, and the direction interpreted. The *Euglena* then bends towards the light by using its contractile myonemes.

MUSCLE

Muscle movement is based on the properties of contractile proteins, as are all animal movements, including that of pseudopodia, cilia, and flagella. Muscle tissue is about 80 per cent water, which plays a vital role in contraction in addition to its functions in maintaining the life of the muscle fibres. The remaining 20 per cent consists of the proteins myosin, actin, and tropomyosin which make up the contractile mechanism.

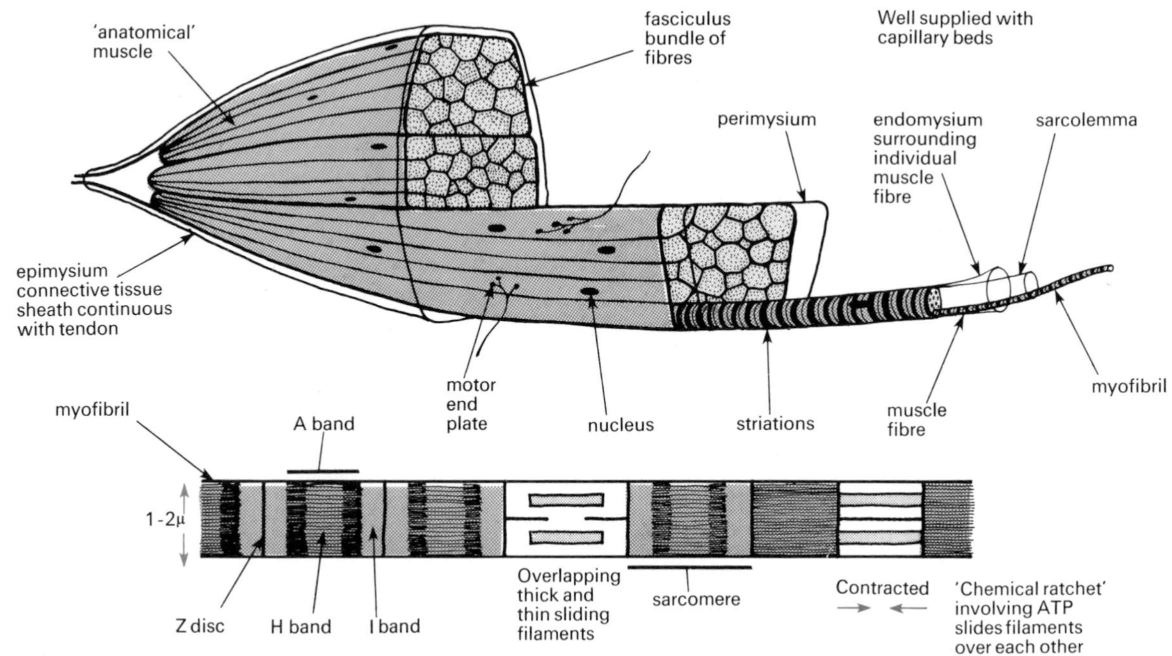

Figure 11.7 *Structure and function of striated muscle*

Myosin molecules have one part which contributes to the physical structure of the muscle, and two chemically active sites, one of which binds to actin, the other of which is enzymic, catalysing the breakdown of ATP to release energy used in muscle contraction. **Actin** is also part of the structure of the muscle and each molecule has one molecule of ATP bound to it. **Tropomyosin** is thought to be important in starting and stopping contractions, and it possibly sensitizes the contractile proteins to calcium which is important in contraction (muscle also contains as much zinc as calcium, but no clear function of the zinc has yet been identified).

These proteins make up the muscle fibres and interact to produce contraction.

Sarcoplasmic reticulum

In striated muscle fibres the smooth endoplasmic reticulum is well-developed and forms a plexus around each myofibril which is known as the sarcoplasmic reticulum. This conducts the excitatory impulses from the surface of the fibre to each individual myofibril and thus speeds contraction.

The impulse also causes calcium ions to be released into the reticulum. These are necessary for the splitting of ATP.[367]

Energy

The contraction of muscle is dependent upon energy from ATP which is closely bound to the contractile proteins. Measurements indicate that in a frog muscle there is enough ATP for about eight contractions, but this store is continually regenerated from another energy store material, phosphocreatine, which provides enough energy for about another 100 contractions. Phosphocreatine is in turn regenerated by energy from the muscle glycogen stores which contain energy sufficient for about 20 000 contractions under aerobic conditions.

Under anaerobic conditions the glycogen store only yields energy for about 600 contractions.

Different types of muscle fibre

White fibres These have little or no myoglobin or cytochrome pigments, and very few mitochondria. They carry out anaerobic respiration, converting glycogen to lactic acid and building up an oxygen debt.

They are used to provide immediate muscle contractions during the period when the circulatory system has not had time to increase the supply of oxygen to the fibres, and when the muscle output exceeds the input. The heat produced is absorbed by a temporary rise in body temperature as the blood circulation cannot carry it away rapidly enough.

Red fibres These are narrower in diameter than white fibres, have myoglobin, cytochromes, and many mitochondria, and carry out aerobic respiration.

They are used in long sustained activity, when the circulation has adjusted to the demands of contraction and the oxygen is absorbed and transported to the muscles at the same rate as it is used up by respiration. Moreover, the respiratory substrate is mobilized from storage at the same rate as it is oxidized, and the heat is transported away at the same rate as it is produced. Thus a steady state exists in these fibres, whereby the output equals the input.

As they are adapted for 'long running', the energy substrate is often indirectly fat which can only be utilized by aerobic oxidation. These fibres also have a network of capillaries in close contact to facilitate all the exchanges involved between the fibres and the blood.

Occurrence of fibres Sometimes fibres occur separately in red and white 'muscles', thus in man the locomotor muscles are predominantly white, and the long-acting postural muscles are predominantly red. Similarly in dogfish the locomotory myotomes which are in virtually continuous action are red. Some birds have the two types of fibres intermixed together, using the white fibres for the sudden contractions required on takeoff, and using the red fibres during long flights. Rabbits have mainly white muscles, whereas hares, which are capable of longer bouts of running, have red.

The white fibres are sometimes referred to as 'sprint' fibres and the red as 'cruise' fibres, but the speed of contraction of different fibres is a property of contractile proteins and not of the colour.

Stimulation Voluntary muscles are stimulated to contract by the spindle reflex, and by direct motor nerve fibres. The motor nerve axon loses its myelin sheath, the nerve fibre comes into direct contact with the muscle fibre membrane, and special motor end plates are formed.

Slow fibres Some muscles are supplied by smaller axons with simple nerve-muscle junctions which lack the special type of synapse of the end plates. Several stimulations of these are required for a contraction, which is long and slow instead of short and quick. Such slow fibres are found where prolonged contractions are required, as in the maintenance of posture.

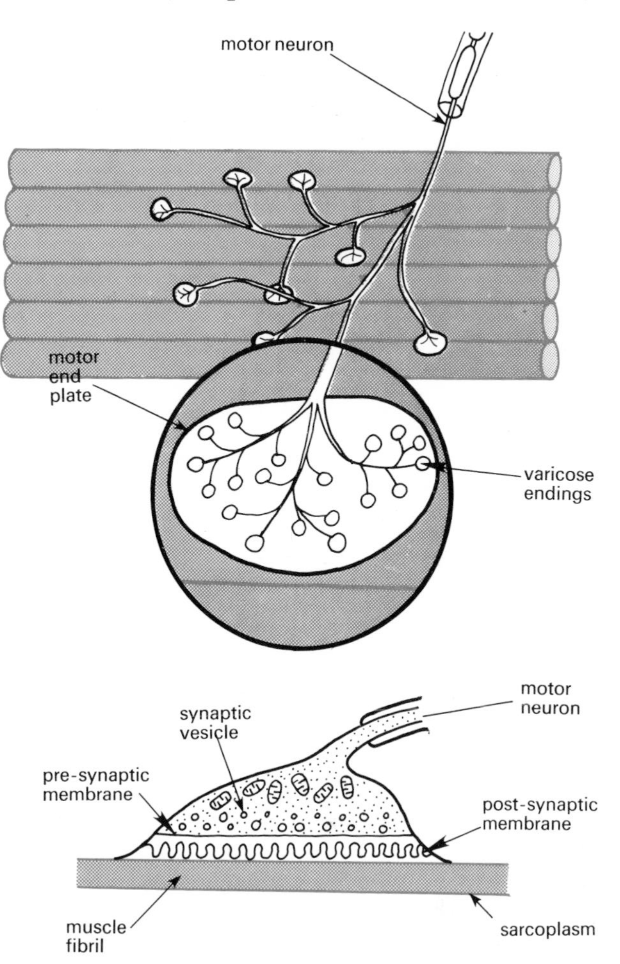

Figure 11.8 *Innervation of striated muscle*

In humans, both fast and slow muscle fibres are found. Males show a wide range of mixtures of these fibres, varying from 90 per cent fast and 10 per cent slow in Olympic standard sprinters, to 10 per cent fast and 90 per cent slow. Females do not show such a wide range of mixtures of these fibres and will never be able to equal the physical performance of males at the extremes of the ranges.

Muscle contraction

Each fibre, when stimulated, either contracts maximally or not at all—the so-called **all or nothing** response. The graded response of whole muscles is brought about by different numbers of muscle fibres being stimulated.

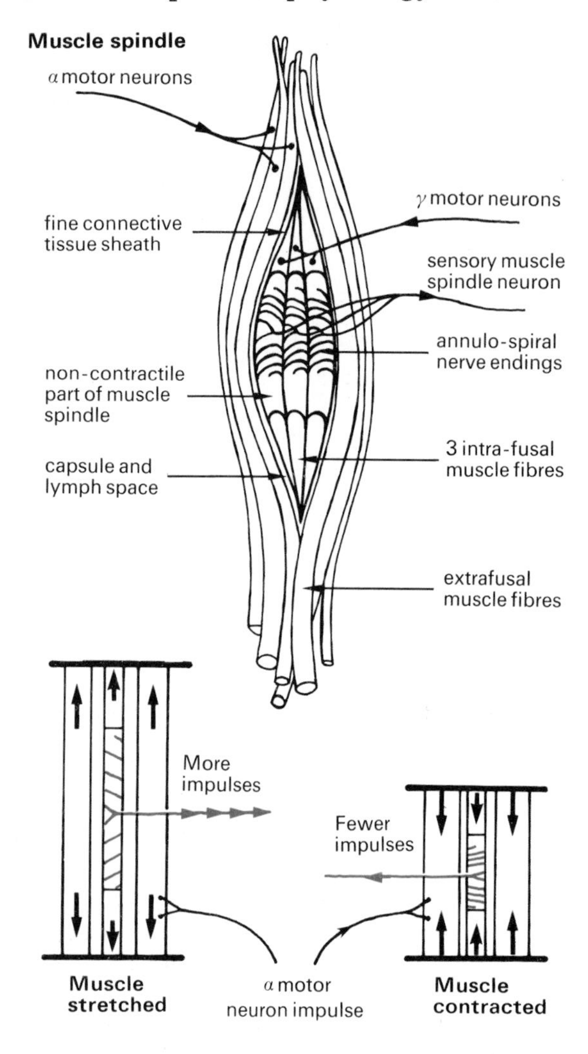

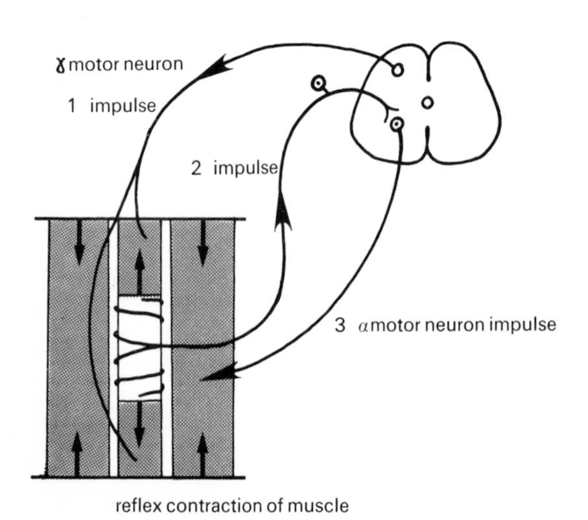

Figure 11.9 *The mechanisms of stretch receptors*

Muscle fibres are grouped functionally into **motor units**, a motor unit being those muscle fibres innervated by a single nerve fibre. The number of muscle fibres in a motor unit can vary from ten to several hundred, and determines the amount of fine control that can be achieved.

Muscle spindles

The degree of contraction of muscles is detected by the special sense organs, the muscle spindles. Muscles can contain up to sixty of these. If the muscle is stretched, extension of the spindle triggers a reflex contraction which attempts to maintain a constant length.

However, the spindles are also used in graded voluntary movements. The ends of the spindle fibres are contractile and by stimulating these to contract, the central receptors are stretched and trigger off the reflex contraction of the muscle. This mechanism is used in maintaining the muscle tone and in the fine control of muscles. Direct stimulation of the muscle by the CNS is found only in sudden and strong contractions.

Skeletons

Muscles produce movement and locomotion by reacting against each other and against some type of skeleton. There are three main types: the **hydrostatic skeleton**, the **exoskeleton**, and the **endoskeleton**.

MUSCLE ATTACHMENT

The shape of a particular skeleton reflects the demands of the musculature. As muscles are only effective during contraction they are usually found in antagonistic pairs, in which the contraction of one stretches the other. They are attached to the long bones in such a way that they work at a mechanical disadvantage of about 1:10. This means that to lift 10 kg by bending the arm the biceps must generate a 100 kg effort.

However, this mode of attachment gives a wide arc of movement at the end of the limb for a relatively short distance of muscle contraction. Also the concentration of muscles at the end of the limb nearest the body (proximal end) lightens the end furthest from the body (distal end) where changes in its velocity are greatest. Furthermore, the muscles carry out their contractions where changes in the velocity of the limb are least. Both of these points reduce the effort required from the muscles, and explain why most muscles are found above the joints which they work.

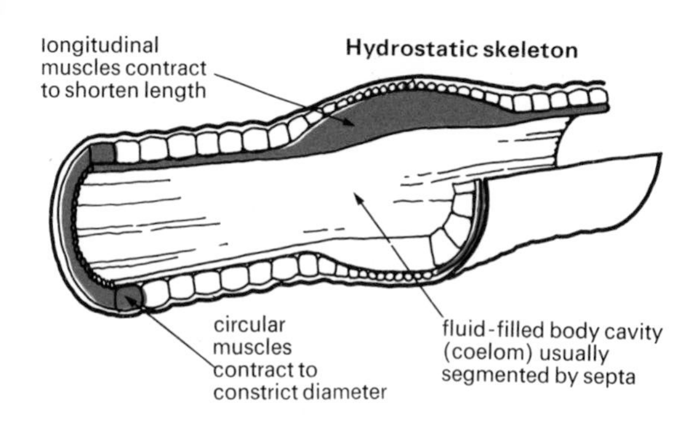

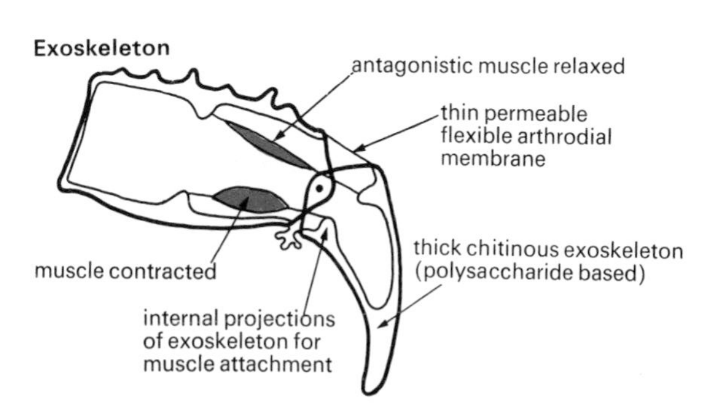

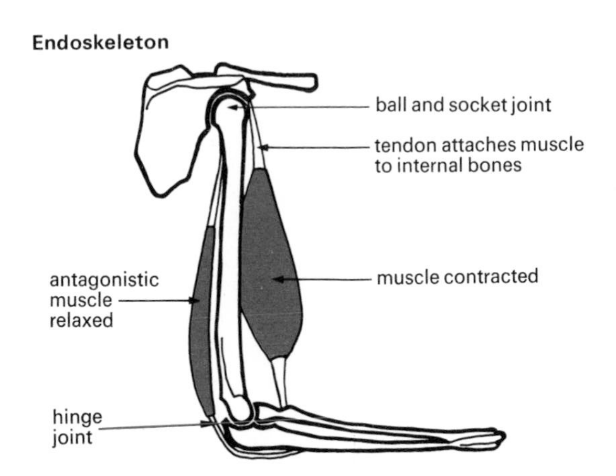

Figure 11.10 *Different types of skeleton*
With the hydrostatic skeleton the muscles react against the fluid and the circular and longitudinal muscles usually co-ordinate to produce peristaltic waves along the length of the animal. An exoskeleton limits the size of animals due to weight problems; it also necessitates moulting (ecdysis) during growth. With a bony endoskeleton the shapes of the bones are related to muscle attachment, leverage, and jointing.

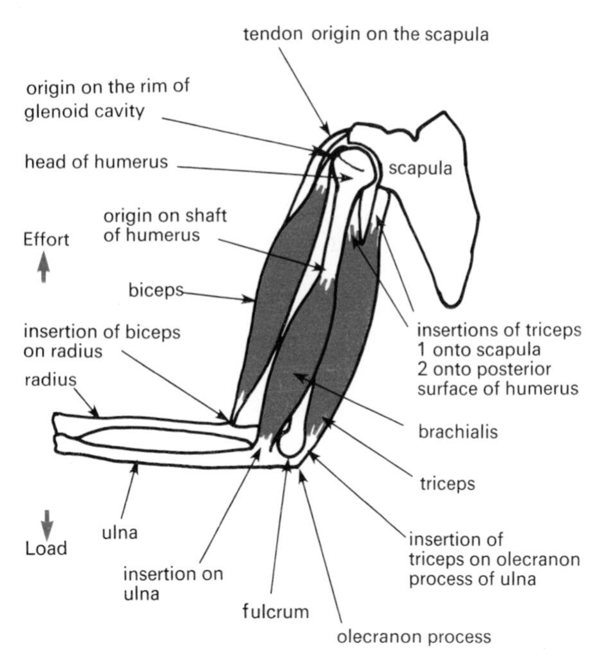

Figure 11.11 *Functional skeleton (joints, levers, and muscles)*

MOVEMENT AND LOCOMOTION IN VERTEBRATES

There is a fantastic variety of movement and locomotion in the vertebrates, adapted to every conceivable habitat and environmental niche: aquatic, terrestrial, and aerial.

Fast movement through water requires streamlining and powerful musculature to overcome the resistance of the water; movement on land presents problems of balance, support, and shock absorption; and flight requires lightness, efficient metabolism, powerful flight muscles, streamlining, a much modified skeleton, and fine co-ordination.

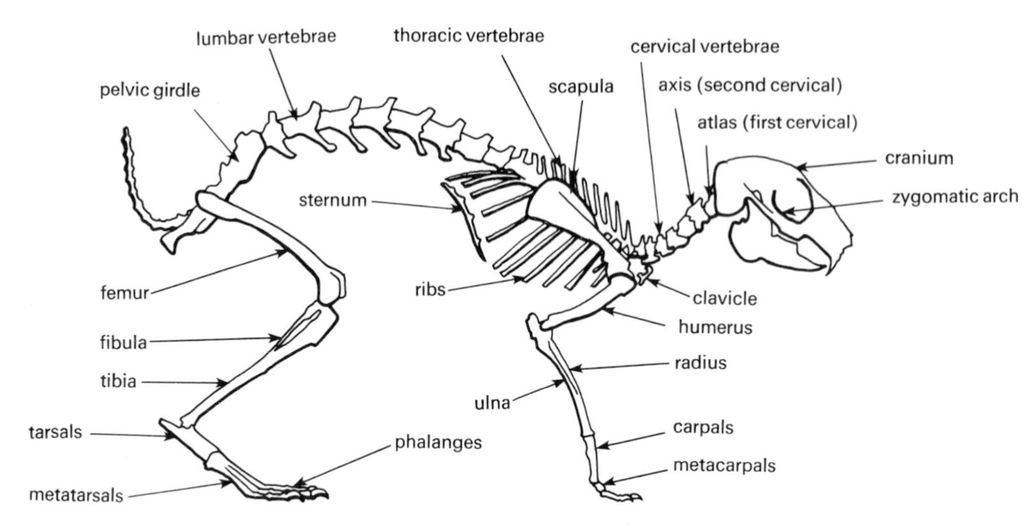

Figure 11.12. *A mammalian endoskeleton [rabbit]*

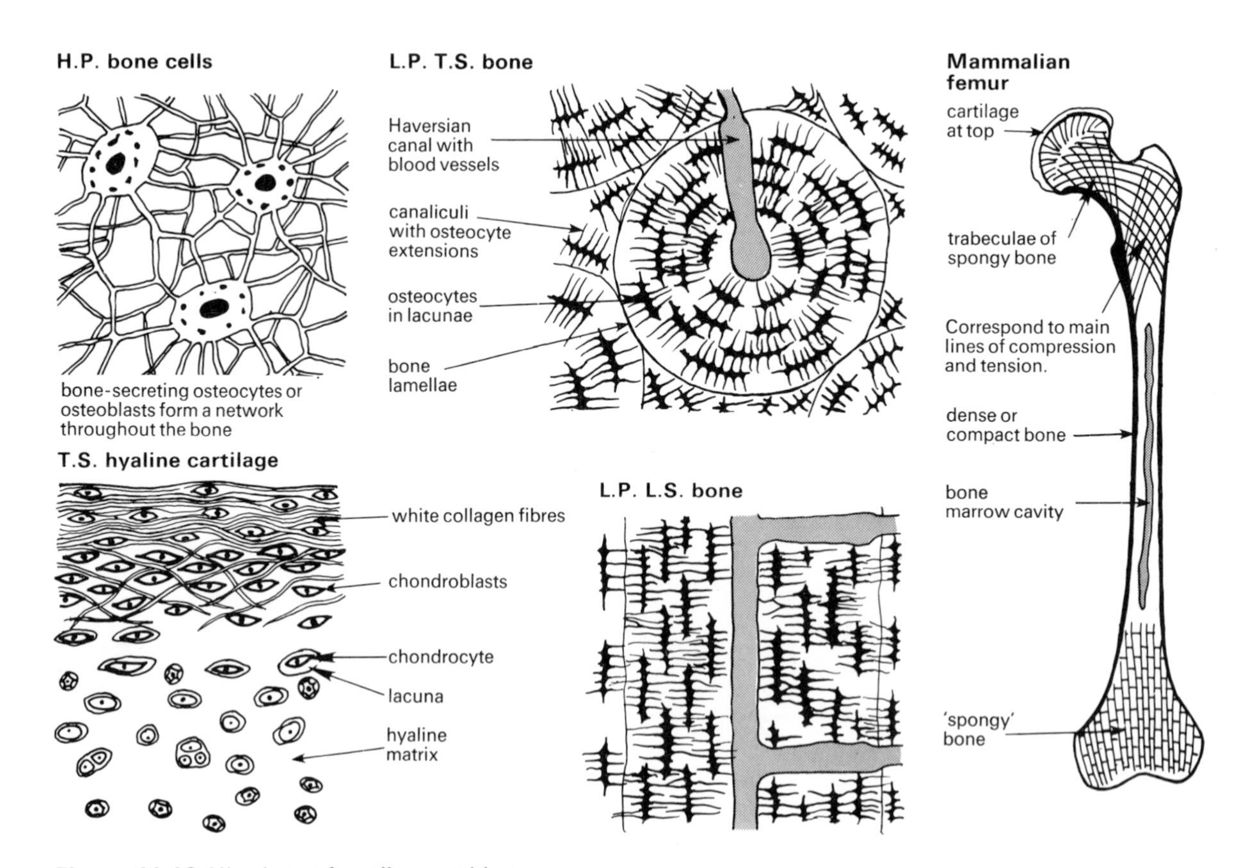

Figure 11.13 *Histology of cartilage and bone*

So-called 'spongy' bone has the same hard composition as dense bone but has sponge-like irregular spaces or lacunae. Bone is a stiff, rigid, elastic, strong, tough, yet light material. It offers as much resistance to stretching as to compression and is therefore difficult to bend or twist. Collagen fibres confer tensile strength; hydroxyapatite crystals and calcium phosphate renders it stiff and strong. Parathormone and calcitonin regulate the deposition of calcium in the bone from the blood and the release of calcium from the bone into the blood, so that the blood calcium level is held at 9–11 mg per 100 cm^3.

Figure 11.14 *Slow flapping flight in the pigeon*
The sequence shown covers about 6/100 second. The 'arm' part of the wing provides lift, and the 'hand' part provides lift and forward thrust. At most instants one part of the wing is moving down and providing lift. Feathers interlock on the downstroke, increasing air resistance and therefore upward thrust. On the upstroke sequence the feathers 'open', lessening the air resistance and therefore decreasing any down thrust.

Figure 11.15 *Swimming in fish* (*bony fish*)

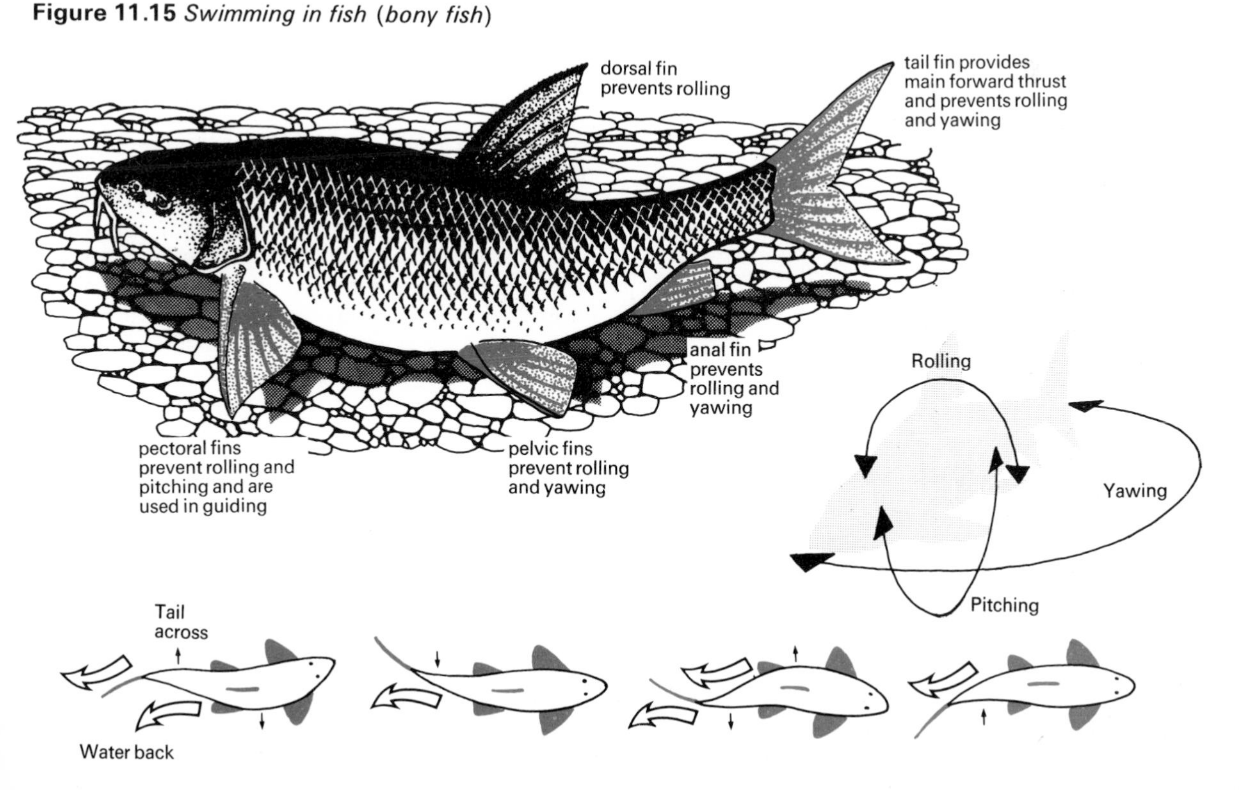

RESPONSE AND BEHAVIOUR

Patterns of behaviour

The response of animals to their environment is more complex than that of plants. It is usually expressed as movement but can include any change in their activity, and thus involves all types of effectors including glands.

Behaviour is an expression of the activity of the co-ordinating systems of the animal, the nervous and the hormonal systems. Therefore many factors operate between the stimulus and the response so that there is seldom a direct cause and effect relationship between the two. Also the complexity of response and behaviour increases with the increasing complexity of the nervous and hormonal systems.

There are many approaches to the study of behaviour, including those which attempt to study isolated components of behaviour under experimental conditions. Such studies may obscure the overall picture of an animal's behaviour in relation to its natural habitat. The approach that attempts to study the behaviour of an animal in its natural habitat is known as **ethology**.

Many attempts have been made to classify behaviour into convenient categories for study, such as reflex, instinctive, learned; but there are no clear distinctions as all behaviour has genetically-determined and acquired or learned components in varying proportions.

The study of behaviour is one of the major branches of modern biology. It is still in the analytical and descriptive stage of development but will ultimately lead to a synthesis of neurophysiology, endocrinology, genetics, ecology, evolution, psychology, and, some even suggest, sociology, economics, and politics! Such a synthesis is attempted in the field of sociobiology, but this remains an area of great controversy.

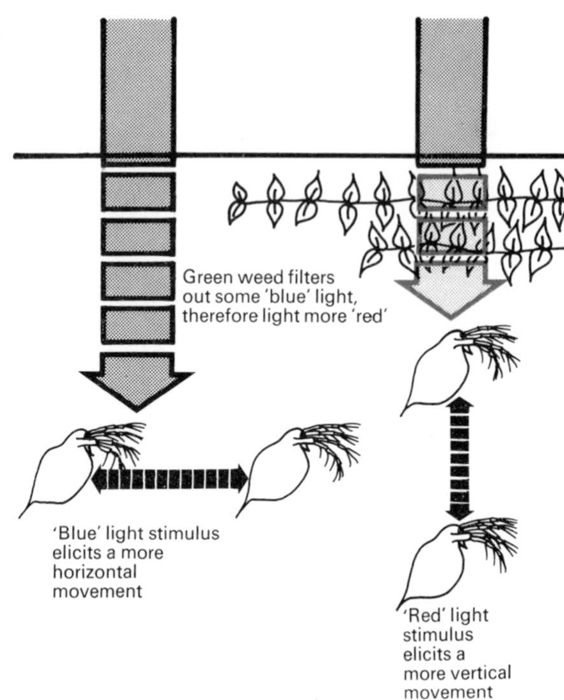

Figure 11.15 *Movement, response, and behaviour in Daphnia*

Daphnia are positively phototactic to weak light, rising to the surface at dawn and dusk. The horizontal movement in 'blue' light will result in the *Daphnia* finding the weed patches, and the vertical movment in 'red' light will result in the *Daphnia* remaining under weed, which is rich in oxygen, algae, and bacteria upon which they feed.

Simpler patterns of behaviour

The movements of animals in response to a stimulus represent some of the simpler types of behaviour, although the complexity of such a response varies with the complexity of the animal. In the less complex animals these movements can be classified as taxes or kineses, but in the more complex they become part of very complex behavioural patterns, including migrations.

Taxes are movements, the direction of which is determined by the stimulus, and include both positive and negative phototaxis (light), chemotaxis (chemicals), geotaxis (gravity), and rheotaxis (water currents). For example, woodlice show negative phototaxis as they move away from bright light.

Kineses are non-directional changes in movement as a result of a stimulus, for example woodlice move faster in drier surroundings than in damp ones, and daphnia move more vertically under green weed.

Complex behavioural patterns

So difficult is it to describe and explain an animal's overall behaviour that several attempts have been made to analyse complex behavioural patterns into simple components. None of these attempts are entirely successful, but they do provide convenient and more easily understood explanations than would be possible without them.

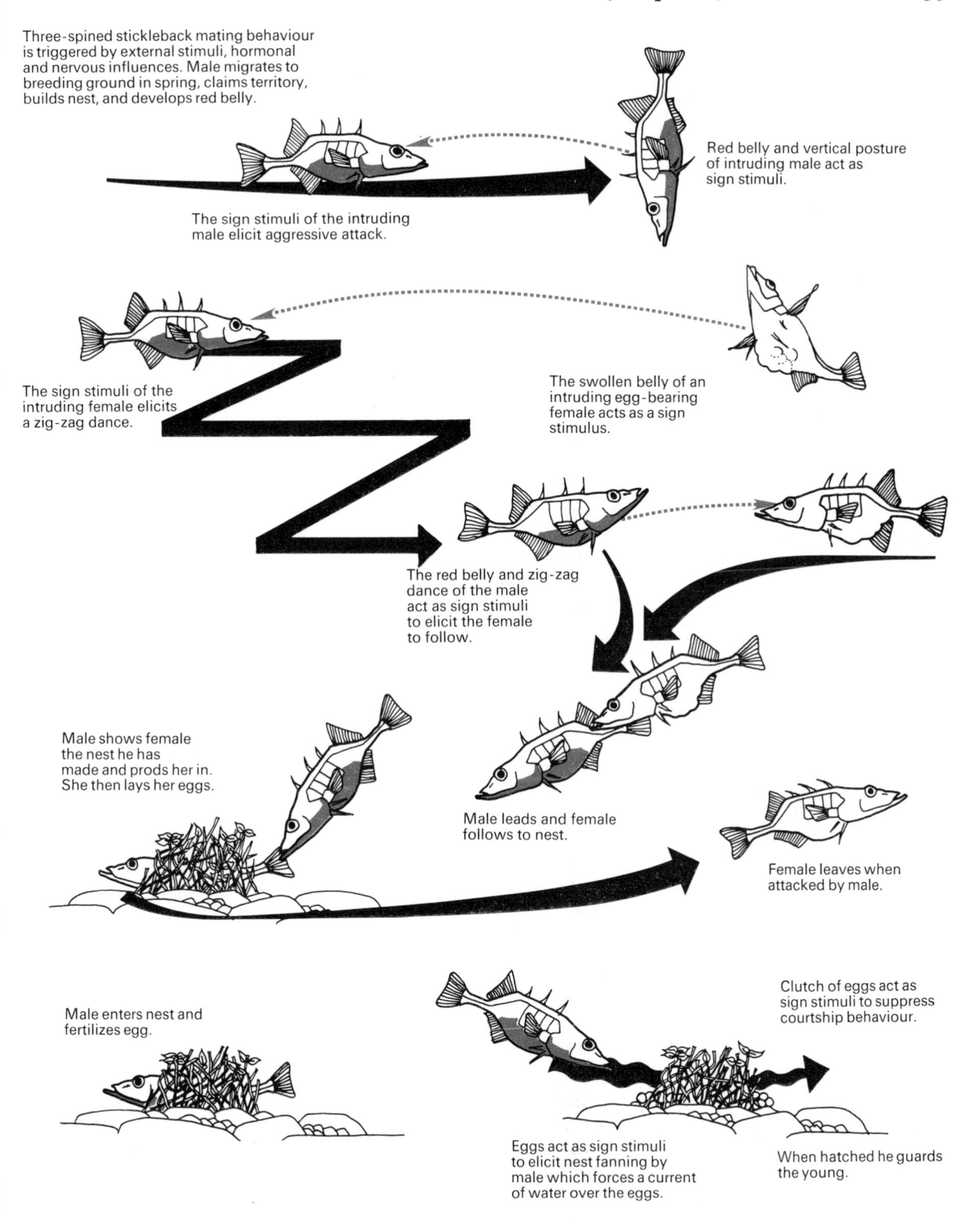

Figure 11.16 *Courtship and mating behaviour in the three-spined stickleback*

INNATE FIXED-ACTION PATTERNS

Many movements, once elicited by a particular stimulus, seem to be independent of further environmental stimuli for their completion; these are referred to as fixed-action patterns (FAP). Each animal has a particular set of FAPs and apparently only a limited ability to develop new ones, thus these patterns are described as being innate and species characteristic (typical for each species).

However, the apparently stereotyped behavioural responses of animals always have a variable component. Also, animals may be 'freed' from their responses from time to time when it is necessary for their survival. Thus, when the temperature drops, woodlice are 'freed' from the behavioural responses that keep them in dark, humid conditions so that they can then emerge to feed when the cool conditions reduce the danger of water loss by evaporation.

SIGN STIMULI

From the great complexity of sensory information to which it is exposed, an animal usually responds particularly to certain key 'sign stimuli'. These 'sign stimuli' are important in displays which animals use to communicate with each other, particularly in complex mating rituals where the 'sign stimuli' are thought to release the appropriate responses in the partner.

INNATE RELEASING MECHANISMS

The release of the FAP in response to specific sign stimuli is thought of by some as being controlled by an innate releasing mechanism, as if some inborn 'switch' were activated to release the stereotyped behaviour in its entirety. However, in most cases the reaction to the stimulus is, in fact, influenced by learning and, in addition to the release of responses, the mechanism may also involve the inhibition of other responses.

DISPLACEMENT ACTIVITIES

Conflict arises when two or more stimuli compete for the expression of responses which are mutually exclusive. This usually results in the suppression of all but one of the responses. There is normally a hierarchy of response in which the higher ones inhibit the lower ones but, if there is competition for expression between two higher ones, neither may occur and a lower one may be disinhibited and appear as a displacement activity (if the necessary stimulus for its appearance is present). For example, if birds have conflict between fighting and fleeing responses, it is not unusual for them to start preening.

REDIRECTIONAL ACTIVITIES

Conflict situations may also give rise to redirectional activities, whereby an appropriate response is made to one of the conflicting stimuli, but is redirected. For example, an angry person faced with the conflicting responses of aggressive physical attack on the opponent and normal social constraints may thump the table instead of the opponent.

HORMONAL INFLUENCES

The influence of hormones on behaviour is particularly well displayed in the birds. **Leuteotrophic hormone** (LTH) is responsible for broodiness in hens and pigeons and is involved in the formation of the incubation patch in perching birds. This is interesting as it is an example of a trophic hormone (one that normally stimulates another endocrine gland) stimulating non-endocrine tissues directly.

Thyroxine is also involved in broodiness, and exercises some control over pre-migration behaviour. **Oestrogens** in birds stimulate nest-building activity and are partly responsible for behavioural characters at breeding time. **Progesterone** is also thought to be involved in incubation behaviour. **Male hormone** levels are linked with dominance amongst males; seasonal rises result in an increase in aggression, the enlargement of territory in territorial types, and the onset of sexual behaviour.

RHYTHMS

Rhythms are behavioural patterns which appear to be initiated by some internal rhythmic activity of the central nervous and hormonal systems. Such rhythms include annual, lunar, and daily or circadian cycles of behaviour. These internal rhythms do not depend on external stimuli for their continuation, but do require external stimuli to keep them 'geared' to the external cycles.

Circadian rhythms

At one time these were known as endogenous diurnal rhythms. The word 'diurnal' implies a cycle of exactly twenty-four hours, but in almost all cases these 'rhythms', when 'free-running' under constant experimental conditions, vary between twenty-one and twenty-eight hours. Therefore these rhythms are now referred to as circadian (*circa* = about, *diem* = day).

Under natural conditions circadian rhythms are coupled to the precise twenty-four-hour period of the environment in a process known as **entrainment**. The rhythms are independent of temperature, which is necessary to prevent them changing with fluctuations in daily and seasonal temperatures. This is unusual for any metabolic activity.

Circadian rhythms occur in both plants and animals. In plants they include spore discharges,

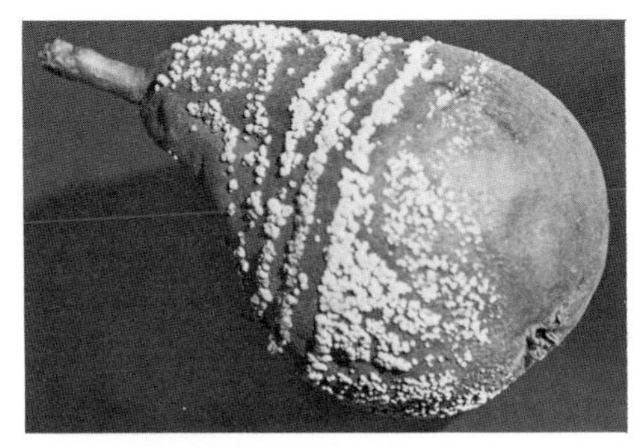

Circadian rhythm of fungal spore discharge is shown by the concentric rings on this pear

leaf movements, various metabolic processes, and general growth rates; but are less obviously related to behaviour, essentially a term applied to animals. Circadian rhythms in animals are widespread and include various metabolic processes, such as the functioning of the endocrine glands, which in turn have a profound effect on the behaviour.

Figure 11.17 *Circadian rhythm of activity in the cockroach*

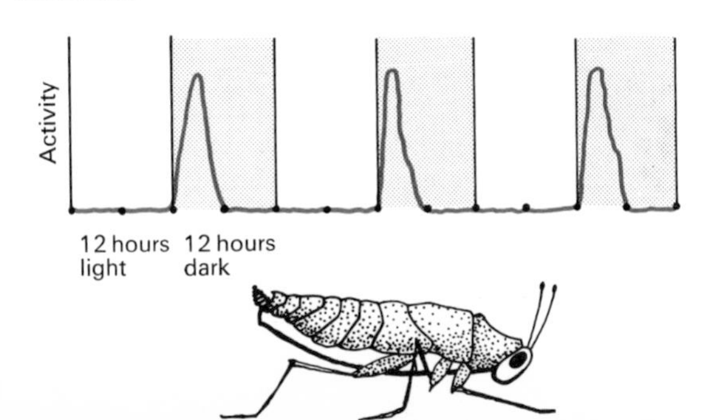

Learning

Learning is difficult to define in any precise sense and cannot be separated clearly from so-called innate behaviour. It is more a term of convenience used to cover a diverse group of behaviours that appear to be based mainly on past experience, and which are thus characteristic of individuals rather than of the species as a whole. It therefore follows that there can be no clear subdivisions of such an ill-defined area. However, again for the sake of convenience, several subdivisions are suggested, including habituation, classical conditioning, operant conditioning, imprinting, and insight learning.

HABITUATION

Habituation occurs when the animal becomes so 'used to' a repetitive stimulus that it learns to 'ignore it' and no longer produces the expected behavioural response. The adaptive advantage is that if the animal survives the conditions that are associated with this stimulus for long enough for habituation to occur, then they are almost certainly harmless.

CLASSICAL CONDITIONING

In classical conditioning a basic reflex is conditioned to be elicited or brought about by a stimulus other than that which would normally elicit that reflex. The classic experiments in this field were performed by Pavlov working with the saliva-producing reflex in dogs.

Dogs were conditioned to associate the appearance of food with the ringing of a bell. After a certain number of repetitions the dogs would salivate on hearing the bell even if no food was presented.

Such conditioned reflexes play a part in the learning process of most animals. However, this particular approach to the study of behaviour and learning is more useful to sensory physiologists than to ethologists, since it does not provide much understanding of the animal's behaviour in its natural environment.

OPERANT CONDITIONING

A response is made more likely to occur by associating some immediately-received reward or reinforcement with that response. The reinforcement does not elicit the response, as in a conditioned reflex, but makes that response more likely. Such responses appear to have no eliciting stimulus (and are thus described as being emitted rather than elicited) and as they always involve the animal actually doing something they are also known as operants. The reinforcing of these is known as operant conditioning. Thus an emitted operant, like leaf flicking in a pigeon, will be reinforced by the discovery of food organisms beneath the leaves; and the pigeon thus learns by operant conditioning. This explanation removes the need to suggest the existence of some innate fixed-action pattern of leaf flicking.

Work pioneered by Skinner led to the idea of behavioural objectives in education, whereby the outcome of education is expressed in terms of particular required behaviours which must be reinforced. This approach found its main application in programmed learning in which the correct behavioural outcome—the selection of the correct answer—is immediately reinforced by progress through the programme. An important aspect of the approach is that punishment has no part in it. Indeed, punishment, by creating anxiety, may be harmful to the learning process. Incorrect responses are simply not reinforced. Needless to say this is again an area of great controversy.

IMPRINTING

During certain brief critical periods animals are susceptible to imprinting. Thus in ducks, geese, and chickens, newly hatched birds become imprinted by the first large moving object that they see, and will follow that object, normally the mother, from then on. If the first large moving object happens to be a man or any other such object, they will still follow it!

There is an interesting example of the significance of imprinting in the herring gull[91] and the lesser black-backed gull. The herring gull has a yellow ring around each eye and a yellow patch at the corners of the beak; the lesser black-backed gull has a redder colouration at these points. 'Changelings', birds that have hatched from eggs swapped between these two species, adopt the behaviour of the surrounding population, mate, and produce fertile offspring. The females are imprinted by the colour of the eye ring of the 'parents', and it would therefore appear that one of the main isolating mechanisms[89] between these two species is the female's preference for birds resembling her parents. (The males are apparently not choosey.)

Such imprinting is usually very persistent, and can be put to good use in the obedience training of dogs which have a critical period for imprinting at about 6–7 weeks.

INSIGHT LEARNING

Insight learning occurs when there appears to be an immediate understanding of a new problem, without any trial and error experimentation, based on what is known as intelligence and reasoning in the accepted sense. This type of learning is accepted as being characteristic of man and of some of the higher primates, but such is the complexity of the workings of the brain that this term is more of a label than an explanation of brain function.

Social behaviour

Social behaviour involves communication between individuals whereby the mutual exchange of stimuli results in the desired behavioural response. Social groups have advantages in obtaining food, in gaining protection from predators, and in reproducing successfully. In many cases, particularly in aquatic types, the conditioning of the environment by body secretions means that members of a group have a more efficient metabolism and growth rate than isolated individuals.

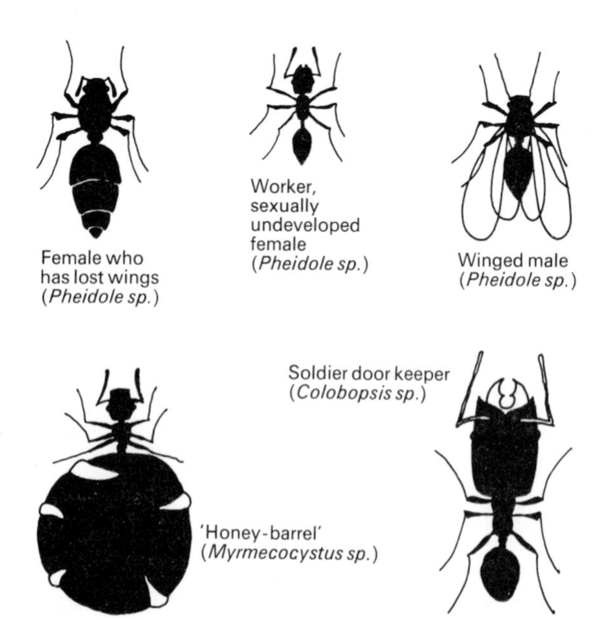

Figure 11.18 *Morphological specialization in ant castes*

INVERTEBRATE SOCIETIES

Invertebrate societies tend to be based on a rigid inherited caste system founded on morphological differentiation, as in the ants, termites, and bees. Reproduction is limited to a single female, the queen, and the social group consists entirely of her own offspring. These, apart from a few fertile males and females, are all sterile, so removing the need for reproductive competition.

VERTEBRATE SOCIETIES

Vertebrate societies are not so fixed. Each individual can reproduce and lead a separate existence, although usually much less successfully than within a group. With this individual identity there exists a measure of aggression, therefore there must be a system of social constraints to reduce the competition and consequent disruption that this would generate. These constraining behavioural responses have a genetic basis, but are mainly learned by individuals within the framework of the society.

The competition between group members for food, space, sex, etc., is often resolved by the establishment of **dominance hierarchies**; once established these reduce aggression, conserve energy, and promote the survival of the individuals and the species. Dominance hierarchies are a typical feature of all vertebrate societies, but can only develop in groups of limited size since animals must be able to recognize other group members to know their place. This in turn depends on the animal's ability to learn.

Group cohesion is encouraged by sexual attraction, infant attraction, mutual grooming, mutual secretions, and antagonism to non-group members; thus neighbouring groups are kept distinct. Group cohesion forces must be stronger than the potential group disruptive forces.

The advantages of dominance hierarchies in addition to the control of disruptive aggression are that, if food is short, the dominant animals will survive, whereas, if all had a share, all would die. In addition the dominant members will be the healthier and best able to reproduce and care for the young.

The term dominance hierarchy can be misleading, however, as it stresses domination as the important factor. Equally necessary for the maintenance of the group structure is the appropriate subordinate behaviour in the low-ranking members. Subordination is not the same as lack of dominance, it implies a particular behaviour pattern of its own.

Pecking orders

The domestic fowl (*Gallus domesticus*) was the first vertebrate in which dominance hierarchies were studied. That this behaviour is not a product of the domestication of the fowl and its restriction to relatively small areas is shown by the fact that pecking orders are found in its wild ancestor, the red jungle fowl (*Gallus gallus*) of India and Indochina. These form flocks of 5–20 birds with dominance hierarchies within a territory centred on the roosting sites.

Cocks and hens form their own pecking orders with the male hierarchy above that of the female, so that even the dominant hens should submit and mate with the male, although this sometimes does not happen.

Once the hierarchy established, pecking declines in frequency as the rank order is recognized by individuals. As the hierarchy is based on aggression, which in turn depends on the level of sex hormones, it becomes stronger in the breeding season.

Baboon troops

These are held together by the strong social relationships that exist between the troop members. Sexual attraction is not a main binding factor, as for most of her life the female baboon is not sexually receptive. Baboons have highly-developed social natures independent of sex. The males form a dominance hierarchy which not only depends on their relative physical condition and abilities, but also on the formation of 'gangs', the members of which support each other in various disputes.

Territory

Territorial behaviour, in which a certain area is claimed and defended against intruders, serves to control population density, and ultimately population size as the territories spread and reach the edge of suitable areas. The primary claimant is usually a male and the boundaries are defined by patrolling, vocalizing, scent markings, and aggression against intruders. Sometimes territorial behaviour is found only during the mating season. In social animals it is more typical of those forming large groups. It serves to distribute individuals of a social group, reducing conflict and 'psychological castration' (such as impotence in males) which is often seen in dominance hierarchies.

Rabbit dominance hierarchies develop within a well-defined territory. Males and females form two separate hierarchies and the dominant ones of both breed preferentially in the central warren. Their young have a higher survival rate than those of the lower ranks, as those of the latter are born and raised at the edge of the territory, and are chased out of the central warren if they attempt to enter. They are thus much more vulnerable to predators.

Human behaviour

Any attempt at interpreting human behaviour on the basis of observations of animal behaviour should be treated with caution.

There are often large differences in behaviour between closely related species, let alone between animals such as the primates and man. Conversely, when interpreting animal behaviour, anthropomorphism (the attribution of human characteristics to animals) must also be avoided.

Figure 11.19 *Social behaviour in baboons*

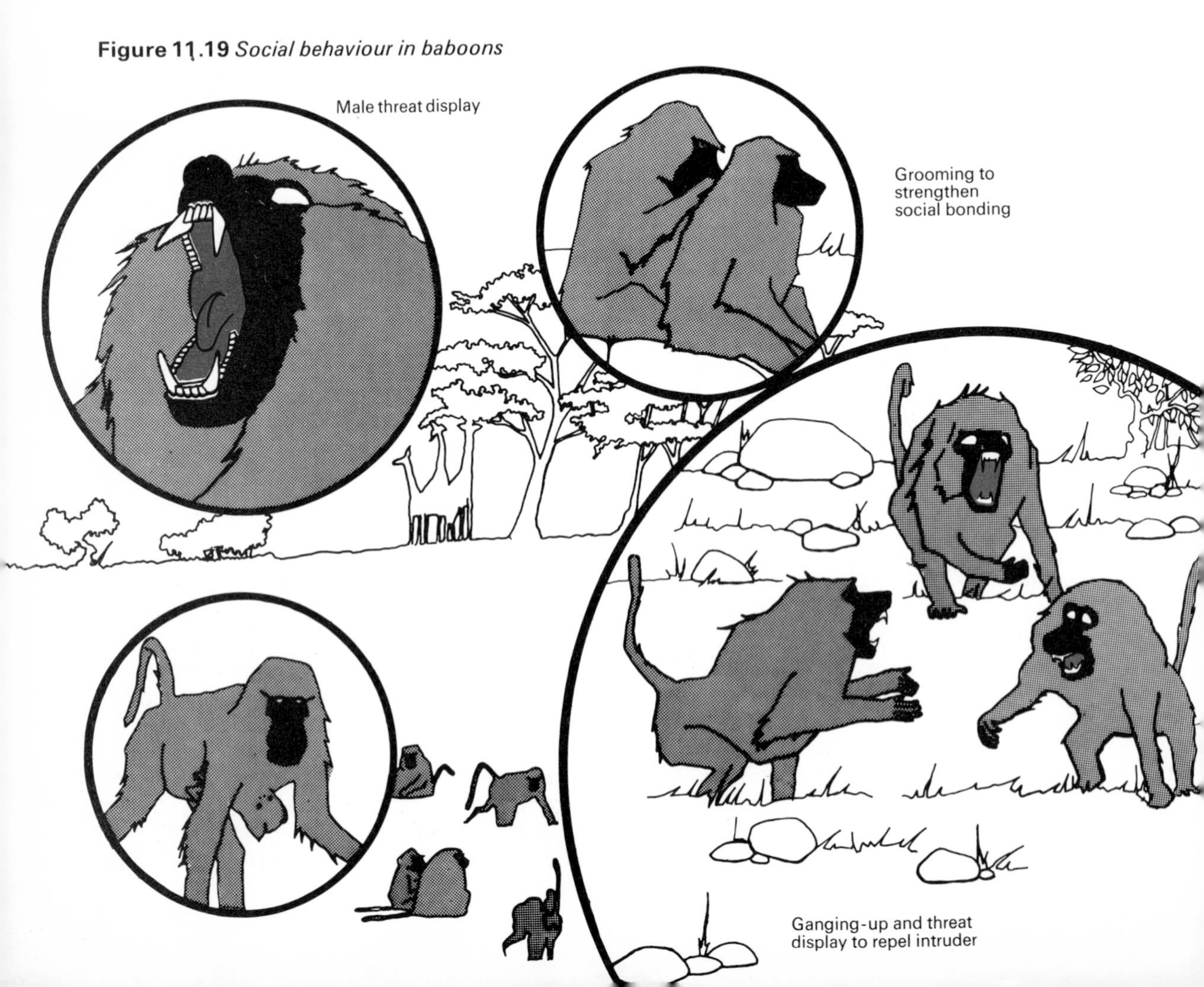

Experimental and applied aspects

One experimental procedure in the investigation of animal behaviour is to present animals with either other individuals or models under carefully controlled conditions and to observe their subsequent behaviour. Two 'classic' experiments of this kind involve the stickleback and the herring gull.

In investigations of the behaviour of a male three-spined stickleback, a nest-building male guarding territory is presented with a sequence of intruders; some models and some live fish held in tubes; and the subsequent behaviour observed.

Guided example

Newly-hatched chicks of the herring gull stimulate the parent to regurgitate food which it has collected, by pecking at the tip of its beak.

Tinbergen investigated the nature of the sign stimuli that elicited or triggered the pecking response in the chicks. He presented a series of life-sized, flat cardboard replicas of a parent's head, on which various features had been altered, to the chicks, and noted the number of times they pecked at each model.

Some simplified accounts of the experiments are provided.

1 The experiments were carried out under carefully controlled conditions. What conditions can you think of that would need to be controlled so that the investigator could be sure that the results were due to the altered features of the replica head?

The more controlled the conditions the better, but some aspects would be more important than others. Obviously the surrounding conditions, for example of temperature and light, would have to be within certain optimum limits so that the behaviour of the chicks could be as natural as possible. Also all the chicks would have to be equally as hungry, and it is for this reason that, in the original experiment, they were taken for the experiment before they had been fed by their parents.

Figure 11.20 *Investigation of the behaviour of a male three-spined stickleback*

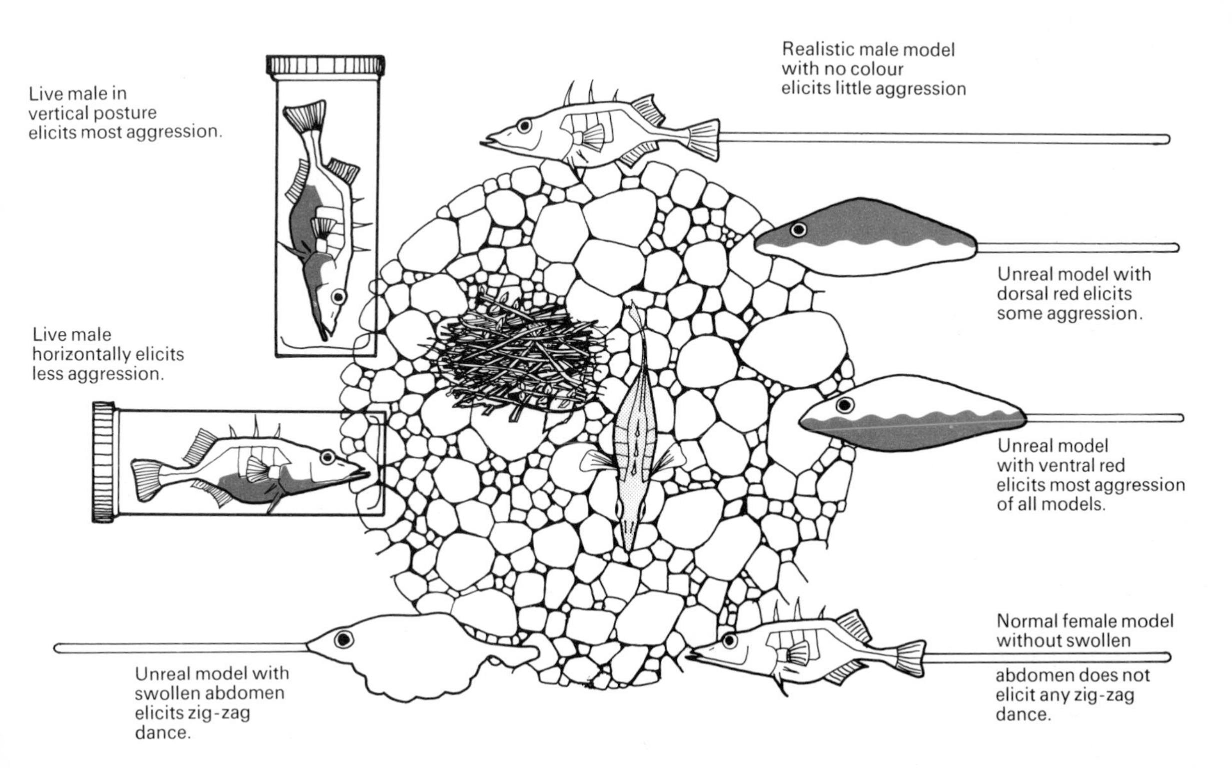

Each model was presented for the same length of time and the different models were presented in various orders, so that the sequence of presentation could be eliminated as a factor influencing the chicks' response; for example it might be that the first one presented triggered the most responses.

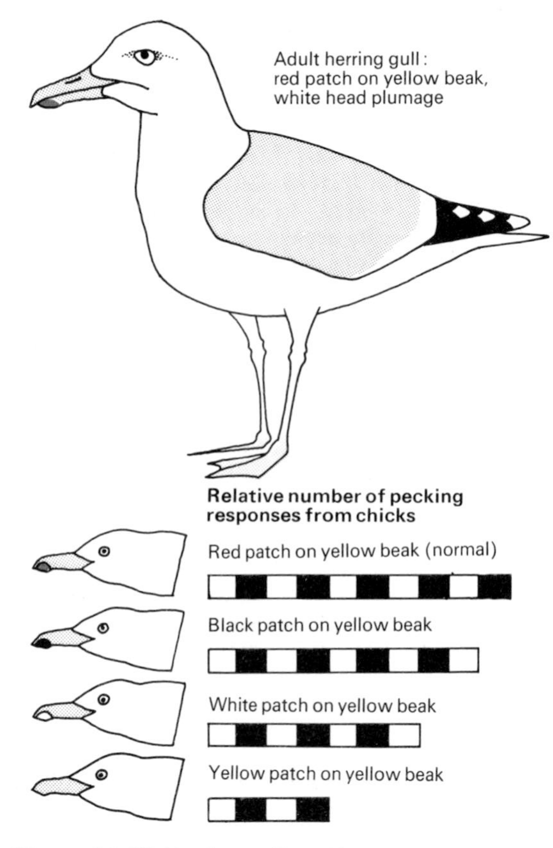

Figure 11.21 *Herring gull pecking response Experiment 1*
All models had white heads and yellow bills, each with an equal-sized patch on the lower bill. Each patch, however, was of a different colour.

2(a) The results of Experiment 1 show that the beaks with white, red, or black patches all elicit more than twice the number of responses than does that with the yellow patch. This suggests a feature to which the chicks might be responding. What do you think this is?
The yellow patch contrasts least with the rest of the beak, therefore one sign stimulus to elicit pecking might be the contrast of the patch to the rest of the beak.

2(b) In Experiment 1 the red patch elicits the most responses. This suggests another feature to which the chicks might be responding. What do you think this is?
The colour red.

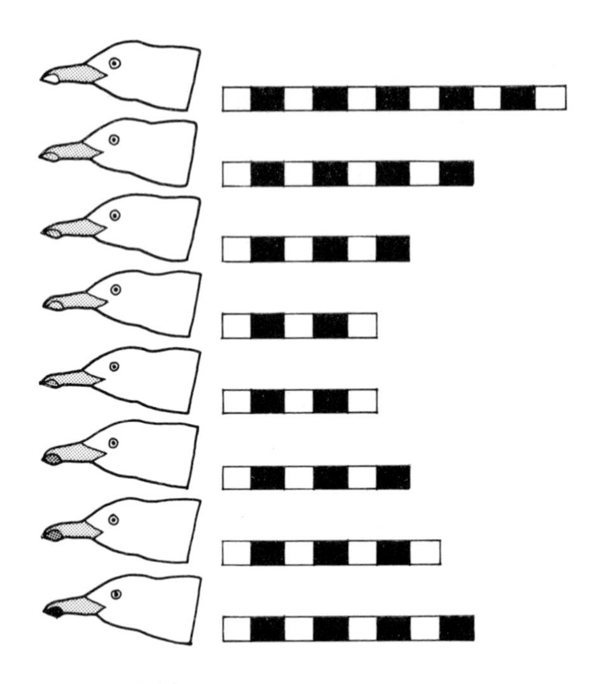

Figure 11.22 *Herring gull pecking response Experiment 2*
All models presented had beaks of a uniform grey with patches varying from white to black.

3 Do the results of Experiment 2 support the idea that contrast of the patch is a sign stimulus?
The results of Experiment 2 support the idea that the contrast of the patch with the beak is a sign stimulus since the models differ only in the degree of contrast between the patch and the beak. The greater the contrast, the greater the number of responses, irrespective of whether the patch was lighter or darker than the beak.

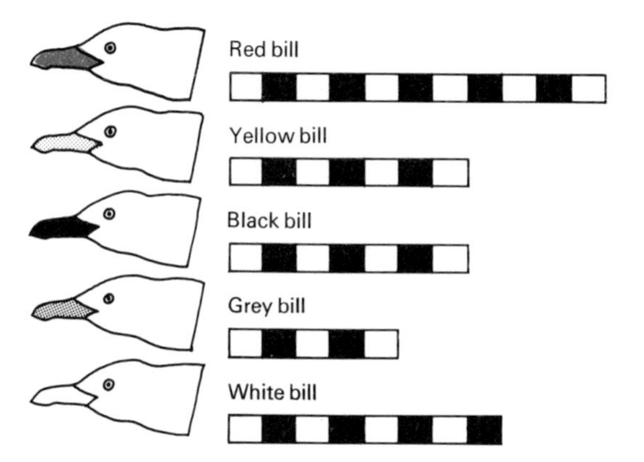

Figure 11.23 *Herring gull pecking response Experiment 3*
All models presented had uniformly-coloured beaks with no patches, but each was a different colour.

4 What conclusion can be drawn from Experiment 3?
The results of Experiment 3 indicate that the colour red is also a sign stimulus for pecking, as there are no contrasting patches. However, they do not indicate whether the stimulus is the colour red alone, or red on the beak.

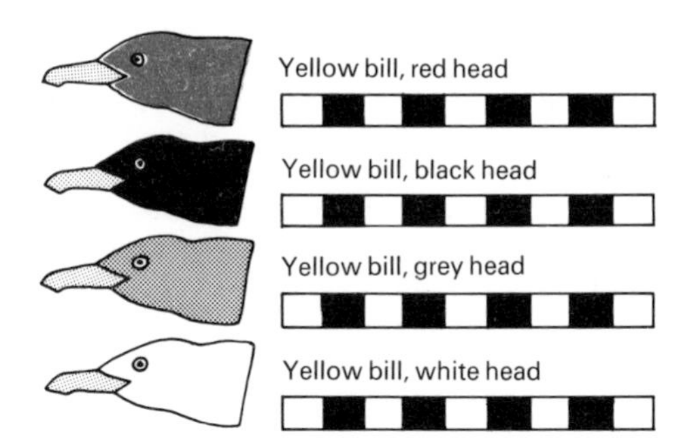

Figure 11.24 *Herring gull pecking response Experiment 4*
All models presented were identical except for the colour of the head. The beaks were all of a uniform yellow with no patch.

5 What do the results of Experiment 4 indicate about the position of the colour red in eliciting pecking responses?
The results of Experiment 4 indicate that the colour of the head does not influence the response and that therefore it is not the colour red alone which is a sign stimulus but the red on the beak.

6 What conclusions can you draw from the results of Experiment 5? Do they differ from those of Experiment 4 in any unexpected way?

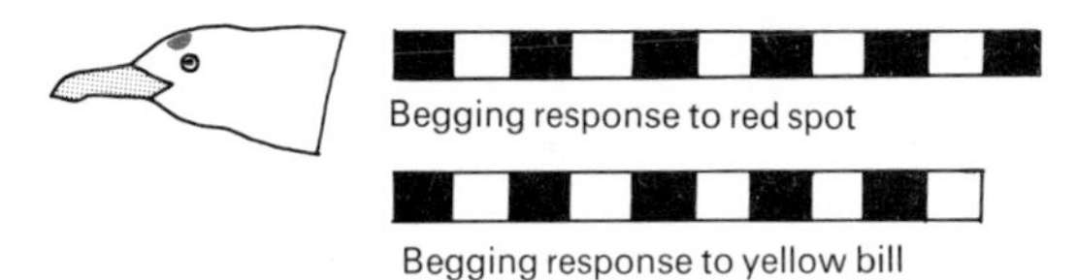

Figure 11.25 *Herring gull pecking response Experiment 5*
In another experiment a model was presented which had a uniform-coloured yellow bill and a red patch on the head.

The results of Experiment 5 could indicate that the patch is involved in directing the pecking response of the chicks. That there were an almost equal number of responses to the beak and the patch on the head indicates that the shape or position of the beak might also be involved in directing the response, and that when the patch is shifted to the head then there is confusion as to where to direct the response.
On the basis of the results of Experiment 5 one might have expected the model with the red head in Experiment 4 to have elicited more responses. However, in Experiment 5 there was a patch on the head, and it could be that it is the contrast of the patch rather than the colour that is important in directing the response.

7 What criticisms could you make of any of these experiments and what further experiments could you suggest that might be carried out to investigate the pecking response of Herring gull chicks? (These are open-ended questions intended to stimulate thought about experimental method, and no 'answer' is given here.)

Questions

1 (a) Including relevant examples, distinguish between the following terms:
(i) kinesis and taxis;
(ii) phototropism and photoperiodism.
(b) Give explanations for the following:
(i) When two established groups of hens are mixed there is a great deal of fighting at first and some birds stop laying.
(ii) Auxins are essential for the growth of flowering plants but may be used for weed control.
(JMB)

2 How would you test the hypothesis that the aggressive behaviour of the male stickleback is **not** related to the concentration of thyroid hormone in the water? (JMB)

3 (a) Describe carefully how movement takes place in
(i) an amoeboid protozoan,
(ii) a ciliate,
(iii) the main stem (or coleoptile) of a flowering plant.
(b) Explain why the two animals need to move from place to place whereas the flowering plant does not. (L)

5 Twenty woodlice were placed in a choice chamber, illustrated below, ten in the humid half and ten in the dry half.

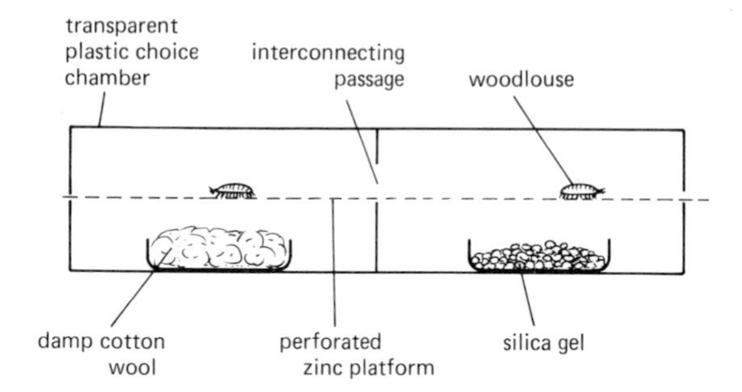

After two minutes the numbers of woodlice in each half of the choice chamber were counted and their positions drawn. Then the numbers were restored to ten in each half. After another two minutes a further record of numbers and positions was made. This technique was repeated four more times. The complete results are shown in the following diagram.

Answer the following questions:

(a) Which of the two choices apparently attracted the larger number of woodlice? Justify your conclusion by some simple mathematical treatment of the results (*graphs and histograms are not required*).

Test	Positions (choice chambers drawn from above) (o = one woodlouse) Humid Dry	Number in humid half	Number in dry half
1		15	5
2		13	7
3		16	4
4		10	10
5		14	6
6		7	13

(b) Suggest two hypotheses to explain the results from tests 4 and 6.

(c) What do your conclusions from (a) and (b) above indicate concerning the behaviour of woodlice? What value could such behaviour have in the normal life of woodlice?

(d) Criticize the experimental method used and suggest how it might be modified to investigate the behaviour of woodlice under other conditions of humidity. (L)

6 Male zebra finches show courtship behaviour towards female zebra finches. Experiments were performed in each of which a male zebra finch was caged with

(i) a female zebra finch with a red beak,

(ii) a female zebra finch with a black beak,

(iii) a simple model of a female zebra finch with a grey beak.

The results obtained are expressed in the following table.

	Percentage response by males	*Mean time spent courting (s)*
Females with red beak	87.5	6.3
Females with black beak	62.5	4.0
Models of female with grey beak	43.8	0.7

(a) Comment on and explain these results as fully as you can.

(b) Explain the significance of the performance by the males of apparently meaningless acts, such as preening, when caged with models of females. Give *one* other example of this form of behaviour.

(c) What are the differences between instinctive and learned behaviour? (L)

Further reading

Carthy J. D., *The Study of Behaviour*, Institute of Biology Studies in Biology No. 3 (London: Arnold, 1967).

Dethier V. G. and Stellar E., *Animal Behaviour*, Foundations of Modern Biology Series (New Jersey: Prentice-Hall, 1964).

Hill, T. A., *Endogenous Plant Growth Substances*, Institute of Biology Studies in Biology No. 40 (London: Arnold, 1973).

Manning, A., *An Introduction to Animal Behaviour* (London: Arnold, 1967).

Pennycuick C. J., *Animal Flight*, Institute of Biology Studies in Biology No. 33 (London: Arnold, 1972).

Tinbergen N., *The Herring Gull's World* (London: Collins, 1953).

12 Homeostasis

Cells require certain optimum conditions for their correct functioning. These conditions are achieved by maintaining the state of the tissue fluids surrounding the cells within certain limits. The maintenance of these constant internal conditions in the face of changes in the external environment is known as homeostasis. Many major homeostatic mechanisms are concerned with maintaining the constancy of the internal fluid environment surrounding the cells, but homeostatic mechanisms also occur within the cells.

The homeostatic state is not fixed but is dynamic, continually responding to both external changes in the environment and to differing demands made on the body, as during exercise, for example. Homeostasis is most developed in the higher animals, particularly the homoiotherms (warm-blooded animals), and involves all body systems. However, some attempt can be made to illustrate the principles of homeostasis by considering how particular systems control specific aspects of homeostasis such as **temperature regulation**, **circulation**, **respiration**, **blood glucose**, and **osmoregulation**.

Larger, less lobed shade leaves and deeper-lobed sun leaves of oak

Temperature regulation

Plants

Plants have lower metabolic rates than animals and produce much less heat as a result. Their particular problem is the absorption of heat from direct sunlight. Leaves have a large surface area exposed to the Sun for the absorption of light for photosynthesis, but at the same time heat is also absorbed and in direct sunlight can raise the temperature of leaves up to 50 °C.

In some plants (e.g. the oak) deeply-lobed sun leaves are adapted to lose heat more quickly than the larger, less lobed shade leaves. Wilting also serves a role in temperature control by decreasing the surface area of the leaves exposed to the Sun. Transpiration[168] in plants lowers their temperature, although the maximum effect rarely exceeds 5 °C. Some plants have stomata that open with increase in temperature to increase the rate of evaporative heat loss. More usually, however, stomata close with increase in temperature to conserve water.

Animals

POIKILOTHERMS (ECTOTHERMS)

Animals other than the mammals and the birds cannot control their temperature, and are unable to maintain it between narrow limits. They produce little internal heat and have little insulation, therefore their body temperature equilibriates with the external temperature and shows the same fluctuations.

Water has a high specific heat and its temperature does not fluctuate rapidly, therefore aquatic poikilotherms do not face the same problems as those on land, where the air temperature can change rapidly over a wide range.

Terrestrial poikilotherms achieve some measure of temperature control by means of their behaviour. By burrowing, being nocturnal, or exposing themselves to the Sun, according to their environment and their particular needs, they can regulate their temperature to some extent and avoid the extremes which would interfere with their metabolism.

HOMOIOTHERMS (ENDOTHERMS)

Mammals and birds maintain their body temperature, within certain limits, at a constant optimum, relatively independent of the external temperature. This provides constant optimum conditions for the enzyme-controlled metabolic processes, and greatly extends the geographical range of these animals. Homoiotherms generate much internal heat and have good insulation. Once the correct body temperature is attained, heat gain must equal heat loss if this optimum temperature is to be maintained.

Under normal conditions, the body temperature in man can be maintained at 37 °C without shivering or sweating. This is achieved by controlling the amount of heat lost via conduction, convection, and radiation, and by shunting blood to and away from the skin as necessary. However, if the conditions are more extreme in either direction, then additional mechanisms come into play.

Overcooling

If the body is losing too much heat then a variety of mechanisms act to prevent this loss and to generate more heat.

Vasoconstriction of the blood vessels in the dermis of the skin cuts down heat loss by radiation (which can be as much as 60 per cent of the heat loss when at rest in a temperate climate), and heat loss by convection and conduction. Vasoconstriction of the blood vessels supplying the sweat glands stops sweat secretion and prevents loss of heat due to its evaporation. However, the skin is not completely waterproof, and a certain amount of water passes by diffusion through the keratinized layers of the epidermis. This is known as insensible sweating and can be as much as 800 ml a day in man.

Reflex muscle twitches leading to shivering generate more heat, as does an increase in the secretion of hormones from the adrenal cortex and medulla, and from the thyroid gland, which stimulate metabolism.

The hair or feathers are raised to trap a thicker layer of insulating air between the body and the external air; and over a long period, if sufficient food is available, the insulating layers of subcutaneous fat are increased in thickness.

Overheating

If the body is gaining too much heat then a variety of mechanisms act to prevent this gain.

Vasodilation of the blood vessels in the dermis of the skin increase the amount of heat lost by radiation, convection, and conduction. Vasodilation of the blood vessels supplying the sweat glands increases the sweat secretion and allows the loss of heat via its evaporation. Energy is required in the form of heat to change the state of water from liquid to vapour and the amount of heat required is known as the latent heat of evaporation. When sweat evaporates from the skin this heat is taken from the body and the body temperature is lowered. One litre of sweat removes about 580 kcals (2424.4 kJ) from the body.

Man has about two million sweat glands, which are of two types: eccrine and apocrine. (All glands are present at birth, therefore there are more per unit surface area in children.)

Eccrine glands occur all over the body. They are innervated by the sympathetic nervous system and, when the body temperature in man rises from 0.2–0.5 °C above normal, they secrete a dilute solution of extremely variable composition known as sweat. Each gland can produce about 150 ml sweat a day. The composition of sweat depends on a complex of factors including the external temperature, degree of activity, state of hydration of the body, diet, and even emotions. Those glands of the palms, soles, and forehead are particularly affected by emotions, for example sweat is secreted there within two minutes of the presence of adrenaline in the blood.

Sweat is 99 per cent water and 1 per cent substances in solution; the main one is sodium chloride, which can vary from 0.1 to 0.4 per cent. The pH can vary between 4.0 and 6.8. Sweat also serves an excretory function by carrying urea, ammonia, and lactic acid to the exterior, but this excretory function is secondary to that of temperature control. Sweat contains traces of a large number of substances, including urocanic acid which is thought to help protect the body from that part of the sunlight that can damage cells.

Apocrine glands occur mainly in the axillae. They develop from hair follicles and have no nervous control, but are stimulated by adrenalin to secrete a milky fluid, which is acted upon by bacteria to produce substances with strong odours. They play no part in temperature control.

Table 12.1 *Average concentration of sweat compared to that of plasma (mg %)*

Substance	*Plasma*	*Sweat*
Chloride	360	320
Sodium	340	200
Potassium	18	20
Lactic acid	15	35
Urea	15	15
Ammonia	0.5	5
Glucose	100	2

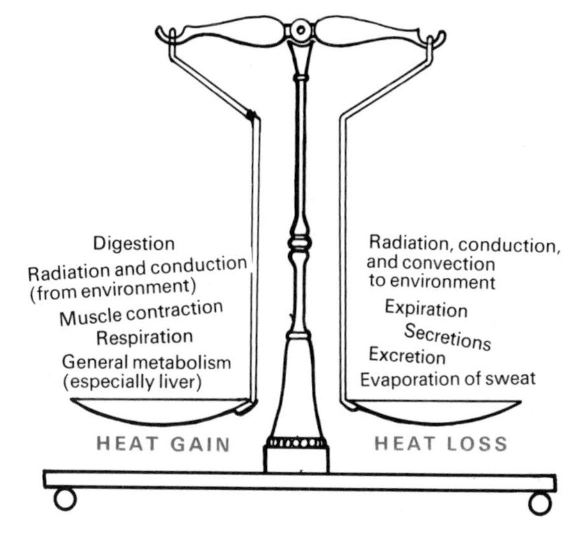

Figure 12.1 *Heat balance in mammals*

Figure 12.2 *Mammalian (human) skin structure*
Skin functions to protect against mechanical damage, bacterial infection, u.v. light, water loss, etc. It is a major sense 'organ', produces vitamin D on exposure to sunlight, and is involved in temperature regulation.

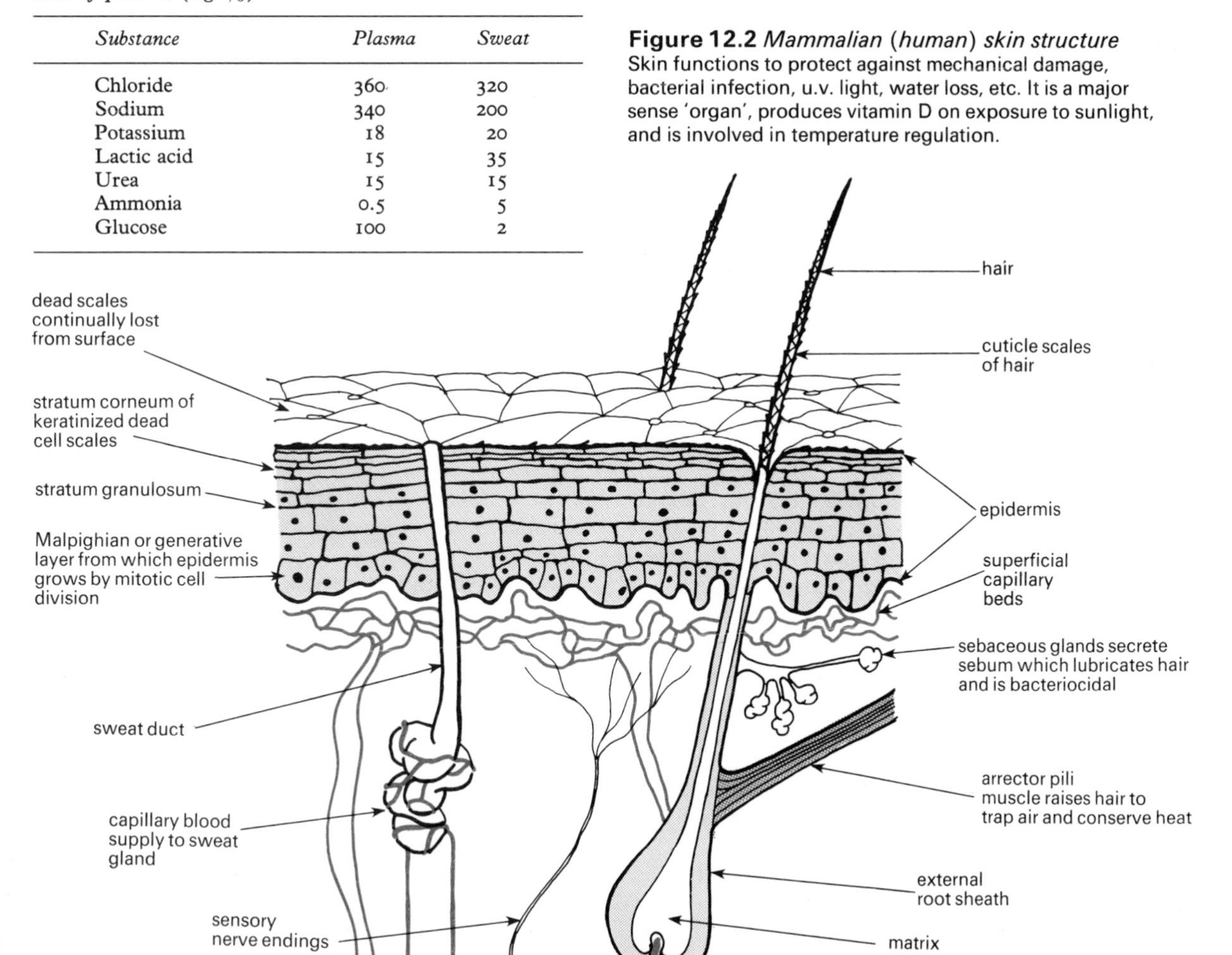

The control centre

The control centre of the temperature-regulating reflexes is the hypothalamus.[200] Here there is some form of physiological 'thermostat' which is normally set to maintain the core temperature at the optimum 37 °C. The core temperature at which sweating starts is influenced by the temperature of the skin. When the skin is cold, the sweating threshold, that is the core temperature at which sweating normally begins, is raised.

With certain infections, the hypothalamic thermostat becomes set at a higher point therefore, although the body temperature is normal, there is a subjective feeling of being cold. The body temperature is raised by shivering, and by an increase in metabolic rate to the new level determined by the higher set point of the thermostat. This results in a fever. When the infection is over the thermostat is reset at the normal temperature and the body temperature drops to normal by sweating.

Under constant conditions the temperature of the body in man shows a circadian rhythm,[235] being at a minimum near midnight and at a maximum in the afternoon.

Countercurrent heat exchange

The core temperature is 37 °C, but the peripheral temperature varies widely, particularly in the extremities of the limbs which may have temperatures well below that of the core. The arrangement of the blood vessels in the limbs forms a countercurrent heat exchanger which warms this colder blood from the limbs on its return to the body, thus lessening its cooling effect on the core temperature.

Efficiency range

The range of external temperature that can be tolerated by different mammals varies considerably. This range of external temperatures, within which the body temperature can be maintained at a constant optimum without changing the metabolic rate, is known as the efficiency range. Below the low critical temperature the metabolic rate rises to generate more heat. Above the high critical temperature the metabolic rate also rises, resulting eventually in heat death, due mainly to different biochemical reactions responding differently to the increase in temperature and becoming increasingly out of phase.

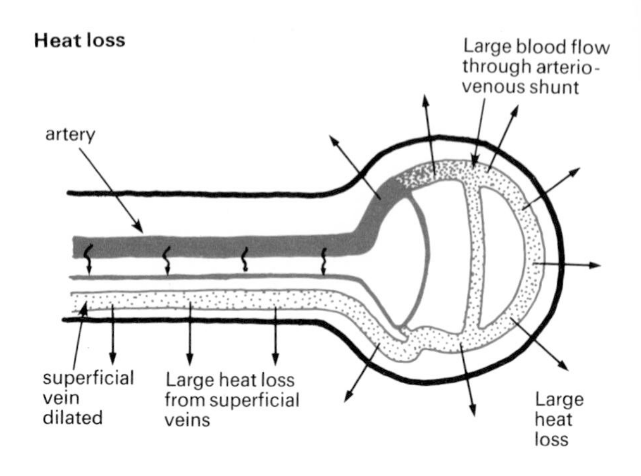

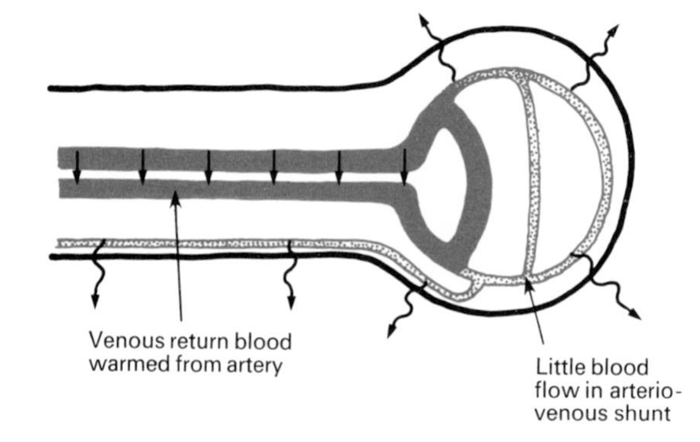

Figure 12.3 *Countercurrent heat exchange in a limb*

Temperature detection

The 'free nerve endings' in the skin detect changes in skin temperature. The various encapsulated nerve endings in the skin which have been given special functions in temperature detection, such as 'Krause end bulbs' (cold) and 'Ruffini corpuscles' (hot), are now considered to be artefacts of skin ageing.

Deeper-seated thermoreceptors in the hypothalamus, medulla oblongata, and spinal cord detect changes in the core temperature of the body. It appears that there are also thermoreceptors in the femoral vein in man, as the onset of sweating is influenced by the femoral temperature.

Problems of size

Smaller animals have larger surface areas, compared to their volumes, than do larger animals. They therefore lose more heat per unit volume, and problems of heat loss set a lower limit to the

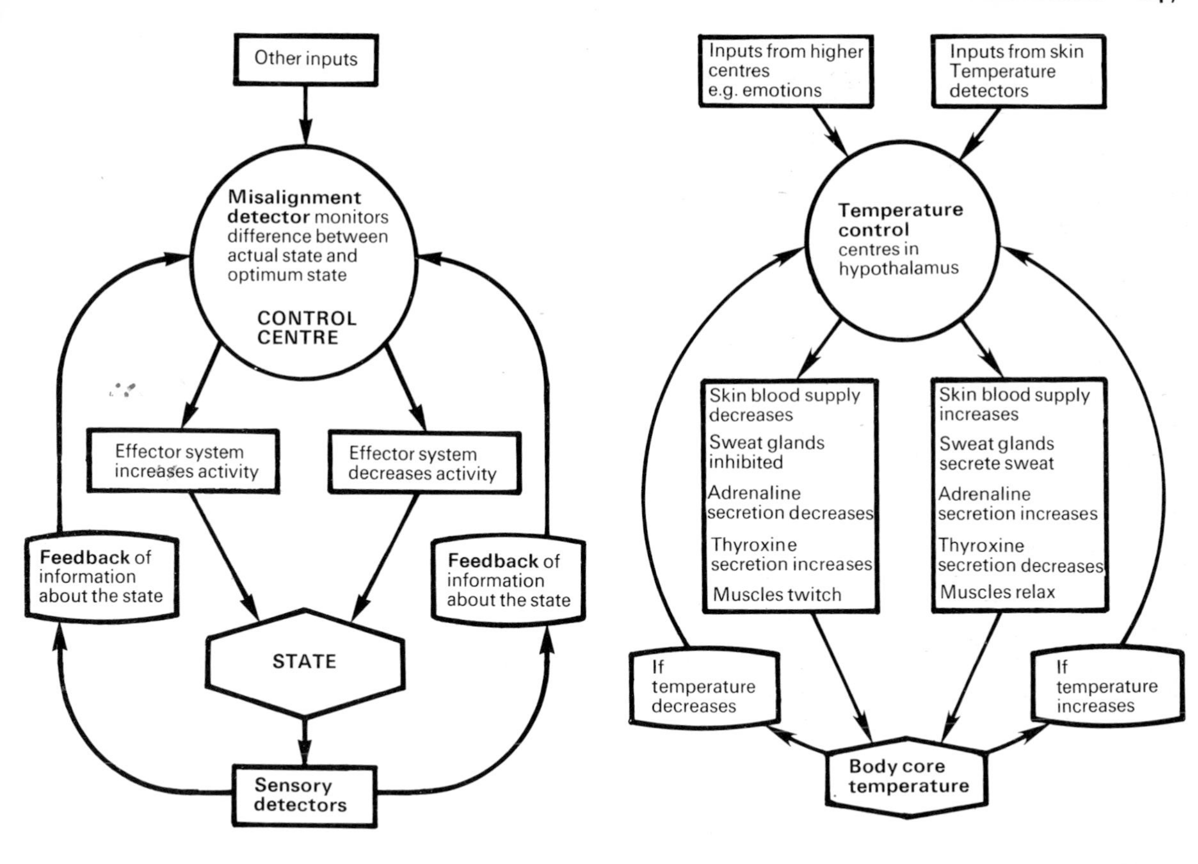

Figure 12.4 *Closed loop homeostatic control system*
There are five fundamental points that need identifying in closed looped control systems:

(a) the particular state being held constant;
(b) the feedback of information about this state to the control centre;
(c) the sensory input from other receptors relating to the state;
(d) the control centre;
(e) the effector system, the action of which is controlled by the control centre to maintain the particular state.

Figure 12.5 *Homeostatic control of body temperature*

size ranges of birds and mammals (which must maintain their temperatures at 40 °C and 37 °C respectively).

Small animals cannot carry thick insulation and sometimes additional metabolism is necessary to produce the required heat. Both the smallest birds (the humming birds) and the smallest mammals (the shrews) weigh about 2 g, which presumably represents the smallest size for homoiotherms.

The young of mammals also have particular problems in maintaining a high body temperature, due in part to their large surface area to volume ratio. The newborn of many species, including man, have regions of special adipose tissue known as brown fat. These are innervated by sympathetic nerves and are involved in the rapid production of heat by the special metabolism of free fatty acids. These pads of brown fat are also found in some hibernating animals.

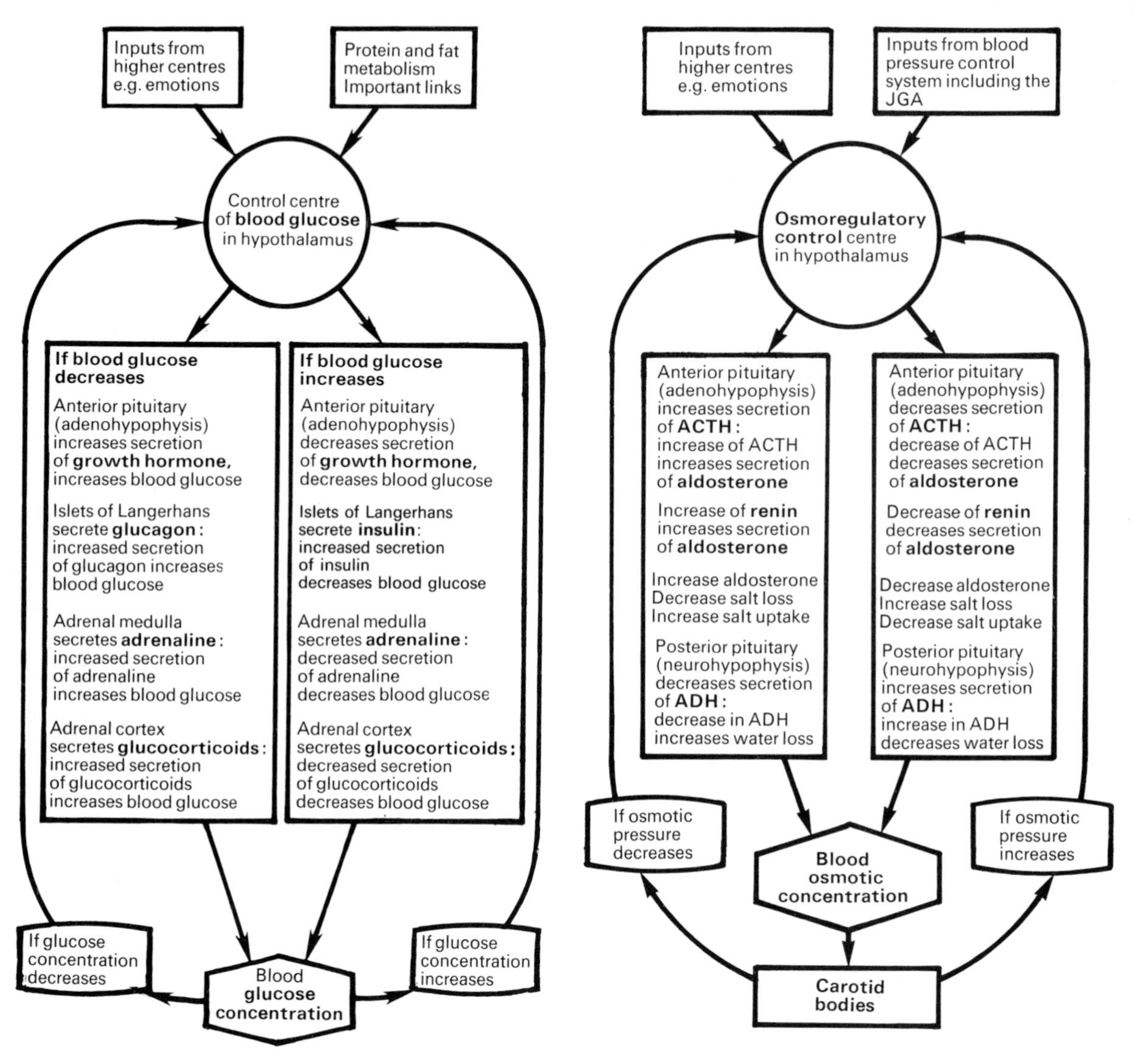

Figure 12.6 *Homeostatic control of blood glucose levels*

Figure 12.7 *Homeostatic control of blood osmotic concentration*

Hibernation

Hibernation refers to the state in which warm-blooded animals undergo a marked drop in the core temperature of the body, and a corresponding reduction in the metabolic rate. (The term heterotherms can be used to describe these animals.) True hibernators can rewarm to the original temperature without absorbing heat from the environment. This definition excludes **torpid poikilotherms**, and amongst the mammals it excludes the bears as they do not undergo a temperature drop. Hibernation is usually preceded by extra feeding and the laying down of fat stores in the body.

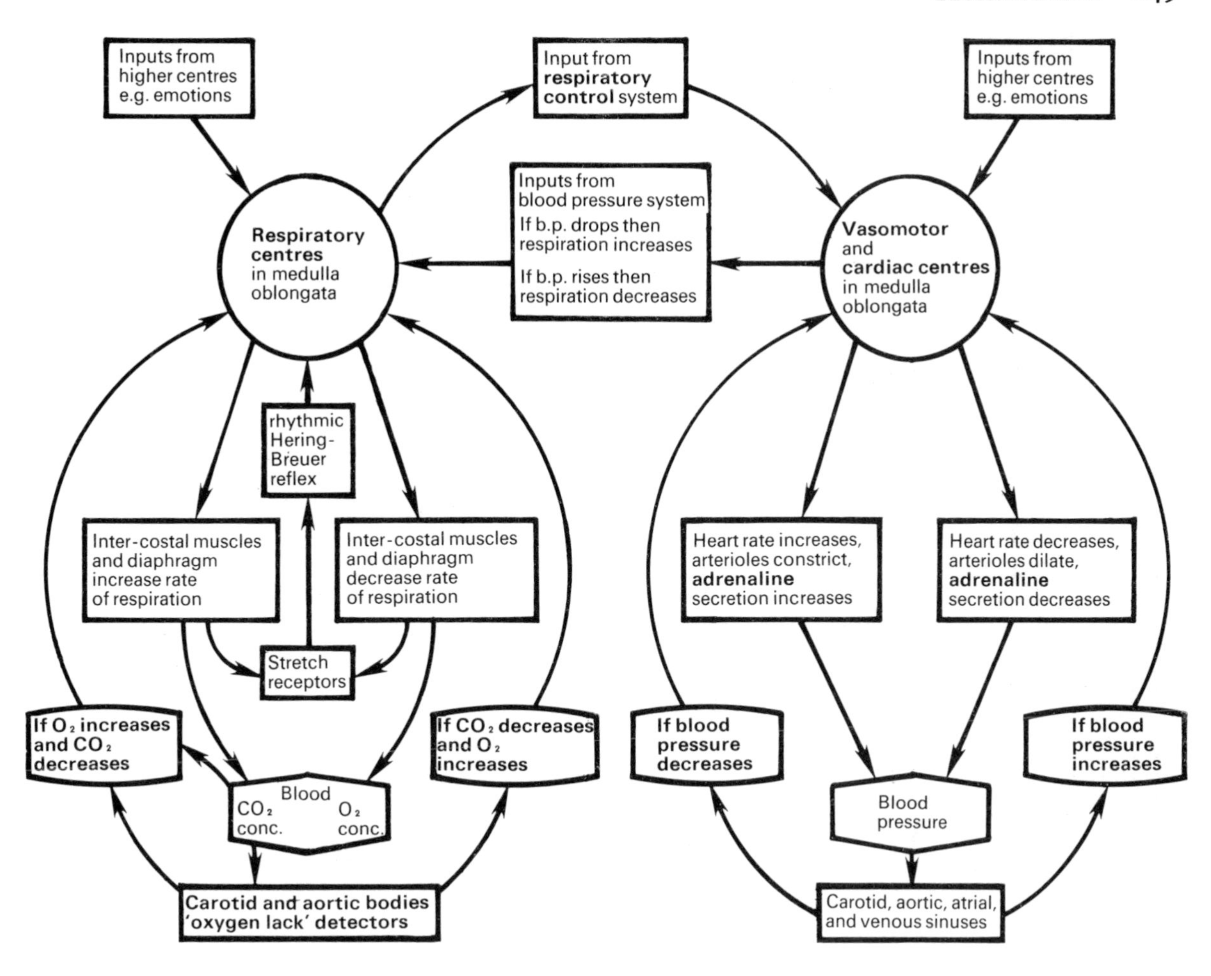

Figure 12.8 *Control of respiration and blood pressure*

In some species, hibernation is triggered by an external environmental stimulus such as food shortage or a drop in temperature; in others hibernation is rhythmic and occurs at fairly set times in the year; and in yet others it is not clear what the stimulus to hibernate is. In all cases hormonal influences are involved.

Small mammals especially are prone to hibernate, as they have a high basal metabolic rate and few food stores; thus when food is short they cannot maintain their metabolic rate and have to drop their temperature and basal metabolic rate in order to survive.

In some ways the hibernating animal is in a state comparable to a poikilotherm; its body temperature is close to that of the surroundings and rises and falls with it. However, if the external temperature increases sufficiently, the hibernator warms up and returns to normal, and if the external temperature drops too far, the hibernator increases its basal metabolic rate to prevent a fatal drop in its body temperature. In this respect the hibernator does not act as a poikilotherm, and in fact hibernation is a well-regulated physiological state.

During arousal from hibernation the body is warmed by an increase in the metabolic rate to up to six times the basal metabolic rate. The body temperature in some mammals can increase by 30 °C in under two hours. Brown adipose tissue around the thorax and main blood vessels (once known as the hibernating gland) is involved in arousal in many species. Under stimulation by the sympathetic nervous system it is involved in the rapid production of heat by the special metabolism of free fatty acids.

Dormancy

Forms of dormancy other than true hibernation are found throughout the animal and plant kingdoms in response to a variety of environmental stresses, including low temperature, food shortage, and drought.

Animals

Poikilotherms, including reptiles, amphibia, and many invertebrates, enter a state of **torpor** when the external temperature drops, during which their metabolic rate falls dramatically. Such torpid animals only recover when the external temperature rewarms them. Low temperature is not the only trigger, as can be seen in some caterpillars which become torpid (in insects known as diapause) immediately after hatching in August and do not start their activity until the following spring.

Some moths that do not feed in the adult state are often active in the winter, which may indicate that food availability is more important to those that do become torpid than is temperature level.

Aestivation is the term used to describe the state of torpor entered during periods of heat or drought. It is found in all vertebrate groups except the birds, and in many invertebrates. It is especially prevalent in fish, amphibia, and reptiles that inhabit freshwater habitats which periodically dry up. These animals then aestivate in the mud and only reactivate when the water returns.

Animals that cannot survive adverse conditions in the adult or larval forms produce special resistant structures such as spores in many protozoa, and resting eggs in many crustacea such as *Daphnia* (the water flea).

Daphnia produces special resting eggs which can survive extreme adverse conditions to hatch months or even years later. Usually they carry the population over winter although a few females may also survive. They hatch, when conditions improve, to give females only; these reproduce parthenogenetically to produce more females. With the onset of adverse conditions again, males are hatched from the unfertilized eggs. When mature, these males fertilize females, and fewer, larger, thick-shelled fertilized resting eggs are produced to survive the adverse period.

Plants

Although plants are autotrophic and have a low metabolic rate when compared to animals, they still utilize various types of dormancy to survive adverse conditions. Such conditions include extremes of temperature and drought. These are closely related since low temperatures decrease the availability of the soil water, and high dry air temperatures cause water losses by evaporation from the plant and the soil.

Plants withstand adverse conditions in a variety of ways. Resistant spores, seeds, fruit, and resting propagules such as winterbuds are often used. These dormant stages can also be used to distribute the species over an area, their dormant condition allowing time for this to occur.

Deciduous leaf fall can occur in response to drought conditions due to either low or high temperatures. The abscission of the leaves reduces water loss by transpiration and prevents photosynthesis. New leaves emerge from dormant buds when water is readily available again. This would be in the spring in temperate climates, when the rise in temperature renders the large volumes of soil water available to the roots; and in the rainy season in hot, dry climates.

In some plants food is stored in underground vegetative storage organs prior to the dying down of the aerial parts. Such perennating organs include modified stems such as **corms**,

tubers, and **rhizomes**; modified roots such as **swollen tap** and **adventitious roots**; and modified leaves such as are found in **bulbs**.

Physiology of exercise

During exercise the body systems make continual adjustments in response to the various demands made upon them in attempting to achieve homeostasis. These physiological adjustments during exercise involve particularly the circulatory system, the respiratory system, and the regulation of body temperature. The ways in which these systems adapt illustrate the dynamic nature of the adjustments made to maintain the constancy of the internal environment in the face of changing demands.

As well as depending upon homeostatic mechanisms to maintain the functioning of the body at an optimum during exercise, the body depends on a certain amount of exercise to be able to maintain homeostasis. For example, when the body is immobilized, body protein can be lost, due mainly to muscle wastage, at rates of up to 1 kg a week. Compact bone becomes less dense, due to demineralization, and the resultant release of calcium into the blood can damage the kidneys. The circulatory and respiratory systems become less efficient and less able to fulfil their roles in maintaining homeostasis.

Circulatory system

During exercise the circulatory system must adjust to the extra demands of the increased metabolism. Extra oxygen must be supplied to the contracting skeletal muscles, waste products must be removed and the excess heat must be transported to the skin.

SUPPLY OF OXYGEN

During exercise a muscle requires on average about fifty times the amount of oxygen than it does at rest. To achieve this tremendous increase, the cardiac output can increase six times from 5 l per min to 30 l per min, the redistribution of this blood to the muscles increases their supply a further three times, and about three times more oxygen is given up by the blood to the tissues. The combination of all these increases results in the necessary increase in supply of oxygen to the tissues, an increase of $6 \times 3 \times 3 = 54$.

ACTION OF THE HEART

With the body at rest the left side of the heart pumps about 5–6 l of blood per minute. During heavy exercise this can increase to 30 l per minute, or even to 35 l per minute in a top athlete. This increase is achieved by an increase in the rate of contraction, and by an increased stroke volume, that is the amount of blood pumped out at each contraction.

During exercise the increased stroke volume (from 70 ml to up to 200 ml of blood from each ventricle) is due to a more complete emptying of the ventricles, and not due to their filling with more blood. When the heart is beating very fast the stroke volume decreases due to the lack of time available for the heart to fill completely between contractions.

The heart rate increases very rapidly at the onset of exercise. In a trained athlete it can double within one minute of starting exercise. This increase is too rapid to be the result of the activities of the metabolic mechanisms concerned with cardiac acceleration, such as increased carbon dioxide, lack of oxygen, increased temperature, and secretion of adrenalin and thyroxine; and must be the result of stimulation by the sympathetic nervous system which in turn must be responding to some unknown stimulus. Indeed, the heart rate can increase in anticipation, before the start of exercise.

During a continuing period of exercise, however, the rate is maintained by nervous and hormonal factors, especially adrenalin.

BLOOD SUPPLY TO ACTIVE MUSCLES

The resting blood flow in human muscle is very low, about 4 ml per 100 g of muscle per minute. During exercise this can increase up to 120 ml per 100 g of muscle per minute. The cardiac coronary blood flow to the heart muscles under conditions of rest is about 200 ml per min, but during exercise this can increase to up to 2.5 l per minute. Therefore there must be wide and rapid adjustments of the circulatory system to achieve the necessary flow to the muscles, and blood must be withdrawn from supplying other regions. This redistribution is under the control of the vasomotor nerves of the sympathetic nervous system.

At the start of exercise the small arteries and arterioles dilate and precapillary sphincters open, increasing the flow of blood to the muscles. In addition, exercising muscle liberates chemicals which act locally to dilate the blood vessels.

Since the majority of the energy produced in respiration is lost as heat, there is an increase in the blood flow to the skin during exercise which increases the heat loss by convection, radiation, conduction, and sweating. However, during maximum exercise most of this blood to the skin is withdrawn to supply the skeletal muscles, with consequent complications in the regulation of body temperature.

Blood is also diverted from the gut and liver to the contracting muscles with a consequent decrease in digestive and absorptive functions and a decrease in the tone of the muscles of the gut wall which interferes with peristalsis.

EXCHANGES WITHIN MUSCLES

Blood flow through the muscles slows as more capillaries open; this allows more efficient exchanges between muscles and the blood.

The low oxygen concentration in the muscles during exercise increases the diffusion gradient between the muscles and the oxyhaemoglobin in the red blood corpuscles, thus encouraging dissociation of the oxyhaemoglobin and the release of oxygen to the tissues. Also the higher temperature of contracting muscle and the higher levels of carbon dioxide encourage dissociation of the oxyhaemoglobin by moving the dissociation curve[157] to the right. This results in more oxygen being given up to the tissues. For example, 100 ml of blood carries 20 ml of oxygen when fully saturated. At rest it gives up 25 per cent of this carried oxygen to the muscles, but during exercise it can give up as much as 75 per cent.

VENOUS RETURN

The contraction of muscles in rhythmic exercise massages the veins and promotes a faster return of blood in the veins to the heart. Venous return is further aided by the increased respiratory ventilation rate which increases the effect of the 'respiratory pump', thus decreasing the pressure in the thoracic cavity and increasing the pressure in the abdominal cavity. The same effects are exerted upon the lymphatic vessels, and aid the return of the lymph to the circulating system.

BLOOD PRESSURE

With the increased cardiac output during exercise a rise in blood pressure might be expected, but this tendency is opposed by the dilation of the blood vessels supplying the muscles and the opening of more capillary beds. However, when the muscles are actually contracted, the increased blood flow to the muscles is resisted by mechanical compression so that there can be a temporary rise in blood pressure, which usually drops when the muscles relax.

Respiratory system

The extra oxygen needed during exercise to saturate the extra cardiac output is obtained by an increase in lung ventilation by increasing the rate and the depth of breathing, and by an increase in the rate of gaseous exchange.

VENTILATION

At rest the normal ventilation rate is about 16 breaths per minute, and the tidal volume or amount of air breathed each time is about 500 ml. The volume of air breathed per minute is therefore about 8 l. During exercise this can increase to up to 200 litres by increasing the rate and particularly the depth of breathing.

However, during all-out exercise the maximum cardiac output is reached before the lungs reach their calculated effective functional limit. Therefore respiratory factors do not usually limit athletic performance, although an increase in the efficiency and comfort of ventilation will reduce the effort involved.

GASEOUS EXCHANGE

Gaseous exchange is aided during exercise. This is due to a steeper concentration gradient between the alveolar air and the blood in the pulmonary capillaries caused by blood entering the lungs which is more depleted of oxygen than normal. Indeed, the difference in oxygen concentration between the blood in the pulmonary artery and that in the pulmonary vein is an important determinant of the amount of physical activity possible.

Gaseous exchange between the air in the alveoli and the blood in the capillaries is so fast that the amounts of oxygen and carbon dioxide exchanged in a certain volume of blood remains the same as normal during exercise, even though the transit time through the capillaries may be halved. The amount of carbon dioxide expired is closely related to the ventilation rate, but the amount of oxygen taken up is more related to the blood flow through the lungs.

OXYGEN DEBT

The respiratory system cannot respond instantaneously to the demands of exercise, particularly at the start of vigorous muscular contractions, therefore sufficient oxygen for aerobic respiration may not reach the muscles. During these periods of undersupply the muscle fibres carry out anaerobic respiration which leads to the production of lactic acid. This lactic acid accumulates and must be oxidized as soon as sufficient oxygen is available.

A certain amount of lactic acid requires a certain amount of oxygen for its oxidation. The amount of lactic acid that accumulates therefore represents an **oxygen debt** of a certain volume of oxygen that must be 'paid off' at a later time.

Sudden vigorous exercise, as in a 100 m sprint race, can be entirely anaerobic. If an athlete runs 100 m in about 10 s he can accumulate 40 g of lactic acid, which represents an oxygen debt of 6.7 l of oxygen. As the maximum rate of oxygen utilization is between 4–6 l per min, then he must breathe at his maximum for about $1\frac{1}{2}$ minutes after the end of the race to repay this debt.

The maximum accumulation of lactic acid that can be tolerated without too much impairment of function is about 120 g. This can be produced within 30 seconds and represents an oxygen debt of about 20 l of oxygen. The increasing acidity of the blood as a result of the accumulation of lactic acid also helps to stimulate respiration. However, much of the acidity due to lactic acid accumulation is buffered[363] by bicarbonate ions in the plasma.

There is no fully satisfactory explanation of the so-called 'second wind' by which the breathing returns almost to normal with continuing exercise within certain limits. However, the adaptive changes in the circulatory and respiratory systems, which enable any oxygen debt built up at the start to be paid off on the run, must be important factors.

The ability to sustain exercise at a high rate is mainly dependent on the rate at which oxygen can be taken up and used in relation to body weight (except over periods of several hours, when the availability of oxidizable substrate becomes limiting). This can be expressed as a ratio of oxygen consumption per minute to the body weight. The figures produced show that there are unbridgeable gaps between certain types in relation to athletic ability.

Thus the average figure for a sedentary female is about 30 cm^3/kg, and for a sedentary male about 48 cm^3/kg. If the average sedentary male in his prime was to undergo strenuous training he could raise this figure to an absolute maximum of just under 70 cm^3/kg; the figure for an Olympic standard male distance runner will be about 85 cm^3/kg.

Males and females have equal powers of oxygen consumption when young, but this ability falls off more rapidly with age in the female than in the male, to the extent that a male of 65 has about the same powers of oxygen utilization as a female in her prime.

Table 12.2 *Oxygen and energy utilization over various distances*

Distance (m)	*Time*	*Speed* ($k\,h^{-1}$)	*Total energy expended* (kJ)	*Rate of energy usage* (kJ/min)	*Oxygen needed* (dm^3)	*Oxygen breathed* (dm^3)	*Oxygen debt* (dm^3)	*% aerobic respiration*	*% anaerobic respiration*
100	10 s	37	200	1200	10	0–0.5	9.5–10	0–5.0	95–100
800	1 min 45 s	27	520	296	26	9	17	35	65
1500	3 min 35 s	25	720	204	36	19	17	55	45
10 000	27 min 30 s	21.5	3000	109	150	133	17	90	10
Marathon (42 186 m)	2 h 10 min	20	14 000	108	700	685	15	98	2

METABOLIC RATE

The metabolic rate increases during exercise, and this increased metabolism above basal metabolic rate is termed the excess metabolism of exercise. For example, heat production can increase during exercise to more than twenty times that produced by the basal metabolism. The increased metabolic rate persists at a high level for a period after exercise to an extent that the energy liberated during this recovery period may be equal to that liberated during the period of exercise.

Energy supply

During exercise carbohydrates, fats, and proteins are oxidized in respiration to supply the necessary energy. A consideration of these reactions can provide some idea of the complexities of the interlinking pathways that exist in metabolism.

Although only catabolic or breakdown processes will be dealt with here (with a few exceptions), most of the reactions are reversible, either directly or via alternative pathways. This leads to the synthesis or anabolism of carbohydrates, fats, and proteins from many of the intermediates. For anabolism to occur there must be a supply of organic materials in the diet in excess of that required to provide the necessary energy.

The liver is the centre for the regulation of carbohydrate, fat, and protein metabolism, and these complex reactions are controlled according to the requirements of the body. Information about these ever-changing requirements is detected and relayed to the liver by a combination of nervous and hormonal pathways. The liver is also directly stimulated by the level of certain metabolites in the blood, for example glucose.

BLOOD GLUCOSE

A variety of homeostatic mechanisms involving a complex of nervous, hormonal, and metabolic pathways attempts to maintain a constant blood glucose level, with the liver acting as the main blood 'glucostat'. Fluctuations in the blood glucose level caused by absorption of glucose from the gut or by utilization of glucose in respiration, are minimized by a complex network of reactions involving protein and fat metabolism, in addition to the metabolism of carbohydrates.

After the glucose has been absorbed into the blood stream from the gut there are three main pathways by which it can supply energy to the tissues. It may be utilized directly; or it may first be converted into glycogen in the liver and the muscles, and into fat in the liver which may then be transported in the blood and stored in adipose tissue fat depots for subsequent use when required.

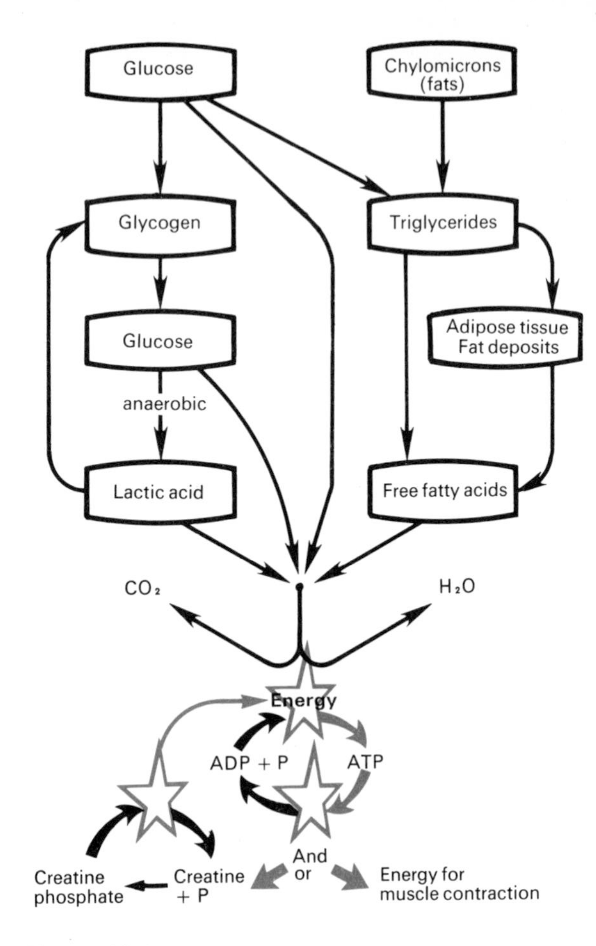

Figure 12.9 *Sources of substrates for respiring tissues*

One of the main reasons why various homeostatic mechanisms attempt to maintain the blood glucose level at or near its optimum, in the face of fluctuating supply and demand, is concerned with nervous tissue respiration. The brain contains no energy reserves and thus depends on a continual supply of respiratory substrate in the blood. The brain cannot utilize ketone bodies from fats, and therefore the blood glucose is its only source of respiratory substrate.

During starvation the blood glucose level is maintained at its optimum, until almost the end,

Marathon runners

by replacement, at first from the liver glycogen and then from glucose-producing or glucogenic amino acids in a process known as **neoglucogenesis** (gluconeogenesis) or new glucose production. This process is speeded up by cortisol from the adrenal cortex. Although the liver can convert excess glucose more readily into fats than into glycogen, very little glucose, if any, is generated from fat metabolism.

During exercise the blood glucose is depleted at a faster rate than it can be replenished, and it therefore drops gradually, resulting in **hypoglycaemia**. This contributes to the onset of fatigue and produces psychological effects by affecting brain function at and below a level of about 75 mg/cm^3 of blood.

Liver glycogen is the only immediately available reserve of blood glucose, as muscle glycogen is not readily converted back to blood glucose. However, the muscle glycogen stores are a source of rapidly available glucose for the contracting muscle fibres during exercise. Indeed, with all the body systems at their optimum, the limiting factor in long strenuous activity appears to be the muscle glycogen reserves. Endurance event athletes often undertake a special 'carbohydrate loading diet' to boost their glycogen reserves. By completely depleting their glycogen stores about six days before the event and then eating only proteins and fats for three days, followed by three days of excessive carbohydrate intake, the body overloads with glycogen as a result of a sort of metabolic 'overshoot'. In this way the normal body stores of glycogen can be increased from about 17 g/kg of muscle to between 30–60 g/kg of muscle.

FAT METABOLISM

Fats occur mainly as triglycerides, which can be hydrolysed to glycerol and free fatty acids, either in the adipose tissue fat 'depots' or in the liver. This utilization of the fats is stimulated by the hormones adrenalin, noradrenalin, and growth hormone. Indeed the level of growth hormone in the blood can increase up to twenty times within thirty minutes of moderate exercise.

Noradrenalin, released at the end of the sympathetic nerves supplying the adipose tissue, mobilises fats into free fatty acids and glycerol. Indeed, contrary to what was once believed, adipose tissue shows much metabolic activity.

During exercise the plasma levels of free fatty acids can increase threefold above the resting level as a result of the mobilisation of the fat reserves and of their metabolism in the liver, which is always involved in the utilization of fats as a respiratory substrate.

When the triglycerides are broken down they produce glycerol and free fatty acids (FFA). The glycerol is further broken down in the normal carbohydrate pathways via pyruvic acid, releasing energy. The fatty acids are converted by the liver to active acetate or acetyl coenzyme A. These can enter the respiratory pathways of the Kreb's cycle if there is sufficient carbohydrate breakdown to supply sufficient amounts of an intermediate (oxaloacetic acid) with which they combine. This is the biochemical explanation of the phrase that 'fats only burn in the flames of the carbohydrates'.

The energy from the complete oxidation of fats in the liver can be used to resynthesize circulating blood lactic acid to glycogen. Under conditions of carbohydrate depletion the acetyl coenzyme A in the liver cannot enter the Kreb's cycle pathway, and these two-carbon compounds join together to form four-carbon ketone bodies (e.g. aceto-acetic acid). These are circulated to the respiring tissues where they are resplit into acetyl coenzyme A and utilized via the Kreb's cycle pathway in the cells, again if sufficient carbohydrate breakdown is occurring.

The ketone bodies formed, although they require a certain amount of carbohydrate metabolism for their utilization, are used to reduce the use of the more valuable blood glucose. Indeed, after 2–3 hours of exercise the R.Q. (respiratory quotient) falls to 0.75, showing that most of the energy is being derived from fats.

In the absence of sufficient carbohydrate breakdown (when the R.Q. is less than 0.75) these ketone bodies, which can also be produced from ketogenic amino acids, accumulate in the blood and lead to the condition known as ketosis. In ketosis the blood becomes very acid (acidaemia) and there is breathlessness and vomiting, with a consequent dehydration and ion imbalance. As a result of these checks on fat utilization as a respiratory substrate there is a maximum amount of fat which the tissues can utilize as a respiratory substrate of about 2.5 g fat per kg body mass per day.

Water loss

At the start of exercise an increased number of sweat glands come into action and, when all the glands are functioning, each one increases its rate of secretion.

During exercise there can be large losses of water as sweat, and even minor dehydration impairs muscle action. The volume of water lost depends on the external temperature and humidity, the amount of exercise, and the volume of water drunk during the exercise. Figures quoted can vary from 1–4 l/h. The water is lost first from the extra-cellular fluids and can, in extreme cases, lead to decreased plasma volume and consequent blood pressure problems. This water loss is monitored by the body's osmoregulatory mechanisms and there is a decrease in urine production if the losses via the sweat are too high.

Salts are also lost in the sweat, the most important of which is sodium chloride. The loss of sodium ions is most important physiologically; the chloride ions are just accompanying anions. Generally there is a lower sodium concentration in the sweat than in the plasma, therefore sweating does not decrease the plasma sodium concentration. However, if the fluid losses are replaced by drinking pure water, the plasma sodium concentration then decreases, and this can result in muscular cramps. Also, as a result of this decrease in concentration of the extracellular fluids, water is drawn by osmosis into the cells and causes swelling of the tissues.

Long-term effects of exercise

As a result of exercise the cardio-vascular system becomes more efficient in its general functioning. There is a slower heart rate; a lowering of blood pressure; a decrease in the speed of the blood flow; an increase in blood volume, and in the number of red blood corpuscles in a given volume (the red blood cell count); and a more efficient use of oxygen. With heavy exercise over long periods there can be an enlargement or hypertrophy of the heart.

Summary

The nature of the stimulus for the increased circulation, increased respiratory ventilation, and the higher setting of the body thermostat during exercise is still not understood. The carbon dioxide concentration of the blood is the main stimulus in circulatory and respiratory rhythms, and there are chemoreceptors sensitive to carbon dioxide in the brain and the carotid bodies.

However, although extreme exercise can produce up to 6 l of carbon dioxide per minute, during this exercise little or no change in oxygen and carbon dioxide concentration occurs in the arterial blood. Therefore there appears to be nothing to stimulate the chemoreceptors, and a different mechanism must be responsible for the control of respiratory and circulatory changes during exercise. It is probable that some part of the brain initiates the messages to the respiratory and circulatory control centres to increase the rates of respiration and circulation.

The precise co-ordination between the systems, particularly between the ventilation of the alveoli and the flow of the blood through the pulmonary capillaries during the varying demands of exercise, is one of the most remarkable features of muscular exercise.

At what point in an individual the body is unable to maintain homeostasis during exercise, so that the effort can no longer be sustained, is very difficult to judge. Both physiological and psychological components contribute to the subjective sensation of fatigue. Physiological factors include low blood sugar, accumulation of lactic acid, cortico-steroid hormone depletion, water loss, and increased body temperature. Psychological factors are certainly important, both directly by determining the amount of discomfort that can be tolerated, and indirectly by the influence that emotion can exert on the physiological state.

Although there is great variation between individuals, the age of 40 years marks the end of the plateau of adult vigour in most. There follows a steady decline of powers, as a result of ageing, which can be considered as a decline in the ability to maintain homeostasis over such a wide range of demands as before. This decline is demonstrated clearly in the maximum attainable heart rate which on average is 200 beats per minute at age 25, 180 beats per minute at age 40, and 150 beats per minute at age 65. This decline in the maximum heart rate of 5–10 beats per min each decade is the key to decreasing athletic ability with age.

Experimental and applied aspects

Homeostasis depends on the use of control systems, and many analogies have been drawn between biological systems and the general study of control theory in nonliving systems known as **cybernetics**.

One aspect of biological systems of particular importance is closed loop control. In this the output is continually monitored and compared with the optimum goal state, and information on any difference is fed back to adjust the control inputs, so that the output leads to the attainment of the optimum goal state. Hence the term feedback control.

Negative feedback opposes any detected tendency away from the steady state and returns it to normal. This is the commonest type of system found in living systems.

Positive feedback reinforces the particular tendency, so that it moves further and further away from the starting point. This type of feedback is not common in living organisms as it moves a system away from the steady state.

Hunting occurs when the output fluctuates about the desired value. The amount of hunting depends on many things: the degree of fine or proportional control in the system (as opposed to a simple all or nothing on/off response); the amount of lag between monitoring the output and changing it; and the 'stiffness' of the system, that is its degree of resistance to change.

Indeed, some feedback systems are not homeostatic, due to pronounced hunting. (Homeostatic systems are a special class of stable feedback systems.) Most biological control systems operate with proportional control, that is as the error decreases, so does the output which is attempting to correct that error. This reduces hunting to a minimum. Needless to say, there are many limitations to the application of control system analysis to living organisms. The interconnections between the neurones of the nervous system are very complex and cannot be clearly described in simple terms; nervous systems adjust to changes in the environment and are capable of learning and reinforcing innate patterns as a result of experience. Therefore control system analysis can only serve as a crude description of the way in which the nervous system functions. Moreover, all the systems of the body interact to maintain the overall homeostasis of the body in an incredibly complex manner.

Guided example

Consider these questions after referring to Table 12.3.

1 How is the increased blood flow, from 5800 cm^3/min at rest to 25 000 cm^3/min during exercise, attained?
The increased blood flow is attained by an increased heart rate and stroke volume.

2 Why does the blood flow to the heart muscles remain a constant proportion of the output in all the states from rest to maximum exercise?
The constant proportion of the blood flow passing to the heart muscles reflects the fact that it is the increased work rate of the heart which is generating the extra flow to all parts during exercise. The heart muscles require a blood flow in proportion to the work they are carrying out, which is in turn proportional to the total increased blood flow.

3 Why does the blood flow to the skin increase during light and strenuous exercise?
The blood flow to the skin increases with exercise in order to dissipate the excess heat generated; by a combination of conduction, convection, radiation, and sweating.

4 Why does the blood flow to the skin decrease during maximum exercise and what will be the outcome of this?
The blood flow to the skin decreases during maximum exercise as even more blood is diverted to the now maximally contracting muscles. As a result the body temperature will rise, leading to overheating and eventual collapse.

5 Why is the blood flow to the brain constant throughout all stages from rest to the maximum exercise?
The blood flow to the brain must remain at a constant optimum for the brain to function correctly. The brain's requirements do not change during exercise, neither therefore does the blood flow supplying it. Should the supply decrease, there would be a disturbance of brain function leading to a breakdown in the complex integration and control of all the systems involved in exercise.

Table 12.3 *Blood flow in different part of the body (cm^3/min)*

	Rest	*% of total (approx)*	*Light exercise*	*Strenuous exercise*	*Maximum exercise*	*% of total (approx)*
Heart muscle	250	4	350	750	1000	4
Skin	500	9	1500	1900	600	2
Brain	750	13	750	750	750	3
Kidneys	1100	19	900	600	250	1
Skeletal muscle	1200	21	4500	12 500	22 000	88
Gut	1400	24	1100	600	300	1
Other	600	10	400	400	100	1
Total	5800	100	9500	17 500	25 000	

6 What will be the result of the decreasing blood supply to the kidneys with increasing exercise?
The decreased supply to the kidneys will result in a smaller proportion of the cardiac output being monitored and adjusted at a time when the increasing metabolic activity is raising the level of nitrogenous waste and increasingly disturbing the composition and reaction of the blood. As a result there will be an increasing disturbance of the homeostasis of the body fluids, and a consequent increasing inefficiency of metabolism.

7 What do the figures suggest could be one reason for not partaking of vigorous exercise soon after eating?
With increasing activity, the blood supply to the gut is decreased as more and more blood is diverted to the contracting muscles. This decreased supply to the gut will interfere with digestion, decrease absorption, and decrease the tone of the muscles of the gut, which in turn will interfere with peristalsis.

Questions

1 Three different types of animal, a lizard, a duck-billed platypus, and a cat were subjected to a range of environmental temperatures between 5 °C and 40 °C. After two hours at a particular temperature each animal's body temperature was recorded. The results are given below:

Environmental temperature (°C)	*Body temperature* (°C) Lizard	Duck-billed platypus	Cat
5	5	31	38
10	10	32	38.5
15	15	32.5	39
20	20	33	39.5
25	25	33.5	40
30	30	34	40
35	35	37	40.5
40	40	40	41

(a) Plot the above data so that you can compare the relationships between the body temperature of each animal and that of the environment.
(b) Comment on and explain the significance of these results.
(c) What changes in physiology and behaviour would you expect to occur when each animal is subjected to the extreme high or low temperatures shown in the table above? (L)

2 For **each** of the following situations, describe the physiological responses of the human body and explain how they are brought about.
(a) A diabetic receives a slight overdose of insulin.
(b) A healthy person drinks three pints of water in quick succession.
(c) In an experimental investigation, a subject with his nose closed by a clip can only breathe air from and into a gas-tight bag connected to his mouth. The subject breathes and rebreathes air from this bag for a few minutes. (JMB)

3 Under what conditions and in what ways are changes brought in the rate of (a) lung ventilation, (b) heart beat, and (c) insulin secretion in a mammal? (L)

4 An athlete has an oxygen intake of 0.25 litre per minute when completely at rest. He rested completely for 2 minutes, spent one minute at the starting blocks, ran as fast as he could for 2 minutes, and then stopped. The starting gun went off at 10.00 hours. The following timed sequence of measurements of oxygen intake were made.

Time	O_2 *intake in litres per minute*
09.57–9.59	0.25
09.59–10.00	0.45
10.00–10.01	3.00
10.01–10.02	3.00
10.02–10.03	2.80
10.03–10.04	2.25
10.04–10.05	1.50
10.05–10.06	1.00
10.06–10.07	0.75
10.07–10.08	0.50
10.08–10.09	0.25
10.09–10.10	0.25

(a) Present the above data in graphical form.
(b) Discuss the results. (O&C)

5 Recent studies of temperature regulation in animals have led to the proposal that terms other than *homoiothermic* and *poikilothermic* should be employed to classify thermoregulatory mechanisms. Examples are *ectothermic*, *endothermic*, and *heterothermic*.

An ectotherm is an animal of small size and large surface area which has a high rate of heat exchange with the environment. The heat which determines its body temperature is derived from the environment.

An endotherm is an animal which is usually of larger size, is well insulated and has a low rate of heat exchange with the environment. The heat which determines its body temperature is internally produced.
(a) Hedgehogs, humming birds, bats, and ground squirrels are considered to be heterothermic. Using these examples, construct a definition of the term heterotherm.
(b) Pythons and other large snakes can maintain their eggs at a constant temperature which may be several degrees higher than the ambient temperature during incubation. State which of the three terms would best describe the python and give reasons for your decision.

(c) Houseflies are usually very difficult to catch, but under cold conditions they may be captured easily. Explain this observation in physiological terms and state which of the three terms should be used to describe houseflies.

(d) Many moths are nocturnal and are active at low temperatures. Certain large sphinx moths (mass 2 to 6 grams) are very similar in appearance to humming birds. They visit flowers at night, flying steadily when the temperature is between 12 °C and 35 °C, but they remain inactive when the temperature is above 35 °C. The thorax is covered with long, dense scales and they vibrate their wings for varying periods of time prior to take-off.

(i) Which of the three terms best describes these moths? Give reasons based on the information given above.

(ii) Explain how you would expect the duration of the pre-flight wing vibration period to vary with the ambient temperature. (AEB)

6 (a) What are the essential features of homeostatic (steady state) control mechanisms?

(b) A most important supracellular control mechanism involves the blood acting as a sensory and motor pathway. Using appropriate examples, discuss this statement with reference to the steady state control processes operating in vertebrates. (JMB)

7 The diagram summarizes the ways in which an animal can gain heat from or lose heat to its environment.

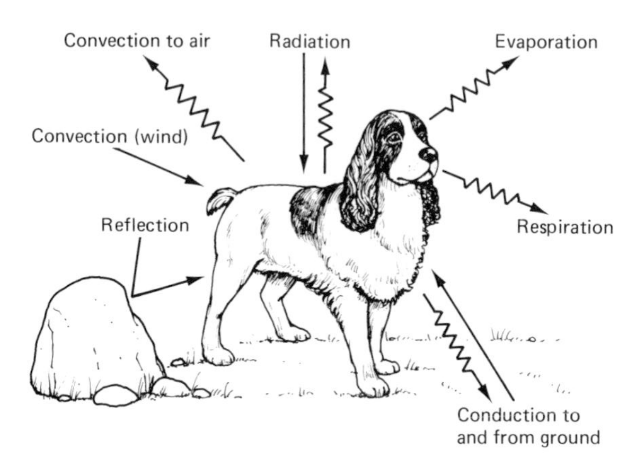

(a) With the aid of the information provided in the diagram, explain how an ectothermic animal (i.e. one with a large surface area which has a high rate of exchange with the environment) such as a lizard can

(i) raise its body temperature to that of the environment,

(ii) lower its body temperature below that of the environment.

(b) Carefully explain how an increase in air currents can influence the amount of heat a mammal loses to the environment.

(c) Explain the connection between problems of heat regulation and the fact that there are no adult birds or mammals less than 2 grams in mass.

(d) Heat loss from the skin of a person sitting in an environment at a temperature above that of his body drops soon after he drinks a large mug of iced water.

What does this information tell you about the way in which the thermoregulatory centre in the hypothalamus region of the brain functions? Explain your answer.

(e) When the body temperature of the marine crustacean *Ligia* approaches too high a level it moves out of its normal damp environment beneath stones into direct sunlight and a higher environmental temperature. Explain how this apparently strange behaviour benefits the animal. (AEB)

Further reading

Hardy R. N., *Homeostasis*, Institute of Biology Studies in Biology No. 63 (London: Arnold, 1976).

Hardy R. N., *Temperature and Animal Life*, Institute of Biology Studies in Biology No. 35 (London: Arnold, 1972).

General further reading

Arthur D. R., *Looking at Animals Again* (London: Freeman, 1966).

Ramsay J. A., *The Experimental Basis of Modern Biology* (Cambridge: Cambridge University Press, 1970).

Simpkins J., *Investigations in Animal Physiology* (London: Heinemann, 1973).

PART THREE

Man and the environment

13 Population and resources

The ideas of Thomas Malthus (1766–1834), typified by the phrase, 'The power of population is infinitely greater than the power in the earth to produce subsistence for man', become increasingly relevant to man's situation as time goes by. It is self-evident that, in a finite world, there is an absolute upper limit to the size of the population that can be supported, no matter what advances are made in food production.

Between 1950 and 1960 the growth rate for the world's population was 2.14 per cent, giving a doubling time of about 33 years. This is almost certainly the fastest growth rate the world will ever see. However, even at the present rate of growth the population of the world is expected to have doubled by some time around the year 2000 AD to an estimated 6000 million. At present the world's population is growing at a rate by which, at every second of the day and night, there are four births, as compared to two deaths. This means that over every 24 hours there are about 173 000 more births than deaths. This represents an annual increase of about 63 million and is the equivalent to adding another population of the USA to the world about every three years. Another striking comparison is that an equivalent number of people to those killed in the Second World War (an estimated 22 million) are born in just over four months.

Table 13.1 *World population estimates for 2000 AD*

UN 'low' variant	5 500 000
UN 'medium' variant	6 100 000
UN 'high' variant	7 000 000

Table 13.2 *Doubling times for the world population*

From	*To*	*Period in years*
0	1650	1650
1650	1850	200
1850	1930	80
1930	1970	40
1970	2000	30

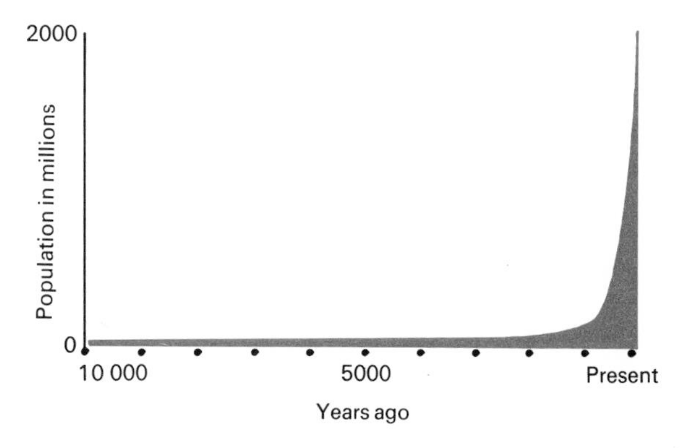

Figure 13.1 *Human population growth*

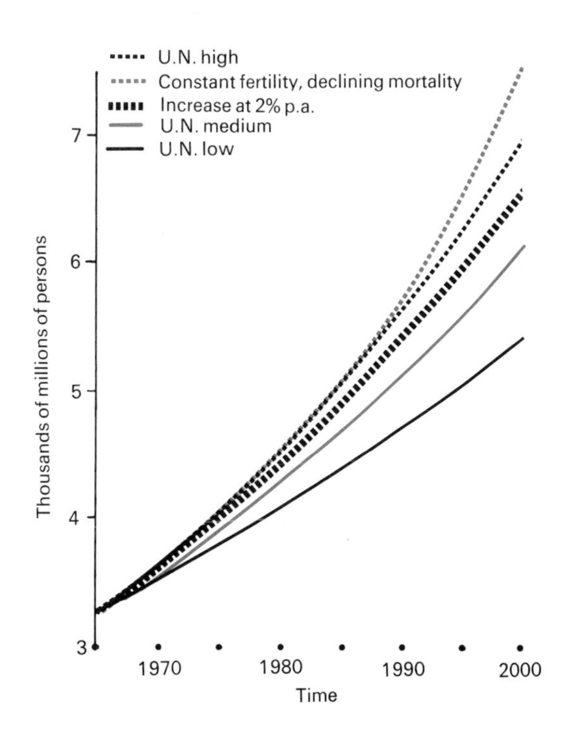

Figure 13.2 *Some estimates of future population growth*

Table 13.3 *Populations worldwide*

	Estimated 1969 population (millions)	*Number of years to double*	*Percentage population under 15 years old*
China	740	50	Unknown
India	537	28	41
USSR	241	70	32
USA	203	70	30
Indonesia	115	29	42
Japan	102	63	25
'Europe'	456	88	25
'Africa'	344	28	43
'Latin America'	276	24	43

The reasons for the world population explosion are thought to be mainly related to the decrease in infant mortality; a result of the prevention of infectious diseases. However, it is not necessarily that simple since many other factors, such as nutrition, customs, mobility, and fertility, enter into the calculations in a way which is not always clear.

Some maintain that such are the numbers alive today, and such are the problems of achieving the necessary world-wide perfect birth control, that the battle against future overpopulation has already been lost. Indeed some would assert that the Earth is already overpopulated, as is witnessed by the present widespread poverty and hunger in the undeveloped countries.

Others would argue that the condition of the undeveloped countries is largely the result of excessive consumption and waste by the developed countries, and consequently a more equitable distribution and use of the Earth's resources would alleviate the problem. This disparity between the developed and the undeveloped countries in terms of materials used is illustrated by the estimate that the average Briton consumes in a lifetime about ten to twenty times as much of the Earth's resources as the average Indian.

However, some argue that even with zero population growth from this moment on, there still appears to be no way in which the undeveloped countries could reach the standard of living experienced in the developed countries as there are simply not enough resources to go round. Furthermore, any advances in food production methods, technology, and international co-operation will be outweighed by those factors that decrease soil production, such as soil erosion, water depletion, pests, pollution, and the exhaustion of oil and minerals needed for modern agriculture.

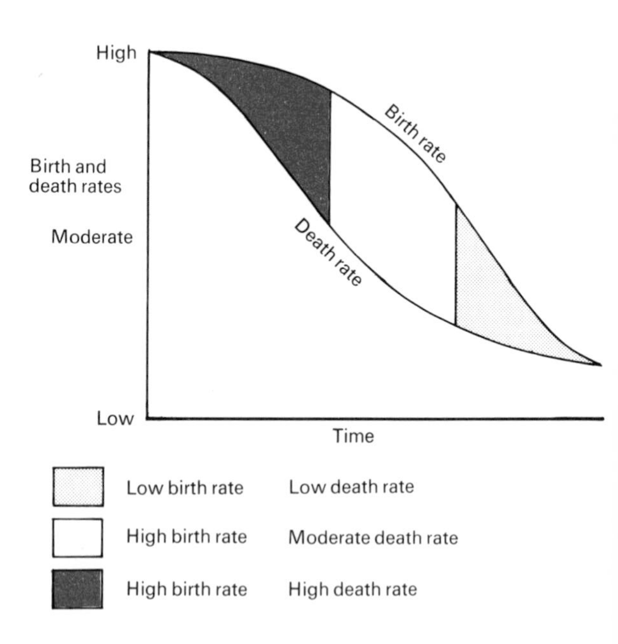

Figure 13.3 *Transition theory*
This postulates that nations pass from an agricultural to an industrial society through a transition phase characterized by high birth rate and decreasing death rate; and that, on the attainment of a certain level of industrialization, the birth rate declines. This theory suggests that the present population explosion could be a transient phenomenon. (After Hay, *Human Populations*, (Harmondsworth: Penguin Books, 1972).)

The population of England at the time of the Domesday Book (1086) was put at 2 million. The population of England in 1976 was about 46 million, and of the UK about 56 million, making the British Isles one of the most densely-populated countries of the world with about 229 inhabitants per square kilometre or 500 per square mile. That is about 1.5 the density of India and ten times that of the USA. Some consider that this is about twice the optimum population for these Islands and that, although there is virtually a zero growth rate, a negative population growth rate is required to prevent an inevitable crisis of food, materials, and space.

Certainly it is overpopulation that compounds all the environmental problems of this Earth, and threatens to overwhelm any improvements in food production, technology, and international co-operation; leading eventually to world famine and a violent competition for the Earth's dwindling resources.

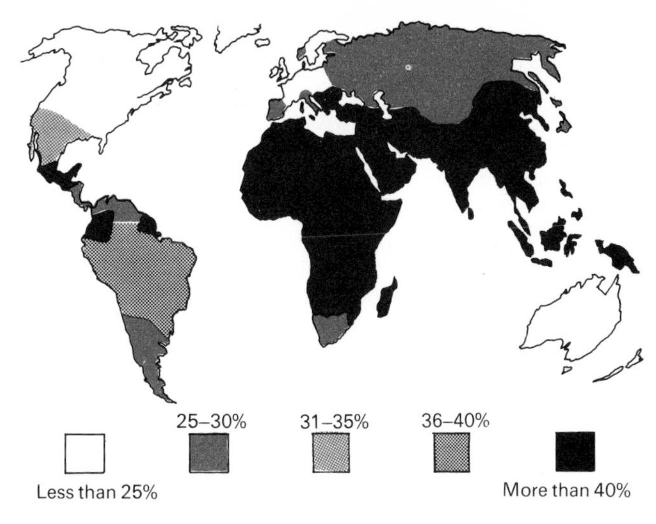

Figure 13.5 *Percentage of the population under 15 years of age*

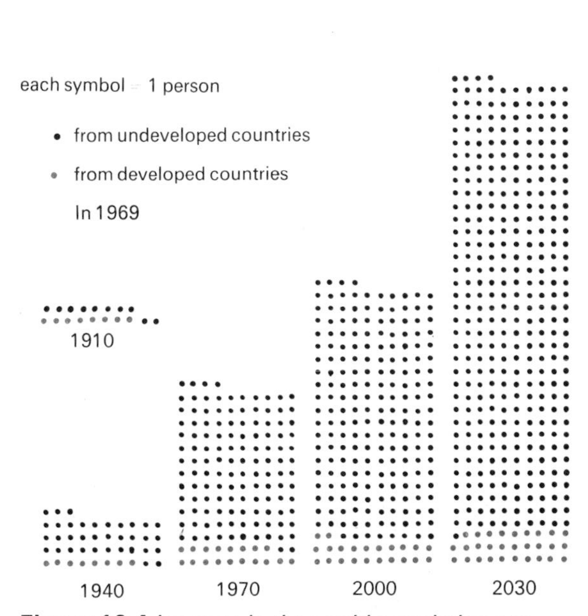

Figure 13.4 *Increase in the world population per minute*

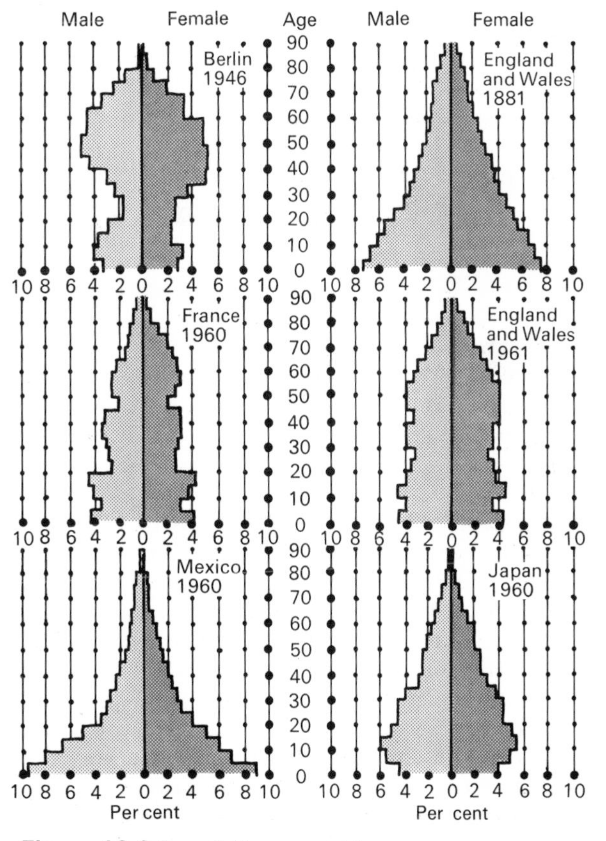

Figure 13.6 *Population pyramids*
These show the age structure of populations very clearly.

Food from the sea

Fishing

Estimates vary, but protein from fish makes up about 25 per cent of the world's total supply of animal protein, with about 5 per cent from freshwater fisheries and 20 per cent from the oceans. About half of the protein obtained from the oceans is used as animal feed in the developed countries, so that protein from the oceans makes up about 10 per cent of man's direct intake of animal protein. Since animal proteins account for about 30–35 per cent of man's intake of all protein, both plant and animal, this means that protein from the oceans makes up about 3.5 per cent of man's total direct protein consumption. Food from the sea also provides about 2 per cent of the world's supply of calories.

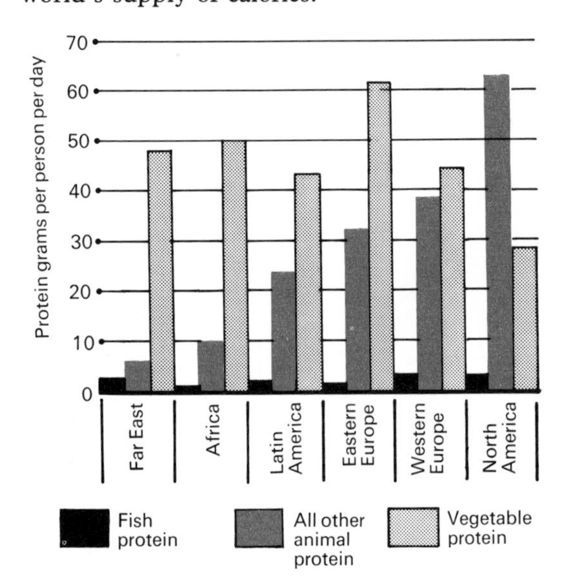

Figure 13.7 *Protein consumption*

Commercial fish stocks can be divided into two categories according to their mode of life and consequently to their method of capture.

Demersal fish live and feed on the bottom and are caught by trawls. Examples are cod, hake, haddock, and all flat fish.

Pelagic fish live and feed at mid-depths and are caught by drift nets. Examples are herring, mackerel, pilchards, and sardines.

The irresponsible use of trawls can cause great damage to the ecology of the sea floor and exert a disrupting effect on the ecosystem that supports the existence of the 'target' fish. Drift and seine nets generally cause less fundamental damage to the ecosystem than do trawls; however, great damage can be inflicted on species associated with the 'target' fish, which could subsequently have far-reaching effects. For example, an estimated two hundred thousand porpoises are destroyed in the nets of the USA tuna fisheries each year.

The shallower seas of the Continental Shelf regions are more productive as a source of protein for man than are the deeper waters of the open oceans. In shallower seas, sinking plankton and organic detritus are utilized by bottom-dwelling animals, which in turn are fished and used as a food source for man. In fact, bottom-dwelling species make up about one-third of the total yield of shallow seas. These regions are also the site of coastal upwellings which bring nutrient-rich water from the deeper water into the phytoplankton zone, thus increasing the primary production which is the basis of all food chains. In addition they receive nutrient-rich run-off from the land, especially via river estuaries.

Deeper waters are less productive as a source of food for man, even though about two-thirds of the ocean's primary production of phytoplankton occurs in deep waters. In these waters the sedimentary organic materials from the surface sink to great depths where light is the limiting factor, and where the deep-water fish which utilize the organic sediments, either directly or indirectly, are bizarre and beyond the effective depth of present fishing techniques. Only in regions of upwelling currents, which bring nutrients back to the plankton zone, are deep waters as productive as shallower seas.

A classic example of the productivity of such regions is seen in the Peruvian 'anchoveta' fisheries of the Humboldt Current in the Pacific. These at one time provided huge catches of fish, and also supported large bird populations, the excreta of which (known as guano) is an ideal fertilizer. However, the future of these fisheries is uncertain as overfishing, natural cycles of abundances, and the shifting of ocean currents lead to dramatically fluctuating catches.

The tremendous primary production by phytoplankton needed to support certain fish stocks is well-illustrated by the estimate that 1 kg of cod is based on a primary production of 50 000 kg of phytoplankton.

All the world's fisheries are threatened by overfishing and pollution. Inland waters are particularly at risk; for example the freshwater fisheries of the Great Lakes of Canada are threatened in this way; as are the fisheries of the Caspian Sea, where the water level is dropping dramatically and compounding problems of increasing salination and pollution. Even such a large body of water as the Mediterranean is at risk, particularly from pollution by domestic sewage and industrial wastes.

The estimates for the future productivity of the sea are based on sound ecological principles, and therefore there is less controversy here than with projections in other areas of food production. For example, rates of photosynthesis can actually be measured at various sampling points and used as evidence leading to estimates of net plant production. This forms the base of the foodweb and from this figure potential yield as a resource for man can be estimated by considering the various steps leading to organisms harvested by man. For example, the transfer efficiency gives a measure of the percentage of the prey's annual production incorporated into the tissues of the consumer species.

Table 13.4 *Estimated productivities at different trophic levels (in millions of tonnes)*

Primary production	130 000	
Trophic level 2	13 000	
Trophic level 3	2000	
Trophic level 4	300	—resource for man, mostly fish
Trophic level 5	45	

Making allowances for the difficulties in harvesting, this suggests that there is an upper limit of about 150 million tonnes of production available to man each year. An independent line of evidence based on a comparison of the actual harvest with its required primary production corresponds closely to these figures.

Estimates once used past trends in landed catch as their evidence, and were frequently revised upwards as technology increased catches. However, these are now very close to the maximum limits calculated as being biologically possible.

With a greater knowledge of the ecology of the seas, a greater international co-operation and control of fishing and pollution, increased technology, and less waste, the oceans can remain an important source of protein for man's growing populations. However, the oceans are close to the limit of exploitation and hold no great untapped reserves of food for the future. Indeed, the open sea which covers 75 per cent of the Earth's surface produces only a fraction of the world's fish catch and has little or no potential.

Whales and whaling

Introduction

The story of whaling provides a classic example of the over-exploitation and cruel destruction of animal species unfortunate enough to be considered as a valuable source of food and materials for man. It further dramatically illustrates the problems of attempting to overcome short term economic expediency with balanced long term arguments. In many ways the fight to 'save the whale' by the conservation movement may be considered as a test-case of man's commitment to the future of his environment.

Whales do not suffer direct competition from man for their habitat and many believe that if the whales cannot be saved from extinction then there is little, if any, hope in the long term for any land-based threatened species. On the other hand, however, the freedom of the oceans is one source of the whale's destruction, as there is as yet no way of controlling the exploitation of international waters, and any protective measures must be adopted by voluntary international agreement. The remoteness of the whale from people's direct experience renders their plight less imaginable than that of animals with which they are more familiar.

Classification

ORDER CETACEA

This is an extremely ancient order of completely aquatic placental mammals. Members of this Order have many special features related to their aquatic existence. They have developed the typical streamlined shape of aquatic animals; the fore-limbs are paddle-like flippers; the hind-limbs are absent; the pelvic girdle is greatly reduced; the fleshy tail flukes provide forward propulsion by their up and down movements; and the skin is devoid of glands and very smooth. Beneath the skin are thick layers of subcutaneous fat or blubber which act as a food store and insulatory layer. The nostrils are positioned dorsally and open upwards to form the characteristic 'blow-hole'. Their physiology shows specializations related to diving for extended periods, often to great depths. Other striking features are the degree of development of the brain, especially the cerebral cortex, and the associated degree of intelligence, communication, and behaviour. Among the whales are the largest animals ever known to have lived.

There are three distinct sub-orders of the Cetacea: the **Archaeoceti**, composed of families of extinct forms; the **Odontoceti**, or toothed whales, which include the majority of living cetaceans; and the **Mysticeti**, or whalebone or baleen whales.

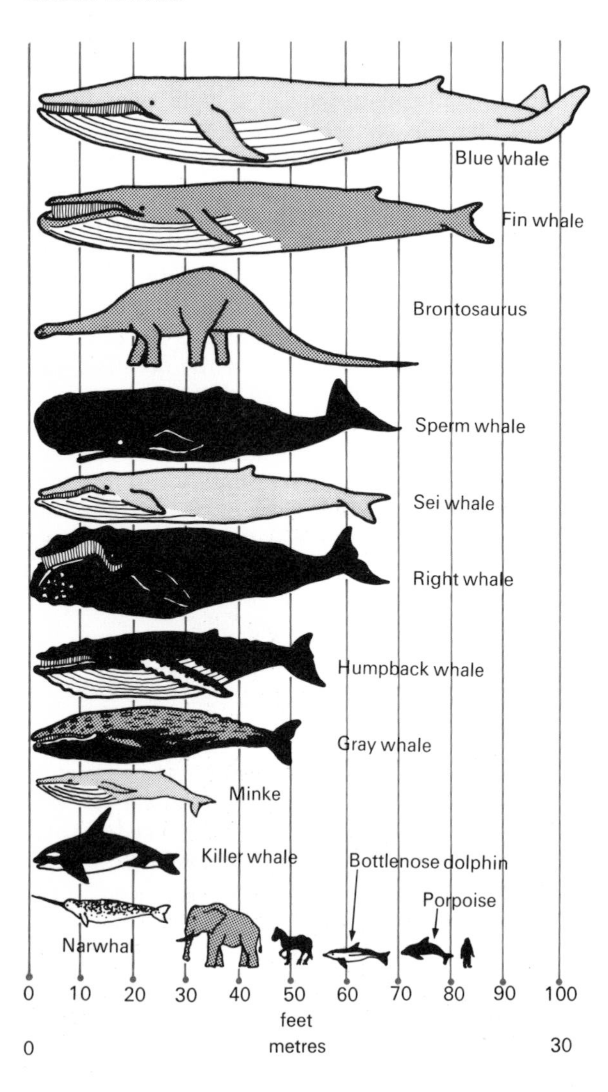

Figure 13.8 *Sizes of whales*

Life cycles

The **great blue whale** feeds on the rich swarms of krill in the polar seas, mainly in the Antarctic. The whales migrate in April and May to the warmer, equatorial waters to breed, during which time they mobilize their great food reserves of blubber, since the warmer waters are not so rich in plankton, and food is scarce. They are monogamous; the gestation period is from 9–11 months; the young are 8 m long and up to 3 tonnes in weight at birth, and grow to a length of 19 m in two years. Usually one calf is born every two years. The natural lifespan is about 50 years, during which time they can grow to 30 m in length and 160 tonnes in weight (equivalent to the weight of 60 elephants or 3200 men).

In contrast to the blue whale, the male **sperm whale** (of Moby Dick fame), which can grow up to 27 m in length and 50 tonnes in weight, herds and defends a harem of the much smaller females which seldom exceed 12 m in length.

Generally speaking, migratory whales tend to move to and from the equatorial regions with very little east–west movement, and very rarely do they cross the equator to mix with those whales of the other hemisphere.

Feeding

The Cetaceans can be divided into three groups according to their feeding habits: the plankton-feeders, the fish-eaters, and the squid-eaters.

The **plankton-feeders** include the largest animals that have ever lived: the blue whales. Plankton-feeders have no teeth but have sheets of fine 'toothed' keratin known as 'whalebone' or baleen, hanging down from the upper jaw. Massive volumes of water are strained through these and vast quantities of plankton are filtered out and swallowed.

The blue whales seek out, and feed mainly on, a specific member of the zooplankton, namely *Euphasia superba* or 'krill', of which they can swallow up to four tonnes in a day. These are shrimp-like Crustacea (about 3.5–6.5 cm long) which occur in densely-packed swarms in the colder, mineral-rich seas of the polar regions. Both the blue and fin whales congregate in regions of great swarms of krill, whereas sei whales have finer horny filters for capturing copepods, and congregate in regions of the greatest *Calanus*

Table 13.5 *Classification of Cetaceans*

Classification	Species	Common name
Order Cetacea		
Sub-order Mysticeti (baleen whales)		
Family Balaenidae ('right whales')	*Eubalaena australis*	*Southern right whale
	Eubalaena sieboldi	*North Pacific right whale
	Eubalaena glacialis	*Black right whale (North Atlantic right whale or Biscayan right whale)
	Balaena mysticetus	*Greenland right whale (bowhead)
	Caperea marginata	*Pygmy right whale
Family Eschrichtiidae	*Eschrictius gibbosus*	*Gray whale { *Pacific gray whale; *Atlantic gray whale }
Family Balaenopteridae	*Balaenoptera acutorostrata*	Minke whale
	B. edeni	Bryde's whale
	B. borealis	*Sei whale
	B. physalus	*Fin whale
	B. musculus	*Blue whale
	(Genus *Balaenoptera* known as the Rorquals, from the Norwegian word 'rorhval' meaning pleated throat)	
	Megaptera novaengliae	*Humpback whale
Sub-order Odontoceti (toothed whales)		
Family Platanisidae (river dolphins)		
Family Stenidae (long-beaked dolphins)		
Family Delphinidae (marine dolphins and killer 'whales')		
Family Phocaenidae (porpoises)		
Family Monodontidae	*Delphinapterus leucas*	White whale (beluga)
	Monodon monoceros	Narwal
Family Physeteridae	*Physeter catodon*	*Sperm whale
	Kogia breviceps	Pygmy sperm whale
Family Ziphidae (including beaked and bottle-nosed whales)		

*Endangered species. *Extinct as a result of hunting.

species production. The free-swimming mollusc *Clione limacina* is another important constituent of many plankton-feeding whales.

A measure of the efficiency of the larger whales in converting zooplankton into body mass is given by their growth, which can be as high as 30 tons a year. Such a growth rate indicates a metabolic efficiency unequalled in the animal kingdom.

The direct exploitation of krill is now occurring on a large scale in the Antarctic and could eventually further undermine the remnants of the whale population, so that those not exterminated by hunting could be finished off by an attack on their food supply. As more conventional fisheries become over-exploited, commercial interest in krill is likely to increase and, as with the whaling industry, once investment in large numbers of krill ships has occurred, the industry will resist constraints.

The baleen whales were key factors in the regulation of the plankton economy of the seas and, before they were decimated by whaling, the

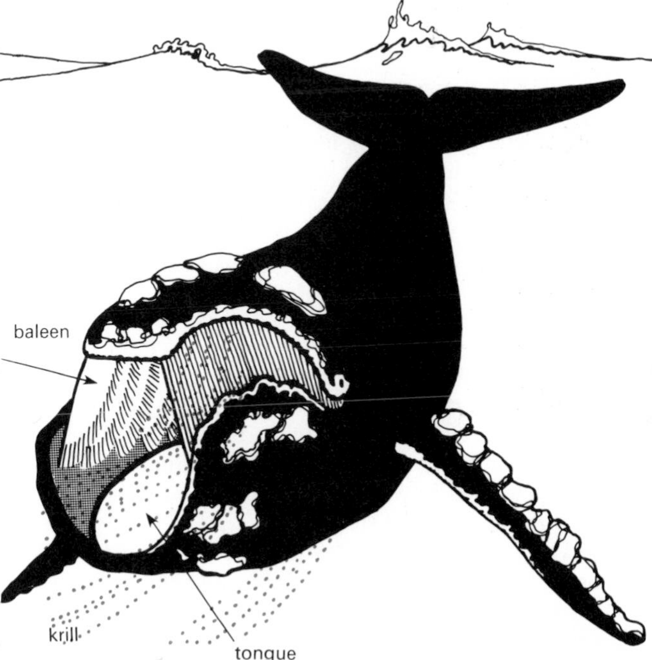

Figure 13.9 *Baleen plates*

Antarctic whale herds are estimated to have consumed about 150 million tonnes of krill each day. However, the materials involved were permanently recycled via the natural cycles. This is not the case with krill ships, which remove the materials from the ecosystem. There can be little doubt that, should man further develop the direct exploitation of krill, it will not be in the balanced way of the self-regulating whale populations and there will be a great danger of permanently disrupting the ocean ecosystems on which so much depends. There is very little knowledge of the systems which are being interfered with, and even less of the long-term outcomes.

The **fish-feeders** include the bottle-nosed whale, the porpoises, and the dolphins, and in addition some plankton-feeders will swallow fish.

The only squid-eaters are the sperm whales, which are the only species to exploit the deep waters of the open oceans, including the blue waters of the tropics. They are found mainly in the Pacific, off the coast of Peru, and around the Galapagos Islands, although they are also hunted in the Atlantic and Indian Oceans. Formidably toothed, with teeth up to 13 cm long, and capable of diving down to 3000 m, they feed mainly on squid up to 1.2 m long; but occasionally they hunt the giant squid of the depths which can reach lengths of up to 16 m.

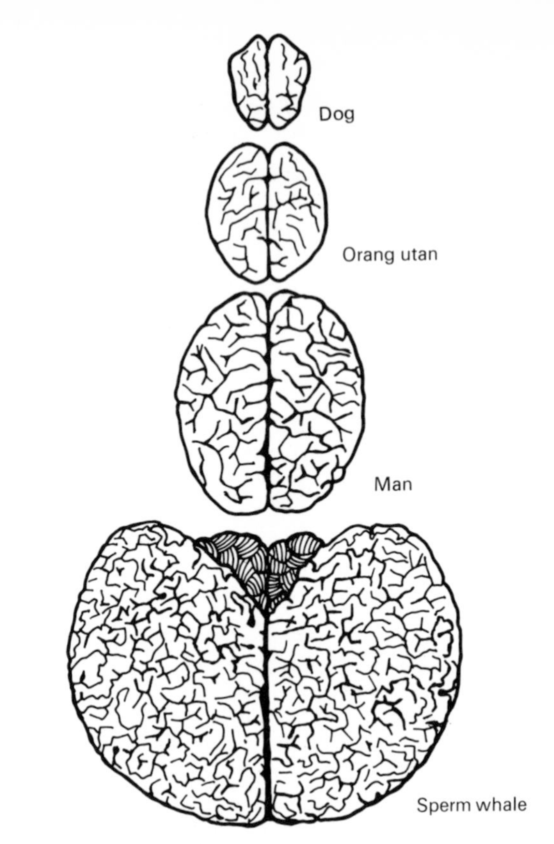

Figure 13.10 *Relative brain sizes (to scale)*
(After Fichtelius and Sjolander, *Man's Place* (London: Gollancz, 1973).)

Brain complexity

The Cetacean brain is large and complex and is at present the subject of much research. The sperm whale's brain can weigh up to 9 kg, which is about 0.03 per cent of its body weight; the brain of the bottle-nosed dolphin can weigh up to 1.3 kg, or 1.2 per cent of the body weight. These figures compare to those of the human brain which weighs up to 1.3 kg, or 1.93 per cent of the body weight.

However, brain weight as a percentage of body weight is not an accurate indication of brain functional ability, especially in an animal such as the sperm whale where much of the body weight is composed of food stores (oil).

The characteristic feature of the Cetacean brain is the striking development and the marked convolutions of the cerebral cortex, which in complexity and depth surpass those of the human brain. There is increasing evidence that such a large and complex brain must harbour a

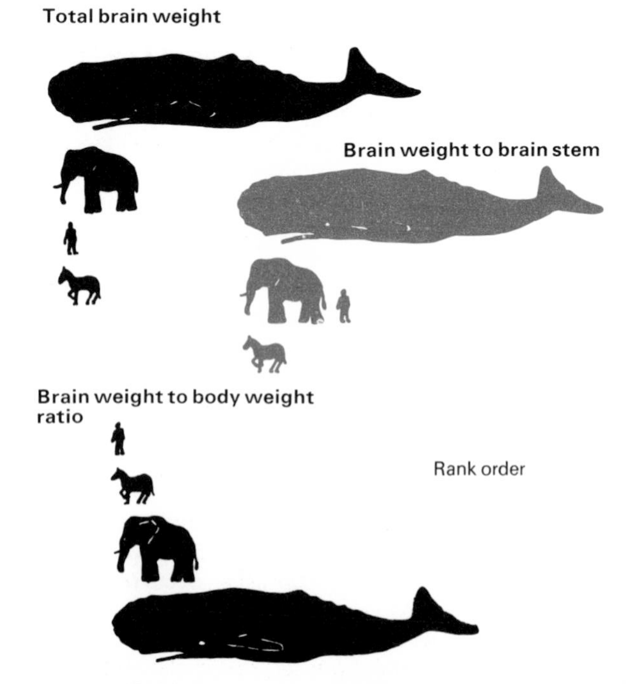

Figure 13.11 *Brain ratio comparisons*

degree of intelligence and consciousness second only to that of man himself; a view reinforced by consideration of the whales' complex powers of communication and co-operation. Such considerations add a further weight to the arguments against the inhumane and indiscriminate slaughter of members of a group of such magnificent animals.

Whale products

All baleen whales yield the same range of products: baleen, meat, oil (from blubber and bones), liver oil, and gelatine materials from the bones. The meat is either used as dog food, cattle food, or as human food, mainly in Japan. The oil, of which a single blue whale can yield up to 40 000 l, is used in the manufacture of soaps, lino, synthetic resins, and at one time half Europe's margarine came from whale oil. Indeed it was the development of methods for hardening oils in the production of margarine that led to the renewal of the onslaught on the whale population in 1905, after a relative lull in the slaughter in the late 1800s. Even so, whales were virtually eradicated in the North Atlantic by then, and this led to the expansion of the Antarctic phase. The liver oil is a rich source of vitamin A, and the gelatine materials are used in glues, films, jellies, and sweets.

The sperm whale yields spermaceti, ambergris, and meat for animal feeds. Spermaceti oil is stored in the spermaceti organ in the front part of the head of the sperm whale; it is a clear oil which, on exposure to air, solidifies into spermaceti wax. This is used in a wide variety of preparations, including certain ointments, smokeless candles, and cosmetics. Spermaceti oil is mainly used as a component of the oils used for tanning leather, and in some lubricating oils. Ambergris is a complex mixture of substances that form great plugs of solid material in the intestine of the sperm whale. Its function in the animal is not clear, but it is used by man as a 'fixative' for the scents of expensive perfumes. At one time ambergris was literally 'worth its weight in gold'.

The role of spermaceti wax, oil, and ambergris in all these materials and processes can easily be substituted for by other materials (e.g. oil from the jojoba cactus is a substitute for sperm oil).

On a world scale, whales supply 5 per cent of the animal fats and only 2 per cent of all fats; with both figures declining rapidly. Indeed, the contribution made by the whaling industry to the world's food supplies is quite literally a drop in the ocean in the face of the relentless tide of the human population explosion.

Control

The collapse of certain whale stocks in the Antarctic in the 1930s led to the ratification of the International Convention for the Regulation of Whales in 1935. (Between 1910 and 1939 it is estimated that about 215 000 blue whales died.) Although this gave so-called protection to the already decimated right whales, it failed to deal with the fundamental problems of long term conservation.

The International Whaling Commission (IWC) was formed in 1946, and held its first meeting in 1949. Its Charter stated that 'The history of whaling has seen overfishing [sic] of one area after another to such a degree that it is essential to protect all species of whales from overfishing.' However, at the same time its stated purpose was 'to regulate the orderly development of the whaling industry'.

Inevitably development meant growth, and growth meant the destruction of the remaining whale populations; this is precisely what has taken place since the Second World War—culminating in the wholesale slaughter of the 1960s when upwards of 50 000 whales a year were being killed, with a peak of nearly 70 000 in 1962. By 1962 the blue whales had been reduced to about 10 per cent of their 1946 levels, and they are still taken on the rare occasions they are encountered.

In fact the only forces ever to come to the aid of the whales have been either economic, or man's preoccupation with warfare. Oil from the great sperm whale virtually subsidized the development of the USA, and they would certainly have been hunted to extinction were it not for the discovery of petroleum in 1859 and the invention of the incandescent bulb in 1879. During both World Wars the whaling fleets were cleared from the seas, and all whale stocks had a brief respite from the relentless hunt.

Now the 'whales' face a deadly race against time, as their survival depends upon the remnants of the population outlasting the ageing pelagic whaling fleets, rather than on the so-called new management policy of maintaining a maximum sustainable yield.

Table 13.6 *Some past and present commercial uses of the whale*

Ambergris (from the intestine of the sperm whale)	Fixative for scent Used in high-quality scented soap
Baleen	Bones for corsets, bustles, and collars Whips and riding crops Umbrellas Brooms Brushes
Blood	Added to adhesives in plywood manufacture Fertilizers
Chemical salts	Creatine used in soaps
Collagens (present in bone, skin, and tendons) which are boiled to yield gelatine	Gelatine is used for: Photographic film Edible jellies Confectionery
Endocrine gland (e.g. pituitary gland)	Medicines and pharmaceutics ACTH source
Liver	Whale oil from liver yields vitamin A
Skin (from toothed and white whales)	Leather for bicycle saddles Handbags and shoes
Spermaceti wax	Ceremonial candles Cold creams Lipsticks Brushless shaving creams Ointments
Unrefined spermaceti oil	Mixed with mineral and other oils for dressing hides in leather industry
Filtered spermaceti oil Sulphurized spermaceti oil	Ingredients of lubricating oil for light machinery Emulsifying agent in compounded oils Cutting oil Textile lubricants Dressing hides
Saponified (spermaceti oil alcohols)	Wide range of industrial uses, cosmetics, detergents, etc.
Tendons	Tennis racket strings Surgical stitches
Whale bone	Bone meal for fertilizers Shoe horns Chess sets Toys
Whale meat	Human food Ingredient in pet foods and animal feed
Whale oil	Glyceridic oil of the baleen whales is used primarily in the production of glycerine, margarine, and soaps Oil cloth and linoleum Printing ink Candles Crayons
Whale teeth (toothed whales such as sperm)	Ivory for carvings and souvenirs Piano keys

species	numbers before commercial whaling	estimated numbers today	percentage remaining today
Blue	210 000	13 000	**7**
Humpback	100 000	7000	**6**
Right	50 000 ?	4000 ?	?
Bowhead	10 000	2000 ?	?
Fin	450 000	100 000	**22**
Sei	200 000	75 000	**38**
Bryde's	100 000	40 000	?
Sperm, male	530 000	230 000	**43**
Sperm, female	570 000	390 000	**68**
Gray	15 000	11 000	**73**
Minke	360 000	300 000	**83**

So-called fully protected species

Figure 13.12 *Whale populations*
(After Victor Scheffer.)

Concept of maximum sustainable yield

When a population has expanded to the limits of its space and food supply it tends to remain stable. It should then show a fairly consistent structure with the proportion of the different ages tending to remain constant, and with surviving young replacing those that die. The so-called maximum sustainable yield (MSY) is calculated on the assumption that if a proportion of a population is killed, the increased availability of food for the remainder can result in an increase in the surviving young. The rate at which the greatest number can be removed without further lowering the population is known as the MSY.

The concept of the MSY is valuable in the management of fish stocks, but the MSY model applied to whales generally, and to sperm whales in particular, takes no account of the population dynamics and social structure of a highly evolved, intelligent, communicative mammal. Also, since no-one knows what constitutes a stable population as far as any species of whales are concerned, it is impossible to apply the concept in this case.

Hunting and killing

In the early days of whaling, the whales were hunted in open boats, killed with hand-thrown harpoons, and towed to a land-based whaling station for processing. Even such simple techniques led to the destruction of those whale populations that regularly migrated close inshore to inhabited coastlines. For example, the population of the Biscayan right whale or the baleine de Basques (*Eubalaena glacialis*) hunted by the Basques in the Bay of Biscay, was extinct as early as AD 1600.

The British despatched their first Spitsbergen Whaling Expedition in 1611, followed by the Dutch in 1612. By 1632 there were signs of a serious decline in stocks of the valuable Greenland right whale (bowhead) which had the longest baleen and the thickest blubber, and by 1720 the whaling off East Greenland was finished. The 'right' whales, slow swimmers with thick blubber and a long valuable baleen, were always the 'right' choice for the early whalers. They were easy to catch and floated when killed. In the nineteenth century the Yankee whaling fleets alone killed 100 000 right whales, reducing stocks to levels from which they have never recovered.

'Classical' whaling essentially ended at the beginning of the twentieth century, although the last Windjammers lingered on until 1925. With the development of powered ships, back-loading factory ships in 1924, fast hunter boats, harpoon guns, explosive harpoons, sonar, and even spotter planes and helicopters, the pursuit has been pressed home with relentless and mindless efficiency.

With the increasing understanding of the degree of complexity of the Cetacean brain, there is increasing concern over methods of killing. The 2 m long, 70 kg explosive harpoon hits the whale at 96 km/h; seven seconds after impact the explosive detonates causing devastating injuries. If the shot is placed correctly death can occur in seconds; however death is usually from internal bleeding and the death time is usually between 5 and 10 minutes, and in some cases several hours.

It would be unthinkable to slaughter other mammals in this way, and indeed other mammals are covered by Humane Killing Legislation. The fate of the harpooned whale has been likened to the firing of two or three explosive spears into a horse, and allowing it to drag a heavy butcher's cart with blood pouring over the road until the animal collapses minutes or even hours later.

Conclusion

Notwithstanding the increasing publicity and pressures against whaling, it appears to many that the fight to 'save the whale' is doomed to failure. Inevitably economic advantage and short-term expediency has over-ridden the most important and obvious long-term conservation measures. Should the fight be lost, the fate of 'the whale' should act as an ominous warning for the future of other endangered species, and in the longer term for the future of the intricate web of life on this Earth.

Land and food production

Under present conditions every human requires about 1 hectare of fertile land to provide an adequate diet. In the USA there are 5 Ha of fertile land per person, in France 1.3 Ha per person, in the UK only 0.4 Ha per person, and in Japan less than 0.4 Ha per person (with a density of about 300 per square kilometre). At the moment Canada and the USA are the only countries to produce food surplus to their own requirements, and massive crop failure in the USA would lead to world-wide famine.

If the protein requirements of modern livestock feeding are taken into consideration, and the number of 'average standard men' that this would feed are calculated, it is possible to consider the livestock of the world in terms of human 'population equivalents'. On the basis of such estimates it has been calculated that the world is carrying a livestock population equivalent in protein demand to a human population of about 13 000–14 000 million.

At present 30–40 per cent of the world's grain production is used as animal feed to produce animal products with an average conversion efficiency of only about 10 per cent. The inefficient use of grain as animal feed on such a large scale as now exists is certain to be decreased in the future as the pressure on food supplies builds up.

A measure of the disparity between the developed and undeveloped countries in terms of food consumption is well-illustrated by the fact that at present one-third of the world's population consumes three-quarters of the world's food supplies. For example, India has a population three times that of the USA, and yet has only half the protein demand. With increased competition for the Earth's limited resources the over-developed, resource-poor countries are the most vulnerable, and will not be able to continue to import the materials and food, especially protein, at anywhere near the present levels. Indeed Britain and Japan are amongst the major recipients of the world's flow of food, especially fishmeal protein, and soya bean protein for animal food.

Japan in particular has acute problems of land, food production, and population. It has a population of around 100 million, equal to about half that of the USA, living in an area only 90 per cent of the size of California with a tilled hectareage less than 5 per cent of the cropland of the USA. There are about seventeen people per arable hectare, and 57 million people depend critically upon food from the oceans.

The UK produces about 50 per cent of its present food requirements, which covers about 75 per cent of the requirements for temperate food (food that can be grown in a temperate climate). Thus, of the 50 per cent that is imported, about 16 per cent is temperate food and the remaining 34 per cent is food that could not be grown in our climate.

Increasing population and decreasing availability of food-producing land will further compound the problems. For example, the population of the UK could reach 60 million by the year 2000 and, since the time of maximum area of agricultural land in 1892, there has been a net loss of 15 per cent. By the year 2000 there will be something like 20 per cent less productive land than existed in 1900. For self-sufficiency in the UK in the future, each acre of productive land will have to feed nearly two people, though more

Table 13.7 *Population in relation to tilled acreage*

Country	*Number of people per acre tilled land*	*Tilled acreage in acres per person*
Japan	7.1	0.14
Netherlands	5.6	0.18
Egypt	4.5	0.22
Jamaica	3.1	0.32
UK	3.0	0.33
China	2.9	0.34
India	1.28	0.78
France	1.0	1.0
World	0.98	1.02
USA	0.46	2.19
USSR	0.39	2.54
Canada	0.21	5.25
Australia	0.12	8.37

optimistic estimates suggest that the UK could still become self-sufficient in food. (It must also be remembered that for every unit of food energy produced, 2.5 units of fossil fuel energy are expended.)

Possibly the greatest increases in food availability could come from the **prevention of waste**. For example, an estimate based on several sampling surveys indicates that about 25 per cent of both home-grown and imported food is wasted in the UK.

Needless to say, productive land is a non-renewable resource which is absolutely beyond value, and which should be recognized as an **irreplaceable national asset**. However, in spite of the increasing importance of productive land, an estimated 40 000–60 000 hectares are lost to urbanization and roads each year in the UK. This represents a loss of land roughly the size of Oxfordshire every 10–15 years. In addition further losses occur via the dereliction of land which occurs as a result of excavations such as open cast mining and gravel pits, spoil heaps, tips from underground mining and industry, and disused industrial sites. There are about 60 000 hectares of such derelict land in the UK, about 40 per cent of which are spoil heaps, 34 per cent excavations, and 26 per cent industrial sites.

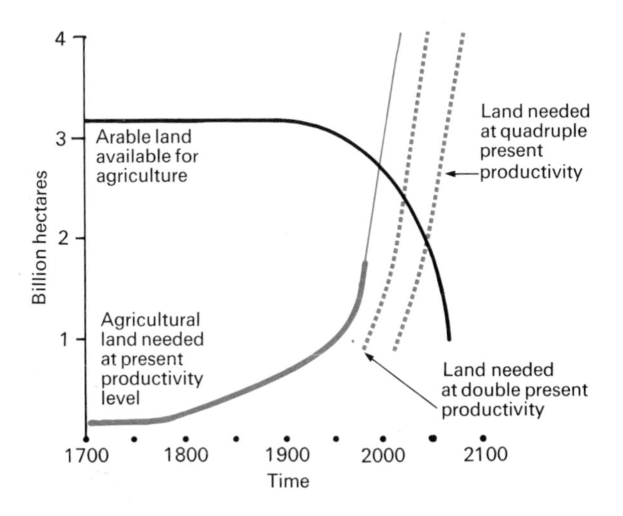

Figure 13.13 *Land requirements and land availability*
Note that a doubling of present productivity only defers the 'cross-over' point of the two curves by about thirty years, and a quadrupling only defers it by about fifty years.

Green revolution

The development and introduction of new, high-yield varieties of cereal crops, especially wheat and rice, has had such an impact on food production that it has been referred to as the 'Green Revolution'. For example, in Columbia new dwarf varieties of rice raised the yield from 2.7 tonnes per hectare in 1961 to 5.4 tonnes per hectare in 1975.

However, these new varieties inevitably require high fertilizer and pesticide inputs which are often not available in the areas where the food is needed most. In fact in the absence of the political will and economic aid necessary for its success, there are signs that the Green Revolution is fast turning brown at the edges. N. E. Borlaug, Nobel Prizewinner and 'father' of the Green Revolution, considers that these developments provide about one generation's grace for the efforts to check the growth of populations.

It would appear that between 10 to 20 years 'lead' time is needed for the successful development and introduction of a new variety of cereal, but the International Rice Research Institute in the Phillipines did produce and introduce a high-yield strain of rice (IR8) in just six years. However, there are signs that the IR8 strain was introduced prematurely, as it is very susceptible to pests, is less palatable, and has poor milling qualities.

Table 13.8 *Comparison of grain yields of two strains of rice with different amounts of fertilizer*

Nitrogen applied (kg/Ha)	*Grain yield (tonnes/Ha)* Traditional rice—'peta'	*IR8*
0	5.2	4.8
20	5.4	5.6
40	5.2	6.8
60	5.0	7.2
80	4.8	7.7
100	4.4	8.4
120	4.2	9.0

Most new, high-yield varieties of cereal crops are either dwarf or semi-dwarf varieties. This decreases storm damage and avoids the losses suffered by tall wheats which tend to collapse when heavily fertilized.

Two major problems facing agriculture are the increasing demand for pesticides and fertilizers in the face of the diminishing oil supplies, on which the production of both are heavily dependent.

The breeding-in of disease resistance to major crop plants has been achieved with great success, despite the great difficulties involved. However, such is the rate of mutation of many major pathogens that this a feature of agriculture that will never be overcome.

A major advance in plant breeding would be the establishment in major crop plants of the ability to develop root nodules containing symbiotic nitrogen-fixing bacteria, such as are found in leguminous plants. This breakthrough would lead to a great reduction in the need for nitrate fertilizers, which are heavily dependent on energy, and the mining and transport of non-renewable mineral supplies. There are also hopes of breeding the C4 photosynthetic[130] ability into a wider range of crop plants. C4 plants, such as maize, have a greater productivity than the more usual C3 plants.

Further improvements may be achieved as a result of plant tissue culture methods by which the best varieties may be cloned.[58] Plant meristems are often free of viruses, and the virus-free cloning of important food crops such as the potato will contribute to increased crop yields.

A further challenge to plant breeders is to increase the nutritional quality of many crops, especially in relation to amino acid content and balance.

Timber

Timber represents another major biological resource. Wood is still an important constructional material, despite the use of many modern substitutes in developed countries. Since many of these substitutes, such as metals and plastics, are derived from finite resources which are close to their depletion point, wood could increase rather than decrease in importance as a constructional material. In addition to its use in building, wood is important in many parts of the world as a fuel for which there is often no local substitute. It is also used on a huge scale as a source of wood pulp for the paper industry, but much of this could be recycled with great efficiency.

Under the pressure of present-day demand in both developed and undeveloped countries it is unlikely that timber can be considered as a renewable source (as strictly it is), either now or at any time in the future. The deafforestation of the Earth's surface is now in its final phases, continuing as it is at a rate far in excess of any replanting schemes. The annual deficit of trees is now running at about 11 million hectares.

Land loss through soil erosion

The relentless process of soil loss over the years is dramatically illustrated by the following passage, written about 2350 years ago.

PLATO (428–347 BC) CRITIAS III

'What proof then can we offer that it is fair to call it now a mere remnant of what it once was? It runs out like a long peninsula from the mainland into the sea, and the sea basin around it is very deep. So the result of the many great floods that have taken place in the last nine thousand years (the time that has elapsed since then) is that the soil washed away from the high land in these periodical catastrophes forms no alluvial deposit of consequence as in other places, but is carried out and lost in the deeps. You are left (as with little islands) with something rather like the skeleton of a body wasted by disease; the rich, soft soil has all run away leaving the land nothing but skin and bone. But in those days the damage had not taken place, the hills had high crests, the rocky plain of phelleus was covered with rich soil, and the mountains were covered by thick woods, of which there are some traces today. For some mountains which today will only support bees produced not so long ago trees which when cut provided roof beams for huge buildings whose roofs are still standing. And there were a lot of tall cultivated trees which bore unlimited quantities of fodder for beasts. The soil benefited from an annual rainfall which did not run to waste off the bare earth as it does today, but was absorbed in large quantities and stored in retentive layers of clay, so that what was drunk down by the higher regions flowed downwards into the valleys and appeared everywhere in a multitude of rivers and springs. And the shrines which still survive at these former springs are proof of the truth of our present account of the country. The acropolis was different from what it is now. Today it is quite bare of soil which was all washed away in one appalling night of flood, by a combination of earthquakes and the third terrible deluge before that of Deucalion. Before that, in earlier days, it extended to the Eridanus and Ilisus, it included the Pnyx and was bounded on the opposite by the Lycabettos; it was covered with soil and for the most part level'.

Rill erosion in Oregon. Soil loss was 62 tonnes per hectare

Gully erosion in Oklahoma caused by a progressive loss of valuable agricultural soil, which vegetation alone is unable to stabilize

That this process continues at an accelerating rate is hard to picture, and the consequences in the face of the population explosion are hard to imagine.

The extent of soil erosion in Latin America, for example, is colossal, estimated at a rate equivalent to the loss of 202 000 hectares of soil to a depth of 1.5 m each year. There is also no doubt that the rate will increase as the great tropical rain forests of the Amazon Basin are cleared. Such losses of soil will compound the problems of food supply in South America which, with its annual population net growth rate of about 2.9 per cent giving a doubling time of only 24 years, will be where the population explosion will reach its highest intensity.

Another example illustrating the scale of the problem is provided by the relentless spread of deserts which is occurring as a result of land misuse and climatic changes. Thus in 1900 an estimated 10 per cent of the Earth's surface was classified as desert, but by 1970 this figure had doubled to 20 per cent.

Any removal of vegetation will increase the rate of soil erosion. Deafforestation, the removal of hedgerows, overgrazing, and overcropping can all lead to soil loss by wind or water erosion. The continuous use of land for the monoculture of a single crop can lead not only to the build-up of pests and diseases, and pollution due to the increased use of inorganic fertilizers and pesticides; but also to the loss of soil structure, which can render it vulnerable to erosion.

In addition the concentration of populations into urban areas, which results in greater environmental impact in terms of water extraction, pollution, and waste disposal, can also lead to an undermining of soil stability. The flow of organic material and minerals in food to the urban centres and their disposal in sludge and sewage effluent, with very little being returned to the soil, results in a gradual depletion of the soils, an increasing need for artificial fertilizers, and again a loss of structure which renders the soil liable to erosion. A similar trend in the concentration of livestock into 'factory' farms, and the disposal of their excreta in the same way as that of humans, further increases the problem.

Although some estimates indicate that only half the world's potentially arable land is being cultivated, such are the problems of bringing this land into production and such are the rates of soil loss by erosion that it would appear that the world's land bank is almost already fully exploited.

The largest unexploited productive land mass in the world is Siberia which, with its present population of only about 18 million, in many ways represents the last frontier of mankind. In potential it can be considered as a New America; possibly capable of supporting a population of up to 500 million.

This contrasts strongly with that other vastly underpopulated area, Australia. Australia has a land mass greater than that of the USA, but has a population of only about 12 million, which is less even than that of Greater Tokyo. However, as a result of soil and climate, the maximum population of Australia has been estimated to be as low as 50 million, less than that of the British Isles. There have, even so, been some striking successes in soil improvement in Australia. For example, the barren 90-mile desert in South Australia has been rendered productive by the addition of minute quantities of zinc and copper trace elements which were previously deficient.

Water

Water shortage

Despite the losses due to soil erosion, pests, and diseases, water shortage is the major single factor in producing famine in the world today. The ground water reserves are shrinking rapidly as a result of their depletion proceeding faster than their replenishment, and inevitably the ground water supply will fail. In 1963 the US Geological Survey estimated that it would take over a hundred years to restore the groundwater reserves in Texas by natural replenishment to the levels that existed in 1900. Depletion of ground water reserves in coastal regions can also result in seepage of sea water under the land mass, causing salination of the remaining water.

Water supply

Large-scale sources of fresh water include upland lakes (both natural and man-made), wells, and rivers.

Upland lakes are generally used to supply urban areas at some distance. The catchment and drainage area must be kept free from pollutants, such as nitrate fertilizers, persistent pesticides, and animal and human wastes, that might drain into the body of the water.

Wells exploit underground sources of water. Shallow wells utilize water relatively near the surface of the land, and are becoming increasingly polluted with substances such as nitrate fertilizers which are leached or washed down through the soil. Deep wells penetrate deeper-lying water-bearing layers of rock or aquifers. Usually such water is hard due to the dissolved elements that it has accumulated whilst percolating down to the aquifers; and it is usually of good quality as a result of the natural filtration and purification effect that such percolation has.

Artesian wells are deep wells which utilize the natural pressure of deep-lying water to raise itself to the surface without pumping. Such wells once provided the bulk of London's water supply, but the supplying aquifers have become severely depleted so that they now supply only about 10 per cent of London's requirements.

Rivers are the major source of water in Great Britain, and about two-thirds of London's supply comes from the Thames. The extracted water is usually returned to the river as sewage effluent, and may be extracted again further down stream. It has been estimated that some of the London water supply has already passed through an average of seven sets of kidneys!

Water purification

STRAINING

If drawn from a river, the first stage in the process of water purification involves straining the water through a series of graded screens to remove the larger suspended materials.

STORAGE

The water is then held in a storage reservoir which allows suspended solids that pass through the screens to settle. Flocculating agents such as aluminium sulphate ($Al_2(SO_4)_3$) may be added to clump the suspended organic matter and bacteria together into lumps which then settle faster. Also, whilst in the reservoir, the water is exposed to u.v. light, which destroys many pathogenic micro-organisms, as well as to the natural purification processes that are carried out by bacteria, protozoa, and algae in the water. At the same time the water is exposed to airborne pollutants and, if high in nutrients, to the danger of algal blooms.

FILTRATION

After being held in the storage reservoir for a period of time the water is filtered to remove any remaining suspended solids and pathogenic micro-organisms. The water is trickled down through gravity filter beds of about 1 m of sand and gravel, grading from sand at the top to coarse gravel at the bottom.

Beds which use fine sand as the top layer do not allow the water to filter through them very rapidly and are referred to as **slow gravity beds**. As the water percolates down through a slow gravity bed a biologically active zone develops in the upper layers. The first few millimetres are dominated by algae which take up nitrates, phosphates, and carbon dioxide, and release oxygen due to their photosynthesis.

This upper, or autotrophic, layer traps a thin skin of inorganic particles which acts as an extremely fine-meshed filter. Below the autotrophic layer is a heterotrophic layer about 30 cm deep dominated by saprophytic bacteria. These break down any organic matter in their nutrition and respiration.

The surface scum that forms must be removed every few months, and this contributes to the high operating costs of slow gravity beds. They also occupy a large area of land.

Rapid gravity sand filters have coarser sand and allow a faster rate of filtration over a smaller surface area. However, they do not build up a system of purifying micro-organisms to the same extent as the slow beds, and as a result do not purify the water to the same high standard. When these filtering methods are used, the organic matter in the water is first precipitated by chemicals, such as ammonium aluminium sulphate, and allowed to sediment in settling tanks.

Another type of rapid sand filter has the water forced through under pressure, and yet other methods of filtration employ cylindrical micro-strainers and filters of aluminium silicate.

STERILIZATION

Any bacteria remaining after filtration and aeration are removed by sterilization with chlorine or ozone. Chlorine is not as efficient as ozone and may not destroy all pathogenic bacteria. Higher levels of chlorine may therefore be used and, after holding in a contact tank to ensure mixing, the excess can be removed by treatment with sulphur dioxide. Ozone at 1 ppm destroys all bacteria, bacterial spores, and viruses within 10 minutes.

HARDNESS ADJUSTMENT

The so-called 'hardness' of water is caused by the presence of dissolved magnesium and calcium salts. Natural waters can vary between 50–400 ppm of dissolved elements, and adjustment can be made during water treatment to bring it to about 100 ppm of dissolved calcium. Some areas have natural levels of fluoride of up to 4 ppm, and in those areas lacking fluoride, fluoridation to levels of 1 ppm is sometimes carried out in an attempt to reduce the level of tooth decay in the population.

WATER QUALITY CONTROL

Testing the water quality at the end of the purification process involves both chemical and biological tests. Chemical tests include those for nitrates, phosphates, lead, and pesticides, etc. Biological tests include a coliform count and a measurement of the biochemical oxygen demand (BOD).

The coliform count measures the number of the bacteria *Escherichia coli* (which come from human and animal intestines). A count of one or more per mililitre of water is taken as a positive indication of pollution by human and animal excreta.

The biochemical oxygen demand (BOD) is defined as the amount of dissolved oxygen (in mg per litre) consumed by microbiological action when a sample is incubated for 5 days at 20 °C. This normally gives a rough indication of the organic matter present in the sample as it is a measure of the activity of the aerobic bacteria in oxidizing that organic matter.

Clean, safe, drinking water should not contain any pathogenic micro-organisms, any suspended solids or toxic chemicals. In addition it should have no definite taste, smell, or colour.

Recycling

Sewage

Sewage contains both so-called 'foul sewage', and storm water which drains from all the hard-covered surfaces of built-up areas. In combined sewage systems both components run into the same pipe system, and the stormwater helps to dilute and transport the 'foul sewage' from domestic and industrial sources. However, the stormwater causes large and variable fluctuations in the flow reaching the sewage works which must therefore be 'buffered' to prevent disruption of the treatment process. The minimum flow must always be fast enough to prevent solids from settling and silting up the system of sewers.

Any flow above the capacity of the sewage works is released untreated on the basis that the large volume of storm water dilutes the 'foul sewage' to tolerable limits. If organic waste is sufficiently diluted, it is broken down by organisms in the river or sea water and does not become a pollutant. However, should a body of water become overloaded with organic waste, the natural purification process breaks down, due mainly to the depletion of the oxygen content.

The organic wastes from urban areas with their high density populations pose considerable disposal problems. Not only is there the problem of oxygen depletion of the water, but also the problems of the spread of pathogenic disease organisms and their eggs; eutrophication[322] from synthetic detergents and inorganic fertilizers; and the poisoning of the water with toxic materials such as lead and persistent pesticides, all of which can find their way into domestic sewage.

SEWAGE TREATMENT

Sewage treatment is essentially a process of water recycling, and indeed water withdrawn from some rivers for domestic use will have already been so used, perhaps more than once.

The basic principles underlying sewage treatment are to remove solid organic matter as a sludge (primary treatment); to supply sufficient oxygen for micro-organisms to break down the organic matter left in solution and suspension in the water after settlement of the sludge (secondary treatment); and, where possible, to prevent toxic materials and pathogenic organisms from reaching natural waterways on the soil (tertiary treatment). Basically, sewage works receive strong sewage and discharge very weak sewage as effluent.

Important measures of the strength of sewage are the BOD[279]; the amount of suspended solids (SS); and the coliform count.[279] However, large volumes of sewage and industrial effluent are still discharged to the sea without treatment; for example 90 per cent of the sewage poured into the Mediterranean is untreated. The discharge of untreated sewage into the sea is less dangerous than into rivers. The volume into which it can be diluted is greater, the salt water kills practically all fresh water micro-organisms, and the marine organisms break down the organic matter.

Table 13.9 *Pollution content of sewage*

	Average untreated sewage	*Average treated sewage*
BOD	350 ppm	20 ppm
SS	850 ppm	30 ppm
E. coli	10^6 cm^{-3}	5×10^4 cm^{-3}

BOD of 20 ppm and SS of 30 ppm, the Royal Commission Standard 20:30 Limit, are the normal minimum criteria for effluent discharged into inland water.

There are many variations of the process of sewage treatment, but all follow basically the same sequence of events.

SCREENING

The incoming sewage has the larger debris removed by coarse mesh filters, and the solids filtered out can be passed to sludge grinders, ground down, and added to the flow. Comminuters screen and macerate the solids at the same point. In addition booms may be used to remove floating grease and oil. (Bradford sewage was once a valuable source of oils (lanolin) recovered from the effluent from the wool industry.)

GRIT REMOVAL

The sewage is then passed along channels about 20 m long at a speed of about 0.3 m/s. This allows grit to settle, which is then removed, washed and used for road in-fill, etc.

SEDIMENTATION

The sewage passes to sedimentation tanks, which are from 1.5–3 m deep, where up to about 50 per cent of the suspended solids are settled out as primary sludge over a period of about 10 hours. Sometimes chemicals can be added to coagulate the suspended solids and speed their sedimentation. This primary sludge is rapidly removed to prevent its decomposition in the tanks. The sedimentation tanks also act as a buffer system to regulate the even flow of sewage through the works.

BIOLOGICAL PURIFICATION

The effluent from the sedimentation tanks is next subjected to biological oxidation. This process exploits the purifying activities of micro-organisms, such as bacteria, protozoa, fungi, and to a lesser extent algae. The bacteria are the most successful in degrading organic wastes, mainly as a result of their small size and consequent large surface area to volume ratio, which facilitates the exchange of materials with the effluent.

Saprophytic(zoic) organisms, mainly the bacteria, but also to a lesser extent the protozoa and fungi, utilize the organic matter in the effluent as nutrients. They secrete enzymes, which carry out extra-cellular digestion of the organic wastes, and reabsorb the less complex products. These are used as a source of carbon and other elements for their synthetic reactions, and as a respiratory substrate in aerobic respiration which provides the energy for such synthetic reactions.

Another oxygen-demanding process during the purification is the oxidation of ammoniacal compounds and nitrites to nitrates by the chemosynthetic nitrifying[111] bacteria.

Sewage is a rich source of bacteriophage[289] viruses which may play an important part in reducing the number of bacteria during the purification process. The major role of the protozoa (mainly ciliates) in the purification process is to act as predators upon the bacteria. By 'grazing' upon the saprophytic bacteria the protozoa prevent them from reaching self-limiting numbers, and the bacterial populations are kept in a prolonged state of physiological vigour. As a result the rate of assimilation of organic materials is greatly increased. In addition the protozoa prey upon any pathogenic bacteria in the effluent, such as diphtheria-, cholera-, and typhus-causing bacteria, *Streptococci*, and *Escherichia coli*. Compared to their role as predators, their contribution to effluent purification by absorbing organic matter saprozoically, engulfing organic particles holozoically, and secreting substances that flocculate suspended colloidal particles, appear to be relatively minor.

The fungi act in the same way as the bacteria, and can be equally efficient in removing organic matter from the effluent. However, they produce more biomass for equal amounts of nutrients absorbed than the bacteria, and their rapid growth and filamentous structure can physically interfere with the movement of effluent through the works. Some predaceous fungi are also found which feed on Nematodes, Rotifers, and protozoa.

The algae, when present, will absorb some mineral elements from the effluent and contribute some oxygen from their photosynthesis, but they play a minor role in the purification process.

The micro-organisms responsible for this biological purification require oxygen for their respiration, and there are two main methods of effluent treatment which ensure adequate supplies.

Percolating filter method

In this process the effluent is trickled down from slowly-rotating arms through about 2 m of graded coke, slag, or stones, with the finer 'grades' at the top. The term 'filter' is a misnomer since no filtering action as such is carried out. These graded particles provide a large surface area for the micro-organismic film which forms naturally upon them, and the spaces between the particles allow the easy penetration of air containing oxygen.

A properly maintained percolating filter bed behaves as a **balanced ecosystem**, with many complex interactions between the organisms and between the organisms and their environment; there being an ecological succession of different members of the community, varying both with depth in the filter and with the seasons. On average it takes about 60 days for the balanced system to become established.

The bacteria adsorbed on to the surface of the bed purify the effluent and are kept from blocking the spaces by the 'grazing' protozoa, especially in the upper zones. Fungi are undesirable as dominant members of this film community as they quickly block the beds and support a large population of fly larvae. Algae usually form a thin film over the surface of the bed and this may interfere with its efficient operation.

In addition to the protozoa, Annelid worms, nematode worms, insects such as springtails, and various fly larvae, and other invertebrates all graze on the bacterial film. Colder weather causes these so-called 'scouring' organisms to retreat deeper into the filter and as a result some surface blocking can occur. The activity of the protozoa is also depressed more by low temperature than is the activity of the bacteria, which further encourages blocking.

In warmer weather the grazers return to attack the surface layers, producing large amounts of 'humus' which is washed out of the beds in what is known as the 'spring-unloading'.

Activated sludge

In this method of biological oxidation the level of dissolved oxygen in the effluent is maintained by rapid stirring or by the bubbling of air through the effluent. Such treatment leads to the build-up of a thin, sponge-like network of activated sludge containing bacteria, protozoa, and fungi. There is not such a great diversity of organisms present as in the percolating beds. Samples of this activated sludge are removed and used to seed tanks containing raw effluent. The activated sludge method uses less land area than the percolating filters and does not involve a 2 m drop from top to bottom; it also avoids any fly nuisance. However, it is expensive to operate, more sensitive to toxic materials, and requires more skilled attention.

Both methods are at risk from physiological poisons such as detergents, cyanides, etc., which destroy the micro-organisms. It has been estimated that about 25 per cent more sewage works than were necessary in the pre-detergent era are now required to purify the same volume of sewage.

HUMUS TANKS

After biological oxidation, the effluent passes to humus tanks where less noxious solids settle out readily. Such tanks are desludged frequently, and are another source of activated sludge for the reseeding of raw effluent.

TERTIARY TREATMENT

If the effluent is still of inadequate quality it can undergo micro-straining through a very fine stainless steel mesh. It can also be 'polished' by being held in lagoons, passed through sand filters, or being allowed to percolate through grass plots before entering the body of water into which it is to be discharged. The term 'sewage farm' derives from the older method of irrigating land with effluent, integrating it with the soil and farming the land on a rotational basis.

When discharged into a river the effluent should not contain oil or grease, raise the BOD of the river above 4 ppm, contain suspended solids that could settle on the river bed, interfere with water abstraction, prevent agricultural use, or raise the temperature of the river unduly. The effluent flow into rivers from cities can distort natural hydrological cycles, for example the flow of water in the river Trent is increasing due to effluent from Birmingham, which gets its water supply from North Wales. Thus large volumes of water from North Wales end up in the Trent.

SLUDGE TREATMENT

When first removed, the sludge has a high water content of up to 90–99 per cent. This can be dumped as it is, especially at sea, but it is better to reduce the water content for ease of transport, and to reduce the risk of spreading pathogenic organisms.

The commonest method of treatment employs anaerobic sludge digestion. In this process anaerobic bacteria digest the sludge, producing gas which is up to 75 per cent methane and can be used as a fuel to drive electricity generators which supply electricity to the sewage works. More than a third of the solid content of sludge can be converted to gas in this way. The process also destroys many pathogens and gets rid of much of the grease which originates from soaps and renders the sludge difficult to be assimilated in the soil if used as a fertilizer.

The digested sludge is dried by spreading on drying beds and, if free from pollutants and pathogens, can be used as fertilizer. About 1 million tonnes of dry sludge is produced in Britain each year; about 46 per cent of it is spread on the land.

Animal manure can be a rich source of protein, for example poultry manure dries down to about 30 per cent crude protein. This can be milled to a fine powder and added as a protein supplement to animal feeds. However it must be correctly sterilized as otherwise it can be the vehicle of many infectious disease-causing organisms.

Urban refuse

It is estimated that each household turns out about a tonne of rubbish a year, with paper and cardboard being the main component.

Timber is theoretically a renewable resource, but such is the rate at which forests are being destroyed that as far as man's future is concerned, timber may be considered to be a finite non-renewable resource. For example, every year a forest the size of Wales is destroyed to provide paper for the UK, with one fully grown tree providing pulp for about 400 copies of a newspaper.

The UK is one of the world's largest importers of woodpulp, importing up to 80 per cent of its requirements for the paper and board industry; yet large amounts of easily-recoverable paper and board are wasted each year. That such paper is recoverable is shown by the fact that in war time about 60 per cent of refuse paper was recycled. However, since that time the figure has dropped to 25 per cent, with only about 10 per cent of household waste paper being collected separately for recycling.

A similar situation exists with waste glass. Of the 6500 million bottles made in the UK each year, only about 15 per cent are returnable. The raw materials for glass-making are cheap and locally available, therefore recycling is given an even lower priority than in other industries. However, recycled glass represents a saving of energy and could gradually become more attractive economically.

Cars represent a tremendous source of waste. About 750 000 cars are scrapped along with about 10 000 million tyres each year in Britain, but only a limited amount of their materials are recycled at present. This is due mainly to the processing costs, which can amount to 90 per cent of the value of the recovered materials. The scrapped cars represent an annual loss of about 190 000 tonnes of iron and steel, 3300 tonnes of aluminium, 3300 tonnes of zinc, and 2500 tonnes of copper. However, due to its value and its ease of recovery, almost all of the lead in the batteries is recovered.

The manufacture of an average car in Britain consumes 20 000 kWh of electrical energy. If the iron and steel is recycled some of this energy can be saved, since steel made from scrap requires about 30 per cent less energy than steel made by refining new iron. Apart from the wasted resources involved, the non-recycling of scrapped cars and their components represents a source of environmental pollution. Also an estimated 30 000 tonnes of used sump oil are dumped annually by do-it-yourself motorists. This is very toxic to soil organisms and is only slowly broken down in the soil. If poured down drains it can be a serious contaminant of sewage treatment plants where it interferes with the micro-organisms involved in the purification of the effluent.

It is difficult to estimate the amount of irreplaceable metals that are wasted after only one use, but it is certainly more than 50 per cent. Even given the necessary encouragement, there are difficulties in recovery. For example, very few of the 7000 million cans thrown away each year in the UK are recovered, due to the difficulties of separating the mixture of metals used in their manufacture. Also, although the record for the recovery of ferrous scrap is among the best, with at least 90 per cent of the potentially recoverable scrap iron and steel being reused, there is still only an overall recovery rate of about 60 per cent.

Recycling of waste

The idea of waste is a human concept, a product of short-term economic pressures and of a disregard for the finite nature of the Earth's non-renewable resources. Exponential[357] growth of population and resource consumption in the face of the finite nature of the world's resources is leading to the foreseeable end of accessible supplies of many of the materials on which present day technology depends. Most of these materials are non-renewable and therefore dependent upon recovery and recycling if their use is to be continued.

The recycling of waste material also alleviates some of the problems of pollution and dereliction of the environment. Where the recycling of a material may not appear to be economic at a particular time, it may become so when the cost of disposal and resultant pollution borne by the general public are taken into account.

For example, the recovery and re-use by industry of its 'waste' products can be encouraged by the levying of a charge by local councils for the reception of trade and industrial wastes into the sewers, and by limiting the discharge of certain substances. At some time in the future, as the pressure on the Earth's dwindling resources continues to increase, recycling will become a central organizing principle for the entire economy, just as it is in the economy of the natural cycles of materials.

Recycling depends for its success on the ease of reclamation and reuse of the particular materials. Successful recycling involves not only separate waste collection of the various materials, but also the manufacture of goods in such a way as to avoid the irreversible combination of materials that are subsequently impossible to reclaim separately.

Even today it is technically possible to recover at least two-thirds of the wasted resources of the world and, with manufacture and the economy geared to recycling, considerable success could be achieved.

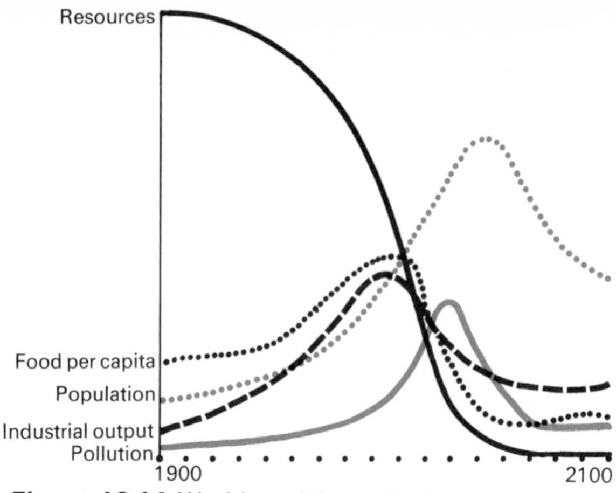

Figure 13.14 *World model standard run*
(After Meadows *et al.*, *The Limits to Growth*, (London: Earth Island, 1972).) This model run assumes no major change in the physical, economic, or social relationships that govern the present world system. The population eventually drops, due mainly to food shortage and resource depletion.

Experimental and applied aspects

When dealing with such large-scale systems, with such complex interactions as are involved in a consideration of human populations and resources, it is necessary to attempt to establish 'models', or representations by analogy, with which the behaviour of such systems can be simulated. Such 'models' can be of varying complexity, ranging from the relatively simple modelling procedure involved in plotting graphs to the construction of complex computer-based models.

Graphs are widely used as models to estimate the upper limit of the availability of resources. Most such estimates are based on the extrapolation of past and present trends into the future, and are not strictly predictions, but **projections** into the future.

The future projection of trends on the basis of past trends involves many difficulties. One such involves finding the best straight line to join various points on graphs, as only slight deviations can result in very large differences in projection, especially when using log scales.[351] The problem of limited and inaccurate data is another major difficulty, and the continuous monitoring of developments to detect any deviations is essential to improve the accuracy of such projections. Also, the concept of resources arises from an interaction between man and his environment; and, by having certain requirements for certain materials, man confers upon them the role of resources; therefore estimates of the future availability of resources require some estimate of future requirements and technology.

The most ambitious computer-based model that has been attempted is that of the 'Club of Rome'/Massachusetts Institute of Technology (MIT) Project on the Predicament of Mankind, a computer-based global model, the results of which were first presented in 1970, and published as the now famous *Limits to Growth*. Such models obviously have serious limitations, but they do provide a starting point for the con-

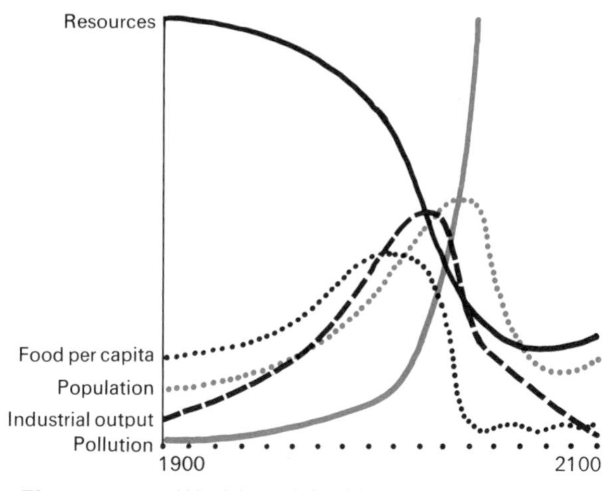

Figure 13.15 *World model with natural resource reserves doubled*
(After Meadows *et al.*, ibid.) The increased availability of resources allows greater industrialization, but pollution causes a rise in the death rate and a decrease in food production. Resources are still depleted.

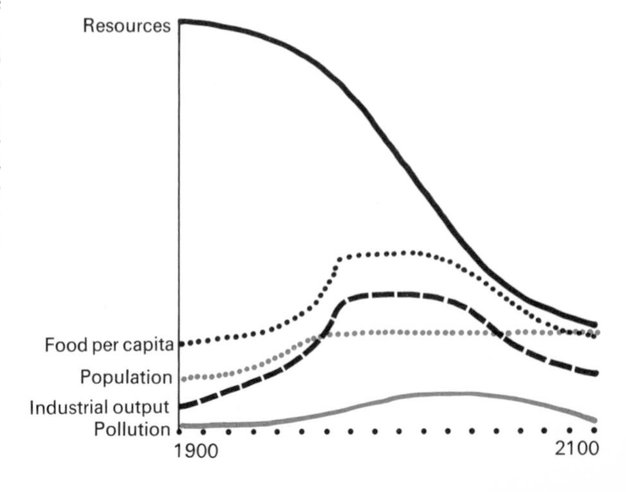

Figure 13.16 *World model with stabilized population and capital*
(After Meadows *et al.*, ibid.) A temporary stable state is attained; however, resources are still eventually depleted.

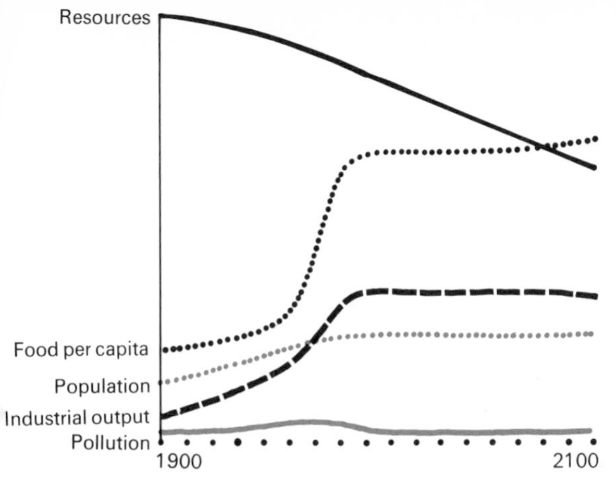

Figure 13.17 *Stabilized world model*
(After Meadows *et al.*, ibid.) Resource recycling, pollution control, methods of restoring eroded and infertile soil, decreased emphasis on industrial production, and other measures are added to the stabilization of population and capital to produce an equilibrium state far into the future.

sideration of these large-scale problems, and serve to stimulate research and debate. However, no matter how sophisticated these models become they cannot encompass the quality of the human resource itself, and will thus always lack a critical component.

Guided example

1 What ecological technique[123] could be employed to estimate the size of fish stocks and the proportion of those stocks that were being exploited?
One method would be to tag and release a sample of those fish caught and subsequently record the numbers recaptured.

Tagging, release, and recapture experiments in the North Sea between 1929 and 1932 showed that about 75 per cent of the released fish were recaptured; this indicated that most adult fish were in fact being caught and that only about 25 per cent had any chance of escape from the nets.

2 What would be the first obvious signs of overfishing of a particular fish stock?
The first obvious signs of overfishing would be a decrease in the size of catches, and in the size of the largest fish caught, leading to smaller and smaller catches from greater and greater efforts, following the Law of Diminishing Returns. Eventually overfishing leads to the collapse of fish stocks, and of course to the collapse of the associated industries.

The story of the Pacific sardine (*Sardinops caerula*) provides a classic example of the results of the over-exploitation of fish stocks. The first cannery opened in California in 1899, and there was subsequently a great expansion until in 1936 about 800 000 tons of fish were unloaded. However, symptoms of overfishing were apparent in 1941, with fewer and fewer adult fish being caught, and there was a catastrophic collapse of the industry as catches fell to less than 50 000 tonnes in 1952.

Derelict buildings abandoned as a result of the collapse of the Californian fisheries—Cannery Row, California

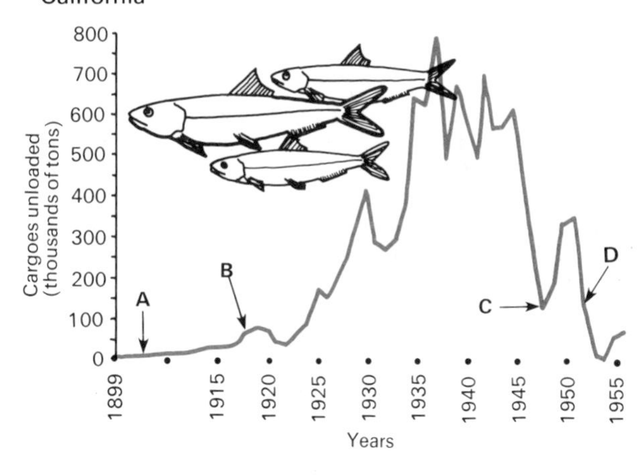

Figure 13.18 *Fluctuations in the catch of pacific sardines (Sardinops caerula) unloaded in American Pacific coast ports*
(From California Department of Fish and Game Report, 1957.)
(A) First cannery opened in California, 1899.
(B) Start of fisheries in British Columbia.
(C) Abandoning of fisheries in British Columbia.
(D) Abandoning of fisheries in San Francisco.

3 What effect does the removal of large amounts of fish have on the balance of the marine ecosystem?
The removal of large amounts of fish affects the balance of the marine ecosystem by affecting the state of the plankton either directly or indirectly, depending upon the

trophic level of the fish in the food chain; and by reducing the number of predators that live on the fish.

4 With strict control of breeding grounds, fishing seasons, and mesh size, fishing can actually result in the increase and improvement of fish stocks. How could this come about?
If adults dominate in a population they compete with and take food from the young, and the food that they consume does not increase the biomass of fish; for example, of the food absorbed by a 20-year-old bass, only 1 per cent is used to increase its weight. If the adults are removed by fishing, there is an increase in the survival of the young and a greater gain in biomass for the same amount of food consumed by the population; for example, of the food absorbed by a 4-year-old bass, as much as 25 per cent can be used to increase its weight.

In this way, as long as fishing does not exceed a certain point, production will be higher than in unfished waters. It is this maintenance of a **maximum sustainable yield** *which is the aim of fishing controls. However, many important fish stocks have already been depleted past the limit of their sustainable yield, and even past their point of recovery; this leads to the exploitation of other stocks which are necessarily becoming progressively fewer.*

Questions

1 Write an essay on **one** of the following:
'The Green Revolution';
Food from the sea;
Recycling. (O)

2 What are the causes of the human population explosion which is at present occurring in various parts of the world? Give a detailed description of the more important biological effects of this explosion and suggest any methods which might be taken to mitigate them. (O)

3 What is sewage? Discuss **two** different methods by which sewage is purified and rendered harmless. What recycling processes are involved in these methods? (O)

4 What do you understand by a safe water supply and what tests are used to demonstrate that water is suitable for drinking? Describe the chief processes involved in the conversion of polluted river water, coming from an industrial town, into water fit for domestic use. (O)

5 (a) Explain why a knowledge of the 'Biochemical Oxygen Demand' of a sample is important in assessing the effectiveness of purification treatment of sewage.
(b) By means of a diagram and notes, outline the principal steps in one method by which grossly polluted water may be rendered relatively pure in a typical sewage works.
(c) Using the headings 'Aerobic organisms' and 'Anaerobic organisms', explain the role of micro-organisms in the processes referred to in (b) and in the subsequent treatment of the solid wastes. Give named examples. (JMB)

Further reading

Borgstrom G., *The Hungry Planet* (London: Collier-Macmillan, 1972).

Committee on Resources and Man, National Academy of Science, *Resources and Man* (San Francisco: W. H. Freeman, 1969).

Dorst J., *Before Nature Dies* (London: Collins, 1970).

Ehrlich P. R. and Ehrlich A. H., *Population, Resources, Environment* (San Francisco: W. H. Freeman, 1970).

Friends of the Earth, *Whale Manual '78* (London: FOE, 1978).

Higgins R., *The Seventh Enemy* (London: Hodder and Stoughton, 1979).

Lauwerys J. A., *Man's Impact on Nature* (London: Aldus, 1969).

Massachusetts Institute of Technology, *Man's Impact on the Global Environment* (Harvard, MIT Press, 1971).

Meadows D. *et al.*, *The Limits to Growth: A report for the Club of Rome's project on the predicament of mankind* (London: Earth Island, 1972).

Overman M., *Water*, Aldus Science and Technology Series (London: Aldus, 1968).

Townsend W. N., *An Introduction to the Scientific Study of the Soil* (London: Arnold, 1977).

14 Bacteria, viruses, and disease

Bacteria

Structure

'Bacteria' are typically unicellular and may form colonies. They can be grouped and named according to their shape; there thus are rod-shaped **bacilli**, spiral-shaped **spirilla**, comma-shaped **vibrios**, and spherical **cocci**. The cocci can be further subgrouped into the streptococci, which associate in chains, staphylococci, which associate in branches, and others. The size of individual bacterial cells ranges from spheres of 0.1 μm in diameter up to rods about 20 μm long.

Each bacteria has a polysaccharide cell wall (rarely of pure cellulose) surrounded by a slime layer which may be sufficiently thick to form a mucilagenous capsule. Such capsules can sometimes run together to form a jelly-like mass of bacteria known as **zoogloea**. Most bacteria are colourless, although pigmentation is not uncommon. Some of the pigmented ones contain bacteriochlorophyll which has similarities with the chlorophyll of higher plants, but which is distributed evenly throughout the cell and not located in chloroplastids. The colourless types are very susceptible to damage by the ultra-violet light in direct sunlight.

There is no true nucleus, only an ill-defined nuclear area containing genetic material; for this reason they are referred to as prokaryotic (or akaryotic). Partition of the genetic material (DNA) during cell division is accomplished without the formation of a nuclear spindle. There are no mitochondria, even though most carry out aerobic respiration. Food is stored mainly in the form of glycogen granules, which is more typical of animal cells than plant cells.

Some bacteria possess flagella at some stage in their life cycle, but bacterial flagella have a completely different structure from those occurring in eukaryotes (cells with a true nucleus).

Respiration

Bacteria can be either aerobic or anaerobic. Aerobic bacteria require a supply of oxygen to enable them to oxidize their respiratory substrates, but anaerobic bacteria can carry out their respiration in the absence of oxygen. Anaerobic respiration leads to the production and accumulation of fermentation products such as ethanol (alcohol), butyric acid (spoils butter), lactic acid (sours milk), and acetic acid (in vinegar). Some bacteria are obligate anaerobes, that is they can only survive in anaerobic conditions, and others are facultative, being able to respire both aerobically and anaerobically.

Autotrophic bacteria

Autotrophic bacteria are capable of synthesizing complex organic compounds from simple inorganic substances, either utilizing light energy in photosynthesis or chemical energy in chemosynthesis.

PHOTOSYNTHETIC BACTERIA

Photosynthetic bacteria contain photosynthetic pigments and carry out photosynthesis. However, this differs from the photosynthesis of higher plants in that bacteria do not use water as a source of hydrogen for the reduction processes involved in the synthesis of fixed carbon dioxide into complex organic substances.

Photosynthesis in a green plant

$$2H_2O + CO_2 \longrightarrow (CH_2O) + O_2 + H_2O$$

Photosynthesis in green bacteria

$$2H_2S + CO_2 \longrightarrow (CH_2O) + 2S + H_2O$$

Some purple and brown bacteria use organic acids as their hydrogen source and, in as much as are utilizing an external source of organic compounds, could be confused with heterotrophic bacteria. However, unlike the heterotrophic bacteria, they require light to be able to utilize the organic compounds.

CHEMOSYNTHETIC BACTERIA

Chemosynthetic bacteria synthesize organic compounds from simple inorganic compounds using energy from inorganic chemical reactions. Some of these bacteria have key roles in the nitrogen cycle, for example *Nitrosomonas* and *Nitrobacter*.

Nitrosomonas

$$NH_3 + 3O \longrightarrow HNO_2 + H_2O + \text{Energy (79 kcals)}$$

Nitrobacter

$$HNO_2 + O \longrightarrow HNO_3 + \text{Energy (22 kcals)}$$

This differs from respiration because in respiration the energy is released as a result of the oxidation of organic substrates. Chemosynthetic bacteria utilize energy from inorganic chemical reactions to synthesize organic compounds, some of which are subsequently oxidized in respiration to yield energy for metabolism.

Heterotrophic bacteria

Heterotrophic bacteria cannot synthesize organic compounds from simple inorganic substances and they therefore require a supply of organic material as food. Some of this organic matter is oxidized during respiration to release energy for metabolism, and some is used as material for synthetic reactions.

Most bacteria are heterotrophic, utilizing either a non-living source of organic matter (saprophytic) or a living source of organic matter (parasitic, and therefore disease-producing or pathogenic). Some heterotrophic bacteria merely require any organic compound which is capable of being metabolized, but some have very exacting requirements for one or more specific organic compounds; for example *Salmonella typhosa* requires the amino acid tryptophan before growth can occur.

Reproduction

Reproduction is mainly by asexual binary fission, the products of which often accumulate to form characteristic colonies by which the bacteria may be identified. However, sexual reproduction, by fusion between members of different strains, and the transfer of genetic material does occur.

Certain bacteria (of the bacillus type only) are capable of forming highly-resistant spores which can withstand adverse conditions. Spores are formed as conditions gradually become unsuitable for vegetative growth.

Harmful bacteria

The natural activities of some bacteria have an adverse effect on many of man's activities. Denitrifying bacteria result in the loss of valuable organic nitrogen from the soil to the air. Saprophytic bacteria cause spoilage of food, spoilage of many stored products, and the destruction of many materials used by man, for example textiles and timber especially. Parasitic bacteria cause diseases of man, animals, and, to a lesser extent, plants.

Contrary to popular belief, only a small proportion of bacterial species are disease-producing or pathogenic. Pathogenic bacteria may cause the disease by their presence alone, but in most cases the disease is caused by the production of poisons or toxins by the bacteria. Such toxins can either be exotoxins, secreted by the living bacteria; or endotoxins, liberated on the death and lysis of the bacteria. Most toxins are protein in nature, and can be either tissue-destroying enzymes or complex metabolic poisons.

Examples of diseases caused by bacteria in man include tuberculosis, tetanus, and most forms of food poisoning; in plants they include most forms of leaf wilt and root rot disease.

Viruses

The existence of viruses was first inferred from work on pathogenic agents which could pass through bacterial-proof filters, but it was not until the development of the electron microscope that viruses were actually seen; because they are too small to be seen under the light microscope. They range in size from 6 μm to 400 μm, and may be smaller than some protein molecules.

Although there are a variety of structures, all viruses basically consist of a protein coat surrounding a strand of DNA or RNA. Unlike all other organisms, viruses never contain both nucleic acids.

Viruses do not possess any of the normal cell organelles or any metabolic enzymes, and their status is questionable; indeed they can be obtained in the crystalline form and still retain their powers of infection.

They are all obligate parasitic pathogens, unable to multiply outside the host cell. When infecting a cell the protein coat attaches to and penetrates the host cell membrane, and the nucleic acid (DNA in most animal viruses and RNA in most plant viruses) is then injected into the cell, leaving the protein 'ghost' outside. Once inside the cell the viral nucleic acid directs the host cell cytoplasm to construct new viruses which are eventually liberated by the bursting or lysis of the host cell.

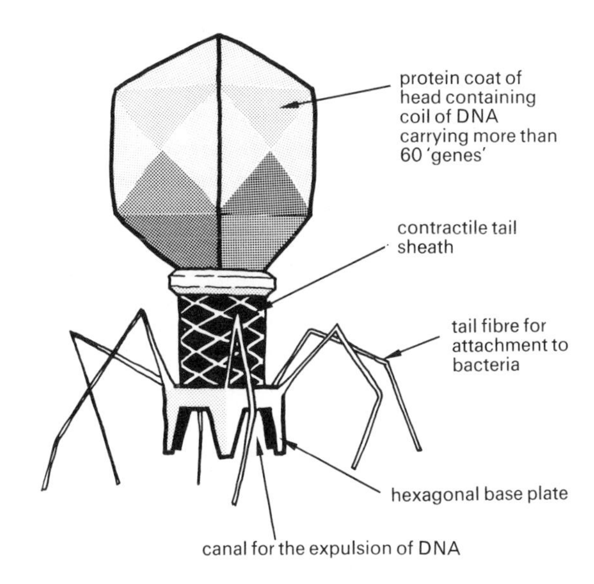

Figure 14.1 *Bacteriophage virus*

A particular group of viruses, known as **bacteriophages**, specifically attack bacteria and are found in great numbers in the gut and in sewage, where bacteria are plentiful.

There are two general categories of phage viruses.

Virulent phages reproduce inside and destroy every bacterium they enter.

Temperate phages only destroy some of those bacteria they enter; those that survive continue to contain the phage genetic material until such time as it becomes active again, reproducing phage particles and destroying the cell by lysis. Bacteria carrying temperate phage are said to be **lysogenic**.

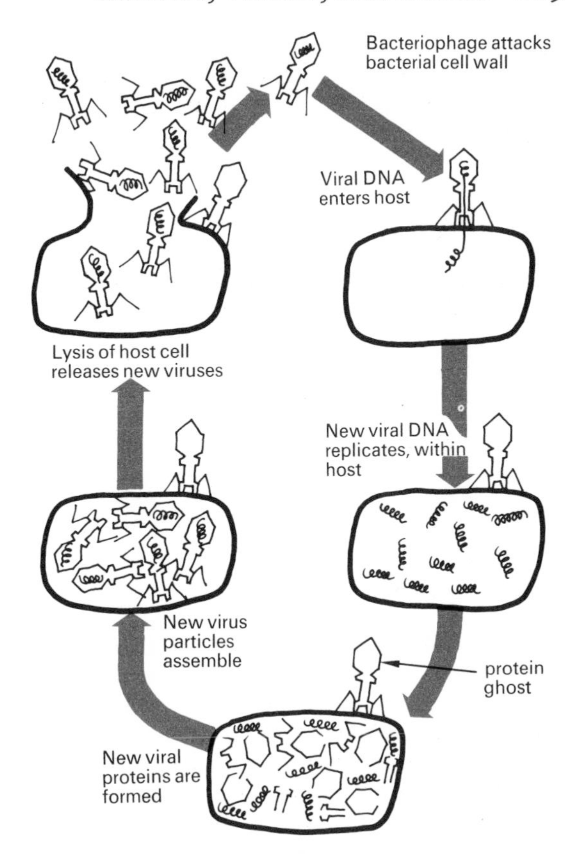

Figure 14.2 *Life cycle of virulent bacteriophage*

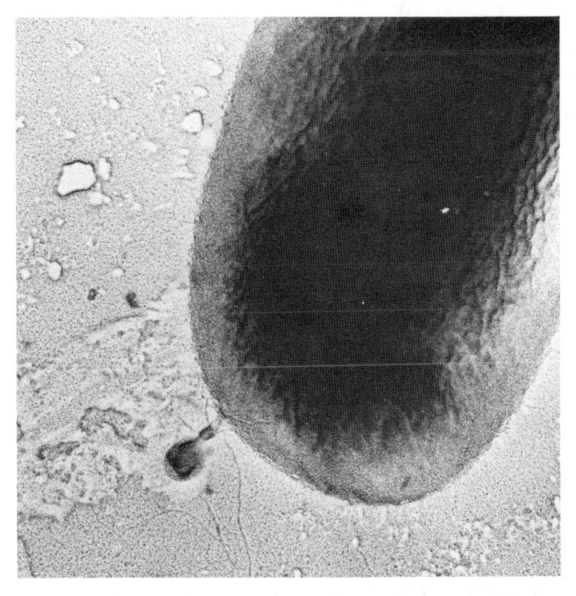

Bacteriophage virus on the surface of a bacterium (× 88 000)

Viruses of animals

Virus diseases of higher animals, including man, are restricted by the body's normal defences against disease, however, a substance known as **interferon** is also produced by infected cells, which, although it is effective against all types of virus, is only effective in the tissues of the same species as produced it. The rapid rate of mutation of certain strains of virus makes it difficult for the body to develop a lasting immunity with antibodies, either as a result of infection or vaccination. Viruses are unaffected by antibiotics. Due to their size and peculiar nature, they are difficult to culture and therefore difficult to study; there are therefore many aspects of virus infection which are not fully understood.

Influenza is caused by a group of myxoviruses which can be divided into three types. Virus A undergoes frequent mutations and causes virulent forms of 'flu' which can start pandemics; virus B may produce severe symptoms but usually only causes local outbreaks; and virus C rarely occurs in adults and only produces mild symptoms in children. The influenza pandemic of 1919 caused more deaths than all those killed in the First World War, and this same strain (which appears to have occurred only in pigs since that time) could emerge again to sweep the world on the same scale. However, the factors that govern these changes in 'behaviour' are not understood.

The enteroviruses include three types of poliomyelitis virus which can be found occurring naturally and harmlessly in the gut. But should these viruses break out from the gut and invade the nervous system they can cause paralysis. It is not known what factors help this migration, but it could involve genetic susceptibilities, and possibly carriage by the migrating larvae of *Toxacara canis* and *Toxacara catii*. There can be many symptomless carriers of this disease, which is spread from the faeces to the mouth. About 80 per cent of the population have a natural immunity to poliomyelitis. This could be the result of genetic factors or of the previous exposure to a mild form of the disease. The oral vaccine for polio contains the live viruses of all three strains.

Some viruses, for example the *Raus carcoma* virus of chickens, are known to cause the growth of tumours, and the polyoma virus is known to cause a wide variety of cancers in mice, hamsters, and rats. In humans viruses are involved in the growth of veruccas and warts, but have yet to be definitely identified as a cause of any specific type of cancer.

Many viruses still remain a mystery, and the term virus disease is often used to cover a multiplicity of conditions which are not fully understood. Even when a specific virus is identified and linked to a condition it is difficult to ascertain its exact behaviour. An interesting example is provided by the *Herpes* virus, which causes 'cold sores'. More than 90 per cent of the population carry the virus which is generally caught in childhood and, once caught, carried for life. It starts as a skin infection, the symptoms of which soon subside, only to reappear in response to any lowering of the resistance due to infection or stress. It is thought that the viruses migrate up the nerve fibres which supply the skin, thus causing the symptoms to subside, and then migrate down the fibres periodically to cause the reappearance of the symptoms. The general effect of the virus in the body is unknown.

Other examples of diseases caused by viruses in man include the common cold, chicken pox, smallpox, mumps, German measles, infective hepatitis, influenza, measles, and rabies. Examples of animal diseases caused by viruses include foot-and-mouth disease and myxomatosis.

Viruses of plants

Apart from a few viruses which attack fungi, plant viruses only infect Angiosperms. They are mainly transmitted by plant-sucking insects, especially aphids, but can be transmitted by virtually any contact between plants such as by soil nematodes and fungi, abrasions, or natural and artificial grafting. Some viruses are transmitted from generation to generation in the plant's seeds.

Virus diseases of crop plants can cause tremendous damage and are therefore of great economic significance, often having a profound effect on the affairs of man. Entire industries associated with certain crops have been wiped out in some regions, enforcing rapid and dramatic changes in the local human population. Examples include tobacco mosaic virus (TMV), potato leafroll virus, and sugar beet yellow virus.

Infectious diseases

Diseases which are caused by micro-organisms and can be passed from one individual to another are known as **infectious** diseases. Infectious diseases that are readily spread by contact are known as **contagious** diseases.

If there are an unusual number of cases of such diseases within a population at a certain time, the situation is described as an **epidemic**; an epidemic of world wide proportions is known as a **pandemic**. Indeed, with an exponentially increasing population, increasing urbanization, and ease of international travel, the potential for a world-wide pandemic has never been greater.

When a disease is usually found within a population at certain levels it is then said to be **endemic**. (The study of the distribution of diseases of mankind is known as epidemiology.)

In any outbreak of infectious disease there is a variety of responses shown by members of a population. Some will succumb to the acute illness while some will only show mild symptoms; during their recovery these latter will act as 'convalescent carriers' and continue to excrete the organism after recovery. Others will be 'symptomless carriers' or 'excreters' who may harbour the pathogen for varying periods of time without showing any symptoms of the disease. Even when identified, such carriers do not usually respond readily to treatment with drugs or antibodies.

Immunology

Immunity is the ability of organisms to resist infectious diseases and, although it can involve many of the body's defences against infection, most attention is concentrated on the antigen–antibody reaction.

An **antigen** is a complex molecule, usually a protein or polysaccharide, that stimulates the production of an opposing **antibody** which leads to the destruction of the antigen.

There are five main types of antibodies: **antitoxins** adsorb toxins and render them harmless; **agglutinins** (the commonest variety) agglutinate or 'clump' particles such as bacteria together; **precipitins** precipitate soluble antigens out of solution; **cytolysins** lead to the breakdown of foreign cells; and **opsonins** render particles available for phagocytosis.

Immunity can be attained in a variety of ways. **Hereditary immunity** occurs passively as a result of the inheritance of some genetically determined resistance.

Naturally acquired immunity can be active or passive. Active naturally acquired immunity can occur as a result of natural exposure to the infectious agent, sometimes in amounts too small to actually cause the disease. If the immunity is passive the body does not produce its own antibodies, for example this may be due to the transfer of antibodies from the mother to the offspring via the placenta or the first secretions of the mammary glands (colostrum).

Artificially acquired immunity may also be active or passive. Active artificially acquired immunity occurs as a result of **vaccination**. The principle behind vaccination is that by introducing a non-disease-producing dose of a disease organism or its toxin into the body, the body's defence mechanisms against the disease are stimulated and a later invasion by the disease can be successfully resisted.

Vaccines can be prepared from living attenuated (diluted) organisms, dead organisms, or their deteriorated toxins which, although no longer toxic, are still antigenic. Living attenuated vaccines include those for rabies, polio, TB, German measles, measles, and yellow fever. Vaccines containing dead organisms include those for influenza, whooping-cough, typhoid, paratyphoid, cholera, plague, and leptospirosis. Vaccines containing toxoids include those for diphtheria and tetanus.

Following the injection of the antigen there is a latent period of one week before the specific antibody appears in the blood. Its concentration rises slowly to a peak in what is known as the primary response. If the antigen is subsequently introduced, either as a result of an occurrence of the disease or of a second 'booster' injection, there is a much more rapid, vigorous, and longer-lasting secondary response which protects against the disease.

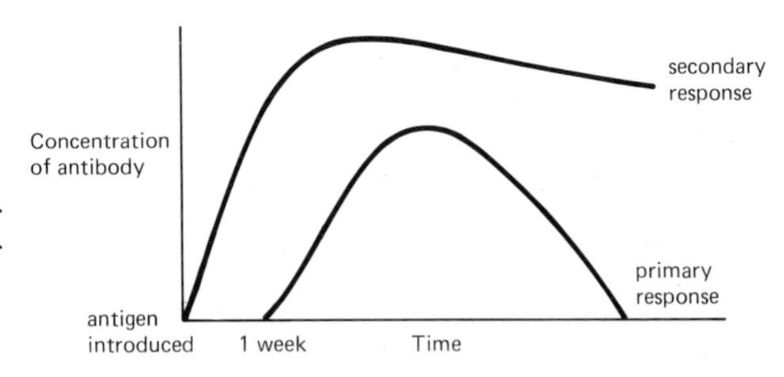

Passive artificially acquired immunity occurs when ready-made antibodies are injected into the body. This can be therapeutic, as in the case of a treatment for an infection that has occurred and which is too dangerous to leave to be dealt with by the body's natural defences, for example with tetanus or rabies. Or it can be carried out as a prophylactive or preventative measure to prevent the contraction of a disease it is difficult to immunize against.

AUTO-IMMUNITY

The mechanism by which the body 'recognizes' its own or 'self' protein and tissues is not clearly understood, although this ability is a prerequisite for the identification of foreign or 'non-self' protein and tissues, and the subsequent production of antibodies against them.

Under certain circumstances the body does 'turn on itself' and begin to produce antibodies against its own proteins and tissues in a process known as auto-immunity. Self-destructive diseases such as rheumatoid arthritis and Hashimoto's disease of the thyroid are thought to arise in this way.

MECHANISMS OF IMMUNITY

There are two main theories to explain the way in which cells involved in the immunological response, or immunocytes, can recognize and respond to 'non-self' antigens by producing antibodies.

The **Classical Instructive Theory** suggests that all immunocytes are capable of responding to all foreign antigens by using the invading antigen as a template or pattern for the construction of the requisite antibody.

The **Clonal Selection Theory** suggests that only one type of immunocyte is capable of responding to one type of antigen to produce the necessary antibody. Thus this theory postulates that an individual possesses the complete range of antibody-producing cells against every possible type of antigen, even though these may never be encountered. This would be a very large number indeed.

ROLE OF THE THYMUS

The thymus, which is found close to the ventral side of the heart, continues to grow and enlarge until the onset of sexual maturity, after which time it gradually atrophies with age. During its growth phase it receives precursor cells from the bone marrow which divide and differentiate into the so-called **T-lymphocytes**. These in turn leave the thymus to 'seed' the lymph nodes and spleen, where they multiply further and develop their full immunological powers. The T-lymphocytes have antibodies attached to their surface and circulate in the body fluids where they destroy any 'non-self' cells such as micro-organisms, grafts, or transplants of foreign tissues. This mechanism of defence is referred to as **cellular immunity**.

The bone marrow also produces the so-called **B-lymphocytes** which pass in the blood stream

Humoral immunity

Soluble antigen
Antibody circulates to site of antigen in blood
Drained off to lymph nodes
Plasma cells produce and release antibody
Activates B-lymphocytes
Some constitute memory cells
Others become plasma cells

Cellular immunity

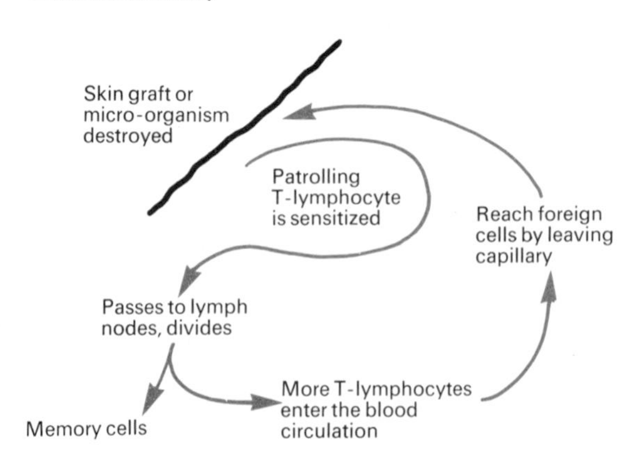

Figure 14.3 *Types of immunity*
(After *Open University S321 Unit 8*, p. 41.)

directly to the lymph nodes, spleen, and Peyer's patches of the gut. Once in the lymphoid tissue the B-lymphocytes differentiate into plasma cells which, when stimulated by the necessary antigen, produce and release antibodies into the blood stream. This mechanism of defence is referred to as **humoral immunity**.

In both cellular and humoral immunity, cells stimulated or sensitized by the presence of an antigen divide to form further lymphocytes or plasma cells of the required type, and also form 'memory' cells which enable the system to provide a rapid secondary response, should the same infection occur again.

Channels branch throughout the lymph nodes and the spleen which are lined with large phagocytic cells or macrophages. These filter out and engulf any micro-organisms or cell debris that are circulating in the body fluids. These 'fixed' macrophages are also thought to process antigens in some way into a form which stimulates the activity of the lymphocytes.

INFLAMMATION REACTION

Damage to tissues and the presence of foreign antigens causes certain white blood cells and tissues to release a local hormone called **histamine**. This causes inflammation by vaso-dilation; and an increased permeability of the capillary walls leads to a greater tissue fluid exudation. The increased flow of fluids dilutes any toxins and brings more blood-clotting materials and white blood cells to the site of the damage. White blood cells reach the site both passively and by active chemotactic movement or diapedesis.

PHAGOCYTOSIS

The stimulus to extend pseudopodia depends upon the surface properties of the particle to be engulfed, and upon the presence of a certain type of antibody known as opsonin. The engulfed particles are not always destroyed, and some pathogens such as the TB bacterium may even multiply within them and be spread further around the body.

HYPERSENSITIVITY

Normally the antigen–antibody reaction occurs in the blood, tissue fluids, and lymph. If, however, the antigen–antibody reaction occurs within the cells of a tissue, more histamine is released and the tissues are inflamed in an **allergic reaction**.

Histamine accounts for many, if not most, of the symptoms of bronchial asthma, hay fever, and other allergies. It causes the dilation of capillaries, and makes them more permeable to tissue fluids; it causes spasms of smooth muscles, skin swellings, and stimulates glands that secrete watery nasal fluids, mucus, tears, and saliva. Anti-histamine drugs compete with histamine for the receptor sites on cells, and therefore block the histamine action. There are, however, side effects.

Summary of the body's defences against infection

1 PHYSICAL BARRIERS

Skin—impermeable surface layer of keratinized dead skin scales.
Enamel—forms a hard covering on the teeth.
Ciliated epithelial cells—trap and remove unwanted particles.
Mucus—secreted by many glandular epithelia; protects surfaces and traps particles.
Blood clotting—damaged cells release an enzyme, thrombokinase or thromboplastin, which, in the presence of calcium ions and vitamin K, converts plasma prothrombin into the enzyme thrombin which converts the soluble plasma protein fibrinogen into insoluble fibrous fibrin. These fibres entangle red blood cells to form the clot.

2 ANTIMICROBIAL SUBSTANCES

Tears, saliva, sweat, and **sebum**—contain bacteriocidal substances.
Phagocytic white blood cells—secrete phagocytin and leukin to destroy bacteria.
Stomach acid—bacteriocidal

3 RETICULO-ENDOTHELIAL OR MACROPHAGE SYSTEM

Fixed phagocytic macrophage cells are found lining the lymph nodes, the spleen, the connective tissue (especially subcutaneous), the liver (Kupffer cells), and the bone marrow (littoral cells). Although widely scattered, they respond to infection as a system.

4 BLOOD AND LYMPH

Phagocytic white blood cells (W.B.C.) (mobile macrophage cells) engulf micro-organisms and cell debris.
W.B.C. and damaged cells secrete **histamine** which triggers off inflammation which is the first stage of the body's defences.
T-lymphocytes circulate in plasma to attack foreign cells.
B-lymphocytes produce antibodies which circulate in the plasma.

5 INTERFERON

Interferon is produced by cells infected by viruses, and can diffuse to other cells and confer immunity to further viral attacks. It is little understood.

Diseases of urbanization

Not all diseases are caused by infectious agents such as bacteria, viruses, protozoa, and fungi. Indeed, even diseases which do have a causal organism may also have several other factors which contribute to their development. For example, rheumatic fever is caused initially by a bacterium infection. But whether or not this infection leads to rheumatic fever also depends on bad living conditions and a genetic predisposition.

With the success of vaccination programmes and the continual development of antibodies, infectious diseases are not the immediate threat that they once were in areas of high population density. The main threat today in developed countries comes from what may be termed the diseases of urbanization; those diseases that are related to the way of life of modern man.

Such factors as overcrowding, disrupted personal relations, lack of social structure and natural rhythm of daily activities, and noise, can lead to **stress**. Despite a high level of physical and material security, stress can result in severe neuroses, nervous breakdowns, lethargy, tension, enlargement of the adrenal glands, and a shrinkage of the lymphatic tissue which increases the risk of contracting infectious diseases. In addition, many serious psychosomatic diseases such as asthma, duodenal ulcers, ulcerative colitis, and neuro-dermatitis may occur.

Stress, lack of exercise, an unbalanced diet, drug dependence (especially smoking and excessive drinking), and exposure to pollution all contribute to a greater or lesser extent to one or more of the three main diseases of industrial societies, namely those of the heart and circulatory system, cancers, and mental disorders.

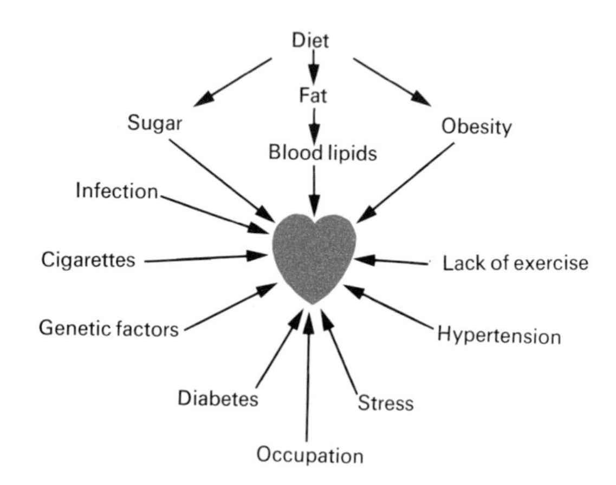

Figure 14.4 *Contributory factors in heart disease*

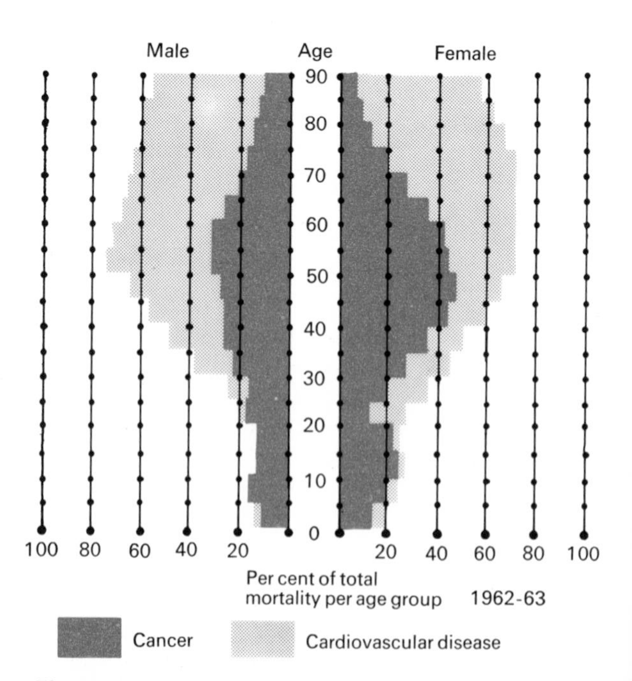

Figure 14.5 *Cardiovascular disease and cancer mortalities*

Diseases of the heart and circulatory system account for 35 per cent of annual deaths in the UK, and cancers for about 20 per cent. Close to 50 per cent of hospital beds in the UK are occupied by the mentally ill, and a further 25 per cent are occupied by those whose condition may have resulted from mental stress, such as sufferers from obesity, accidents, alcoholism, and smoking-related diseases.

All societies in the world have some chemical substance, either natural or synthetic, which is used by a section of the community to alter their psychological state. However the use or abuse of drugs, both legal and illegal, is a particular characteristic of most urban societies, possibly also as a result of mental stress. The most widely used include hallucinogens (e.g. LSD); stimulants (e.g. amphetamines and cocaine); morphine-type drugs (e.g. morphine, opium, and heroin); barbiturate-type drugs (e.g. pentobarbital); marijuana; and nicotine obtained from tobacco smoke. In all cases the ultimate effect of drug use depends on a complex interaction between the chemical nature of the drug and the metabolism and mental state of the user. The most important drugs, in terms of being the most widely used and the most widely abused, are alcohol and nicotine.

Alcohol abuse causes untold damage, even though only about 1 in 500 of those who drink become acute alcoholics, and tobacco smoking may indirectly account for almost half the number of annual deaths in the UK.

In addition to these conditions there are occupational diseases, including all forms of damage to the human body and mind that arise from a person's occupation. Needless to say, the incidence of such diseases is at a peak in industrialized, and therefore urban, areas, where the range and complexity of jobs and materials handled increase the risk of damage to health. Particularly bad cases often occur as a result of exposure to carcinogenic chemicals, dust particles such as coal dust and asbestos (particularly 'blue' asbestos), ionizing radiations such as X-rays, and excessive noise.

Tobacco smoking

Epidemiological studies, both retrospective (those examining the smoking habits of patients already suffering certain diseases) and prospective studies (those examining the occurrence of certain diseases amongst groups of smokers and non-smokers over a period of time); leave little doubt that smoking is the primary cause of many debilitating and fatal diseases. Indeed smoking is known to be the largest single cause of preventable death and ill-health in the UK, even though less than 50 per cent of the adult population are smokers. There is no lower threshold below which smoking is known to be safe, and even passive smoking of other people's smoke carries with it an increased risk of smoking-related disease.

CONSTITUENTS OF TOBACCO SMOKE

Tobacco smoke is a mixture of carbon particles, tarry droplets, and hot gases. At least 1000 constituents have been identified, many of which are known to be harmful. Its composition varies depending on the type of tobacco, the method of curing, and how it is smoked. Cigarette smoke is acidic and can be inhaled straight into the lungs. Cigar and pipe smoke is more alkaline and more irritant: its nicotine can be absorbed from the mouth and it is not usually inhaled.

The constituents of tobacco smoke are carcinogenic (cancer-producing) substances, irritant substances, nicotine, and carbon monoxide and other gases.

Carcinogenic substances are present in the tars (responsible for the brown stains on smokers' fingers). These contain at least seventeen compounds which can cause cancer when applied to the skin of mice. Many laboratory experiments have also shown that animals can develop cancer of the larynx and lung from inhaling cigarette smoke. Tar yields of British cigarettes vary from under 4 mg to 34 mg per cigarette. The main carcinogenic substances are the polycyclic hydrocarbons of which benz(a)-pyrene is an important member. Polonium (a radio-active element which could cause cancer) is also present in tobacco smoke.

Irritant substances in the tobacco smoke cause the familiar smoker's cough. They inhibit the action of the minute protective beating hairs (known as cilia) in the bronchial tubes. These normally help to remove dust and particles from the lungs. More than ten substances—notably acrolein—that paralyse cilia have been identified in tobacco smoke. The irritant activity caused by acrolein probably leads to the destruction of the fine structure of the alveoli in the lung, a condition called emphysema. Tobacco smoke also stimulates the excessive mucus secretion which forms the phlegm of bronchitis.

Nicotine is a colourless poison, about 90 per cent of which is absorbed by smokers who inhale. Cigarette brands sold in Britain yield from less than 0.3 mg to over 2.3 mg of nicotine per cigarette. Nicotine is thought to be the drug which many smokers are dependent on or addicted to, indeed it is one of the most dependence-producing drugs known. It stimulates the release of adrenaline and noradrenaline; two hormones which increase blood pressure, constrict the blood vessels, and increase heart rate and output of blood from the heart. This causes an extra strain on the heart and can be serious for the person who has already had a coronary attack. Nicotine also makes the heart more irritable and can precipitate dangerous irregularities. Nicotine increases the concentration of fatty acids in the blood and the 'stickiness' of blood platelets, the small cell fragments involved in normal blood clotting and thrombosis. This may be one reason why cigarette smokers are more prone than non-smokers to atherosclerosis (fatty deposits in the arteries). Nicotine also reduces ADH secretion, which results in excess urine production.

Carbon monoxide, a colourless, odourless, poisonous gas is also found in tobacco smoke. A smoker who inhales can build up a concentration of 400 ppm carbon monoxide (CO) in the air reaching his lungs. This is eight times the level of carbon monoxide permitted in industry. CO yields of British cigarettes have been found to vary from about 10 ml to 18 ml per cigarette. Cigarettes with ventilated filters have been found to produce about 21 per cent of the CO of ordinary filter cigarettes. Carbon monoxide combines with haemoglobin 200 times more readily than does oxygen. It therefore reduces the blood's ability to carry oxygen. CO also damages the walls of arteries and so probably increases the risk of narrowed coronary arteries leading to heart attack.

Cigarette smoke also contains hydrogen cyanide, butane, and other gases which may damage health, but whose role has not yet been proven. Little is known about the effects of the large number of other substances in tobacco smoke, many of which are considered to be hazardous in industry at certain concentrations.

PASSIVE SMOKING

The contamination of community air by smokers is the most common form of local air pollution and leads to what is known as 'passive smoking'. This is the breathing of both the 'side-stream' smoke direct from a burning cigarette, and the exhaled 'main-stream' smoke. Side-stream smoke has a higher concentration of dangerous substances than main-stream smoke; for example it has twice the concentration of tars and nicotine, three times the concentration of phenols and particulates, four times the concentration of benzyl(a)pyrenes and cadmium, five times the concentration of carbon monoxide, and fifty times the concentration of nitrosamines.

Nitrosamines have been demonstrated to be potent carcinogens in animals and are suspected of having a similar effect on man. Under certain conditions after one hour in a smokey room a non-smoker may inhale quantities of nitrosamines equivalent to that inhaled when smoking about fifteen filter cigarettes. Similarly, the smoking of one cigarette can release 100 mg of another potent carcinogen, benzyl(a)pyrene, into the air.

As the evidence indicates that all carcinogens exert their effect in direct proportion to the level of exposure, and that there is therefore no safe threshold beneath which they have no effect, it follows that exposure to the carcinogens of side-stream smoke must increase the risk of lung cancer in those that breathe it. It might further concern non-smokers to know that after breathing a smoky atmosphere they excrete measurable amounts of nicotine in their urine.

It is argued that the levels of such substances to which non-smokers are exposed are too low to produce measurable increases in smoking-related diseases. However, it has been demonstrated that families in which there is at least one heavy smoker suffer 50 per cent more respiratory illnesses than those coming from a non-smoking family. In addition, a large number of people have allergic reactions to tobacco smoke, which could be dangerous to those at risk from chronic bronchitis and coronary heart disease.

Another example of the passive absorption of harmful substances from smoking occurs across the placenta. One estimate gives as many as 1500 foetal infant deaths each year in the UK associated with women smoking during pregnancy.

LUNG CANCER

Lung cancer has increased from being a rare disease to being a major cause of death over the last fifty years. This is a direct result of the increase in the smoking habit. Estimates vary, but about 30 000–50 000 people die from lung cancer each year in the UK, that is between 5–8 times the number killed on the roads. About 90 per cent of those who die from lung cancer are regular smokers and between 33–50 per cent of smokers die from this disease.

The earlier a person starts smoking, the greater is the risk; a fact which makes the problem of teenage smoking particularly serious. The majority of those who die from lung cancer are males of 55–64 years, and lung cancer accounts for 1 in 7 of the male deaths in this age range. Although the figures are lower for women, they are increasing as a result of an increase in smoking amongst women.

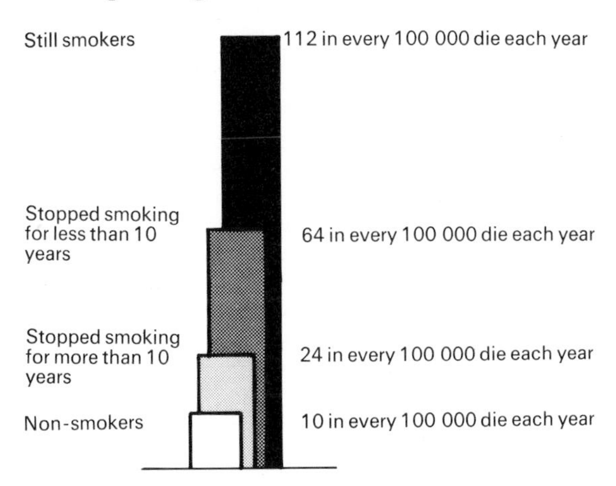

Figure 14.6 *Mortality from lung cancer*
(From *Smoking and Health*, 1962.)

HEART DISEASE

The association of heart disease with smoking is substantially less than the association of smoking with lung cancer and chronic bronchitis. It is frequent amongst non-smokers and is less directly associated with the number of cigarettes smoked. However, smoking roughly doubles the risk of heart disease, and far more smokers die of this cause than of lung cancer.

FURTHER EFFECTS

Smoking causes an increased risk of cancers of the mouth, pharynx, larynx, oesophagus, bladder, and pancreas; and an increased risk of gastric and duodenal ulcers. In addition smoking contributes towards the effects of bronchitis, emphysema, and asthma, which are estimated to cause the premature death of between 30 000–40 000 people each year in the UK.

The blood vitamin C levels of smokers decreases from about 0.9 mg per 100 cm^3 of blood to 0.5 mg per 100 cm^3 and, although the significance of this is not understood, it must relate to some alteration in metabolism.

Table 14.1 *Average number of deaths per year in the UK*

Total deaths per year in UK	500 000
Coronary heart disease (estimated to be doubled by smoking)	150 000
Lung cancer } Smoking-related diseases	30 000
Bronchitis } Smoking-related diseases	30 000
Road accidents	6 000

Food poisoning

Food poisoning is essentially a modern 'disease'. Although its causes are understood and standards of food hygiene are continually being improved, cases of notifiable food poisoning have increased dramatically over the years, especially since the Second World War. With the development of communal feeding and the associated bulk handling of food, retail cooking and packaging, and long storage periods, ideal conditions have been created for the widespread occurrence of food poisoning.

In the past the cause of food poisoning was always attributed to the presence of certain chemicals; for example, alkaloids or ptomaines, formed by the breakdown of putrefying tissues, especially in fish, were always taken as the cause of a certain type of food poisoning called ptomaine poisoning. (The term is still widely used even though such ptomaines are harmless when swallowed.)

Some cases of food poisoning are in fact caused by the presence of chemicals such as the alkaloid solanin in green potatoes, persistent pesticides, and heavy metals such as lead; but the vast majority of all cases of food poisoning are caused by bacteria. The first food-poisoning bacteria were isolated in 1888 and since that time the full significance of their role has been recognized. One of the food-poisoning bacteria, *Clostridium botulinum*, produces one of the most lethal toxins known—as little as 0.0001 g is lethal and just 250 g could kill the entire human race. Bacterial food poisoning causes disturbance of the gastro-intestinal tract, with abdominal pain, diarrhoea, and usually vomiting.

A common misconception is that food poisoning occurs as a result of eating 'bad' food which can be detected by taste and smell. In fact, food-spoilage bacteria rarely cause food poisoning and food-poisoning bacteria rarely cause food spoilage. This means that food that looks, tastes, and smells perfectly normal could still cause food poisoning.

Food poisoning is generally caused by food which has become contaminated by food-poisoning bacteria, which has subsequently been stored under conditions favourable for their multiplication, and then been inadequately cooked. The contaminating bacteria may be present in the raw food or introduced during handling or processing.

One reason for the difficulty in reducing the incidence of food poisoning is that, due to the number of different causal agents, an attack on one front alone can never be successful. For example, a high standard of personal hygiene may reduce food poisoning due to *Staphylococcus* sp., but will have little effect on *Salmonella*. *Salmonella* bacteria can occur in large quantities in contaminated raw food, and therefore food poisoning from this source can only be prevented by a combination of personal hygiene, correct handling, safe storage, and correct cooking.

Experimental and applied aspects

Applied aspects of bacterial activity

Role	*Comments*
(a) Silage	Silage production is a method of animal food preservation. Plant material is cut and packed into a 'silo', in which bacterial fermentation of the carbohydrates occurs. After about a month, the plant material has been changed into a nutritive mass which will keep for years.
(b) Sewage disposal	Aerobic saprophytic bacteria are important in percolating filters and activated sludge, and in breaking down organic wastes. Anaerobic bacteria digest sludge and are also active in cesspools and septic tanks.
(c) Butter	A 'starter' bacterial culture (e.g. *Streptococcus lactis*) is added to pasteurized cream. The bacteria curdle and coagulate the cream, and yield products that impart flavour and aroma. The product is then churned to butter.
(d) Cheese	Milk is coagulated either by rennet or bacteria-produced lactic acid. Bacteria are involved in the ripening of the cheese.
(e) Yoghourts	*Lactobacilli* bacteria and others are used to produce fermented milk beverages.
(f) Vinegar	Bacteria of the genus *Acetobacter* oxidize alcohol to acetic acid. Vinegar contains about 4 per cent acetic acid; traces of other fermentation products give odour and taste.
(g) Cocoa	Bacteria are involved in the fermentation of cocoa beans. This develops the flavour, aroma, and colour prior to roasting.
(h) Sauerkraut	Bacteria of the *Lactobacillus* group ferment shredded cabbage leaves. Sometimes a pure culture is added, usually those already on leaves are allowed to act.
(i) Flax retting	Bacteria break down the pecten binding the flax (hemp) fibres, which, when isolated, are used in weaving.
(j) Leather	Bacteria are used in several stages of the preparation of leather from hides. The removal of the hair can be by carefully-controlled bacterial decomposition, bating (the placing of hides in weak fermenting infusions of dog or bird dung or bran mash) involves a complex of unknown bacterial action, as does the final process of tanning.
(k) Tobacco curing	Tobacco leaves are stacked and allowed to ferment; a process involving a complex of bacterial action.
(l) Lactic acid	This can be made industrially by the action of *Lactobacillus* or *Streptococcus lactis* on a variety of carbohydrates.

(m) Ethyl alcohol	This is prepared mainly by the action of yeasts, but certain anaerobic bacteria can also be used.
(n) Acetone	This can be obtained by the action of anaerobic bacteria on carbohydrates.
(o) Propionic acid	This can be produced by the action of *Propionioacter* on a variety of carbohydrates.
(p) Enzymes	Several bacteria can be used for the commercial production of amylase and protease.
(q) Vitamin assay	The specific nutritional requirements of some bacteria for certain vitamins can be used as a means of determining the presence of these vitamins in food.

Applied aspects of fungal activity

Role	*Comments*
(a) Fermentation	Yeasts (*Saccharomyces* sp.) carrying out anaerobic respiration result in the production of ethyl alcohol and carbon dioxide. The carbon dioxide evolved by yeast is utilized in baking to 'raise' the dough, and accounts for the porous nature of bread. (Between 1920 and 1930 the entire citric acid industry switched from extraction from fruit to fermentation by *Aspergillus niger*.)
(b) Cheese-making	Several fungal moulds are utilized in the maturation and flavouring of cheeses (e.g. *Penicillium camamberti* and *P. roqueforti*).
(c) Antibiotics	*Penicillium* sp. producing penicillin, *Streptomyces griseus* producing streptomycin.
(d) Enzyme production	*Aspergillus niger* is used in the production of amylase, protease, and pectinase.
(e) Hormone production	*Giberella fujikuroi* produces the plant growth hormone giberellin.
(f) Vitamin production	A yeast, *Ashbya gossypii*, is used in the production of riboflavin.
(g) As food	There are many edible fungi, none of which, however, are very nutritious.

Gram staining

In 1884 Gram devised the single most important stain in bacteriology. It divides bacteria into two categories on the basis of whether, after staining with crystal violet and treating with iodine, they can be decolourized with acetone, alcohol or aniline oil. Those that resist decolourizing are termed Gram positive (virtually all the cocci); and those that do not, and which subsequently take up the red counterstain of neutral red or safranin, are termed Gram negative (virtually all the bacilli).

Antibiotics

In 1928 Alexander Fleming discovered the antibacterial effect of a secretion from the mould *Penicillium notatum*, and took the first step in the tale of antibiotics. This strain yielded about 2 μg or 'units'/cm^3 of extract. In the 1940s, Florey and Chain (who shared a Nobel Prize with Fleming) isolated and produced penicillin in quantity. A world-wide search for new strains led to the discovery of *Penicillium chrysogenium*, which gave a yield of up to 200 units/cm^3. X-ray-induced mutations raised the yield to 500 units/cm^3, and u.v. treatment gave the strain Q176 which yields up to 1000 units/cm^3 and is the basis of the present penicillin industry.

One side effect of antibiotics is that, when taken orally, the poorly-absorbed ones such as tetracycline can virtually sterilize the gut of symbiotic bacteria, whilst at the same time allowing the growth of yeasts. This can severely disrupt normal gut function, which depends upon the correct bacterial community in the large intestine. Other side effects include skin rashes, hearing impairment, suppression of bone marrow function, and even sudden death from anaphylactic shock.

Examples of antibiotics are penicillin, produced by *Penicillium notatum*; streptomycin, produced by *Streptomyces grisens*; and aureomycin from *Streptomyces aureofaciens*. As more and more resistant strains of bacteria emerge, there is a continual search for new antibiotic-producing fungi, for example 116 000 soil samples were tested before resulting in the discovery of 'terramycin'. One pharmaceutical firm alone tests about 20 000 fungi each year for new antibiotics, and reports a success rate of about three useful strains every ten years.

The necessity for this relentless search is demonstrated by the fact that by 1946 14 per cent of the infections in a general hospital were by penicillin-resistant strains of bacteria, and that by 1950 the majority of *Staphylococcus* infection in all general hospitals were resistant to penicillin. The use of antibiotics in veterinary practice, agriculture, and food processing should be strictly controlled to prevent this proliferation of resistant strains of pathogenic organisms.

Antibiotics have no effect on viruses.

Guided example

1 Bacteria and fungi are cultured in a variety of ways, but one of the commonest methods is to culture them on agar. Agar is a complex polysaccharide (a polymer of galactose) extracted from seaweeds which are found in Japan, New Zealand, and California. It is chemically inert and is not attacked by most bacteria and fungi. What must therefore be added to the agar to allow the growth of the micro-organism under study?
The necessary nutrients must be added. With most bacteria this is usually some form of meat extract, but in certain cases the micro-organisms will have very exacting nutritional requirements that must be supplied.

2 When agar is heated with excess water to just under 100 °C it forms a liquid sol which, on cooling to about 32 °C, sets to form a stable gel which will not reliquify until reheated to 98 °C. Can you think of any advantage in the fact that the agar sets at one temperature and liquifies at a higher one?
Whilst in the liquid form the agar can be easily poured into Petri dishes or tubes, and, once set and inoculated with the micro-organism under study, it can be incubated above its setting temperature without reliquifying.

3 Once the Petri dish has been poured it is left closed until set (usually about 30 min) and then dried in an incubator for 10–20 min at about 37 °C in the position shown.

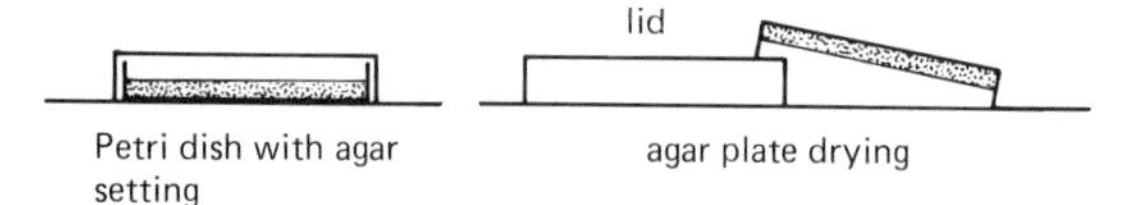

What is the reason for drying the Petri dish 'upside-down', when it would appear to be more efficient to dry them 'right way up' and to allow the surface moisture to evaporate more freely?
There will be many airborne micro-organisms and their spores in the air in the incubator, and to dry the plates the 'right way up' would allow many of these to settle and contaminate the plate prior to its inoculation with the particular organism under study.

4 The sensitivity of bacteria to different antibiotics and drugs can be tested by the use of paper discs impregnated with such anti-bacterial agents. These are placed on an agar plate inoculated with the bacterium under study, and the subsequent pattern of growth of the colonies is observed.

To which agent is the bacterium (a) most resistant, and (b) most sensitive?
(a) Erythromycin; (b) Sulphafurazole.

5 Why are the patches of inhibition of growth circular in each case?
This is due to the radial diffusion of the antibacterial agents from the discs.

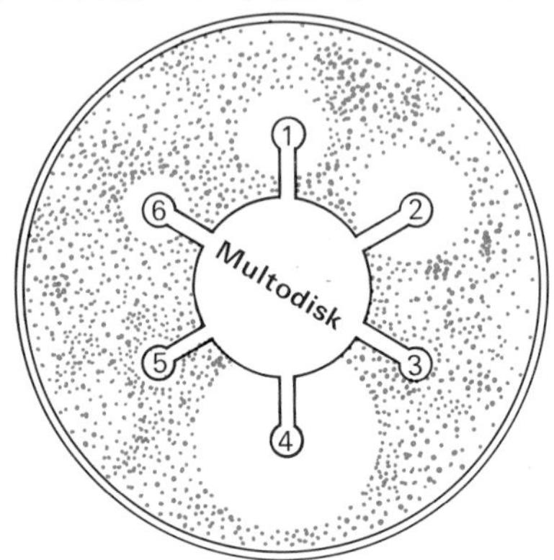

Figure 14.7 *Inhibition of bacterial growth by anti-bacterial agents*
Multodisks are made of sterile filter paper with the tip of each arm impregnated with a different anti-bacterial agent.

6 Although this method indicates the relative sensitivity of the bacterium to different anti-bacterial agents, it is not easy to compare the sensitivities of different bacteria. How might it be possible to get a more accurate comparison of sensitivities to different antibiotics?
This may be achieved by comparing the reaction of different bacteria to those of a known sensitive bacterium such as 'Oxford' staphylococcus on the same agar plate.

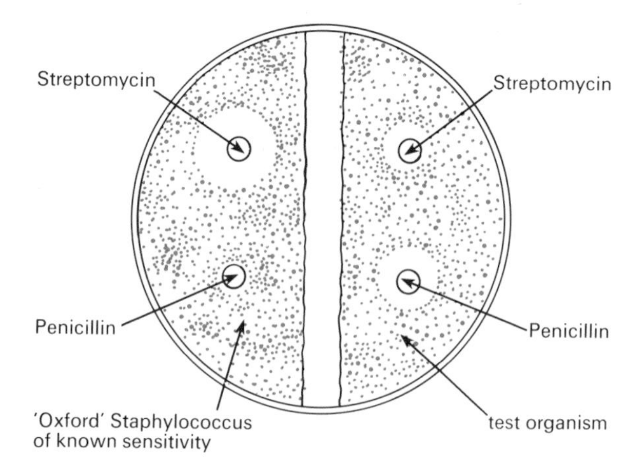

Figure 14.8 *Plate with 'Oxford' staphylococcus*
A zone comparable to, or greater than, that of Oxford staphylococcus is reported as sensitive; those with smaller zone are reported as resistant.

7 To test the sensitivity of bacteria to various disinfectants, a culture of bacteria is established in a nutrient broth and a measured volume of this broth culture is mixed with a measured volume of disinfectant of known strength. The two are mixed thoroughly and then streaked out on to an agar plate after varying periods of time. The plate is incubated and the results observed.

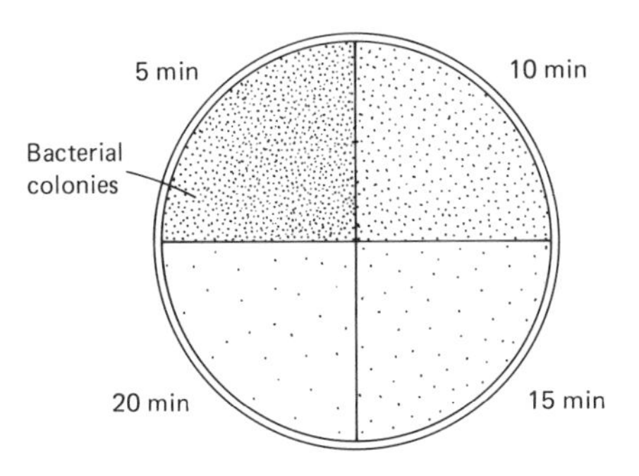

What would be wrong with concluding from this single plate alone that the disinfectant is gradually destroying the bacterium in the broth culture?
There could be another reason for the decline in the growth of the bacterium after 30 min, or it might be that different amounts of bacterial culture are being streaked out each time. To prove that it is the disinfectant that is the effective agent it is necessary to have a control carried out at the same time in which all the conditions are identical except that distilled water is added instead of the disinfectant.

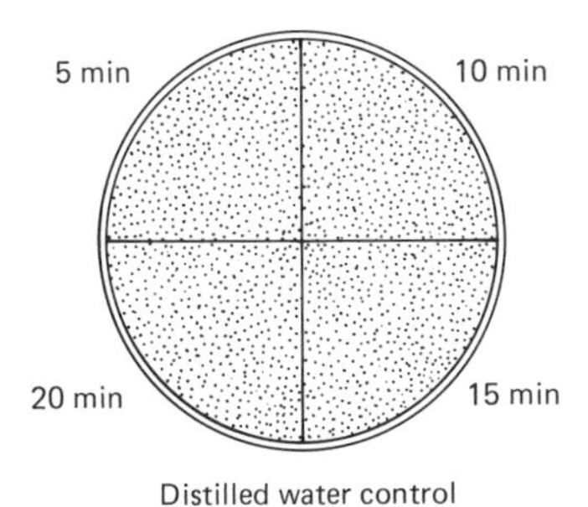

Distilled water control

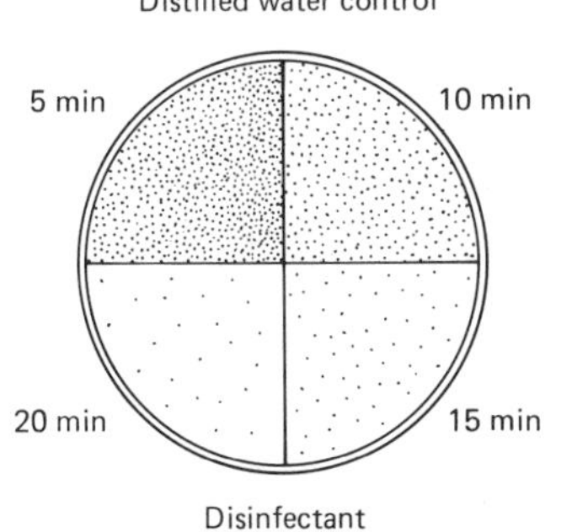

Disinfectant

Questions

1 Discuss the ways in which bacteria, fungi, and viruses may be (a) beneficial, and (b) harmful to man. (L)

2 Describe the main methods for the control of human diseases and comment on the problems which may arise when such control measures are applied. (L)

3 How may organisms that cause disease enter the human body? Discuss the ways in which the spread of diseases may be controlled. (L)

4 What are antibiotics and what are the particular characteristics which make them so important to the medical and veterinary sciences? Discuss the commercial production of any **one named** antibiotic emphasizing biochemical details. What problems have arisen from the widespread use of antibiotics? (O)

5 List the main agents which cause disease in human beings. For each type of agent suggest how the disease spreads and discuss methods by which it may be controlled. (O)

6 What are the main risks to public health in an urban society? How can these risks be minimized? (L)

Further reading

Bergel F. and Davies D. R. A., *All About Drugs* (London: Nelson, 1970).
Boycott J. A., *Natural History of Infectious Diseases*, Institute of Biology Studies in Biology No. 26 (London: Arnold, 1971).
Bryan A. H., Bryan C. A., and Bryan C. G., *Bacteriology* (New York: Barnes and Noble, 1965).
Clegg A. G. and Clegg P. C., *Man Against Disease* (London: Heinemann, 1975).
Postgate J., *Microbes and Man* (Harmondsworth: Penguin Books, 1969).
Royal College of Physicians, *Smoking and Health Now* (London: Pitman Medical, 1971).
Willis J., *Addicts: Drugs and Alcohol Re-examined* (London: Pitman, 1973).

15 Pests and pest control

Introduction

There can be no doubt that the greatest potential for increasing food supplies lies in the ability to control and reduce the losses, due to plant and animal pests, of crops and stored products. These losses are on a collossal scale and, although estimates vary widely, can exceed half the total crop yield world-wide.

In nature a subtle and intricate system of checks and balances prevents the excessive dominance of individual species—so characteristic of those organisms considered to be pests. Complex ecosystems have a greater stability than simpler ones. The typical 'monoculture' of modern agriculture, in which vast areas are covered by a single crop plant, presents ideal conditions for the population 'explosion' of insect pests. This can be further encouraged by the use of 'broad-spectrum' pesticides which also destroy the pests' natural predators and parasites. Needless to say, the greatest danger lies in the transport of a potential pest species to a country where there are no natural enemies and perfect conditions for its rapid spread through massive monocultures.

Insect pests

Insect pests cause untold losses to man's food supply. Their tremendous powers of reproduction, motility, adaptation to every ecological niche, and variability by which strains resistant to pesticides can rapidly arise, means that they can never be defeated but can only be held at bay. For example, the number of strains of pesticide-resistant insects and mites increased from 182 in 1965 to 364 in 1977.

Whilst feeding on plant and animal tissues, insects can act as vectors of many diseases. Perhaps the greatest agricultural losses of all are due to their transmission of plant virus diseases. Direct damage is caused by insects biting, chewing, or sucking plant tissues, in either or both of the young and adult stages. The classic example of direct damage by an insect pest is shown by the locust.

LOCUSTS

The term locust covers a dozen or so species of Orthopteran grasshoppers which have swarming and migrating characteristics. Such insects have two distinct phases: a solitary grasshopper-like phase and a swarming and migrating locust-like phase.

As numbers increase, the solitary phase insect begins to change its behaviour, shape, and colouring and enter the locust-like phase. These changes are under hormonal control and are somehow triggered by the stimulus of overcrowding. When conditions are right, huge swarms of vast numbers of locusts build up; one of the largest swarms recorded covered about 1000 square kilometres and contained an estimated 40 000 million locusts which ate up to 80 000 tonnes of food a day. This swarm had a biomass equivalent to that of about 1 million people. In one day a swarm can eat the amount of food that would feed an estimated 400 000 people for a year.

Swarms can move up to 3000 km in a month, devastating great tracts of land as a result of their periodic feeding stops. As much as 20 per cent of the Earth's land surface is periodically invaded by these swarms. Attempts at control involve international co-operation, and a very early warning system, since once the young hoppers have undergone the final moult into winged adults, it is too late to prevent the swarm damage.

Control of insect pests

There are two main methods used in attempting to control insect pests: chemical control and ecological control.

CHEMICAL CONTROL

Chemical control involves the use of one of a wide variety of different chemicals known as pesticides. As their name implies, pesticides should only kill pests, but some are toxic to such a wide variety of organisms that the term **biocide** would be more appropriate. Such persistent broad-spectrum poisons have many side-effects, including toxic effects on the operators, the destruction of the soil microflora and fauna, outbreaks of other insects as a result of the destruction of predators, have strong toxic effects on the consumers of the food as a result of biological concentration,[315] and cause the development of resistant strains of insects as a result of selection.

The problem of the development of resistance is well-illustrated by the emergence of a strain of 'peach-potato' aphid resistant to the organophosphorous insecticides in 1973. This important pest had been previously efficiently controlled by these insecticides but, since the emergence of resistant strains, carbamate compounds have had to be used instead. Inevitably strains resistant to carbamates are now emerging, and yet another change of pesticide will be necessary. This ability to resist results from the possession of an enzyme which can break the insecticide down to harmless products.

The development of strains of the host that are resistant to the pest is the ideal method of pest control. For example, once soil is infested with the potato root eelworm (*Heterodera rostochiensis*), it is virtually impossible to clear it of this damaging pest. However, resistant potato varieties have been developed which can grow in soil that would otherwise have been rendered unuseable for potato growing.

Another form of chemical control, which avoids all the side effects of insecticides, is the use of insect hormones. Pheromones[217] or 'external hormones' can be used to interfere with the insect pests' normal behavioural patterns and reproduction; or juvenile hormone can be sprayed to prevent the insect from completing its life cycle.

BIOLOGICAL CONTROL

Biological methods of control attempt to reassemble the natural enemies of a pest in an invaded area. If successful, this results in a permanent balanced equilibrium between the

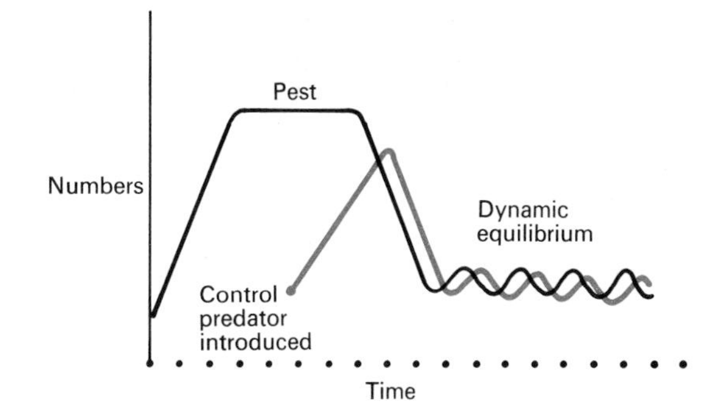

Figure 15.1 *Biological control*

pest and predator or pathogen being established. With effective biological control the characteristic out-of-phase fluctuations in the numbers of the pest and the controlling enemy occurs at such levels that, even at their peak, the pest does not cause significant damage. Control by a predator can be even more effective if the predator also feeds on non-pest species, and thus is maintained in high numbers at all times.

Although some striking successes have been recorded, biological control is not always as simple as that described above because complex interactions can occur. There have been many failures, and biological control cannot yet be regarded as a general alternative to pesticides. However, it can provide an almost perfect way to keep pests in check, and avoids all the undesirable side effects of pesticides.

One of the earliest examples of biological control occurred in the 1890s when the cottony cushion scale (*Icerya purchasi*) was accidentally introduced into California. In the absence of any natural predators it soon became a major pest and caused tremendous damage in the citrus orchards. A search in Australia led to the discovery of a beetle (*Rodalia cardialis*) which was a natural predator upon the scale insect. This was introduced into California and the pest was brought under effective control in about two years.

Pathogenic micro-organisms such as viruses and bacteria can also be used against insect pests. For example, some success has been achieved in the control of the European pine sawfly (*Neodiprion sertifer*) in Canadian forests by aerial spraying with a suspension of virus pathogenic to this insect; similarly the bacterium *Bacillus thuringiensis* can control the tomato moth 'worm' which attacks both foliage and fruits.

A third ecologically safe method of pest control is the sterile male technique, in which large numbers of males, sterilized by exposure to radioactive emissions, are released into the environment to compete with the normal males for females. This results in many failed matings and a consequent drop in the population of the pest insect.

Pests other than insects can also be controlled biologically, two classic examples having occurred in Australia (a country which has seen the introduction of many alien species). These were the control of the prickly pear cactus and the control of the rabbit population.

In 1787 cochineal bugs were introduced into Australia for the production of the red dye carmine. At the same time the prickly pear cactus (*Opuntia*) was introduced from North America to act as a food source. The natural enemies of this plant were absent, and by 1925 an area greater than that of Great Britain was blanketed by the species. The natural enemy of the cactus was the caterpillar of *Cactoblasis cactorum* from North America. Between 1928 and 1930, 3000 million of these insects were released in Australia, and control of the cactus was 95 per cent complete within a few years.

The myxomatosis virus is indigenous to America where it exists as three closely-related strains affecting various species related to the rabbit. The virus was first introduced to Australia in 1950, where it was transmitted through the vast rabbit populations by mosquitos, ticks, and fleas. By 1952–3 about 99.5 per cent of the several hundred million-strong rabbit population of Australia had been killed. The virus had the same devastating effects on the rabbit population of Great Britain in 1953. This led to a change in the vegetation of the countryside as the grazing pressure was removed, allowing the growth of many plants previously cropped by the rabbits.

The gradual reappearance of rabbits after these epidemics is not simply due to the natural selection of resistant strains of rabbits; it also involves the natural selection of less virulent strains of virus. If a virus is too virulent and the host dies too quickly there is less chance for the spread of the disease, and these virulent strains therefore tend to be selected against as the host population decreases. Therefore, as the host population thins out, less virulent strains of virus, which allow the hosts to live for a reasonable length of time, are naturally selected for.

Fungal infections

Fungal infections are responsible for losses of crops and of stored products on a scale that it is hard to imagine. As with plant virus diseases, entire crop-industries have been destroyed with far-reaching repercussions. For example, an outbreak of coffee rust (*Hemileia vastatrix*) in Sri Lanka (Ceylon) led to the development of the tea plantations as a substitute crop. Thus, although there were only about 1000 acres of tea on the island in 1875, by 1880 there were 300 000 acres of tea plantations.

There are many destructive fungal diseases of plants, three dramatic examples of which are potato blight, wheat rust, and those, such as Dutch elm disease, which destroy large trees.

PHYTOPHTHORA INFESTANS (POTATO 'BLIGHT')

Potato blight is caused by an obligate parasite which infects the tissues of the potato plant with intercellular hyphae. These have fine branched haustoria which penetrate the host cells and absorb nutrients from them.

Branched sporangiophores are protruded through the stomata of the stems and leaves. These bear sporangia which are dispersed intact, either by direct contact with nearby leaves or by splash droplets. The sporangia release biflagellate zoospores which can swim in surface moisture for a few hours before settling and producing a germ tube which can penetrate new host plants. (There is a sexual reproductive stage, but this is very rare except in Mexico.) The fungus mycelium overwinters in infected tubers. The severity of the disease is largely determined by local weather conditions. It is most serious in areas of high rainfall and mild temperature.

Potato blight spread throughout Europe, and was first recorded in Britain in 1845, although it was not until 1861 that the causal agent was identified as *Phytophthora infestans*.

From 1800 to 1845 the population of Ireland doubled from four million to eight million, and it became one of the most densely populated countries in Europe. The potato was virtually the sole food source of the mass of people, and a family of ten needed about 12 tonnes of potatoes a year to survive. The blight outbreak in Ireland in 1845 was preceded by a glut, which led to a large number of excess tubers serving as a reservoir of infection. In 1845 and 1846 crops were destroyed by the disease on a massive scale. This lead to the death by starvation of one million people and the migration of two million people to the East Coast of the USA and Canada.

An effective control against the disease was not achieved until the acquisition in 1890 of Bordeaux mixture from the French winegrowers. This mixture of copper sulphate and calcium oxide (quicklime) is sprayed on to the leaves, where the copper exerts a fungicidal effect.

Fungal diseases of wheat

Wheat is the most important food crop in the world, but it is susceptible to a wide range of fungal diseases, such as smuts, bunts, and rusts, of which there are thousands of species, varieties, and strains. These fungal diseases cause colossal losses of crops world-wide, with sometimes more than half the total yield being lost.

One of the most serious diseases is stem rust which is caused by *Puccinia graminis tritici*. *Puccinia* is a Basidiomycete[347] which produces **basidiospores** without producing a large toadstool-like fructification. If a basidiospore settles on the leaf of a Barberry plant, an intercellular septate mycelium ramifies throughout the tissues over a relatively small area of about a few millimetres. This produces unicellular **pycniospores** which are oozed out onto the upper surface of the host leaf in a sugary solution. After a complex sexual phase in which + and − strains of pycniospores form a binucleate dikaryotic state in the mycelium, dikaryotic **aeciospores** are produced which are liberated by an explosive mechanism and carried by the wind. These aeciospores infect the leaves of wheat through stomata, and produce a binucleate dikaryotic septate mycelium which in turn produces binucleate **urediospores**. Masses of these burst out of the epidermis to give 'rust' spots. The urediospores are a repeating-spore stage and can reinfect the wheat host, leading to the build-up of the disease.

In the autumn **teliospores** are produced which remain dormant in the host until they germinate in the spring to give diploid basidia. The diploid nucleus, formed as a result of the fusion of the two nuclei in the dikaryotic state, undergoes meiosis to produce four haploid basidiospores which are capable of infecting the Barberry plant again.

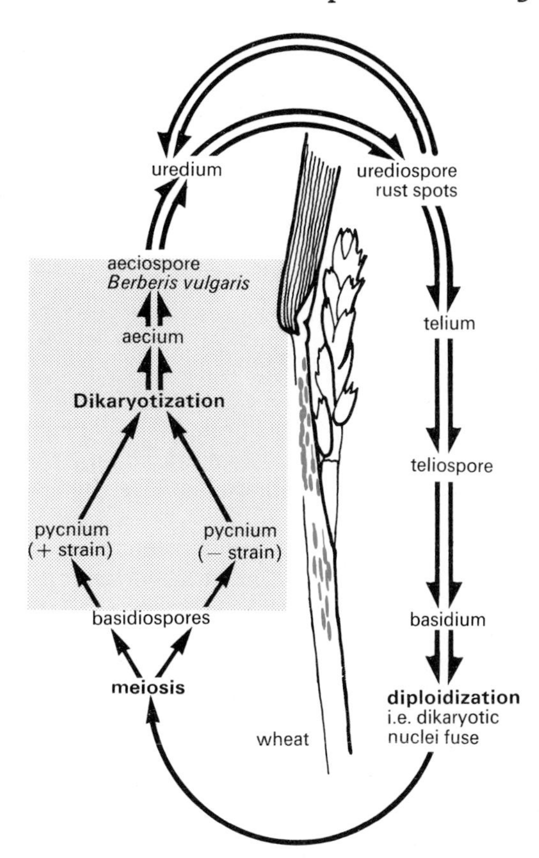

Figure 15.2 *Life cycle of Puccinia graminis*

At the turn of the century William and Charles Saunders developed a new, hard, spring wheat named '**Marquis**' which had all the right qualities and ripened early so that it escaped the worst attacks of rust. The development of Marquis also extended the growing range of hard spring wheats to the North West plains of North America, as it ripened before the onset of the frosts. Marquis also led to the development of the first artificially-bred rust-resistant wheat known as **Ceres** (after the goddess of agriculture, who also 'gave' her name to cereals). This was obtained as a result of crossing Marquis with a rust-resistant Russian wheat known as **Kota**.

However, there is no end to the battle to breed-in resistance to rust diseases because new strains of rust are continually emerging as a result of mutations and sexual reproduction on the Barberry (*Berberis* sp. or *Mahonia* sp.).

Under optimum conditions for the fungus it is estimated that more rust spores are produced in a year than there are grains of sand on all the beaches in the world.

The removal of Barberry reduces the rate of variation in the sexual phase and removes new foci of infection in the spring. For example *P. graminis* would not survive in Britain without Barberry because the urediospores are unable to survive the cold winters.

Fungal diseases of trees

DUTCH ELM DISEASE

Dutch elm disease is caused by the fungus *Ceratocystis ulmi* which is transmitted by the bark beetle *Scolytus* sp. The fungus penetrates the xylem tubes of the functioning sapwood and releases toxins which cause a reaction leading to the plugging of the vessels. This cuts off the transpiration stream to the aerial parts, and can kill a young tree within two weeks. On the death of the tree the fungus can continue to live saprophytically on the dead organic matter.

The disease was first described in Holland in about 1919, and reached England in 1927 and the USA in 1930. The recent outbreak, which threatens to remove elms from the landscape of Britain completely, is due to new strains of fungus and beetle being imported in timber from Canada. The pattern of the spread of the outbreak has been a radiation from around port areas, especially in the South.

There is no doubt that the disease has spread so rapidly in Britain as a result of the unrestricted movement of elm logs from dead infected trees the length and breadth of the country.

The effects of Dutch elm disease

OAK WILT (*Ceratocystis fagacearum*)

This fungus is a relative of the Dutch elm disease fungus, and is transmitted by the oak bark beetle. It penetrates and blocks the sapwood xylem vessels, cutting off the transpiration stream and leading to the death of the tree. However, unlike Dutch elm disease, it is not an epidemic-type disease which destroys vast numbers of trees in a short time. Rather, it is characterized by a slow 'patchy' progress. Nevertheless, it still causes great damage to oak forests, and there is a continuing danger of its spread.

CHESTNUT WILT (*Endothia parasitica*)

Chestnut wilt is a fungus disease which invades the phloem tissue. It has completely destroyed the great chestnut forests of the eastern USA, which at one time contained an estimated 1000 million trees.

Stored products

Usually it is dampness that provides the necessary conditions for the spread of fungi through stored products. Most are harmless saprophytes that simply break down the organic matter, as in natural decomposition; but some fungi can produce extremely poisonous toxins which are a hazard to health. For example, *Aspergillus flavus* is a mould fungus which can infect inadequately stored food products throughout the world, but especially in the tropics. Groundnut meal infected with this fungus has caused disastrous outbreaks of disease in the British and European turkey farming industry. The mould produces the poisonous aflatoxin, an extremely potent liver toxin in birds, which is suspected of being the cause of the high incidence of liver cancer that occurs in parts of Africa.

Summary

With the great increase in volume of international traffic there is an increasing danger of the movement of plant parasites between continents. Such parasites may be introduced into new areas where, in the absence of natural checks and balances and in the presence of extensive single-crop monocultures, ideal conditions for widespread outbreaks exist.

Household pests

Household pests are those animals which either live or thrive under the conditions found in the home; are attracted there by certain stimuli; or find their way in by accident as a result of living in the vicinity.

In contrast to agricultural pests and pests of stored products, which are classified as such as a result of the damage they cause, exactly which animals are considered to be household pests depends as much on tradition and phobias as it does upon the actual amount of damage done.

Houseflies

Houseflies are perhaps the major household pest because, by their feeding habits, they can present a serious hazard to health by transmitting many pathogenic organisms, especially those involved in food poisoning.

They are liquid feeders, feeding by means of a sponge-like proboscis. In order to render the food to be eaten soluble, the contents of the insect's stomach are regurgitated on to the food and then mixed into it. The soluble products of this external digestion are then sucked up. Remnants of the regurgitated stomach contents are frequently left behind, and these provide the major source for the cross-infection of human food; a secondary cause is the material which had adhered to the hairs during the previous meal being left behind by the feet. The main hazard in urban areas comes from cross-infection of food from dog excreta which contains many pathogens, including *Salmonella* sp. and *Toxacara* eggs.

The female housefly lays about 200 eggs in organic matter, upon which the larvae feed when they hatch. The life cycle can take less than two weeks under optimum conditions, and up to several months in cold weather. In temperate climates most flies die with the onset of winter, but some larvae, pupae, and adults overwinter to form the basis of the next year's population.

Although serious household pests, flies and their larvae play a vital role as decomposers in the economy of the natural recycling of organic material.

Common house spiders

All British spiders are harmless, and beneficial as they prey on flies and other insects. The common house spider (*Tegenaria domestica*) is one of the largest European spiders, with a span of up to 6 cm. The spiders prefer humid sites, which explains their regular appearance in the bath; specimens seen in other parts of the home will almost certainly be chance visitors as the atmosphere is usually too dry to allow them to establish themselves.

Gnats and mosquitoes

It is only the female mosquito which sucks blood; she must have a blood meal before she can reproduce. Male mosquitoes feed on nectar and fruit juices and do not bite. The mosquito's eggs are laid in static water, and the larva (or wriggler) feeds on particles suspended in the water whilst attached to the underside of the surface tension film by an air tube through which it breathes.

The life cycle of the mosquito is very variable as the larval, pupal, and adult stages may each last from seven days to several months. Most species do not travel more than about 1.5 km from their breeding place in search of food. They are normally summer insects, but two species can survive the winter in the house: the large grey mosquito (*Theobaldia annulata*), which is nearly always responsible for bites in the winter; and the common gnat (*Culex pipiens*) which rarely bites humans.

Wasps

There are seven different species of social wasps in the UK, all with the same general appearance and habits, but the one usually seen in and around houses is *Vespula vulgaris*, the common wasp.

Unlike the honey bees, wasps do not store food for the winter, and a colony therefore lasts only one season. Towards the end of the season the males and new queens develop and leave the colony to mate. The males soon die, but the fertilized queens overwinter in protected positions. They emerge in mid-April, and each one then searches out a good site in which to establish a new colony.

The queen builds a spherical nest about the size of a walnut; inside it she builds between ten and twenty hexagonal cells and lays an egg in each. She feeds the larvae until, after about a month, the first sterile female workers emerge from their pupae.

The queen continues to lay eggs while the workers enlarge the nest with 'paper' formed from chewed wood, care for the larvae, and defend the developing colony which can contain up to 6000 members. The larvae are fed on food balls of flies and moths which are caught in flight by the workers. The workers themselves feed on liquid food, such as fruit and nectar, and also are fed a sugary liquid which the larvae regurgitate in exchange for a food ball.

During most of the summer the workers are so busy maintaining the colony that they are only infrequently seen in and around the house. However, when they stop rearing larvae later in the summer they search wide areas for sweet substances, and it is at this time that they are often seen in and around the house. Unlike the bees, wasps do not leave their sting in the victim, and therefore sting more readily. The sting is a modified ovipositor, and the venom contains histamine.[293]

Bed bug

The bed bug (*Cimex lectularius*) is perhaps the most unpleasant of all the household pests. Both adults and the young of both sexes feed by piercing the skin and sucking blood through their tubular mouthparts. After feeding, their paper-thin bodies are greatly distended with blood. They are nocturnal, emerging about once a week from cracks and crevices to feed just before dawn. The females cannot breed until they have had a meal of blood, but the adults can survive as long as a year without feeding. The bugs possess stink glands which are used when they are disturbed.

Crane flies

Although crane flies (*Tipula paludosa*) can cause much disturbance by their erratic flight when they are attracted by lights into the house, they are completely harmless and in fact do not even feed as adults. The eggs are laid in the ground, and the larvae or 'leatherjackets' feed on roots and vegetation; they are sometimes so numerous that they kill the grass, causing bald patches to appear.

Hoverflies

Hoverflies (*Syrphus ribesi*) resemble small wasps and are often mistaken for them when they accidentally enter the house. They are distinguishable by their very distinctive flight pattern, hovering motionless in the air and darting off very rapidly at intervals. The adults feed on nectar and pollen, particularly of the Umbelliferae, but the predatory larvae feed voraciously on aphids and are therefore very useful in the garden.

Clothes moth

The clothes moth (*Tineola bisselliella*) is not a native to temperate climates and is therefore only found inside warm buildings. The eggs are laid in and on materials and the larvae can take up to a year to reach the pupal stage. The larvae feed mainly on the keratin content of wool, but they also require dietary supplements from other organic material such as dead insects. The adults, which are about 6–8 mm long and of a golden colour, do not feed and are not attracted to light.

Woolly bears

The so-called 'woolly bears' (*Attagenus pellio*) are the bristly larvae of the carpet beetle, both larvae and adult being about 4 mm long. The larvae feed on and cause damage to woollen carpets, blankets, and any fibrous fabrics, and have overtaken the common clothes moth as the major British textile pest. The hairs of the larvae are easily shed and can cause a skin rash and irritate the lungs if breathed.

The adult beetles fly and are attracted to light so they are often found crawling about window-sills; outside the adults feed on nectar and pollen in the summer.

Cockroaches

Most of the more than 3000 or so species of this insect are tropical or sub-tropical, and the two species which are common household pests in Britain thrive in warm, moist conditions, such as are generally found in kitchens, and are rarely found outdoors. None of the six introduced species in Britain can survive the winter out of doors. Neither of the two common species, the common or oriental cockroach (*Blatta orientalis*) and the German cockroach (*Blattella germanica*), are native to Britain. The three native species, which are very much smaller, are found only in southern England and south Wales.

Cockroaches are nocturnal insects which feed on practically anything organic. They can taint food with their odour, and carry bacteria, parasitic worms and other pathogens.

Ants

Each colony of ants contains many sterile female workers, one or more queen, and, during part of the year, a few fertile males. The workers search for food, including flower nectar and the 'honeydew' secreted by aphids. This consists of the sugary contents of the phloem sieve-tube elements of the plants on which the aphids live. The workers maintain the nest and tend the eggs and developing larvae. They do not have wings, but the males and young queens develop wings at swarming time during which they take off on their nuptial flight. After mating the males soon die, while the queens drop their wings and find a suitable place in which to start a new colony. After mating the queens remain fertile for as long as twenty-five years. Foraging workers often find their way into the house.

Lice

There has been a recent resurgence of these insects in Britain of almost epidemic proportions.

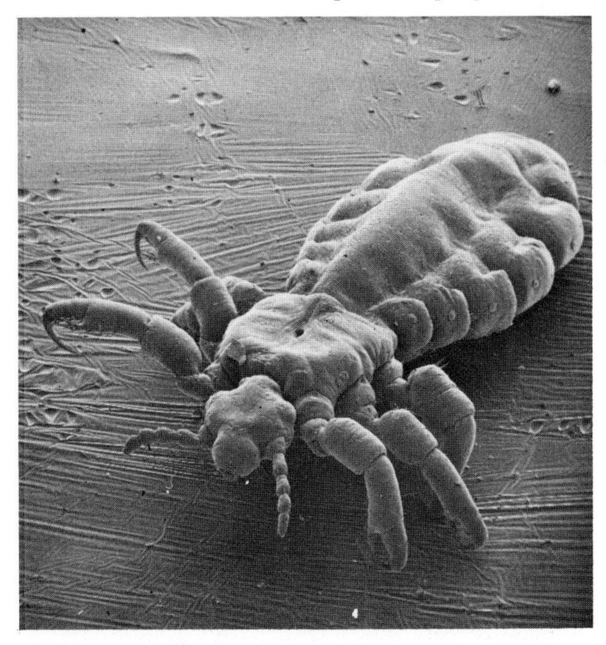

The head louse *Pediculus humanus capitis* (× 88)

Infection by these insects has nothing to do with the standard of personal hygiene, but occurs by direct or indirect contact with infected hair or materials. These ectoparasites spend their whole lives on their host, and have a flattened body shape and strongly hooked legs to help maintain their position.

Humans are infested by three types of lice: the head louse (*Pediculus humanus var. capitis*); the body louse (*Pediculus humanus var. corporis*); and the pubic louse (*Phthirus pubis*). Of the three, it is the head louse which has recently become so common.

Both adults and young of both sexes feed by piercing the skin and sucking blood through their tubular mouthparts. A single female can lay up to 300 eggs in her lifetime of about a month. Each egg or 'nit' is attached singly to the hair or clothes fibres by a tough cement, and the young nymph emerges through a small cap at one end.

Dry rot

Many fungi attack damp timber with a water content of 30–50 per cent; the dry rot fungus (*Serpulum lacrymans*) thrives in timber with a water content as low as 20 per cent and can invade even drier timber from the original centre of infection. Millions of brown spores are produced, which spread by air currents to infect other timbers.

Zoonoses

These are a group of diseases which can be transmitted from other vertebrate animals to man. In the urban environment it is man's close association with pets, rather than with domesticated farm animals or wildlife, which brings the greatest risk. In particular the tremendous rise in recent years in the number of dogs kept in urban areas renders them an increasing hazard to health.

Nematodes (round worms)

Toxacara canis is a nematode gut parasite of dogs and is found mainly in dogs under a year old.

The adult worms virtually disappear from the gut of dogs over a year old in a process of 'self-heal' which is not fully understood. When in the gut, the adult worms produce fertile eggs which pass out of an infected dogs' gut with the excreta. Infective larvae develop within the very resistant egg cases and such eggs can remain infective on the soil for up to ten years. Due to the properties of the egg cases, they are not usually washed away by rain water, and can even be carried as wind-borne dust.

If swallowed by another dog, the larvae hatch out in the gut, bore through the gut wall, migrate through the body in the blood system, break through from the pulmonary capillaries into the alveoli of the lungs, and migrate up the trachea to be reswallowed into the gut where they develop into mature worms up to 17 cm long. In older dogs with acquired immunity the larvae do not complete the migration and become inactive and encapsulated in the tissues of the body.

It has been estimated that about 20 per cent of the adult dogs and 98 per cent of puppies are infected in the USA, and there is no reason to suppose that these figures will differ in other parts of the world. Most puppies are infected across the placenta before birth, either by the larvae migrating from the gut of the mother, or by larvae emerging from encapsulation in the mother's tissues under the stimulation of levels of circulating sex hormones characteristic of pregnancy. Many of these larvae arrive in the gut of the puppy within three days of birth and mature in about another week. The adult worms begin producing fertile eggs in the puppy. The eggs are infective to any mammal swallowing them.

In alternative hosts the larvae hatch in the gut, begin their migration through the body, and become encapsulated in the tissues. If such infected alternative hosts are subsequently eaten by a dog, the larvae undergo their usual full migration round the body and back into the gut where they develop into adult worms.

The fertile eggs are infective to humans, and children are especially vulnerable to the migrating larvae which can encapsulate in the liver, lungs, brain, and retina. As a result, the larvae might be implicated in certain cases of hepatitis, asthma, epilepsy, and retinal damage leading to blindness. In their migrations they could also act as 'carriers' of infective agents such as the polio virus which multiplies in the intestine.

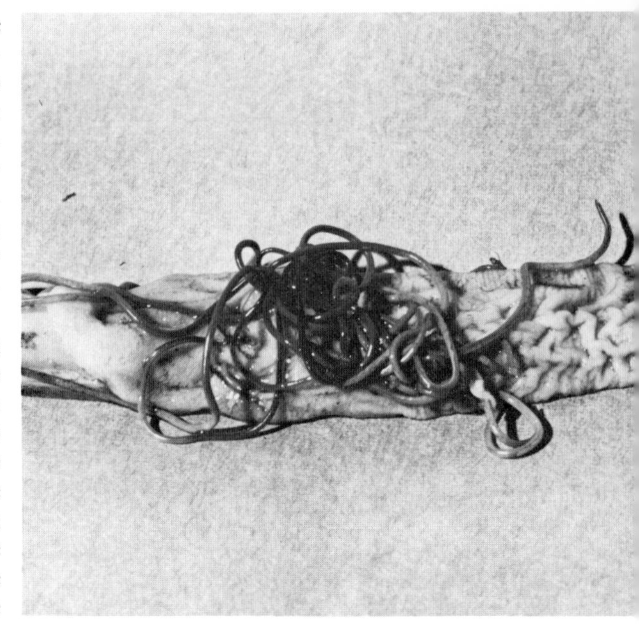

Toxacara cati (actual size) in an opened cat gut

In Britain it would appear that about one in twenty people are, or have recently been, infected with *Toxacara*. However, there seems to be no correlation of the occurrence of toxacariasis with dog ownership, which implies that contamination of the environment (especially where children play) is as serious a source of infection as close contact with an infected dog.

Toxacara cati is a nematode gut parasite of cats and shows many similarities to *Toxacara canis*. However, cats do not develop the same immunity that adult dogs show, and therefore cats can become repeatedly infected, either directly, or from a wide range of alternative hosts, throughout their adult life. For this reason they should be treated regularly and not just when kittens.

Platyhelminthes (tapeworms)

There are many tapeworms found in dogs and cats, some of which carry a risk of cross-infecting humans. A common tapeworm of both cats and dogs is *Diplidium caninum*. This has an interesting life cycle in which the flea or louse acts as the secondary host.

Echinococcus granulosus is a cestode worm which is primarily a parasite of dogs and sheep.

The small adults, which are less than 1 cm long, are found in the gut of the dog, and fertile eggs with very tough resistant shells are passed on to the ground with the dog's faeces, where they can remain infective for long periods.

If taken in by a grazing animal, such as a sheep, the larvae hatch in the gut, where they bore through the wall and migrate all over the body, mainly to the liver and the lungs. Here they form slow-growing cysts which contain large numbers of hexacanth larvae capable of reinfecting another dog, if and when the sheep should be eaten. However, should the eggs from the dog enter a human being instead of a sheep the same migration occurs, but the cysts can grow to an enormous size, containing as much as 10 litres of fluid. These hydatid cysts can form in the brain, liver, kidney, and other parts of the human body.

The worm is most common in parts of Wales where man, dogs, and sheep live in close contact. However, there is a great danger of it being spread throughout the countryside by packs of dogs from hunt kennels which have their own slaughter yards and do not sterilize the meat before feeding it to the dogs.

Insects

Dogs and cats also carry insect pests which can transfer to humans. Examples are the dog mite, which causes canine scabies, and fleas.

There is no true 'human' flea. *Pulex irritans*, which lives on foxes, badgers, and hedghogs, is the only flea in Britain which is able to breed whilst feeding on a diet of human blood. However, it prefers cool damp conditions and has been virtually eliminated by the drier and warmer conditions of more modern housing.

This 'ecological niche' has now been occupied by the cat flea (*Ctenocephalides felis*), which is now possibly the most widely spread domestic insect pest other than flies. The eggs are laid around the cat's nest and the larvae hatch out after 2–12 days and feed on household dust which is mainly shed human skin scales. The larval stage can last up to six months; and the pupal stage can last for up to a year, during which time it is safe from insecticides. On hatching, the fleas are stimulated by vibrations and will feed twice a day on any animal moving by, although a flea has to be starving before it will bite humans, as human blood literally makes it sick before it is conditioned to it. The adults can survive up to four months without feeding.

The tapeworm *Diplidium caninum* can be picked up by humans from the cat or dog flea.

Protozoa

A protozoan parasite, *Toxoplasma gondii*, occurs in both dogs and cats and can cause toxoplasmosis in humans. This varies in severity in adults but can cause damage to the developing embryo by infection across the placenta. Toxoplasmosis is rare in Britain, but quite widespread in other countries.

Bacteria

Dogs suffering from infection by the micro-organism *Leptospira canicula* can pass them in the urine for many years after they have apparently recovered from an acute attack. These micro-organisms can infect man, and cause leptospirosis which can result in serious jaundice conditions, kidney disease, and symptoms similar to meningitis. Dog urine should be avoided as far as possible; untrained puppies are a particular risk.

Dog faeces carry many bacterial pathogens, including *Clostridium tetani*, which can cause tetanus if introduced into a wound; and *Salmonella* sp., which are the main cause of food poisoning in this country. Indeed, the deposition of an estimated two million litres of dog urine and 2500 tonnes of dog faeces each day on the streets of Britain represents a very serious health hazard, and one to which greater public concern should be brought to bear.

Viruses

Rabies is a viral disease which attacks the central nervous system of all mammals, including man, and which is spread by bites and scratches or contact with a skin abrasion. By 1902 the disease was eliminated from Britain, although there have been sporadic outbreaks since.

If bitten by a rabid animal a course of vaccinations can prevent the development of the disease, but once the symptoms appear the disease is usually always fatal. The prevention of the re-entry of this disease to this country depends on the continuing strict application of the quarantine laws.

Practical and applied aspects

Some pesticides in common use

Pesticide	Notes
Synthetic organics	Persistent.
Paris Green (Calcium acetoarsenite)	The first chemical insecticide used successfully on a large scale.
Chlorinated hydrocarbons	Persistent in the environment.
DDT and its metabolites, e.g. DDE	First synthesized in 1874, not recognized as an insecticide until Second World War. Used widely to control malaria mosquitoes and houseflies, but strains of these have evolved with resistance. Low oral toxicity to mammals.
Aldrin and dieldrin	Formerly widely used as seed dressing, but use suspended because of death of birds and other wildlife.
BHC and lindane (Benzenehexachloride)	Used for seed treatment. At one time widely used in gardening products.
Organophosphorus insecticides	Short-lived in the environment.
Parathion and malathion	Developed as poisonous gases for use in Second World War. Extremely dangerous to use—related to nerve gases.
Carbamate insecticides, fungicides and herbicides	Fairly short-lived, also attack nerve function.
Natural organics	Non-persistent.
Pyrethrum	From flowers of several species of chrysanthemum. Fast-acting 'knock-down' agents.
Derris	From dried and ground roots of derris plant. As prepared for an insecticide, it is relatively non-toxic to humans.
Inorganics	
Lead arsenate, Sodium fluoride	Stomach poisons for chewing insects.
Sulphur dusts	Main chemical control for mites, crawling stages of scales, and some caterpillars.
Hydrogen cyanide	Used as a fumigant of stored products.
Abrasives	These scratch the epicuticle and the insect dies of dehydration.
Herbicides	
Phenoxy compounds 2,4-D, MCPA, MCPB	In many common gardening products; widely used in rangelands and woodlands. Causes foetal damage in mammals (teratogenic).
2,45-T	Used as a defoliant. Teratogenic.
Diquat and Paraquat	Becoming widely used in agriculture and gardening. When accidentally drunk, destroys the lungs; there is no antidote.
Fungicides	
Arsenic compounds, Copper compounds (e.g. copper arsenite); Zinc compounds, e.g. zineb; Mercury compounds, e.g. phenyl mercury acetate; Manganese compounds, e.g. maneb	Compounds of toxic heavy metals, widely used in industry and as seed dressings.
Bordeaux mixture (copper sulphate and calcium oxide)	First control of potato blight fungus.
Rodenticides	Highly toxic to mammals.
Warfarin	Introduced in 1950; by 1960 resistant strains emerged, the result of a single dominant gene (linked to coat colour).
Sodium fluoracetate	Cause of an accident a few years ago in which cattle died.

Biological warfare

Much work has been done on developing new and more virulent strains of various diseases, stocks of which exist in various parts of the world. If, as the result of this work, a disease such as pneumonic rabies, which spreads directly from person to person by droplet infection, were to be developed, the results of an outbreak would be unthinkable. Rabies is a dreadful disease, and is virtually always fatal once the symptoms have appeared.

Similarly, any new plant pathogens could have devastating effects on world crops. The particular danger is that, with advances in genetic engineering and the study of micro-organisms, any group with a few well-trained microbiologists and the necessary facilities could build its own biological doomsday weapons.

Guided example

Weeds

Just as animal pests are animals which are successful at exploiting conditions created by man, so weeds are simply plants which have certain characteristics which enable them to colonize and thrive under those conditions created by man's cultivation of the soil. In fact some so-called weeds can be useful even in the cultivated land itself; for example by providing cover for predatory insects which feed on crop pests; by concentrating certain trace elements into their tissues which can subsequently be released into the surface soil; and as indicators of the state of the soil.

The main disadvantages of weeds is that some act as reservoirs of crop diseases, for example club-root is harboured by shepherd's purse;[352] and they can compete with the crop plants for nutrients, water, and space, etc. Study the information provided on the characteristics of some weed species, and consider the questions that follow.

1 In Table 15.1 both weed species are given as being variable; this means that there is a range of variation in the offspring. How would this feature enable such species to thrive as weeds?
Having a range of genetic variability, or genetic 'plasticity' as it is sometimes called, means that there would be a range of genotypes capable of surviving in a range of environmental conditions. In this way weed species can survive in a wide range of different habitats and be ready quickly to exploit cultivated conditions when and if they become available; also they can thrive in a wide variety of cultivated soils, under a wide range of conditions.

2 In Table 15.1 the two weed species are annuals, and yet in Table 15.2 most species quoted are perennials. What do you think are the advantages and disadvantages to weed species of being annuals or perennials?

Table 15.1 *A comparison of the characters of closely-related weedy and non-weedy forms in two genera*

Eupatorium	
E. microstremon (weed) ($n = 4$)	*E. pycnocephalum* (non-weed) ($n = 20$)
Variable	Not very variable
Annual	Perennial
Quick to flower	Slow to flower
Photoperiodically neutral	Short day requirements
Self-compatible	Self-incompatible
Tolerant of drought	Intolerant of drought
Tolerant of waterlogging	Intolerant of waterlogging
Ageratum	
A. conyzoides (weed) ($n = 20$)	*A. microcarpum* (non-weed) ($n = 10$)
Variable	Not very variable
Annual	Perennial
Quick to flower (6–8 weeks)	Slow to flower (in second year)
Flowers in any day length	Flowers better in short days
Self-compatible	Self-incompatible
Tolerant of drought	Intolerant of drought
Tolerant of waterlogging	Intolerant of waterlogging

After T. A. Hill, *The Biology of Weeds* (London: Arnold, 1977).

Table 15.2 *Characteristics of some perennial weeds*

Species	*Reproductive parts: overwintering state*	*Seed production*
Aegopodium podagraria (ground elder)	Shallow rhizomes	Unimportant
Agropyron repens (couch grass)	Shallow rhizomes, aerial shoots persist over winter	Fairly important
Cardaria draba (hoary cress)	Deep creeping roots; small rosettes of leaves over winter	Important
Convolvulus arvenis (field bindweed)	Very deep (> 3 m) creeping roots	Very important
Ranunculus repens (creeping buttercup)	Creeping surface stems, some leaves over winter	Very important
Rumex crispus and *R. obtusifolius* (curled and broad leaved dock)	Tap roots, rosette of leaves over winter	Very important

After T. A. Hill, *The Biology of Weeds* (London: Arnold, 1977).

Annual weeds that are quick to flower have the advantage of avoiding damage by cultivation, and of avoiding too much direct competition with the crop plant. The seeds are produced and dispersed rapidly, overwinter in the soil, and are ready for early germination in the next year. Perennials can be at a disadvantage from certain culture methods, such as their continual disruption by soil cultivation, competition from the crop, and severe weather in the winter. Nevertheless, in certain cases cultivation can actually aid the spread of perennial weeds; for example the smallest section of a damaged couch grass rhizome will quickly develop into an independent plant. However, it is inter-

esting to note that in all those perennials in Table 15.2, except one, seed production is still important.

3 What do you think is the significance of the fact that the two weed species in Table 15.1 flower independently of daylength (day-neutral plants), but that the two non-weed species only flower in short days?
Plants that require short days before they will flower are late summer/early autumn flowering types, and therefore would be at risk on cultivated soil of being damaged or destroyed by cultivation before they had time to flower. The day-neutral weeds, on the other hand, could flower as soon as they were ready.

4 The terms self-compatible and self-incompatible in Table 15.1 refer to the ability to self fertilize. What advantages does self-compatibility confer on weed species?
Self-compatibility removes the requirement for cross-pollination between different plants, and therefore facilitates the ease of reproduction. However, cross-pollination and fertilization can still occur and is important in maintaining the necessary genetic variability.

Questions

1 (a) Explain how the productivity of economically important plants can be increased.
(b) Indicate the problems which are associated with some of these practices. (JMB)

2 The larvae of the small white butterfly (*Pieris rapae*) do much damage to Brussels sprout plants. A plant breeder who was searching for natural resistance to this pest tested a number of types having different leaf colour. His results are summarized in the graph below.

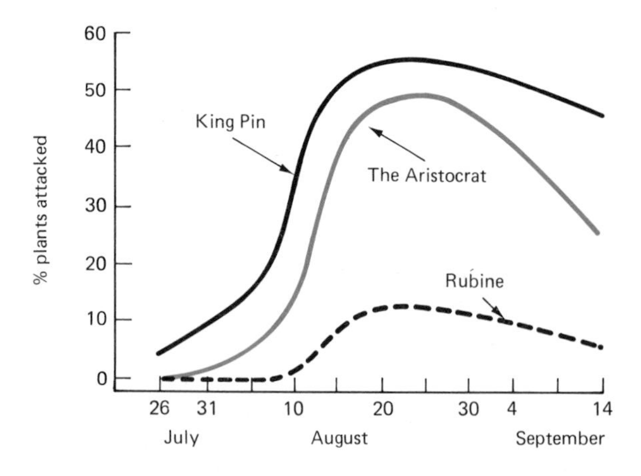

(a) Briefly discuss the relative resistances of the Brussels sprout types to *Pieris rapae*.

(b) In a laboratory experiment, he transferred 20 larvae to a standard quantity of sprout leaves of each of the three types shown in the graph. The larvae used were all taken from a different type of sprout plant. Each day after the transfer he counted the numbers of larvae and pupae on the leaves of each type and when all the larvae had pupated he weighed the pupae. His results are shown in the table below.

Types of sprouts	*No. of larvae* (L) *and pupae* (P) *on days 2 to 8*								
	2	5	6		7		8		Pupal wt.
	L	L	P	L	P	L	P	P	(mg)
Rubine	20	18	2	10	10	4	16	20	185
The Aristocrat	20	16	4	11	9	4	16	20	189
King Pin	20	18	2	8	12	2	18	20	187

Do you consider the differences shown in the table to be due to some factor in the leaves which affects the larvae? Explain your answer. (C)

3 'The success of man in controlling his environment really depends on eliminating those species that adversely affect his crops and live stock.' Discuss this statement.

4 Discuss, with examples, the reasons for preferring biological methods of pest and weed control to chemical methods. (O)

5 Discuss the problems of survival and spread in a *named* parasitic fungus, and relate these to the various stages in its life history. (L)

Further reading

Andrews M., *The Life that Lives on Man* (London: Faber and Faber, 1976).

Carefoot G. L. and Sprott E. R., *Famine in the Wind* (London: Angus and Robertson, 1969).

Carson R., *Silent Spring* (London: Hamish Hamilton, 1963).

Debach P., *Biological Control by Natural Enemies* (Cambridge: Cambridge University Press, 1974).

Graham F., *Since Silent Spring* (London: Hamish Hamilton, 1970).

Gunn D. L. and Stevens J. G. R. (Eds), *Pesticides and Human Welfare* (Oxford: Oxford University Press, 1976).

Mourier and Winding, *Wildlife in House and Home* (London: Collins, 1977).

Owen D., *The Natural History of Britain and N. W. Europe Towns and Gardens* (London: Hodder and Stoughton, 1978).

Wheeler B. E. J., *Diseases in Crops*, Institute of Biology Studies in Biology No. 64 (London: Arnold, 1976).

16 Pollution

Introduction

Wastes, and consequently pollution, are unavoidable products of an industrial society. However, the presence of natural 'background levels' of toxic substances, such as sulphur dioxide from rotting vegetation, and mercury and lead from their natural ores, can complicate the study of the problems caused by substances introduced into the environment by man's activity. For example, normal weathering of rocks results in about 5000 tonnes of mercury being added each year to the oceans; man's activities add about the same amount.

Given the present population, there is no way back to a low environmental impact society. Therefore a balance between maintaining adequate standards of living and the cost in terms of the deterioration of the environment must be aimed at. The problems of gross pollution of the environment will not be resolved until the so-called 'external costs' of pollution are internalized, in other words, when those who produce pollution are made to pay for its prevention, and this cost is 'internalized' into the price of the product or service. In addition, strict accounts could be kept of certain materials, e.g. mercury and plutonium, so that the 'books can be balanced' and any 'losses' accounted for.

Pollutants fall into two main groups: non-persistent and persistent.

Non-persistent pollutants are rapidly broken down and generally it can be said that the 'solution to this problem is dilution'; they are safe until the natural systems become overloaded.

Persistent pollutants are not broken down into harmless substances and some can persist indefinitely. The solution to this type of pollution is not dilution, as many of these substances undergo biological concentration as they move up food-chains.

With all cases of pollution special note must be made of the **synergistic effect**, whereby the total effect of a variety of pollutants can be greater than the sum of the individual effects. This is due to complex interactions between pollutants and between pollutants and living organisms.

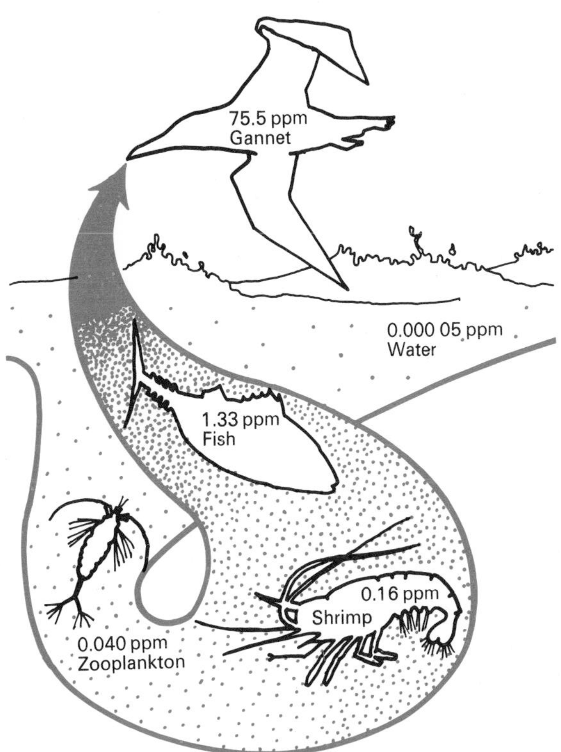

Figure 16.1 *Biological concentrations of DDT residues in a food-chain*

For example, mercury was not considered to be particularly hazardous in the environment as it is relatively inactive and quickly removed from living organisms; however, it can be converted to organic methyl and alkyl mercury compounds by the action of various micro-organisms. These compounds remain in the body for longer and can undergo biological concentration.

Air pollution

Each of us breathes about 83 kg of air a day to obtain about 13 kg of oxygen; in moving this large amount of air over the delicate epithelia of the respiratory tract there is the danger of exposure to harmful pollutants.

The greatest source of air pollution in Britain is the combustion of coal and oil (although on a local scale tobacco smoke can be worse). About eight million tonnes of air pollutants are released in this way each year in Britain. The National Survey of Air Pollution has 1200 monitoring stations, co-ordinated by Warren Spring Laboratory, and in 1972 only thirty-eight were over the unofficial target for clean air using Lawther's 250/500 limit (250 $\mu g/m^3$ smoke, 500 $\mu g/m^3$ sulphur dioxide). These are the limits above which bronchitis deteriorates. (In the great London smog of 1952 the figures reached 400/2000.)

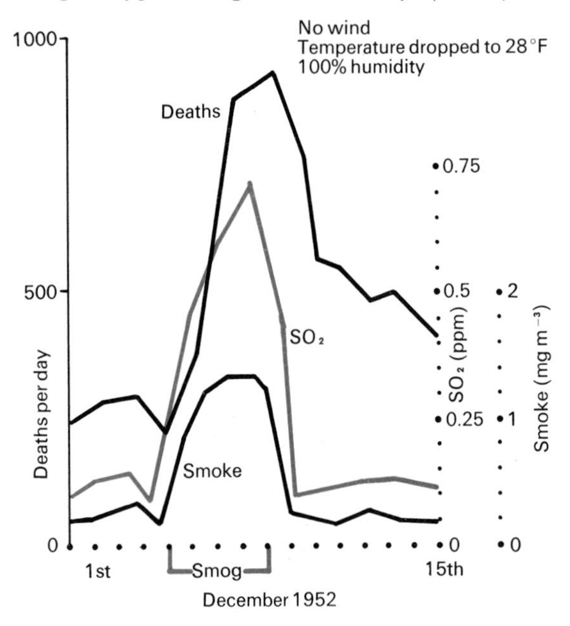

Figure 16.2 *Pollution levels in the London smog of 1952*

Although air pollution in Britain has been greatly reduced, especially since the 1956 Clean Air Act, many problems still remain from the burning of fossil fuels and from the internal combustion engine.

Table 16.1 *Principle air pollutants (millions of tonnes)* (Clean Air Yearbook 1970–71)

Burning of fossil fuels (coal and oil)	
Sulphur oxides	6.14
Smoke	0.84
Grit	0.5
Carbon monoxide	5.0
Internal combustion engine	
Hydrocarbons	0.48
Carbon monoxide	6.2
Lead	0.006
Nitrogen oxides	0.26
Aldehydes	0.26

Combustion also depletes atmospheric oxygen; however, even if oxygen were not being replaced by photosynthesis there would still be several hundred years' supply in the atmosphere at the present rate of usage. In terms of the photosynthetic replacement of oxygen it is interesting to note that on average it takes one oak tree to supply the oxygen needed by two people.

Sulphur dioxide

About 70 per cent of the global production of sulphur dioxide comes from decomposing vegetation, and man's contribution must be seen in relation to this. In Britain about 6 million tonnes of sulphur dioxide are emitted by man's activity each year, but on a global scale the level in the air is well below the known 'threshold' for damage to animals and plants of 0.01 ppm.

Sulphur dioxide is short-lived in the air, about 50 per cent being removed within four days of entering the atmosphere by combining with excess ammonia to give ammonium sulphate (which is not toxic); and about 20 per cent being washed out as sulphuric acid rain.

The toxicity of sulphur dioxide is not due to its acidity but to its activities as a reducing agent. Sulphur dioxide is more damaging to plants than animals because, as sulphurous acid and solid sulphites, it disrupts photosynthesis. The disruption of photosynthesis reduces crop yields and can result in the death of conifers, as is especially clearly seen in the hills around Los Angeles. When rats inhale sulphur dioxide with benzyl-pyrenes tumours are produced, and both these substances are present in smoke.

In assessing the dangers to health it is the peak levels of sulphur dioxide which are more important than the average levels.

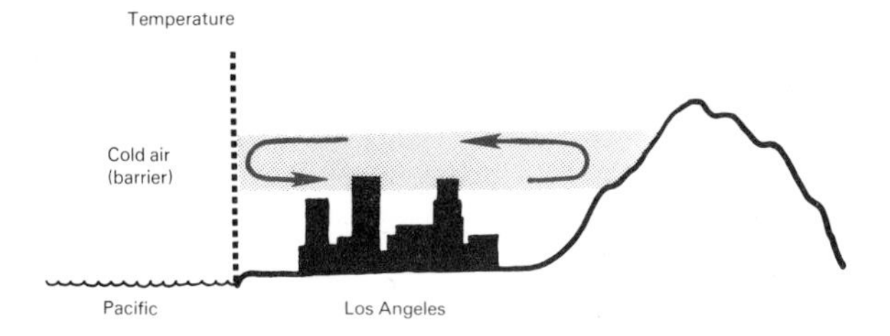

Temperature inversion trapping smog over Los Angeles

The prevention of sulphur dioxide pollution by its removal from waste gas emissions depends on world sulphur prices for its economic viability as far as the industries are concerned.

Under certain circumstances there can be slight advantages to sulphur dioxide pollution, for example it prevents the fungus diseases of roses known as black spot (*Diplocarpon rosae*), and also prevents the formation of PAN (Peroxacyl nitrates) which are formed as a result of the action of intense sunlight on car exhausts trapped by a temperature inversion. The main biological effect of PANs is the damaging of cell membranes. They are extremely active, with concentrations as low as 15 $\mu dm^3/m^3$ (pp thousand million) damaging susceptible plants, and 1 $\mu dm^3/m^3$ causing severe eye irritation.

Rose leaves without and with black spot. Black spot is prevented by a level of atmospheric sulphur dioxide pollution of 100 $\mu g\ m^{-3}$ of air.

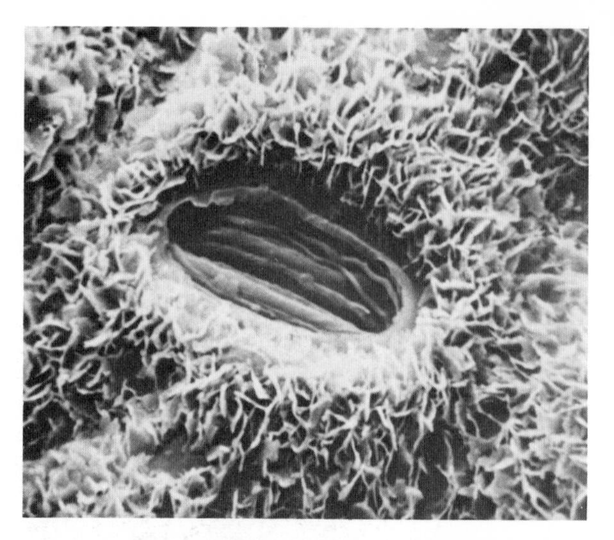

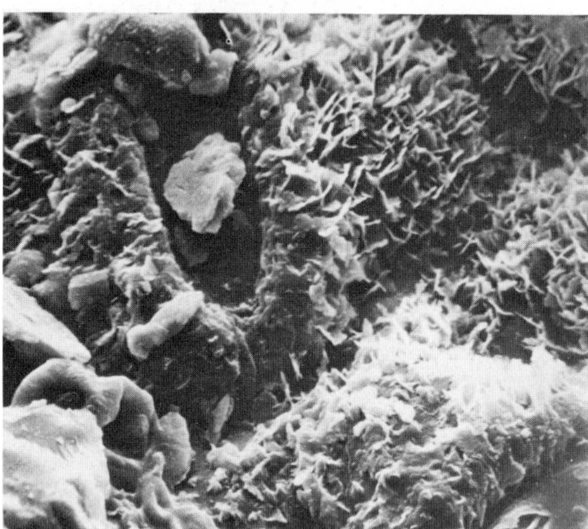

Electronmicrographs of the leaf surface of an oak (*Quercus*) from (a) an unpolluted area, and (b) an area with high fallout of debris which interferes with the opening and closing of stomata (× 24 000)

Smoke and dust

Smoke consists of the visible products of incomplete combustion, namely much carbon and tarry hydrocarbons. Some of these are implicated in diseases of the respiratory tract which, as a delicate, thin, highly vascularized large surface area, is particularly susceptible to damage.

Dust is usually considered as being composed of larger particles than smoke, and thus has a greater tendency to settle out from the air. The particles act as lung irritants and, in the presence of benz-pyrenes, may stimulate cancer growth and thus account for the 'urban factor' in lung cancer. Dust reduces the amount of light penetrating through to leaves and can block up to 60 per cent of the stomata of exposed plants; in these ways it can reduce the rate of photosynthesis.

Carbon dioxide

Pure air contains about 0.03 per cent carbon dioxide and it is calculated that this represents an increase of 15 per cent on that existing in 1900. If fuel combustion continues at the present rate the carbon dioxide level could increase by up to 25 per cent by the year 2000. Increased carbon dioxide aids photosynthesis as, under natural conditions, it is usually the limiting factor, however, at too high a level it can become toxic. There is also speculation about a possible warming effect of the atmosphere due to an altered absorption of the Sun's energy.

Carbon monoxide

Motor vehicles produce up to 80 per cent of all emissions of carbon monoxide in the world. It is a fairly persistent pollutant, lasting for several years in the atmosphere before being converted to carbon dioxide. Its main effect is that it combines irreversibly with haemoglobin in the red blood corpuscles to form stable carboxyhaemoglobin, thus preventing the carriage of oxygen to the tissues as oxyhaemoglobin. Carbon monoxide blood levels of up to 5 per cent are found in those breathing in heavy traffic fumes, and of up to 16 per cent in heavy smokers.

Lead

In some countries, this metal is added as tetraethyl lead to petrol, for reasons which remain controversial. As a result, about 3000 tonnes of lead are emitted via petrol engine exhausts in Britain each year. This 'aerosol' lead in car exhausts represents a danger which is being increasingly recognized after many years of complacency. It is not only breathed in, but also settles and may contaminate the soil, plants, and water. It is estimated that the sea receives one million tonnes of lead a year from car exhausts.

Average figures for daily lead uptake are about 100 μg from solid foods, 100 μg from drinks, and 30 μg from the air. The same amount, that is about 230 μg, is eliminated each day in the urine and the faeces. However, any intake above this level accumulates in the body. The bones will store lead instead of calcium, and it can lead to damage of the nervous system, including the brain, from which children are particularly at risk.

Partial chemotherapy is possible using substances such as EDTA which render the lead soluble and thus excretable. Lead is also deposited in the hair and subsequently lost over a 5–7 year cycle. However, lead poisoning is essentially accumulative and permanent.

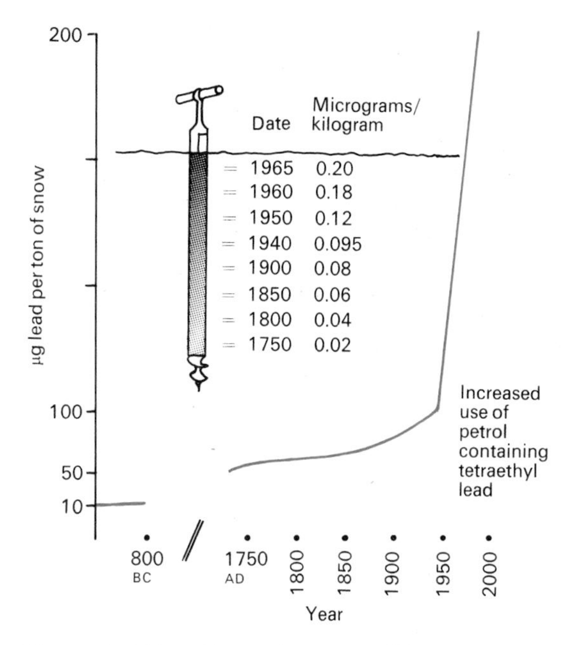

Figure 16.3 *Lead in arctic snow and ice*
The global nature of lead pollution is illustrated by these figures, gained by ice borings in Greenland.

Fluorine

Fluorine is emitted by aluminium, steel, brick, and phosphate fertilizer plants. It is a stable, dangerous, and persistent pollutant which undergoes biological concentration in plants. Some grasses can contain up to 2000 ppm, and high levels of fluorine can reach grazing livestock from this source.

Excessive intake leads to a condition known as fluorosis in which there is a swelling of the joints; and, although fluorine is an essential element of protoplasm, excess inhibits enzyme activity.

FLUORIDATION

In many areas fluoride is added to water supplies at levels of up to 1 ppm in an attempt to reduce the incidence of tooth decay, again especially amongst the young. However, only about 10 per cent of the water supply is consumed directly; the rest is used in a wide variety of ways, including food processing, and food processed with fluoridated water can have its fluoride content increased by as much as three times. Ideally the total average daily intake of fluoride from all sources should be assessed before it is added to the water supply.

Polychlorinated biphenyls (PCBs)

These are added to plastics as plasticizers to improve flame retardance, to increase the stickiness of adhesives, and to increase their resistance to chemical attack.

PCBs are released into the air as pollutants by the incineration of waste plastics. Attention was not drawn to these chemicals until they were found in the fats of eagles and seals at concentrations of up to 240 ppm. These levels had arisen by biological concentration from estuarine levels of about 0.01 ppm.

PCBs are stable pollutants and, although the long-term effects are not clear, some manufacturers have voluntarily banned their use. Certainly, greater care should be taken over the incineration of plastics and the consequent release of toxic substances to the atmosphere, both industrially and domestically.

Water pollution

Such is the burden and complexity of pollutants entering waters today that the sayings 'running water purifies itself in ten miles', and 'the solution to pollution is dilution', no longer generally hold true.

There are five main groups of water pollutants.

(a) Those that reduce dissolved oxygen, e.g. heat, organic materials including sewage, and inorganic reducing agents.
(b) The physiologically toxic substances, e.g. phenols, cyanide, lead, arsenic, pesticides, and herbicides.
(c) Excess nutrients that cause eutrophication, e.g. nitrates and phosphate fertilizers, and phosphate detergents.
(d) Oil.
(e) Suspended solids.

Even the oceans are threatened, for example the Mediterranean receives about 450 000 million tonnes of pollutants from land-based sources each year. Of the sewage expelled by the 120 cities on its shore, 90 per cent reaches the sea untreated, and the river Rhone alone carries down about 500 tonnes of pesticides each year. The Mediterranean also contains about 50 per cent of all the world's floating tar and oil. As it is an enclosed sea; the water takes about eighty years to be renewed.

Pollutants that reduce dissolved oxygen

At 20 °C, a litre (dm^3) of water only contains 6 cm^3 of oxygen, as compared with 200 cm^3 of oxygen in a litre of air at 20 °C, thus any pollutant that decreases this already relatively low level will have a profound effect on water organisms, especially animals, which will have to move an even larger volume of water over their respiratory surfaces to obtain the oxygen they require. They will thus be exposed to a greater risk of contamination from other pollutants.

Domestic sewage is the major pollutant of this type, but animal excreta from intensive farming

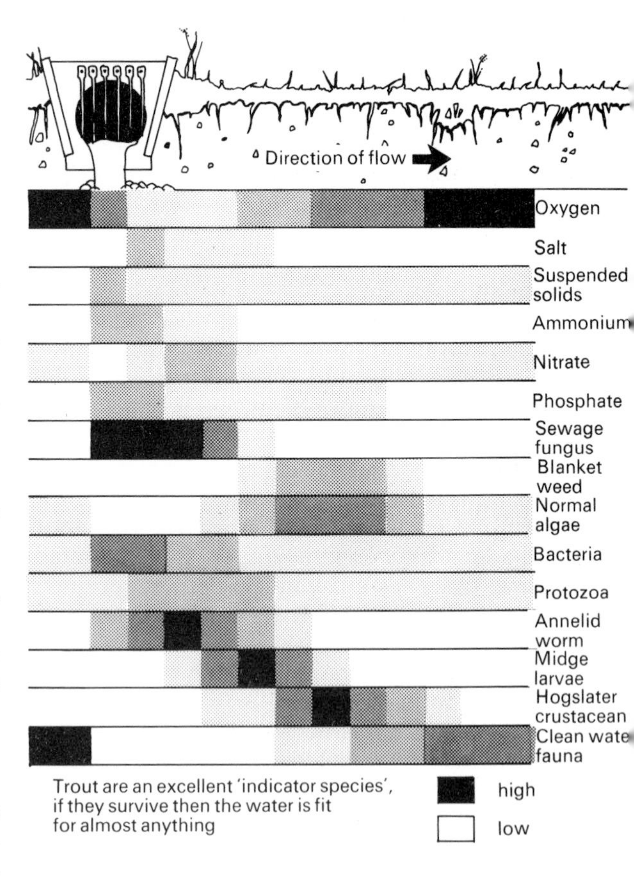

Figure 16.4 *The effect of sewage effluent on river water and its fauna*

* 'Sewage fungus' a green, slimy, cotton wool-like mass of protozoa, algae, fungi, and bacteria (mainly the filamentous bacterium *Sphaerotilus nathans*) is a sensitive indicator of the presence of organic pollutants. The oligochaete annelid worm *Tubifex* can exist in almost O_2 free water. Blanket weed consists of masses of the filamentous green alga *Cladophera*.

lots is an increasing hazard as their potential pollution load is three times that of the human population in Britain. The average sewage output of a single person has a daily requirement of oxygen equal to that contained in 2000 gallons of water, and this amount of oxygen must be supplied by the sewage treatment process.

Sewage also contains synthetic detergents, or 'syndets', of which there are two main types: the so-called 'hard' detergents, which have branched chain molecules and which are not readily biodegradable; and the 'soft' detergents, which have straight chain molecules and are more readily broken down by micro-organisms in sewage plants.

Although detergents are not particularly toxic they can cause foaming, and as little as 0.1 ppm can halve the rate at which oxygen dissolves at the surface, possibly by lowering the surface tension. The lowering of surface tension can also have complex ecological effects on surface water dwelling organisms. Detergents also contain phosphates which contribute to the eutrophication[322] of waters.

Any organic effluent such as that from tanneries, paper mills, food processing, factories, abattoirs, dairies, breweries, silage, etc., will reduce the oxygen content of the water into which it is discharged. It is possible to compare the oxygen requirement of these effluents with that of an individual's sewage contribution to obtain a **population equivalent**. Thus grass silo waste has a population equivalent per cubic metre of waste of 833.3, a dramatic illustration of the importance of proper control of these wastes in addition to domestic sewage.

Thermal pollution decreases the solubility of oxygen in water and also increases the metabolic rate of organisms so that animals require more oxygen. They therefore increase the flow of water over their respiratory surfaces, exposing themselves to an even greater volume of polluted water. Also certain critical temperatures at key points in various animal's life histories may be exceeded.

Physiologically toxic substances

PESTICIDES

Pesticides should be highly specific to their target organisms. Since this is seldom achieved, they become pollutants when they reach and affect the wrong organisms. The persistent pesticides are liable to biological concentration and many have such widespread effects that they could be more accurately known as biocides than pesticides.

Chlorinated hydrocarbons

These include DDT, BHC, dieldrin, and aldrin. They are persistent and undergo biological concentration, particularly in fatty tissue, where they are stored harmlessly until the fat is metabolized, at which time the stored pesticides are released suddenly into the blood stream with often fatal effects. The long-term effects on man are not clear, and fat levels up to 100 ppm have no obvious short term effect.

There is additional concern that DDT depresses photosynthesis in phytoplankton, and could have catastrophic effects on marine food-chains and atmospheric oxygen renewal.

It is for these reasons, and because DDT can be detected in every part of the biosphere, that the use of DDT and many other chlorinated hydrocarbons has now been severely restricted in most countries.

Organophosphorus compounds

These include substances such as parathion and malathion. Parathion is very poisonous and has caused many thousands of accidental deaths throughout the world, but it is non-persistent. Malathion is only one-thousandth as toxic as parathion due to the body's detoxyfication mechanisms, but exposure to apparently harmless doses of parathion can damage these detoxyfication mechanisms and malathion can thus become very poisonous. This again illustrates the complexity of interactions that are possible, and why theoretical 'safe limits' for various substances must be kept continually under review.

HERBICIDES

Herbicides are the most widely used type of biocide. Hormone herbicides mimic the action of naturally-occurring plant hormones and thus interfere with the plant's metabolism. The most common of these is a substance, known as MCPA, which eliminates broad-leaved dicotyledonous[352] weeds in a monocotyledonous[353] crop. Physiologically it has the same effect on all plants but, when sprayed, accumulates on the broad flat leaves of the dicotyledons, but not on the narrow vertical leaves of the monocotyledons. It is non-persistent, and only slight effects on the soil flora and fauna have been recorded.

Another hormone herbicide, known as 2,45-T, which has been widely used as a defoliant, can contain dioxin as an impurity. It therefore has very serious and long-lasting effects on the environment and on man directly.

Other types of herbicides include the compounds Diquat and Paraquat which 'burn' off the aerial parts of plants. These are said to 'disappear' on contact with the soil; however, this must be an assumption because their long-term effects on the soil are not known. They are extremely poisonous if swallowed, and there is no known antidote.

Fungicides can contain copper, as in the copper sulphate used to destroy potato blight fungus (*Phytophthora infestans*), and mercury compounds are also widely used.

Arsenic is acutely poisonous and carcinogenic and persists in the environment for many years. Sodium arsenate was once used extensively to destroy potato plant foliage, and many areas are still contaminated. Lead arsenate, which is extremely poisonous, is still used in fruit tree sprays, and arsenical compounds used on tobacco crops are a further hazard for smokers—and non-smokers who are forced to inhale this local pollution.

Excess nutrients that cause eutrophication

Eutrophication occurs naturally. Upland rivers which are low in nutrients, or oligotrophic, become loaded with nutrients, or **eutrophic**, as they flow into the lowlands. Lakes also undergo a natural ageing process in which their waters become rich in nutrients and very productive, supporting the production of masses of plant material. This in time decays and builds up deposits which eventually obliterate the lake. However, human activities often shorten the time scale of these events. For example, it is estimated that Lake Erie is now in a state that it would have taken another 15000 years to reach under natural conditions.

Substances that contain organic and inorganic nutrients include those added directly to the water, such as human and animal excreta, waste vegetable matter, and phosphate detergents; and those added indirectly by leaching or washing from the soil, such as inorganic fertilizers.

Inorganic fertilizers contain phosphates and nitrates. The nitrates are more easily leached from the soil than the phosphates, and therefore the nitrates represent the bigger hazard to the artificial eutrophication of water. Of the nitrogen in fertilizers added to the soil, only about 50 per cent is utilized by the crops, some re-enters the atmosphere via denitrification, and the rest is leached. The amount of run-off from the soil varies with the type of soil, the season, the weather, and the time of application. Potentially, nitrates are one of the most serious of all pollutants of drinking water, as the soluble forms are not removed by the usual treatment process.

There are vast amounts of nitrates slowly percolating down towards the underground water stores or aquifers which provide about 30 per cent of the water supplies in Britain. As more and more aquifers become polluted with dangerous levels of nitrate, increasing attention will have to be paid to the problems of nitrate removal at water purification works. Although of little apparent danger to adults, babies have a different gut flora of bacteria which convert the nitrates to nitrites. These then combine with haemoglobin to form methaemoglobin, which does not carry oxygen.

One of the ecological results of artificial eutrophication is a sudden increase in algal growth known as an **algal bloom**. These algae generally consist of masses of unicellular *Monodus* species and filamentous *Cladophora* species. They are a great problem to water purification plants as they block the filters. It is not clear what actually triggers a bloom off in the first place, but in some cases it is caused by pesticides destroying algal 'grazers' such as *Daphnia* and other zooplankton, so that there is an initial increase of growth which then continues exponentially as the organisms exploit the rich supply of nutrients. The masses of algae are relatively short lived, and their decomposition by aerobic bacteria uses up all the available oxygen, and gives rise to anaerobic conditions under which only anaerobic bacteria survive.

The whole complex web of life of the water subsequently breaks down, and the condition cannot be reversed by a simple process of reoxygenation.

Oil

Oil is the most obvious pollutant of the seas, especially when washed up on beaches, but it is not necessarily the most serious because it is eventually broken down by bacteria.

More ecological damage can be done by the synthetic detergents used in an attempt to disperse the oil but, in contrast to oil which spreads very thinly over the surface of the water, these are not visible.

The increasing off-shore production of oil around the world increases the risks of serious pollution from this source in addition to the ever-present risk from super-tankers. Even when no accidents occur, it is estimated that about two million tonnes of oil are added to the sea each year from shipping.

Table 16.2 Sources of pollution (per annum)

Source	*million tonnes*	%
Tankers	1.09	17.6
Dry docking	0.26	4.0
Terminal Operations	0.003	0.04
Bilges bunkering	0.51	8.1
Tanker accidents	0.2	3.2
Non-tanker accidents	0.1	1.62
Off-shore production	0.08	1.3
Coastal oil refineries	0.2	3.2
Industrial waste	0.3	4.9
Municipal waste	0.3	4.9
Urban runoff	0.3	4.9
River runoff	1.62	26.1
Natural seeps	0.6	9.8
Atmospheric rainout	0.6	9.8
Total	6.163	

Suspended solids

There are always solids naturally suspended in water, but man's activities can raise the levels until they constitute pollution. For example, this can occur as a result of soil erosion from soils degraded by poor agricultural practice.

The suspended solids settle and may destroy, or at least change, the nature of the bottom-living flora and fauna; light penetration can also be interfered with and photosynthesis thus reduced, with a consequent reduction in oxygen production. Eventually, silting up of the water courses may change the entire pattern of drainage of an area and produce far-reaching effects.

Ionizing radiations

Ionizing radiations are very high-energy radiations which can remove electrons from atoms and molecules as they pass through a material. If an electron is removed in this way then the atom or molecule becomes a positively charged ion; and if that electron is added to another atom or molecule this will become a negatively charged ion. Ionizing radiations are either electromagnetic radiations and fast-moving particles are emitted by **radioactive materials**, either light are part of sunlight, and both electromagnetic radiations and fast-moving particles are emitted by **radioactive materials**, either occurring naturally in ores; or produced in nuclear reactors or by nuclear weapons. In addition the Earth receives cosmic rays from the Sun and outer space, which consist mainly of charged particles.

The particles include positively charged protons and alpha (α) particles; negatively charged beta (β) particles; and neutrons which have no electrical charge. The electromagnetic radiations include those wavelengths of the electromagnetic spectrum described as gamma (γ) rays, X-rays, and ultra-violet light.

The α-particles consist of two protons and two neutrons, and are in fact the nuclei of helium atoms. They travel only a few centimetres in air and cannot penetrate clothing or unbroken skin.

The β-particles are high speed electrons. They travel a couple of feet in air and can only penetrate a few millimetres of living tissue.

The α- and β-particles have been described as 'internal emitters' because they have their greatest effect on the body when swallowed or inhaled.

Neutrons, which are uncharged particles, are not capable of causing ionnization. However, they can knock protons out of the nuclei of atoms and, since these protons are positively charged, this can cause ionization in substances through which they pass.

Gamma (γ) rays travel at the speed of light for miles through the air and can penetrate the body completely. They are very energetic and produce much ionization along the route of their passage through the body.

X-rays are electromagnetic radiations of longer wavelength than γ-rays. Like γ-rays, they travel at the speed of light through air and penetrate the body completely, producing much ionization along their route of passage through the body.

Radioactive isotopes with short half-lives soon decay and, although they have a high activity, their short life renders them less important ecologically than the longer-lived isotopes.

Dose

The amount of radiation received by the body of a living organism is referred to as the 'received dose'. The effects of various doses depend on a complex of factors and can vary considerably. Generally, high level exposures of short duration are more dangerous than low level exposures of long duration since, with the latter, the body's cellular repair mechanisms have a chance to operate. However, all doses are damaging and there is no true threshold of tolerance. All doses add to the so-called 'genetic load' of a species by their damaging effect on the genetic material in the gametes.

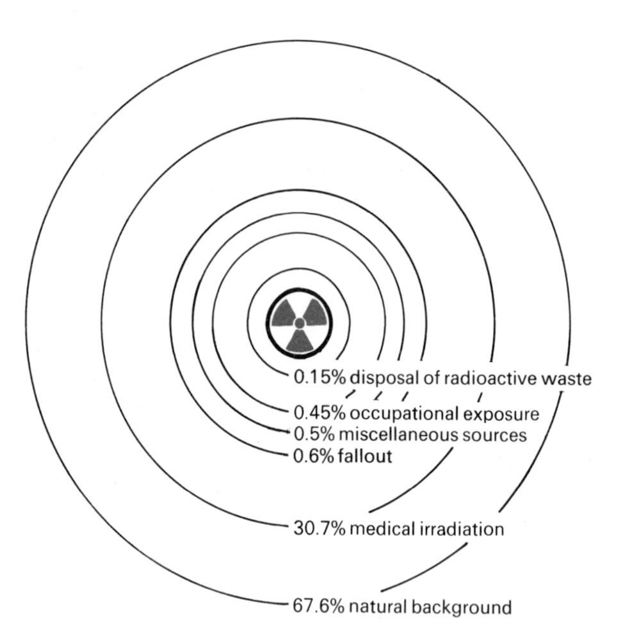

Figure 16.5 *Origins of radiation doses received (measured in annual effective dose equivalents)*

Biological effects of ionizing radiations

The biological effects are not fully understood, but it is known that ions of chemicals in DNA are chemically highly reactive, and their production can lead to both gene and chromosomal mutations. Similarly, the highly reactive ions that are produced by ionizing radiations in the cytoplasm disrupt cell function, probably by interfering with enzyme action. Actively dividing tissues such as the skin are especially at risk, as are also the lymphatics and the gonads.

The effect of the radiations on the bone marrow is to disrupt the production of red blood cells, which leads to anaemia; and to interfere with the production of white blood cells and blood platelets, which, along with the effect on the lymphatic system, reduces the body's resistance to infection to a level which is usually fatal. It is the vulnerability of the bone marrow which renders mammals very sensitive to relatively low doses of radiation.

The effect on the lining of the gut is to induce the so-called gastro-intestinal syndrome of diarrhoea, vomiting, and discharge of blood. Very high radiation doses cause the cerebro-vascular syndrome of damage to the central nervous system and lead to rapid death.

Radioactive isotopes can undergo biological concentration which results in the build-up of high internal levels from relatively low environmental concentrations. The powerfully radioactive strontium 90, which is produced in great quantities by hydrogen bomb explosions, has a half-life of about twenty-eight years and is metabolized similarly to calcium. It can become concentrated up to 500 times the environmental level in bones, where its emissions of β-particles are a danger to the vulnerable dividing cells of the bone marrow.

Caesium 137 has a half-life of thirty-three years and is metabolized similarly to potassium. It tends to spread throughout muscles, and can be concentrated up to 250 times the environmental level. It emits both β-particles and γ-rays.

Iodine 131 has a half-life of eight days and tends to be concentrated in the thyroid, up to 500 times the environmental level, where it emits both β-particles and γ-rays. It is a particular danger to young children.

The knowledge that some of these materials can be excreted from the body gives rise to the concept of the **radiobiological half-life**, which is the time the radioactivity within the body takes to reduce by half.

Long-term effects of low doses of radiation are the possibilities of cancer, leukaemia, a general shortening of the life span, and genetic damage. Increasing levels of ionizing radiation increase the rate of genetic mutations, most of which are harmful.

Table 16.3 *Some significant dose levels to man in rem (r)*

0.1 r	Natural background dose for most tissues in a year, some of which penetrates the body from external sources, but most of which comes from particles swallowed or breathed in
3 r	Received to the gonads by everybody over 30 years from natural background radiation
10 r	The body gradually replaces cells killed by radiation at an estimated rate equivalent to the damage done by an exposure of about 10 r each day
35 r	Doubles the mutation rate of most genes
73 r	Official maximum permitted dose per year for those working with radioactive materials in 1931. Revised down to 5.0 r per year in 1957
350–550 r	If received in 1–2 hours, is the dose at which each individual has a 50:50 chance of survival (LD_{50}) after a period of vomiting, anaemia, haemorrhage, and loss of hair
400–800 r	Fatal within 10 days. Higher doses can be tolerated if the abdomen is well-shielded
5000 r	Rapid death after single exposure

Table 16.4 *Average annual dose to the general population*

	Dose in rems (r)	*Approximate percentages of total from all sources*	*of man-made*
Natural background	0.125	68.0	—
Medical sources	0.055	30.0	94.0
Weapons fallout	0.002	1.0	3.0
Reactor wastes	0.003	1.0	3.0
TV sets	0.001	0.5	1.5
Luminous watch dials	0.001	0.5	1.5
Total	0.187		

Experimental and applied aspects

Radioactive wastes from nuclear reactors

High-level radioactive wastes are those liquids or solids that must be contained because they are too dangerous to be released anywhere in the biosphere. One ton of spent nuclear fuel gives rise to about 440 l of such wastes. By 1969 there were about 350 million litres of such high-level waste stored in 200 underground tanks in the USA; and at least 200 000 cubic metres of new storage space has been required annually since then. Proposals for the glassification of these wastes and their subsequent burial deep underground are being considered.

Most concern about the wastes being produced centres around the man-made element called plutonium. The fissionable isotope plutonium (Pu 239) is created in atomic reactor fuel elements. It is recovered from these by reprocessing and can be used in nuclear weapons, but most of it is ear-marked for the future use as the fuel in so-called breeder reactors. However, very little Pu 239 has been recycled so far, and it is continually accumulating.

Pu 239 emits α-particle radiation and has a half-life of 24 000 years. It has been suggested that no discovery has had a greater military, and therefore political and social, significance than the discovery of plutonium as it is the most convenient material for use for the construction of atomic bombs. There is therefore great concern about the spread of atomic power on the grounds that it will be impossible to keep the accumulating amounts of Pu 239 secure from extremists. In addition there is the problem of keeping large amounts of plutonium isolated from the environment for many thousands of years, which is necessary for a radioactive element with such a long half-life.

By the 1980s, about 350 000–500 000 kg of Pu 239 will have accumulated, and the exponential increase in the amount produced would result in vast amounts being produced by the end of the century.

A particle of Pu 239 oxide, 1 μm in diameter, delivers a dose of 4000 r per year to the 65 μg of surrounding lung tissue by short range α-particle radiation which travels less than 0.1 mm in solids and liquids. This type of irradiation is referred to as 'hot particle' irradiation.

Very large volumes of low-level waste are continually released into the environment, supposedly with little if any effect. It has been calculated that the current disposal of wastes from the use of atomic power for generating electricity is responsible for fewer than ten deaths each year throughout the world, whereas the estimated figure for deaths caused by the burning of fossil fuels is between 10 000 and 20 000 a year. However, the potential dangers of radioactive wastes are serious and long-lasting due to their effect on the genetic material of all living things throughout the entire biosphere.

Nuclear weapons

The power of nuclear detonations is measured in equivalents of tons of TNT. Kiloton (kt) weapons are equivalent to 1000 tons of TNT; those dropped on Hiroshima and Nagasaki in Japan in August 1945 were both 20 kt weapons. Megaton (mt) weapons are equivalent to 1 000 000 tonnes of TNT.

The energy released by a nuclear detonation is distributed approximately in the proportions of 45 per cent as blast and shock; 35 per cent as light and heat; 5 per cent as initial nuclear radiations (within one minute); and 15 per cent as residual radiations from fission products in the radioactive fallout.

The hazards from blast and heat of mt detonations would extend far beyond the range of possible injury from initial nuclear radiations. The heat from a 20 mt blast would char the skin at a distance of about 32 km from the point of detonation or ground zero (GZ), and blister it at a range of about 48 km.

Radioactive fallout results from the condensation of vaporized fission products on dust particles swept up by the hot air produced by the explosion. The typical mushroom-shaped cloud of a kt weapon is caused by the fallout flattening out when it reaches the layer of the atmosphere 11 000–13 000 m above the earth, known as the tropopause, with a constant temperature of $-60\,^{\circ}C$. The cloud from mt weapons, however, passes through the tropopause to heights in excess of 32 km, with a consequently wider dispersal of fallout particles.

In the absence of any wind a circular fallout pattern forms around the point of explosion or ground zero (GZ) with the same diameter as the cloud; in a 10 mt explosion this would be about 90 km. In the presence of wind there is virtually no upwind fallout, but the downwind fallout range can be considerable. Given the average winds in Britain, fallout could extend several hundred miles downwind of GZ. About 60 per cent of the fallout from a surface burst would be deposited within about two days.

Radioactive fallout damage is caused both by contact and whole-body radiation. The contact hazard is due mainly to α- and β-particles emanating from fallout on the skin, or within the body. The whole-body radiation mainly comes from external γ-radiations emanating from fallout in the environment, and is far more dangerous than the contact hazard. γ-radiation can be scattered back to the earth from the atmosphere, and therefore all-round cover by thick shielding is required. The thickness of material needed to reduce the dose rate in a beam of γ-rays by one half is referred to as its 'half-value' thickness.

Material	*Half-value thickness (cm)*
Steel	2.4
Concrete	9.6
Earth	12.0
Water	20.8

A series of assumptions and calculations have been made to produce the following measure of the so-called 'protective factor' of different situations.

	Protective factor
Bungalow	5–10
Trench under a detached two-story house	100
Slit trench in open with 1 m earth cover	200
Lower floors of blocks of buildings	50–500

Guided example

1 With regard to ionizing radiations emanating from man's use of radioactive materials in nuclear reactors and weapons testing, it is argued, as with some other pollutants, that the amount produced is only a fraction of that dose received from the natural background radiation, and is therefore of little significance. What criticism could be levelled against such an argument?
Natural background radiation comes from naturally-occurring radioactive substances in rocks, soil, water, and air; and from mixtures of particles and electromagnetic radiations which enter the Earth's atmosphere from space. As with other pollutants, the existence of a naturally-occurring background level complicates the problems in assessing the significance of the contribution from man's activities. However, although at present the contribution from nuclear tests and reactor wastes is only a fraction of the natural background level, it must be remembered that there is no safe level of radiation, and that damage is directly proportional to the dose received; thus any addition, no matter how small, increases the dangers.

2 The fact that natural background radiation is part of the natural environment has led some to suggest a positive role for it in relation to living organisms. Can you think of any such role?
It has been suggested that such background radiations may be important in maintaining the observed rate of genetic variation in nature, which is of central importance to the theory of evolution.

3 With regard to the atmospheric testing of nuclear weapons, it has been argued that such testing in areas remote from centres of populations is 'safe'. What is wrong with such assurances?
Any testing of nuclear weapons in the atmosphere produces global contamination by world-wide fallout. For example, the large megaton tests of the early 1960s ejected debris into the stratosphere which will continue to have an affect for many years to come. Indeed, everybody in the world has received, and is receiving, some dose of radiation as a result of fallout from explosions.

4 Can you think of any reasons why the Eskimos and Lapps should be more at risk from radioactive fallout than many other populations?
Poor soils with mat-like vegetation, such as are found in the arctic and sub-arctic regions, result in an increased trapping and uptake of fallout by plants. In these regions reindeer and caribou feed on the ground lichens, and Eskimos and Lapps, who feed in turn on these animals, are exposed to internal doses of radiation much higher than the background level.

5 It is argued by some that forms of pollution other than ionizing radiations are responsible for many more deaths annually than those due to radioactivity resulting from man's activities. Others argue, however, that a certain aspect of radioactivity sets it apart from most other types of pollutant. What do you think this special feature could be?
The most serious consequences of polluting the Earth and its atmosphere with radioactive materials are the effects that ionizing radiations have on the genetic material by increasing the genetic load of all living organisms in all parts of the globe. It is this aspect of radioactive pollution which sets it apart from most other types of pollution.

6 Which particular use of radioactive materials carries the greatest threat to the living world?
The danger of nuclear warfare represents the greatest threat to all life on Earth, both short-term and long-term via the effects on the genetic material of all living things.

Questions

1 Explain the importance of *two* of the following:

(a) atmospheric pollution;
(b) overfishing;
(c) pesticides;
(d) plant monoculture. (L)

2 EITHER (a) Write an essay on 'The use and abuse of herbicides and insecticides.'
OR (b) Give a brief account of the nature and effects of *either* atmospheric *or* water pollution, and discuss methods of control. (SUJB)

3 EITHER (a) Describe how you would estimate the effects of a suspected pollutant in a named habitat.
OR (b) Distinguish between contamination and pollution. With reference to a named pollutant, state the effectiveness of legislation in controlling its level in a habitat. (SUJB)

4 EITHER (a) With reference to specific tests and observations, say how you would assess the varying degrees of pollution in the countryside near a large town.
OR (b) Discuss the effects of a named pollutant in *either* the sea *or* in fresh water. What action can be taken at local and national levels to enforce control of the level of pollutant in the habitat? (SUJB)

5 Discuss the effects the following might have if they entered a slow running river:
(a) untreated sewage;
(b) excessive run off from agricultural land;
(c) hot water from the cooling towers of a power station. (L)

6 A river was sampled at a point X along its course and the levels of a number of living and non-living constituents in the water were recorded. Below point X a sewage effluent entered the river at a point designated Y. Samples were again taken at Y and then at intervals downstream from Y until the whole situation was found to be restored to the levels existing at X.
The constituents sampled were:
mineral ions, oxygen concentration, bacterial populations, algal populations, invertebrate animal populations, fish populations.
Suggest the possible fluctuations in the levels of these constituents from values you assign at X until their return to these levels. Explain the reasoning you use in suggesting these values. If you find it helpful construct figures or diagrams.
What measures would you suggest to minimize disruption of the river's ecosystem by such an input? (AEB)

Further reading

Baker R. E. and Bushell J. A., *The Unclean Planet* (London: Ginn, 1971).
Coggle J. E., *Biological Effects of Radiation* (London: Wykeham Publications, 1971).
Lucas J., *Our Polluted Food* (London: Charles Knight, 1975).
Mellanby K., *Pesticides and Pollution* (London: Collins, 1967).
Walker C., *Environmental Pollution by Chemicals* (London: Hutchinson, 1975).

APPENDICES

Appendix 1

Animal taxonomy

Taxonomy

Taxonomy is the study of the principles on which a classification is based. The classification of living organisms is based mainly on similarities in structure, that is from observations of comparative morphology and anatomy, and involves the study of homologous structures.[60] Other evidence for the affinities of group members comes from palaeontology, cytology, genetics, and physiology. Physiology is particularly important in the classification of bacteria and some fungi, where structural similarity can mask great differences in physiology.

The establishing of which groups or taxa are to be used is referred to as 'systematics'. The basic grouping or **taxon** is the species; these are usually fairly well-defined. However, groupings above the level of the species can vary, with many shades of opinion as to the status and size of the various categories. 'Nomenclature' deals with the naming of the groups used and of the individuals in those groups. Some of the main groupings used are: Phylum (Division), Class, Order, Family Genus, species.

The Swedish botanist Linnaeus (1707–78) initiated the **Binomial System of Nomenclature** in 1753, by which every living organism is given a two-word name, the first of the Genus and the second of the species. The name of the Genus is always written with a capital letter, and the name of the species with a small initial letter; for example *Homo sapiens* (man) and *Primula vulgaris* (primrose).

KINGDOM ANIMALIA

Phylum Protozoa

Unicellular; microscopic; about 30 000 species described.

Class Rhizopoda (Sarcodina)

Amoeboid; moving and feeding by ever-changing cytoplasmic extrusions known as pseudopodia, e.g. *Amoeba*.[224] Fresh water; feeding by phagocytosis, i.e. engulfing bacteria, diatoms, etc., with pseudopodia; contractile vacuole for expulsion of excess water that enters by osmosis; reproduction by binary fission.

Class Flagellata (Mastigophora)

Flagellate, moving by one or more flagella; reproduction only by asexual longitudinal binary fission. Includes *Euglena*, *Trypanosoma* (blood parasite of vertebrates, including man, causing sleeping sickness), *Polytoma*.

Euglena[225]
Found in pools and ditches rich in nitrogen. Cell membrane is a firm pellicle; chlorophyll in chloroplastids, therefore photosynthetic; positively phototropic with eye spot; can be saprozoic in dark.

Class Ciliata (Ciliophora)

Ciliated during at least one stage of the life cycle, moving and feeding by means of cilia; usually binucleate with mega- and micro-nucleus; unique form of sexual reproduction known as conjugation; e.g. *Paramecium*,[225] *Zoothamnium*[188] (stalked colonies), *Carchesium* (stalked colonies on the legs of *Gammarus*).

Paramecium
Fresh water. Slipper-shaped; firm pellicle with cilia; mega- and micro-nucleus; strong beating cilia known as the 'undulating membrane' of the oral groove draw in bacteria, etc., which are engulfed in food vacuoles and which then undergo a figure-of-eight passage through the cytoplasm known as cyclosis, undergoing first acid then alkaline digestion; spiralling and gyrating locomotion by metachronal beating of the cilia; reproduction by asexual binary fission, and sexual conjugation.

Class Sporozoa

All parasitic; adults non-motile; usually complex life cycle, and reproduction by means of large numbers of spores; e.g. *Monocystis* (parasitic in the seminal vesicles of earthworm); *Plasmodium* (blood parasite of man causing malaria).

Phylum Porifera

The sponges. Fresh water and marine. Single body cavity connected to exterior by pores, cavity lined with flagellated choanocytes which create water currents; skeleton of calcareous, siliceous spicules, or horny fibres of spongin; no nervous cells, little integration between cells, body acts more as a colony of single cells; asexual reproduction by budding; e.g. *Euspongia* (the bath sponge).

SUB-KINGDOM METAZOA

Truly multicellular; division of labour between groups of cells constituting tissues and organs.

Phylum Coelenterata

Fresh water and marine. Are at the **tissue grade** of organization with muscular tissue and nervous tissue giving co-ordination of reactions and of locomotion; single sac-like body cavity (the enteron) opening to the exterior by a single opening the mouth; **diploblastic**, i.e. body wall of two layers of cells, the outer ectoderm and the inner endoderm, separated by a non-cellular mesogloea; sexual reproduction produces a characteristic ciliated planula larva, asexual reproduction by budding; can exist in two forms: the hydroid polyp and free-swimming sexually reproducing medusa; tentacles bear characteristic poisonous stinging cells known as nematoblasts or nematocysts (seventeen types); characteristic undifferentiated interstitial cells, which can differentiate in to all types of cell (not however in *Obelia*).

Class Hydrozoa

Typically marine and colonial, with polyp colony budding off free-swimming medusae which carry the gonads and produce new polyp colonies as a result of sexual reproduction, e.g. *Obelia* (found in coastal waters to a depth of 100 m), *Physalia* (Portuguese man-of-war), *Hydra* (atypical solitary fresh water form).

Hydra
A typical, fresh water; solitary polyp, no medusa. Feeding by trapping prey with nematocysts and passing food, such as *Daphnia*, to the mouth using tentacles in a co-ordinated manner; can extend by contracting radial muscles in body wall, and contract to a blob using longitudinal muscles; symbiotic green algae (*H. viridis*) or brown algae (*H. fusca*) in endodermal cells.

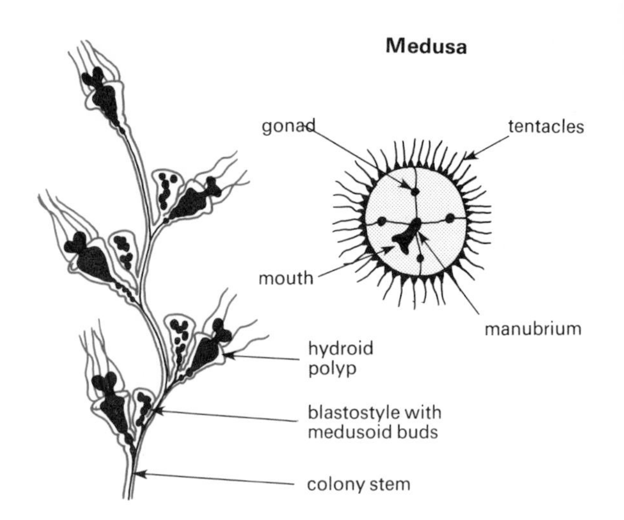

Figure 17.1 *Obelia colony and medusoid (3 mm diameter)*

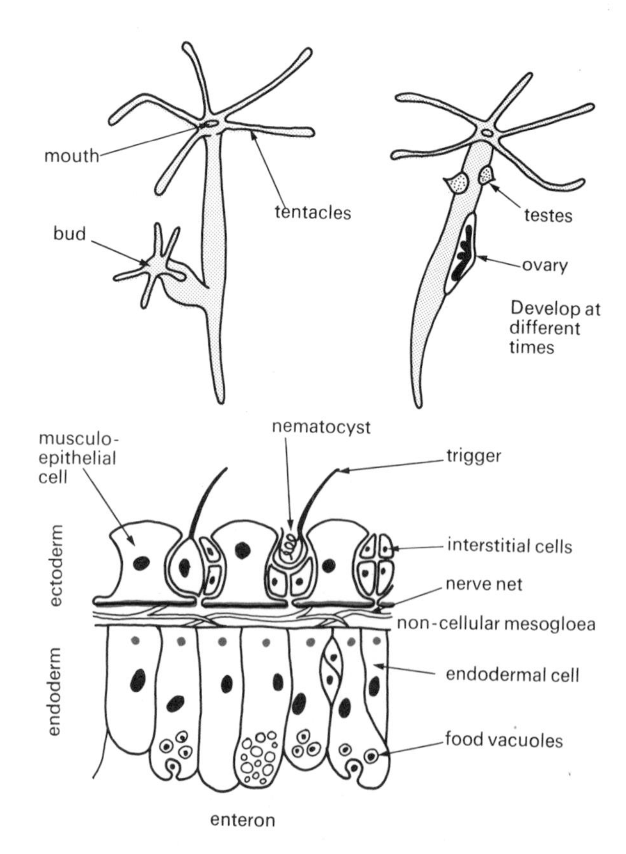

Figure 17.2 *Hydra (about 10 mm long)*
(a) With asexual bud, (b) with gonads, (c) H.P. detail of body wall

Class Scyphozoa

Marine. The jelly-fish. Adults medusae; polyp form can be absent, when present it buds off medusae; e.g. *Aurelia*.

Class Anthozoa

Marine. Polyp form only, solitary or colonial; enteron divided by vertical septa; endodermal gonads; e.g. *Actinia* (sea anemone), *Tubipora* (organ-pipe coral), *Gorgonia* (sea-fan), *Pennatula* (sea pen).

THE TRIPLOBLASTICA

Three-layered body wall, ectoderm, mesoderm, endoderm; general increase in size and complexity.

Phylum Platyhelminthes

The 'flat worms'. Single opening to gut; excretion and osmoregulation by characteristic flame cells; usually hermaphrodite.

Class Turbellaria

Aquatic; ciliated ectoderm; e.g. *Planaria* (fresh water).

Class Trematoda

The 'flukes'. All parasitic. No cilia; suckers for attachment to host; strong powers of reproduction, larval stages and complex life cycle involving a secondary host; e.g. *Fasciola hepatica* (liver fluke of sheep), *Schistosoma* (in veins of man causing schistosomiasis or bilharzia).

Class Cestoda

The 'tapeworms'. All parasitic in the gut of vertebrates. No cilia; no gut; reduced sensitivity and powers of locomotion; hooks and/or suckers for attachment; prolific reproduction involving strobilization and budding-off of mature proglottids full of fertilized eggs; larval stages and complex life cycle involving a secondary host; e.g. *Taenia saginata* (beef tapeworm), *T. solium* (pork tapeworm), *T. serrata* (dog tapeworm), *Diplidium caninum*[310], (dog tapeworm), *Echinococcus granulosus*[310] (dog tapeworm).

Phylum Nematoda (Aschelminthes)

The 'roundworms'. Terrestrial, aquatic, free-living and parasitic on plants and animals. No cilia; alimentary canal a straight tube with mouth and anus; usually separate sexes; e.g. *Ascaris* (common gut round worm), *Ancylostoma* (hookworm), *Heterodera*[303] (root nematodes), *Toxacara*[309] (dog and cat roundworm), *Trichinella spiralis* (man, rat, pig), *Aphelechus* (leaf parasite).

COELOMATA

Possess a fluid-filled cavity in the mesoderm known as the **coelom**. The coelom is always lined with coelomic epithelium which secretes the coelomic fluid. All animals above the Nematoda are coelomate. Coelomate animals show an increase in complexity, and usually size, over the Acoelomates. The coelom allows for the separation of the gut wall from the body wall; the gut can therefore contract independently and there is a corresponding increase in complexity and efficiency of nutrition. Indeed, the gut usually fills the greater part of the coelom. The coelomic fluid provides turgidity for the support, protection, and sometimes the locomotion of the animal, and ducts connect to the outside.

With the general increase in size and decrease in surface area to volume ratio, a blood vascular system with pumping heart carries out internal transport, for which simple diffusion alone is insufficient. The circulation of the blood supplies the complex organ systems.

The overall increase in complexity is reflected in the greater development of the co-ordinating systems, nervous and hormonal, seen in coelomates.

METAMERIC SEGMENTATION

All coelomate animals are also metamerically segmented, that is their bodies are divided into a linear series of segments from the earliest stages of embryological development, so that the body consists of a constant number of segments which are all of the same age (this contrasts with strobilization seen in the tapeworms). As the coelomates increase in complexity, the basic pattern of metameric segmentation becomes increasingly obscured by cephalization (fusion) and specialization, as for example in the head of vertebrates.

Phylum Annelida

The segmented worms. Typically aquatic. Single, pre-oral segment—the prostomium; central nervous system with paired cerebral ganglia joined by circum-oesophageal commissures to the ventral, solid, double nerve cord with segmental ganglia; excretion via characteristic segmental nephridia; body wall bears characteristic bristle-like chaetae.

Class Polychaeta

Marine. Many chaetae borne on lateral extensions of the body wall known as parapodia, sexes usually

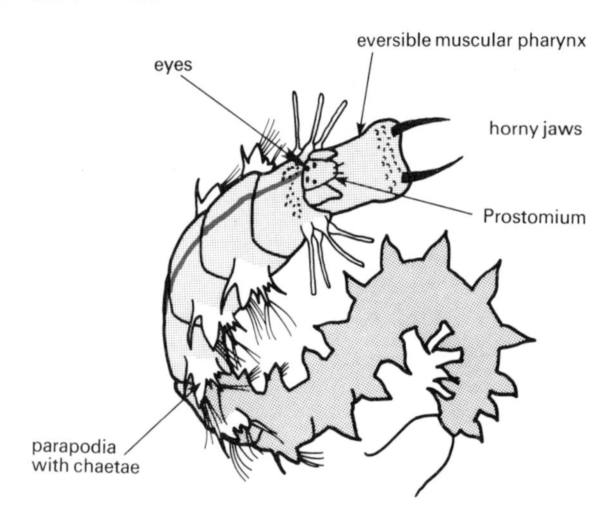

Figure 17.3 *Nereis (ragworm, 15–30 cm long)*

separate; planktonic trochosphere larva; e.g. *Nereis* (ragworm), *Arenicola* (lugworm), *Sabella* (fan-worm).

Class Oligochaeta

Terrestrial (the earthworms), some fresh water. Few chaetae per segment; no parapodia; reduced head region; hermaphrodite, but cross fertilization during copulation; eggs deposited in cocoon formed by a clitellum; direct development, no larval stage; e.g. *Lumbricus terrestris* (garden worm), *Allobophora longa* (long worm), *Eisenia foetida* (brandling), *Octolasion cyaneum* (blue worm).

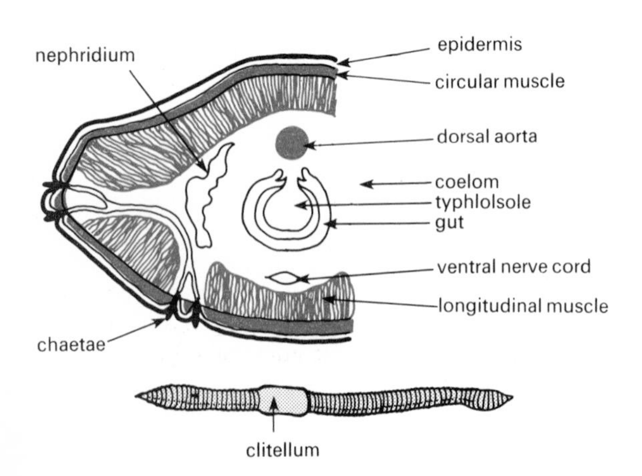

Figure 17.4 *Lumbricus (earthworm)*

Class Hirudinea

Fresh water. No chaetae or parapodia; shortened body with thirty-two segments but many more ring markings on ectoderm; all are ectoparasites—blood suckers, typically have anterior and posterior suckers; hermaphrodite with clitellum producing a cocoon; e.g. *Hirudo* (medicinal leech).

Phylum Mollusca

Fresh water, marine, and terrestrial. Body divided into head, visceral mass, and muscular foot; skin covering visceral mass extends to form a mantle which encloses a mantle cavity, the mantle also secretes an external shell in most cases; open blood system with haemocoel; metameric segmentation much reduced and modified; aquatic forms have a trochophore larva.

Class Gastropoda

Fresh water, marine, and terrestrial. Visceral mass undergoes torsion during development bringing the anus overhead, also shell coiled; large muscular foot; distinct head with eyes on tentacles, feed using radula; e.g. *Planorbis* (fresh-water snail), *Patella* (limpet), *Buccinium* (whelk), *Helix* (common garden snail), *Limax* (slug).

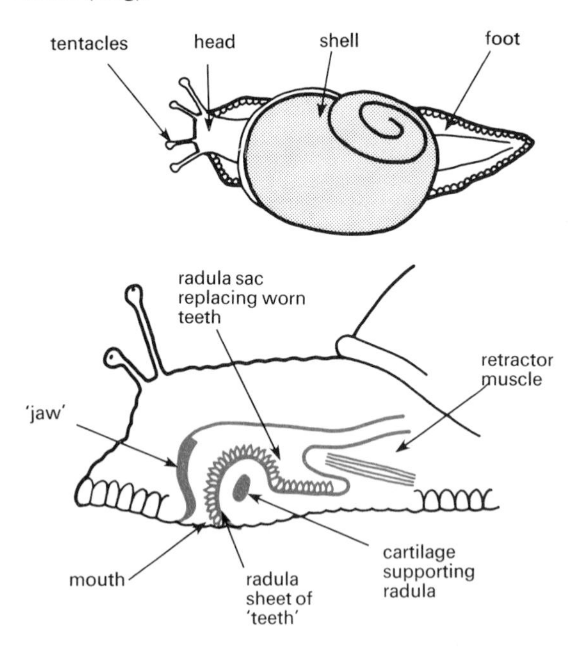

Figure 17.5 *Gastropod mollusc (snail, 6–8 cm long)*

CLASS LAMELLIBRANCHIATA

Fresh water and marine. Body compressed laterally; head reduced; bivalved shells; reduced foot; layered sheets or lamellae of ciliated gills; e.g. *Anodonta* (fresh water mussel), *Mytilis*[142] (marine mussel), *Ostrea* (oyster).

CLASS CEPHALOPODA

Marine. Well developed head region with most complex invertebrate brain and eyes; foot modified to form tentacles with suckers; muscular siphon; shell usually reduced and internal; e.g. *Architeuthis* (giant squid), *Loligo* (squid), *Sepia* ('cuttle fish' squid), *Octopus* (octopus), *Nautilus* (very distinct and ancient survival, chambered shell, links with extinct Ammonites).

Phylum Arthropoda

Fresh water, marine, terrestrial. Chitin cuticle thickened to form strong exoskeleton, flexible at intervals to allow for jointing; paired jointed appendages; exoskeleton shed during ecdysis to allow for growth; coelom much reduced, main body cavity is blood-filled haemocoel which is part of the open circulatory system, contractile dorsal tubular heart; cerebral ganglia and ventral solid nerve cord with typical paired segmental ganglia; no cilia. (Such a large and complex phylum that only some typical sub-groupings are included.)

Class Crustacea

Fresh water, marine, very few partially terrestrial. Two pairs of antennae, three pairs of head appendages, frequently some thoracic ones act as mouthparts; breathe by gills, usually heavy exoskeleton with dorsal shield of carapace. Important members of the zooplankton.

Sub-Class Branchiopoda

Trunk limbs involved in breathing and feeding.

ORDER DIPLOSTRACA

Compressed carapace covering trunk and limbs; compound eyes fused into one; large antennae used for swimming.

SUB-ORDER CLADOCERA

4–6 pairs of trunk limbs form compact and efficient feeding apparatus; e.g. *Daphnia*[232] (water flea).

Sub-Class Copepoda

No compound eyes; no carapace; six pairs of thoracic limbs; no limbs on abdomen; some parasitic; e.g. *Calanus* (marine), *Cyclops* (fresh water).

Sub-Class Cirripedia

Larvae free but adults fixed; hermaphrodite; no compound eyes in adult; carapace as a mantle enclosing the trunk; includes the barnacles; e.g. *Balanus* sp. (common barnacle), *Lepas* (goose barnacle).

Sub-Class Malacostraca

The 'highest' crustacea; wide diversity and large numbers of species; compound eyes, usually stalked; carapace covering thorax of eight segments.

ORDER PERACARIDA

Carapace does not fuse with more than four thoracic segments.

SUB-ORDER ISOPODA

No carapace, dorso-ventrally flattened; e.g. *Armadillium*[242] (pill-bug woodlouse, rolls into a ball), *Oniscus* (woodlouse).

SUB-ORDER AMPHIPODA

No carapace, laterally flattened; e.g. *Gammarus* (fresh-water shrimp).

ORDER EUCARIDA

Carapace fused to all thoracic segments.

SUB-ORDER EUPHAUSIACEA

E.g. *Euphausia*[268] (krill).

SUB-ORDER DECAPODA

Hinder five pairs of thoracic limbs adapted for locomotion; e.g. *Astacus* (fresh-water crayfish), *Carcinus* (common shore crab).

Class Myriapoda

Terrestrial. Elongated bodies with many leg-bearing segments; distinct head, single pair of antennae; internally similar to insects with system of air tubes or trachea, and malpighian tubules; in fact one sub-class, the Chilopoda (centipedes), is more closely related to the Insecta than it is to the other sub-class, the Diplopoda (millipedes).

Sub-Class Chilopoda

Many similar segments with one pair of walking legs, dorso-ventrally flattened, first body segment with a pair of poison claws; carnivorous; e.g. *Lithobius* (stone-dwelling centipede), *Haplophilus* (burrowing centipede).

Sub-Class Diplopoda

Anterior region of four single segments, posterior region of double segments each with two pairs of legs; cylindrical body; short, club-shaped antennae; herbivorous; e.g. *Iulus* (common millipede).

Class Insecta

Typically terrestrial. Body divided into head (six segments, one pair of antennae), thorax (three segments with three pairs of legs and typically two pairs of wings), and abdomen (eleven segments); tracheal system; malpighian tubules for excretion.

More than half the known species of animals are insects, with over one million having been described and thousands being added every year. They are thought to have evolved at about the same time as the Angiosperms in the Cretaceous period 125 million years ago.

The story of the insects is one of fantastic success. Their waterproof covering to the exoskeleton, efficient tracheal respiration, powers of flight, variation, and adaptability, have enabled them to colonize every conceivable terrestrial niche. Their prolific reproduction and astronomic numbers bring them into competition with man for his food supply, and many act as vectors of diseases of man, his crops, and his animals. Some have become secondarily aquatic in fresh water, but only one species (a midge) is entirely marine.

Sub-Class Apterygota (Ametabola)

The primitively wingless insects, as distinct from those wingless insects such as fleas which are thought to be derived from winged ancestors. They show little or no metamorphosis, hence called Ametabola.

ORDER THYSANURA

Bristle tails; e.g. *Lepisma* (the 'silver fish').

ORDER COLLEMBOLA

The spring tails; e.g. *Sminthurus* (lucerne flea).

Sub-Class Pterygota (Metabola)

Winged or secondarily wingless. Either gradual metamorphosis from nymph resembling adult form, and wings developing from external buds (Hemimetabola, Exopterygota); or marked metamorphosis from larval form not resembling adult form, with wings developing from internal buds (Holometabola, Endopterygota).

Hemimetabola; Exopterygota

ORDER EPHEMEROPTERA

Mayflies. Aquatic nymphs, adults short-lived with no mouthparts.

ORDER ODONATA

Damselflies, dragonflies. Aquatic nymphs, adults predaceous with biting mouthparts; large eyes.

ORDER DICTYOPTERA

Mantids and cockroaches.[308] Most ancient group of winged insects still in existence.

ORDER ISOPTERA

Termites. Social and polymorphic (different-shaped individuals).

ORDER ORTHOPTERA

Locusts,[302] grasshoppers, and crickets. Hind limbs adapted for jumping, stridulating (noise) organs.

ORDER DERMAPTERA

Earwigs. Unjointed terminal cerci modified into pincers or forceps.

ORDER MALLOPHAGA

Biting lice.

ORDER ANOPLURA

Sucking lice.[309] Ectoparasites of man.

ORDER THYSANOPTERA

Thrips. Ectoparasites of plants.

ORDER HOMOPTERA

Aphids[303] and cicadas.

ORDER HETEROPTERA

Bed bugs[308] and water bugs.

Holometabola; Endopterygota

ORDER COLEOPTERA

Beetles. Fore wings modified as wing-covers or elytra.

ORDER MEGALOPTERA

Alder flies. Aquatic larvae.

ORDER PLANIPENNIA

Lace-wings. Two long pairs of membranous wings.

ORDER LEPIDOPTERA

Butterflies and moths. Mouthparts as coiled sucking proboscis; scaly wings held vertically at rest in butterflies and flat in moths.

ORDER DIPTERA

Flies,[307] gnats, and mosquitoes.[307] One pair of wings, posterior pair of wings modified as balancing organs or halteres.

ORDER SIPHONAPTERA

Fleas.[311] Piercing and sucking mouthparts; legless larvae; wingless ectoparasites.

ORDER HYMENOPTERA

Ants,[309] bees, and wasps.[307] Polymorphic; first abdominal segment fused to thorax; four membranous wings, anterior wings linked by hooks to posterior wings; female has ovipositor (modified to sting in worker female bees and wasps).

Class Arachnida

Typically terrestrial, some aquatic. Body divided into two: the anterior prosoma, and posterior opisthosoma; four pairs of walking legs; no antennae; many simple eyes.

ORDER ARANEIDA

Spiders. Prosoma and opisthosoma separated by a narrow waist; spinning glands, which about half of the species use for building traps; e.g. *Tegenaria* (common house spider[307]), *Araneus* (garden spider), *Erigone* (money spider).

ORDER ACARINA

Mites and ticks. Rounded body with no division between prosoma and opisthosoma; parasitic with false head or capitulum carrying the mouthparts; e.g. *Ixodes* (sheep tick).

ORDER SCORPIONIDAE

Flexible tail with terminal sting; e.g. *Scorpio* (scorpion).

ORDER XIPHOSURA

Aquatic; e.g. *Limulus* (king crab)—sole living representative of this group.

ORDER PHALANGIDA

Very elongated legs; prosoma and opisthosoma fused, covers whole width; e.g. *Phalanium* (harvestmen).

Phylum Echinodermata

Marine. Larvae usually bilaterally symmetrical and segmented, adult radially symmetrical with five radii and unsegmented; unique water-vascular system involved in movement of characteristic tube feet; no excretory or circulatory system; gut always a simple coiled tube; mesodermal skeleton of calcareous plates.

The body plan of the adult shows no affinities with any other groups, but the larvae show affinities with those of some of the primitive Protochordates.

Class Asteroidea

Star-shaped with five arms; have suckers on tube feet; mouth ventral, anus and external opening of water-vascular system (madreporite) dorsal; e.g. *Asterias* (the starfish).

Class Echinoidea

Globular, cushion, or disc-shaped; no arms; large spines; larvae show affinities to Protochordate larvae; e.g. *Echinus* (the sea urchin).

Class Crinoidea

The feather stars.

Class Ophiuroidea

The brittle stars.

Class Holothuroidea

The sea cucumbers.

Phylum Chordata

Possess a dorsal skeletal notochord, at some stage in their life history, running the length of the body just below a tubular nerve cord. Pharyngeal or visceral clefts at some stage in life history, primitively involved in filter feeding and respiration, develop into gills in fish. Closed blood circulatory system with blood flowing forward ventrally and backwards dorsally. Post-anal matamerically segmented tail.

Sub-Phylum Protochordata (Acrania)

No true cranium, brain, heart, or kidneys.

Class Hemichordata

Short notochord, and nerve cord still in contact with epidermis; body divided into pre-oral proboscis, collar, and trunk; larva shows affinities with that of Echinoderms; e.g. *Balanoglossus* (the acorn worm).

Class Urochordata

Notochord and dorsal hollow tubular nerve cord usually only present in 'fish-like' free swimming larva. Larva settles to from sessile adult and loses typical chordate features during metamorphosis; e.g. *Ciona* (sea squirt). Some retain 'Ascidian tadpole' larval form throughout life; e.g. *Oikopleura* and *Fritillaria.*

Class Cephalochordata

Well-developed notochord in adult extending length of the body, fish-like appearance; e.g. *Amphioxus.*

Sub-Phylum Vertebrata (Craniata)

'The vertebrates.' Notochord is replaced in the adult by the vertebral column, associated with this is a well-developed head region with a well-developed brain protected by the cranium; organs of special sense are well-developed and usually closely associated with the brain; the visceral clefts are generally lost in the adult except in the fish where they persist and are modified as gills; a true muscular heart develops ventrally; true kidneys are formed which utilize the blood pressure produced by the pumping heart; one or more portal systems in the circulation; a well-developed endocrine system; usually two pairs of limbs. All have jaws except members of Class Cyclostomata. The jawed vertebrates are referred to as the **Gnathostomata**, and the jawless ones as the **Agnatha**.

Class Cyclostomata

Round mouthed, without jaws; notochord is retained in the adult; atypical skull; e.g. *Petromyzon* (the lamprey), *Myxine* (the hagfish).

Class Pisces

The fish. Paired limbs are pectoral and pelvic fins; respire by gills; single circulation with blood passing through the heart only once on each complete circulation of the body; lateral line sense organ; possess scales on the skin which are modified to teeth on the jaws; no external or middle ear.

Sub-Class Chondrichthyes (Elasmobranchii)

ORDER SELACHII

All 'sharks'. Entirely marine. The 'cartilagenous fish' with a cartilagenous endoskeleton; placoid scales; hyostylic jaw suspension involving the hyoid arch, with the associated hyoid visceral cleft reduced to a spiracle; no operculum covering the gills, each of which has its own septum; no air-filled swim-bladder, therefore 'heavier' than water and depth in water regulated by fleshy pectoral fins and typical heterocercal tail; spiral valve in the intestine; blood contains high concentration of urea; pelvic fins modified to claspers in male; fertilization internal, and few large-yolked eggs laid in special egg cases known as 'mermaid's purses'; overall tendency to dorso-ventral flattening, and bottom-living flat forms lie on their ventral surface; e.g. *Scyliorhinus* (the dogfish), *Dasyatis* (the sting ray), *Torpedo* (the electric ray), *Raja* (the skate).

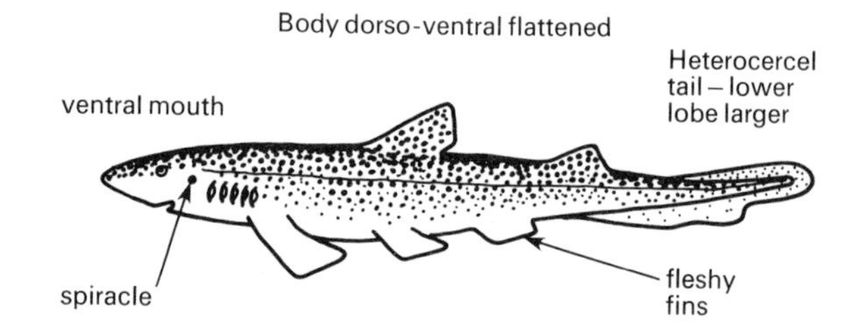

Figure 17.6 *Cartilaginous fish (dog fish)*

Sub-Class Osteichthyes

Fresh water and marine. The bony fish, with a bony endoskeleton, bony scales, and an air-filled bladder or sac derived as a diverticulum or branch of the gut.

ORDER TELEOSTEI

The main group of living bony fish with some 25 000 living species. Bony scales reduced to thin cycloid scales; air sac or swim bladder operated as a hydrostatic organ, by controlling the volume of gas they can adjust their depth in the water; homocercal tail; operculum covering the gills; fins a series of rays supporting a thin membranous skin and with any basal elements retained inside the body; overall tendency to lateral flattening, and bottom living flat forms lie on their side; e.g. *Gadus* (the cod), *Salmo* (the salmon).

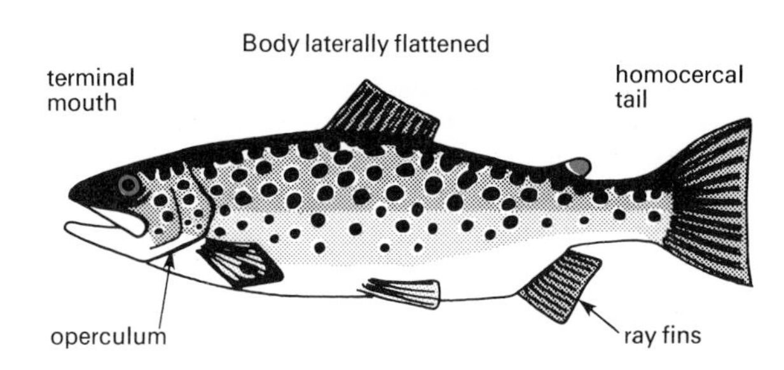

Figure 17.7 *Bony fish (trout)*

ORDER DIPNOI

The 'lung fish'. Air sacs act as lungs, but also still possess gills, have internal nares or nostrils; e.g. *Ceratodus* (in Australia), *Protopterus* (in South Africa), *Lepidosiren* (in South America).

ORDER COELOCANTHINI

'Primitive' order thought to have been extinct since Cretaceous period until rediscovered in Indian Ocean in 1938; e.g. *Latimeria* (the 'coelocanth').

Tetrapods

All the remaining groups of the vertebrates have two pairs of limbs based on the pentadactyl[80] plan.

Class Amphibia

Most members are only partially adapted to a terrestrial existence and are therefore restricted to damp habitats; eggs are not adapted to terrestrial conditions, therefore the adults return to the water to breed (although there are exceptions); the larval forms, known as tadpoles, are aquatic with gills; metamorphosis into the adult results in the loss of the gills and the development of simple sac-like lungs, although in the adult the main respiratory surface is the soft, permeable, well-vascularized skin; three-chambered heart; no external ear; and a single auditory ossicle in the middle ear.

ORDER URODELA

Tail persists in the adult and there is a long vertebral column; gills can persist into the adult stage and if they do the lungs atrophy; limbs are short. Contain about 900 species ranging from terrestrial forms to completely aquatic types; e.g. *Triton* (the newt), *Salamandra* (the salamander), *Amblystoma* (the axolotl, which attains sexual maturity in the larval form).

ORDER ANURA

Distinct metamorphosis into tail-less adult, with no gills and long muscular hind limbs adapted for jumping. Includes the frogs and the more terrestrial toads; e.g. *Rana* (the frog), *Bufo* (the toad), *Xenopus* (the African clawed toad, completely aquatic), *Alytes* (the 'midwife' toad, male carries eggs around on hind legs).

ORDER APODA

No girdles, limbs, or tail in blind burrowing adults; worm-like with external rings; some lay large, yolky eggs on land and development is direct with no larval stages; e.g. *Coecilia*.

ORDER CAUDATA

Only two species in the south east USA; long body; short legs; possess lungs but also retain one pair of gill slits; lay eggs on land; e.g *Amphiuma*.

Class Reptilia

The first group of vertebrates completely adapted to a terrestrial environment. Waterproof keratinized scaly skin; true lungs are the only respiratory surface; heart almost completely divided into two halves to give a double circulation with blood passing through the heart twice on each circulation of the body; large, heavily-yolked 'cleidoic' egg with protective shell, and shell membranes allowing it to be laid on land.

These features enabled the reptiles to exploit the vast, untapped food resources of the land masses and they underwent a wide adaptive radiation, culminating in the great Age of Reptiles in the Mesozoic period. Modern forms are but a relic of this once-dominant terrestrial group of vertebrates.

ORDER CHELONIA

The turtles and tortoises. Dorsal carapace and ventral plastron of bony plates overlaid with horny epidermal plates of 'tortoise shell'. Can be grouped into two on the basis of the articulation of the cervical vertebrae which result in the neck bending sideways or vertically; e.g. *Testudo* (the tortoise).

ORDER RHYNCHOCEPHALIA

Once a more numerous group, now represented by one living type; pineal body shows vestiges of lens and retina but no iris, opening in skull bones over 'pineal eye', but covered with thick skin. Vestigeal 'pineal eye' also seen in many other species of lizard, but not so well-developed as in these; e.g. *Sphenodon* (the tuatara lizard of New Zealand).

ORDER SQUAMATA

The lizards and snakes.

SUB-ORDER LACERTILIA

The lizards.

SUB-ORDER OPHIDIA

The snakes. Limbless with vestiges of girdles; asymmetry of internal anatomy to fit elongated shape, e.g. the left lung is vestigial whilst the right lung is elongated down to the cloaca (the common opening of gut, reproductive, and excretory system), whilst in the vipers the trachea takes on the structure and function of a lung; upper and lower jaws allow tremendous gape, right and left lower jaws not fused and are easily separated; the brain is protected ventrally by the massive sphenoid bone; poison glands, when present, are modified salivary glands; eyes not homologous with other vertebrate eyes, appear to have been 'reconstructed', possibly after re-emerging as a group from a burrowing existence during which the original type were reduced or lost; poisonous types and constrictors; e.g. *Crotalus* (the rattlesnake).

ORDER CROCODILIA

The crocodiles and alligators; e.g. *Crocodilus* (the crocodiles).

Class Aves

Similar in many ways to the Reptilia, and considered by some to be descendants from the Dinosaurs of the Mesozoic period. Feathers, scales on legs; toothless jaws enclosed in a horny beak; many adaptations to flight, anterior limbs form wings, sternum expanded into 'keel' for attachment of powerful pectoral flight muscles, hollow bones with air sacs from lungs, air sacs also in abdomen, no rectum or bladder; well-developed cleidoic egg with rigid calcareous shell, complex mating, nesting and rearing behaviour.

ORDER CARINATAE

The flying birds with well-developed 'keel' or 'carina'.

ORDER RATITAE

The flightless birds; reduced wings, and reduced keel to sternum; e.g. *Apterix* (kiwi), *Dromaeus* (emu), *Struthio* (ostrich).

Class Mamalia

The mammals. Mammary glands secrete milk to nourish the young; skin has sweat or sudorific glands, sebaceous glands, and hair; external ear and three auditory ossicles in middle ear; true diaphragm completely separates thoracic cavity from abdomen; lower jaw consists of a single bone, the dentary on either side; only the left systemic arch present, forming the aorta leaving the heart.

Sub-Class Monotremata

Show many 'reptilian' features, e.g. cloaca; lay large-yolked eggs; relatively low body temperature; a relatively simple brain; have no mammary glands but pour a milky secretion into a groove on the abdomen; the adults lack true teeth but this is a specialization; e.g. *Echidna* (the spiny ant eater), *Ornithorhyncus* (the duck-billed platypus).

Sub-Class Metatheria

The 'marsupials'. More mammal-like than the monotremes but still not considered as 'true' mammals. Viviparous, but the placenta is rarely allantoic,[25] the young are born alive but are poorly developed at birth except for powerful fore-limbs which enable them to reach the pouch in which they complete their development. They are found only in South America and Australia, and are only abundant in the latter; e.g. *Macropus* (the kangaroo).

Sub-Class Eutheria

The 'true' mammals. Completely viviparous with a true allantoic placenta; true mammary glands developed from modified sebaceous glands; roof of forebrain, the neo-pallium, forms the bulk of the extremely well-developed cerebral hemispheres, both cerebellum and cerebrum convoluted, active intelligent and high degree of parental care; diphyodont (not the rat), i.e. milk and permanent dentition, heterodont with different specialized types of teeth, e.g. incisors, canines, premolars, and molars.

ORDER INSECTIVORA

Small nocturnal forms; many primitive characters; full mammalian dentition usually preserved $\frac{3}{3}\ \frac{1}{1}\ \frac{4}{4}\ \frac{3}{3}$; brain has large olfactory lobes, small cerebral hemispheres; some retain the cloaca; numerous young produced—up to 32; most forms are solitary; e.g. *Talpa europaea* (mole).

ORDER CHIROPTERA

More than 800 species, about 14 per cent of all mammal species. The only mammals with flapping flight, the wing is a skin fold which involves all the digits except the first and extends along the body to include the legs but not the feet, great elongation of distal bones of the arms, sternum has a keel, also the clavicles are stout; hang upside down, when hanging metabolic rate drops; wide range of feeding habits within the group, e.g. insectivorous, licking nectar, sucking blood, eating fruit, catching fish; main sensory system is echolocation, high intensity and high frequency vibrations produced by large larynx, and reflection from objects, including prey, detected with great accuracy, echolocation and night flying enabled the bats to exploit the vast untapped food reserves of the night-flying insects; with absence of insects in winter in temperate climates bats must either hibernate or migrate.

ORDER CARNIVORA

Large brain with complicated behaviour for tracking prey; powerful jaws and dentition, with strong canines and some 'chewing' teeth modified as sharp carnassials; e.g. *Canis* (dog), *Felis* (cat).

THE UNGULATES

All have a herbivorous diet; usually large, elongated limbs for rapid locomotion, lateral digits reduced, movement of limbs in fore and aft direction only, formation of hooves; well-developed auditory and olfactory senses; teeth developed for dealing with large volumes of vegetation; alimentary canal with either stomach or caecum, or both, modified to harbour symbiotic cellulose-digesting bacteria.

ORDER PERISSODACTYLA

The 'odd-toed' Ungulates. Main digit is III; e.g. *Equus* (horse), *Tapirus* (tapir).[84]

ORDER ARTIODACTYLA

The 'even-toed' Ungulates, with cloven hooves. Main digits are III and IV; elaboration of stomach into four chambers, the rumen, reticulum, omasum, and abomasum; food cropped and swallowed without chewing, acted on by symbiotic bacteria, regurgitated when softened and chewed ('chewing the cud'); the 'ruminants'; e.g. *Bos* (cattle), *Capra* (goat), *Ovis* (sheep), *Camelus* (camel), *Capreolus* (roe deer).

ORDER PRIMATES

Most features are associated with tree-climbing or 'arboreal' adaptations; ventral elements of pectoral girdle reduced but clavicles act as struts for the mobile arms, development of grasping; good vision with fovea centralis in retina; general reduction of the olfactory sense; typically omnivorous; claws become nails; have four grasping 'hands' (except for man).

SUB-ORDER LEMUROIDEA

E.g. lemurs, bushbabies.

SUB-ORDER ANTHROPOIDEA

Infra-Order Platyrrhina

New World forms. All forest-dwellers, some with prehensile tails.

Infra-Order Catarrhina

Old World forms.

Family Cercopithicidae

Old World monkeys; e.g. baboons,[238] macaques.

Family Simiidae

Great apes, e.g. gorilla, chimpanzee.

Family Hominidae

E.g. *Homo sapiens* (man).
Erect posture in bipedalism involves many modifications of the skeleton and muscles from those arrangements seen in tetrapods. The vertebral column has curves that, with the associated musculature, bring the head and trunk in balance over the hips; and the pelvic girdle is brought into line so that the body weight can be transmitted efficiently through the legs. As a result of the new position of the skull, in relation to the vertebral column, the foramen magnum, by which the spinal cord leaves the skull, is found situated under the brain case. The vertebrae are separated by thick, shock-absorbing discs of fibro-cartilage, which in fact can account for a quarter of the length of the vertebral column. These cushion the quite considerable compression that occurs with the vertical posture (when lying down people are measurably longer). The pelvic girdle is also broad and has a muscular diaphragm to support the weight of the viscera, which now acts downwards instead of 'hanging off' the vertebral column as in tetrapods.

The gluteus maximus muscles of the buttocks are a very human feature and act to straighten the thigh in relation to the vertebral column. The legs are relatively long when compared to other primates and the foot shows many unique features related to bipedalism. In all other primates the foot is a grasping structure with a wide-set 'big-toe', in man the big toe is brought into line with the others (which are reduced) and used as the main lever in locomotion, the heel bone is expanded as a prop at the back, and has attached to it the Achilles tendon from the soleus muscle (which gives the calf muscle its characteristic swelling in man) and the smaller gastrocnemius muscle (which flexes the knee joint). Also the ability to evert the foot, that is to turn the sole to face slightly outwards is a distinctly human feature and is important in balance. The fore-limbs or arms are relatively shortened, and the bipedal habit frees the arms to become highly manipulative. The thumb and digits are opposable: thus the thumb can be rotated so that the palmar surface can be opposed to that of the digits. The radius and ulna are 'crossed', allowing the rotation of the fore-limb, so that the hand can be held with either surface up, and the digits can be controlled individually. Surprisingly, although these features are those upon which man's great manual dexterity and characteristic use of tools are based, they are not unique to man and in fact can be found in other primates as well as in other mammals.

When compared to other primates the face is shortened, the chin is developed, the nose is more prominent, and the facial axis is more vertical with the development of the forehead. Most of these changes, particularly the latter, are associated with the tremendous development of the brain, particularly the frontal lobes of the cerebral hemispheres. The brain volume of between 1200 and 1500 cm^3 is about three times that of the great apes; the human brain is not just a larger version of the ape brain—the patterns of nerve tracts and association centres are distinctly different. The cerebral cortex is very large and convoluted, and large areas are 'uncommitted' to any known motor or sensory function. These areas of uncommitted cortex or 'silent-areas', which occupy up to 90 per cent of the frontal lobes of the hemispheres, are thought to be the higher association centres concerned with intelligence, learning, memory, musicality, and personality. Language and speech are other uniquely human features and in fact the brain at birth is asymmetrical with the left 'language' cerebral hemisphere larger than the right. Characteristically, there is great development, mobility, and control of facial muscles, and of those relating to speech. Indeed, about 25 per cent of the total motor area of the cerebral cortex is devoted to the control of the muscles associated with speech.

Further reading

Buchsbaum R., *Animals Without Backbones, Volumes 1 and 2* (Harmondsworth: Penguin Books, 1966).

Freeman W. H. and Bracegirdle B., *An Atlas of Invertebrate Structure* (London: Heinemann Educational Books, 1971).

The Nuffield Foundation, *Keys to Small Organisms in Soil, Litter, and Water Troughs* (London: Longmans, 1967).

Robinson M. A. and Wiggins J. F., *Animal Types 1: Invertebrates*, and *Animal Types 2: Vertebrates* (London: Hutchinson Educational, 1975).

Romer A. S., *Man and the Vertebrates, Volumes 1 and 2* (Harmondsworth: Penguin Books, 1960).

Appendix 2

Plant taxonomy

There is a greater uniformity in the endings used for plant groupings than there is for the animal groupings; they include Division (-phyta), Class (-atae or -phyceae), Order (-ales), and Family (-aceae).

However, there is greater controversy as to the status and naming of many plant groups, especially those of the algae and fungi. Thus, for example, groupings of the algae, which were once considered as classes of the Division Phycophyta or Algae, are now considered to be sufficiently distinct as to constitute separate Divisions. Also although some groupings are clear, others, again especially in the Algae and Fungi, are very confused, with many types being variously classified by different authorities.

Division Schizophyta

Unicellular or colonial; reproduce by fission; no true nucleus (prokaryotic or akaryotic); if photosynthetic pigment present it is not enclosed in a special organelle; cell-wall structure different from that of higher plants.

Class Schizomycetes

The bacteria. Includes both heterotrophic and autotrophic types, photosynthetic autotrophic types do not evolve oxygen as a result of photosynthesis.

ORDER EUBACTERIALES

The 'true bacteria'; simple forms rod-shaped (bacilli) or spherical (cocci), with no true branching. Unicellular with rigid cell walls; reproduce mainly by binary fission; most are heterotrophic, some autotrophic. This order contains most of the well-known species which cause disease, putrefaction, or fermentation.

Bacteria are classified on the basis of a wide range of characteristics, including morphological, pathological, serological, biochemical, and metabolic features.

ORDER ACTINOMYCETALES

Some members very much like true bacteria; some members more like fungi.

The fungal-like types are filamentous and can form a definite branched mycelium. Some play an important part in soil fertility where they can be as numerous as the true bacteria; many produce important antibiotics; some form symbiotic associations in root nodules of higher plants, e.g. alder.

ORDER CHLAMYDOBACTERIALES

Alga-like, filamentous bacteria, typically aquatic. The Order includes the 'iron bacteria', chains of cells ensheathed in ferric hydroxide, and the 'colourless sulphur bacteria', which grow in water rich in H_2S and obtain energy by the oxidation of the H_2S (chemosynthesis).

ORDER MYXOBACTERIALES

The 'slime bacteria'; have no rigid cell walls and thus in mass appear slimy; cells are motile; many important in the soil in cellulose decomposition.

ORDER SPIROCHAETALES

Protozoan-like, spirally twisted organisms, lacking rigid walls; move by twisting movements of whole body. Many are saprophytic and harmless, but *Treponema pallidum* causes syphilis.

ORDER RICKETTSIALES

Small bacterium-like organisms; can only be cultured outside the host in living tissues (or rarely in body fluids). Most are parasitic on Arthropods; some cause human disease.

Class Cyanophyceae

The 'blue-green algae' (can in fact be classified as algae). Fresh-water, marine, and terrestrial, found in all environments from hot springs to bare rock. Photosynthetic and evolve oxygen; unicellular and filamentous; non-motile; many are symbiotic partners to fungi in lichens; several genera have species which fix atmospheric nitrogen, and are important in soil fertility especially in rice fields; certain planktonic species can cause 'algal' blooms, some of which are poisonous; e.g. *Nostoc*, *Rivularia*, *Anabaena*.

THALLOPHYTA

An 'older' grouping including both the Algae and the Fungi. Have true nuclei (eukaryotic); plant body a simple thallus showing no differentiation into root, stem, and leaves; little if any tissue differentiation; sex organs and spore-producing structures usually unicellular and if multicellular then all cells are fertile; haploid and diploid phases can alternate with each other, in absence of alternation the thallus is usually haploid.

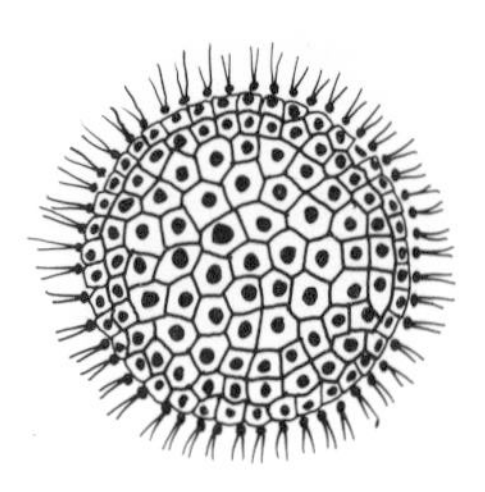

Figure 18.2 *Volvox (diameter about 80 μm)*

Algae

The Algae are mainly aquatic, and are a very large and varied group of photosynthetic plants, all containing chlorophyll a and β-carotene. The marine types are the well-known seaweeds. They were once considered as belonging to a single Division (the Phycophyta), but are now more generally considered as belonging to several distinct Divisions. There is a marked difference between the Divisions, based on morphological and biochemical criteria, such as differences in the type, position, and number of flagella in motile cells; differences in the types of pigments; the nature of the reserve food material; and the composition of the cell wall.

Division Chlorophyta

Class Chlorophyceae

The green Algae. Mainly fresh-water, some marine, and some semi-terrestrial; wide range of vegetative structure from unicellular to thalloid. Pyrenoids usually present in chloroplasts, have starch as the storage product; asexual reproductive cells and male gametes motile, with two apical flagella; adult stage haploid, sometimes alternation of generation between haploid and diploid phases of identical form (isomorphic).

The largest group of Algae with a tremendous range of size and structure, of which only some Orders are described here.

ORDER VOLVOCALES

Fresh-water planktonic. Flagellate, unicellular or colonial forms; asexual reproduction by motile zoospores or daughter colonies; e.g. *Chlamydomonas*, *Volvox*, *Pandorina*, *Eudorina*.

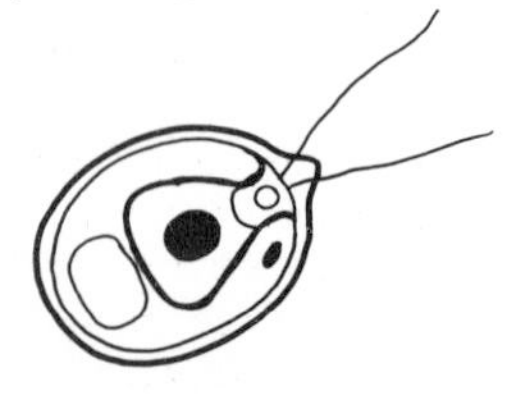

Figure 18.1 *Chlamydomonas (length about 5 μm)*

ORDER CHLOROCOCCALES

Mainly fresh-water plankton, found on moist soil, tree bark, fencing, etc. Unicellular or aggregates; non-motile vegetative cells; cells never undergo binary fission but cytoplasm divides into many motile zoospores; some are symbiotic partners in lichens, and in some invertebrate animals (e.g. *Hydra*); e.g. *Chlorella*.

ORDER ULOTRICHALES

Mainly fresh-water, some marine. Range in structure from simple unbranched filaments to large, 'leafy' thallus as seen in *Ulva lactuca* (the sea lettuce); e.g. *Ulva*, *Ulothrix*.

Figure 18.3 *Ulva (up to 1m long)*

ORDER CHAETOPHORALES

Mainly fresh-water. Branched filaments can form complex pseudoparenchymatous sheets, but some have a very reduced plant body, e.g. *Pleurococcus*, which is found on damp bark, fences, etc., and as a partner in some lichens; e.g. *Pleurococcus*, *Stigeoclonium*.

ORDER CONJUGALES

Almost entirely fresh-water. Unicellular or unbranched filamentous chains of cells; no asexual reproduction, sexual reproduction by amoeboid gametes; includes the planktonic **Desmids** which have a great variety of shapes and a cell wall of two sculptured equal halves; e.g. *Spirogyra*, *Zygnema*, Desmids.

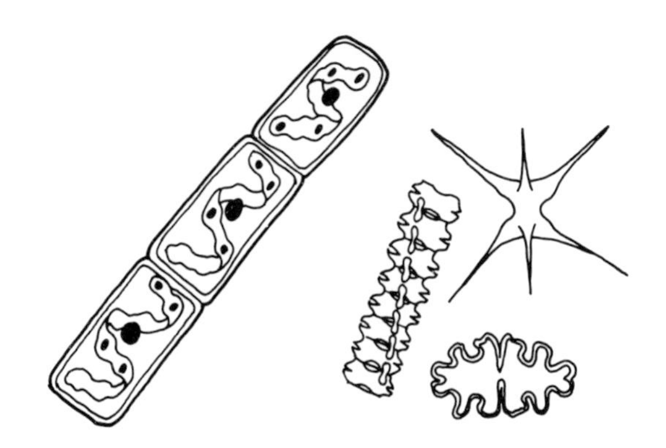

Figure 18.4 *Spirogyra (diameter about 120 μm)*
Figure 18.5 *Desmids (50–150 μm)*

ORDER CHARALES

The 'stoneworts'. Complex structure with whorls of branches separated by long internodes, branches in turn composed of nodes and internodes; colonize sand and mud rather than hard substrates; unique spiral spermatozoids not found in any other algae; show no close relationships to the other green algae; e.g. *Chara, Nitella.*

Division Phaeophyta

Class Phaeophyceae

The brown algae. Marine, the brown 'seaweeds', some rare fresh-water species. Range in structure from minute, simple, branched filaments to massive complex pseudoparenchymatous thalli many metres long; contain chlorophyll, but it is 'covered' by the brown pigment fucoxanthin; contain the polysaccharide laminarin, but never starch. Both asexual zoospores and gametes are motile with two unequal flagella; some show isomorphic and some heteromorphic alternation of generation, with a haploid gametophyte and diploid sporophyte which is usually dominant; a few do not show any alternation of generation.

ORDER ECTOCARPALES

The main group, containing the majority of the Phaeophyceae. Filamentous; some epiphytic, for example on *Fucus*; e.g. *Ectocarpus.*

ORDER LAMINARIALES

Sporophytes can be very large, e.g. up to 100 m long, and can show quite complex morphological and anatomical differentiation. The large complex sporophytes alternate with microscopic female and male gametophytes which grow in felt-like masses; e.g. *Laminaria, Macrocystia.*

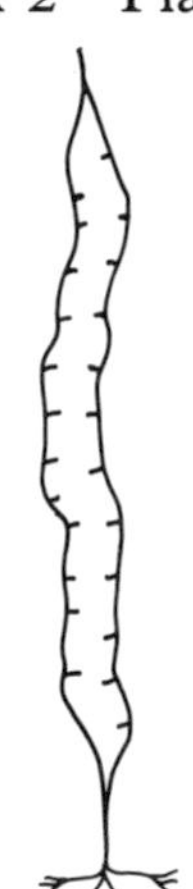

Figure 18.6 *Laminaria (can be several metres long)*

ORDER FUCALES

The commonest inter-tidal algae (along with some of the Laminariales) of the colder seas. Have no clear alternation of generation; no asexual reproduction; diploid thallus produces haploid gametes in special structures known as conceptacles; e.g. *Fucus, Ascophyllum, Sargassum* (*S. muticum* or Japanese seaweed is a recent unwanted introduction into British waters).

Figure 18.7 *Fucus (up to 1 m long)*

Division Rhodophyta

Class Rhodophyceae

The red algae. Mainly marine, some occur in fresh water, but vast majority are the red 'seaweeds' found mainly in deeper water; can be found at depth of 200 m where only the blue light penetrates. The chlorophyll is 'masked' by the red pigment phycoerythrin; true starch is not formed, but so-called floridean starch is found; some are the best source of agar; some are colourless parasites; some are unicellular but others have a complex leaf-like thallus; asexual and sexual reproductive cells are non-motile; most have a life cycle involving three generations, a haploid gametophyte, a diploid-carpo sporophyte on the gametophyte, and an independent diploid sporophyte.

One family, the Corallinaceae, are calcified and are important members of coral reefs. Apart from the Cyanophyceae, the red algae are the only algae that lack motile cells; e.g. *Porphyra*, *Lithothamnium* (reef builder), *Batrachospermum* (fresh-water).

Division Euglenophyta

Class Euglenophyceae

Fresh-water, especially in stagnant water rich in nutrients. Photosynthetic, with chlorophyll a and b, and a unique xanthophyll; motile with one long and one short flagellum, but with only one visible externally; lack firm wall, surrounded by flexible, non-cellulose, polysaccharide pellicle. Often classified as flagellate protozoa; e.g. *Euglena*, *Phacus*, *Colacium*.

Division Pyrrophyta

Class Pyrrophyceae

The dinoflagellates. Few fresh-water, most marine. Important members of the phytoplankton. Yellow in colour; almost all unicellular; two long flagella; cell wall often complex polygonal porous plates of cellulose; fine cytoplasmic threads can extend out through pores and engulf micro-organisms; remarkable flotation processes in many enable them to maintain their position in the plankton; some are symbiotic with certain animals; store starch, fats and oils; e.g. *Ceratium*, *Peridium*, *Gonidium*, and *Gonyaulax* (form periodic 'blooms' in the sea to cause poisonous 'red tides').

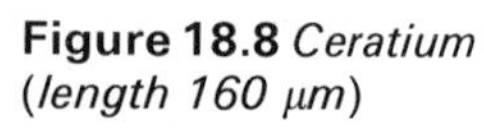

Figure 18.8 *Ceratium (length 160 μm)*

Figure 18.9 *Peridinium (length 40 μm)*

Division Chrysophyta

Class Chrysophyceae

The golden algae. Mainly fresh-water and marine. Important members of the phytoplankton; unicellular; golden-brown due to various xanthophylls; includes the **diatoms** (Diatomales or Bacillariales) with characteristic two silicified 'box and lid' type walls inside the cell membrane, of which there are more than 10 000 species.

Vast deposits of 'Diatomaceous earth' containing 6×10^6 shells per mm^3, deposited in the ancient seas, are now mined commercially as building material, filtering medium, and insulating medium. E.g. *Biddulphia*, *Pleurosigma*, *Dinobryon*.

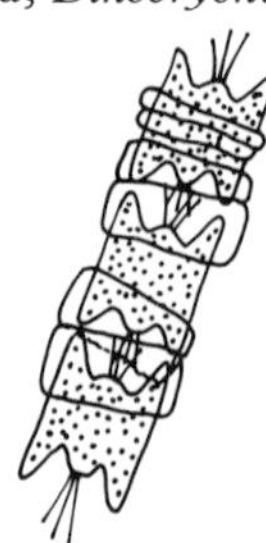

Figure 18.10 *Biddulphia (diatom) (length about 40 μm)*

Division Xanthophyta

Class Xanthophyceae

The yellow-green algae, due to high concentration of β-carotene and xanthophylls. Mainly fresh-water and semi-terrestrial in damp soil; simplest forms always flagellate, some are planktonic, others are attached tubular siphonaceous forms; no starch is produced, but a modified form of laminarin; relatively small group; e.g. *Vaucheria*, *Botrydium*, *Capitulariella*.

FUNGI

The fungi are fresh water or terrestrial, rarely marine. Usually colourless, non-photosynthetic pigments can occur particularly in the spore-bearing fructifications of the higher fungi; heterotrophic, being either saprophytic or parasitic; cell walls not cellulose, but chitinous; vegetative body typically composed of filamentous hyphae known collectively as a mycelium; reproduction mainly asexual via spores of various kinds. The fungi are generally divided into two Divisions, the **Myxomycophyta** or slime moulds, and the **Eumycophyta** or true fungi.

Division Myxomycophyta

Class Myxomycetes

The slime moulds. Semi-terrestrial on the floor of woods, decaying leaves, and timber. Walled spores germinate in wet conditions to produce one or more biflagellate uninucleate 'swarmers', eventually the swarmers lose their flagella to become amoeboid myxamoebae; these reproduce by fission until they fuse in pairs in sexual reproduction; these fused pairs then unite to form a large mass of naked multinucleate cytoplasm known as a plasmodium, some of which can reach over 30 cm in diameter; eventually the plasmodium develops into numerous sporangia which contain the uninucleate spores; e.g. *Cribraria*, *Comatricha*, *Stemonites*, *Plasmodiophora brassicae* (serious parasite of *Brassicae*, e.g. cabbage, causing club-root or 'finger-and-toe' disease).

Division Eumycophyta

Class Phycomycetes

The 'true fungi'. Range in form from microscopic uninucleate unicells to large branched coenocytic (tubular or siphonaceous) filamentous hyphae which form loose mycelia.

ORDER ZYGOMYCETALES

Sometimes considered as a separate class. Mainly terrestrial, saprophytic moulds, some parasitic on insects; with a well-developed coenocytic mycelium; cell walls predominantly chitin; reproduction adapted to terrestrial conditions with multinucleate non-motile spores, with walls resistant to desiccation, being produced in sporangia; in sexual reproduction they never produce gametes, but two multinucleate gametangia fuse forming a zygote; e.g. *Mucor mucedo* (pin mould), *Phycomyces*, *Pilobolus* (on dung), *Rhizopus* (black bread-mould).

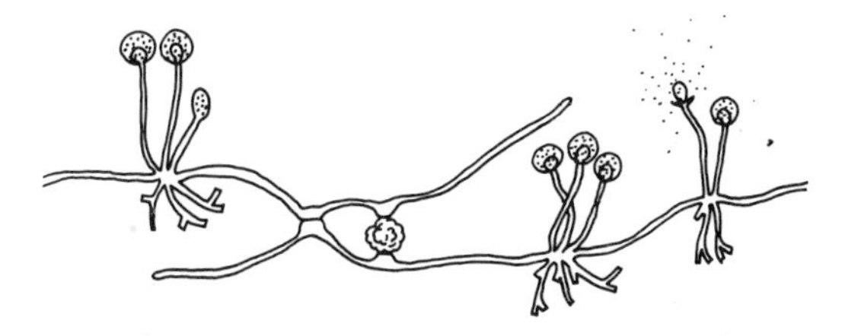

Figure 18.11 *Rhizopus*

ORDER OOMYCETALES

Sometimes considered as a separate class; aquatic and terrestrial. Tubular, non-septate, coenocytic, multinucleate hyphae; cell walls have cellulose but no chitin; reproduction adapted to terrestrial conditions by entire sporangium being detached when ripe, giving rise to biflagellate zoospores when it germinates; both male and female gametes non-motile with the male gamete being carried to the egg by a tube. Two main families.

Family Saprolegniaceae

Mainly aquatic saprophytes, sometimes parasitic on fish. Asexual reproduction by zoospores produced in club-shaped sporangia; e.g. *Saprolegnia*, *Achlya*.

Family Peronosporaceae

Mainly parasitic on higher plants, some of great economic importance; e.g. *Pythium*, *Albugo*, *Phytophthora infestans*[304] (potato blight), *Plasmopara viticola* (downy or false mildew of grapes).

ORDER CHYTRIDIALES

Sometimes considered as a separate class. Saprophytic or parasitic; contains fungi parasitic on plants and protozoa; and the carnivorous predacious fungi, which trap soil animals such as nematodes in fungal loops; some microscopic forms parasitize diatoms; e.g. *Polyphagus* (traps euglenae), *Arthrobotrys* (traps nematodes), *Zoophagus* (traps rotifers), *Zoopage* (traps amoebae).

ORDER BLASTOCLADIALES

Simpler types are superficially similar to the Chytridiales; others, however, have much-branched multinucleate hyphae forming large mycelia; the only Phycomycetes to show sexual and asexual reproduction on different individuals which alternate (irregularly) from generation to generation; e.g. *Blastocladiella*, *Allomyces*.

Class Basidiomycetes

Terrestrial, saprophytic and parasitic. Mycelium of branched septate mycelium with pores in the cross-walls; there are no special sex organs but fusion of uninucleate hyphae of opposite strains results in binucleate or dikaryotic 'fruiting' or secondary mycelia; the hyphae of the secondary mycelia often have characteristic clamp connections; fructifications or fruiting bodies are composed of compacted pseudoparenchymatous hyphae and are usually of distinctive shape; the fruiting body has a fertile layer known as the hymenium bearing special cells, the basidia, in which nuclear fusion and subsequent meiotic division occurs to give characteristically four basidiospores. There are two main sub-classes, the Holobasidiomycetidae containing the well-known mushrooms, toadstools, bracket fungi, and puffballs; and the Phragmobasidiomycetidae containing the important parasites of cereal crops, the rusts, smuts, and bunts.

Sub-Class Holobasidiomycetidae

Have a perennial mycelium, overwintering in the soil or wood, producing new fructifications each year; the fructifications can show a certain degree of differentiation into thick-walled supporting hyphae, branched binding hyphae, water-transporting hyphae with wide lumens, and thin walled hyphae producing basidia; e.g. *Psalliota campestris* (edible mushroom), *Amanita phalloides* (death-cap toadstool), *Serpula lachrymans*[309] (dry rot fungus), *Colvatia gigantea* (giant puffball).

Figure 18.12 *Psalliota*

Sub-Class Phragmobasidiomycetidae

Basidia divided into four cells by transverse septa. Includes the rust fungi of the grasses and cereal crops, which produce no fructifications; e.g. *Puccinia graminis*,[305] *Tilletia* (stinking smut or bunt).

Class Ascomycetes

Mainly terrestrial, some aquatic. Widely varied group, but long-recognized as a natural class mainly on the basis of the production of four or eight characteristic ascospores by meiosis on a specialized structure known as an ascus; the asci are usually borne on or in special structures such as closed, flask-shaped perithecia or disc-shaped apothecia; they frequently produce characteristic fructifications or fruiting bodies of compacted pseudoparenchymatous hyphae; can have specialized hyphae known as conidiophores which cut off asexual conidiospores at their tips; never produce flagellate reproductive cells. Complex systematics with many overlapping characteristics, therefore only a few orders are dealt with here.

ORDER SACCHAROMYCETALES

Much reduced forms; e.g. *Saccharomyces*[299] (yeast).

ORDER TAPHRINALES

Many plant parasites producing various deformities, such as 'witches' brooms on birch trees and others; e.g. *Taphrina*.

ORDER PLECTASCALES

Asexual reproduction usually by bluish-green conidia; e.g. *Penicillium*.[299]

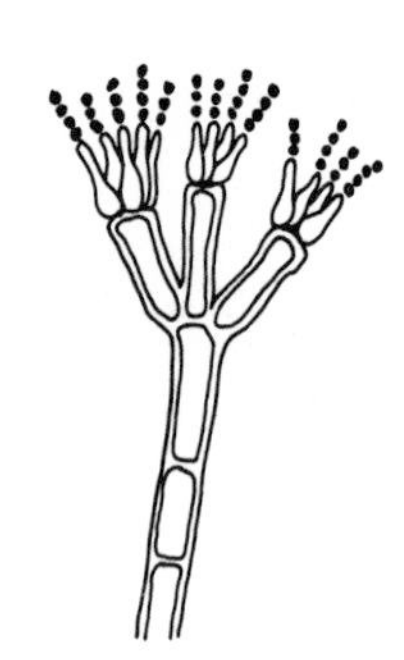

Figure 18.13 *Penicillium*

ORDER ERYSIPHALES

Parasitic powdery mildews on leaves of higher plants; e.g. *Microsphaera* (on oak leaves).

ORDER PSEUDOSPHAERIALES

Includes many plant parasites, e.g. *Venturia* (scab of apples and pears), *Caprodium* (sooty mould of leaves).

ORDER SPHAERIALES

Very large group of saprophytes and parasites; e.g. *Ceratocystis ulmi*[306] (Dutch elm disease), *Gibberella*[299] (source of gibberellins, parasitic on rice), *Neurospora* (widely used in biochemical genetics), *Sordaria* (widely used in biochemical genetics).

ORDER CLAVICEPITALES

Contain many important parasites, e.g. *Claviceps purpurea* (ergot of rye).

ORDER HELOTIALES

Saprophytic and parasitic; e.g. *Sclerotinia* (mould on rotting apples and pears), *Monolinia* (brown rot of some fruit), *Botytris cinerea* ('noble' decay on grapes in dry seasons, responsible for high sugar content).

ORDER TUBERALES

Mycorrhiza, the 'truffles'; e.g. *Tuber* (prized truffle).

Class Deuteromycetes

The 'Fungi Imperfecti'. The natural classification of the Fungi is largely based on their life-cycles, reproductive organs, and sexual reproduction. However, many are known (some 20 000 species) in which there appears to be no sexual reproduction, therefore they cannot be assigned to any particular group and they are all grouped into this artificial class. On structural features alone most would appear to belong to the Ascomycetes. E.g. *Penicillium*[299] (some species), *Dactyella* (traps nematodes in the soil), *Trichophyton interdigitale* (athletes foot), ringworm fungus.

Division Bryophyta

Terrestrial, some secondarily aquatic; all relatively small, widespread from Arctic to tropics. Well-developed heteromorphic alternation of generation with the haploid gametophyte dominant and forming the persistent vegetative body with rhizoids, and sometimes with stem and leaf-like structures. The gametophyte thallus has a delicate waterproof cuticle; typically shows dichotomous branching; bears multicellular sex organs, the male antheridia and the female archegonia; fertilization is achieved by biflagellate male gametes or spermatozoids swimming in a film of water to the female oosphere in the archegonium (zoidogamous fertilization); fertilized diploid zygote grows into the diploid sporophyte generation consisting of a sporogonium of foot, stalk, and capsule, which is completely or partially dependent on the gametophyte generation; produces haploid spores by meiosis; spores are all of one type (homosporous); spores of some form characteristic filamentous protonema before forming gametophyte thallus.

Class Hepaticae

The liverworts. Great variety of gametophyte structure, some flattened and thallose, others leafy or 'foliose'; cells of most contain unique oil-bodies; have unicellular rhizoids; dichotomous branching; sporophyte simple, spore mother cells produce both spores and sterile cells called elators which aid in their dispersal by hygroscopic movements.

ORDER JUNGERMANNIALES

The largest order, about 90 per cent of the species of Hepaticae. Predominantly tropical, relatively simple gametophyte and sporophyte, capsule dehisces by four valves; some are thallose, but majority are foliose with stem and leaf-like structures.

SUB-ORDER ANACROGYNAE

Mostly thallose. Archegonia formed behind growing tip, sporogonium arises from dorsal surface with its base covered by a flap of gametophyte tissue the involucre; e.g. *Pellia*.[30]

SUB-ORDER ACROGYNAE

Mostly foliose; archegonium formed from the apex of the stem; e.g. *Trichocolea*, *Calypogeia*, *Lophozia*.

ORDER MARCHANTIALES

Many have a complex thallus; vegetative reproduction by cup-like gemmae; sex organs (gametangia) borne on special erect branches; antheridia and archegonia mostly occur on different plants; e.g. *Marchantia*.

Class Musci

The mosses. Gametophyte well-developed with spirally arranged leaves arising from a stem; multicellular rhizoids; sex organs in groups at tips of lateral branches of main stem; strong powers of vegetative reproduction either by fragmentation of thallus or gemmae; well-developed sporophyte with photosynthetic tissue and air pores (stomata); some show beginnings of heterospory with two different types of spore giving rise to two separate types of male and female gametophyte; complex dehiscence of capsule with 'teeth' but no elaters; spores always produce well-developed protonema.

Sub-Class Sphagnidae

The 'bog-mosses'. Comprise but one genus, *Sphagnum*;[120] grows in wet, acid areas forming deep accumulative peat forming bogs; spores only germinate in the presence of mycorrhizal fungus; erect stems have no rhizoids.

Sub-Class Bryidae

Largest group with widest variety of form and highest level of differentiation; e.g. *Funaria*,[30] *Bryum* (800 species), *Mnium*, *Polytrichum*.

Sub-Class Andreaeidae

Found in cold regions, forming dark brown cushions on non-calcareous rocks; e.g. *Andreaea*.

TRACHEOPHYTA

All plants above the Bryophytes have a dominant conspicuous sporophyte generation with true roots, stems, and leaves; well-developed vascular tissue of xylem, phloem, and other specialized tissues. They can all be grouped together in this one large Division, however this makes such a huge grouping that some prefer to divide it into a series of smaller Divisions. Other terms which are used in relation to the Tracheophyta are Cormophyta, Vascular cryptogams, and Phanerogams.

Cormophyta is a term used to group all plants with true roots, stems, and leaves together and thus includes the Pteridophyta, Gymnosperms, and Angiosperms.

Vascular cryptogams are those Tracheophyta with 'hidden' or inconspicuous reproductive organs, these are the Pteridophyta which have microscopic reproductive organs carried on a small gametophyte, whereas the conspicuous plant is the sporophyte which produces asexual spores.

Phanerogams are plants with conspicuous reproductive structures in cones or flowers, and which produce seeds; namely the Spermatophyta (Angiosperms and Gymnosperms).

Division Pteridophyta

The 'club-mosses', the horse-tails, and the true ferns, and many extinct groups (especially in the Carboniferous). Well-developed heteromorphic alternation of generations; gametophyte typically a small short-lived thallus; zoidogamous fertilization; sporophyte vascular tissue only contains xylem tracheids; leaves bearing spore-producing sporangia are known as sporophylls; most are homosporous, although some are heterosporous with dioecious gametophytes.

Class Lycopodiatae (Lycopsida)

The 'club-mosses'. Have remained virtually unchanged for about 300 million years; extinct tree-forms were the dominant forms in the Carboniferous flora and some had developed the seed habit; present-day types have herbaceous creeping sporophytes with dichotomously branched stems and roots, and simple unstalked (sessile) leaves; sporophylls aggregated into spike-like cones or strobili.

ORDER LYCOPODIALES

Herbaceous sporophyte with creeping stem; only one main genus *Lycopodium*; in nature spores can take up to seven years to germinate into underground saprophytic lobed tuber-like gametophyte prothallus with mycorrhizal fungus; prothallus can take as long as fifteen years to mature; there are many fossil forms; e.g. *Lycopodium clavatum* [*commonest European species*].

ORDER SELAGINELLALES

Only one living (extant) genus, *Selaginella*; mainly tropical; herbaceous usually prostate stem; small scale-like leaves usually in four rows; heterosporous with very reduced dioecious gametophytes; e.g. *Selaginella*.[33]

Figure 18.14 *Selaginella*

ORDER LEPIDODENDRALES

Extinct club-moss trees; e.g. *Lepidodendron*, *Sigillaria*.

Class Equisetatae (Sphenopsida)

The 'horse-tails'. Deep perennial underground rhizome produces erect annual stems with whorls of branches; distinct sporophylls in strobili born on distinctive sporangiophores with whorls of sheathing scale leaves; homosporous; only one extant genus, *Equisetum*; many fossil forms, many of which were heterosporous; e.g. *Equisetum*, *Calamites* (fossil form).

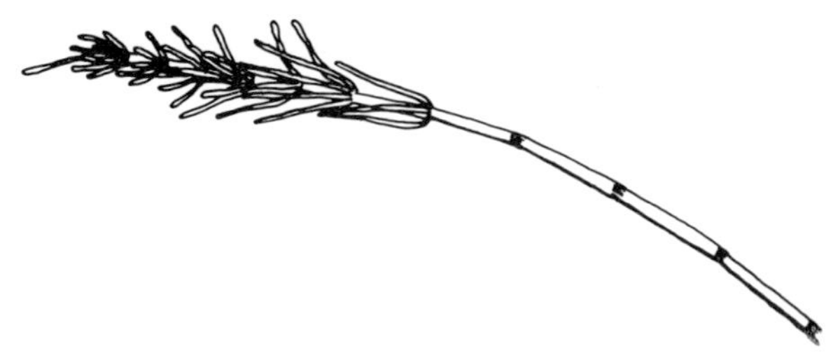

Figure 18.15 *Equisetum*

Class Filicatae

The ferns. Large, stalked leaves known as fronds bear many sporangia usually in clusters or sori on under-surface; generally homosporous but some are heterosporous; many living and fossil forms.

Sub-Class Eusporangiatae

Sporangia with thick walls of several layers of cells; e.g. *Ophioglossum*.

Sub-Class Leptosporangiatae

Large group containing up to 90 per cent of all the Filicatae genera; sporangia have thin walls of only one layer of cells; e.g. *Pteridium aquilinum* (bracken), *Dryopteris filix-mas*[31] (male fern).

Sub-Class Hydropterides

The water-ferns. Small group of ferns living in water or marshy places; all heterosporous; e.g. *Salvinia*.

Class Psilophytatae (Psilopsida)

Only four species of living types, all tropical; herbaceous sporophytes; rhizome with rhizoid but no true roots; minute leaves; colourless underground gametophyte with mycorrhizal fungi; rhizome of sporophyte also has mycorrhiza; fossil forms represent the lowest known Pteridophytes; e.g. *Rhynia* (fossil form, an ancestral land plant), *Psilotum* (two tropical species).

Division Spermatophyta

The seed-bearing plants. The dominant terrestrial plants since the upper Permian in the Mesophytic periods; some are secondarily aquatic; the sporophyte has the highest degree of vascular differentiation; secondary thickening is common; heteromorphic alternation of generations with gametophyte, much reduced and totally dependent on the sporophyte, being retained within the specialized reproductive structures known as cones and flowers; microspores containing the very reduced male gametophytes are released to be carried by wind or insects to the female gametophyte, in the process of pollination; fertilization is via a pollen tube (siphonogamous), and therefore no longer dependent upon the presence of water; after fertilization seeds are formed.

Sub-Division Pteridospermae

The 'seed-ferns'. Extinct group of fern-like plants with ovules (and in turn seeds) borne on special parts of the much branched fronds; wide diversity of form, especially in the Carboniferous period; e.g. *Lyginopteris*.

Sub-Division Gymnospermae

Trees and shrubs. Sporophylls aggregated into cones; seeds exposed on surface of megasporophylls (hence naked seeds); no fruits; xylem elements only tracheids; phloem has no companion cells; many fossil forms.

Class Pinitae

Sub-Class Coniferae

The conifers. Mainly found in the northern hemisphere where they reach the northernmost limits and highest altitudes of all trees; have many xeromorphic features; e.g. *Pinus*[35] (pine), *Araucaria* (monkey puzzle), *Picea* (spruce), *Larix* (larch, deciduous conifers), *Cedrus* (cedar), *Abies* (fir).

Sub-Class Taxidae

Contains a single family.

Family Taxaceae
Separate male and female plants; no female cones, ovules borne singly. e.g. *Taxus* (yew).

Class Ginkgoatae

Represented by a single living species which is recognized as the most ancient type of living seed plant, with the genus *Ginkgo* being represented in the Jurassic by forms similar to today's living species; e.g. *Gingko biloba*.

Class Cycadatae

One main order.

ORDER CYCADALES

World-wide distribution in the Mesozoic period, now a small tropical group with some of palm-like habit; male gametes large and multiflagellate; female cone can weigh up to 35 kg; e.g. *Cycas*.

Class Gnetatae

Have many advanced characteristics, including vessels in the xylem tissue, phloem with sieve tubes and sometimes companion cells; enclosed ovules; and much reduced gametophytes.

Sub-Division Angiospermae

The flowering plants. The dominant group of land plants with close to 25 000 known species; diverse forms adapted to every conceivable terrestrial habitat, with some secondarily aquatic, but only a few species marine (e.g. *Zostera* or eel grass); have the highest degree of cell and tissue differentiation with true vessels in the xylem, and sieve tubes with companion cells in the phloem; microsporophylls (stamens) and megasporophylls (carpels) aggregated into true flowers, flowers contain both sterile and fertile parts; gametophytes extremely reduced and develop entirely within the spore walls; fertilization always siphonogamous, characteristic double fertilization results in diploid zygote and triploid triple-fusion-nucleus which gives rise to the endosperm; carpels enclose ovules in ovaries; after fertilization ovary walls develop into fruits containing seeds.

Includes two classes, the Dicotyledonae and the Monocotyledonae. There are over 450 families of flowering plants, the vast majority being dicotyledons. Some families are natural groupings into which all the members fit quite naturally, as in the Ericaceae; whereas others are very artificial groupings, members of which may be only distantly related to each other, as in the Liliaceae. However, even in the 'natural' families there is still always a range of variation in the criteria used to classify them, such as the number and arrangement of the floral parts. The so-called characteristics of the families therefore only describe the overall trend of members of that group, and the 'typical' member very seldom, if ever, exists.

The families described here contain examples of the typical forms of flowering plants, for example a herbaceous dicotyledon (in the Cruciferae), a dicotyledonous 'evergreen' shrub (in the Ericaceae), a dicotyledonous deciduous tree (in the Fagaceae), and a monocotyledonous plant (in the Liliaceae). These examples also contain illustrations of some of the major trends that are thought to have occurred in the evolution of the flowering plants, such as the reduction in the number of flowering parts, the fusion of floral parts, and the separation of the sexes into either separate male and female flowers on the same plant, or separate male and female flowers on different plants.

Class Dicotyledonae

The larger of the two classes; seeds have two cotyledons or seed-leaves which act as the seed's food store; leaves have net venation; vascular tissue of stem occurs in ring of vascular bundles with cambium, therefore secondary thickening is common and many are woody perennials; flower parts arranged in twos, fours, or usually fives, or multiples of these numbers; root system typically a large primary tap root with lateral branches.

Family Cruciferae
Herbaceous plants; alternate leaves of very variable shape in different species; white or yellow terminal flowers of unusual plan, two pairs of 'clawed' petals, six stamens with two inner pairs, ovary of four carpels, but inner two flattened to form a longitudinal partition down the centre of the ovary, and the outer two forming the valves of the fruit which is a long pod-like siliqua, or a shorter silicula; e.g. *Brassica oleracea* (cabbage), *Raphanus sativus* (radish), *Cheiranthus cheiri* (wallflower), *Capsella bursa-pastoris* (shepherd's purse).

Family Ericaceae
A natural grouping, all of which are woody, either shrubs or perennial herbs; small needle-like leaves and other xeromorphic features; mostly avoiding calcareous soils (calcifuges); mostly form mycorrhizal associations (mycotrophic); flower parts typically in fives or fours, characteristic combination of fused petals and free stamens dehiscing by pores; pollen often forms tetrads; generally ovary superior, axile placentation with many seeds; fruit usually a capsule or berry; e.g. *Erica tetralix* (cross-leaved heather), *Calluna vulgaris* (ling or Scots heather).

Figure 18.17 *Calluna vulgaris or ling*
An evergreen shrub up to 1 m high. The calyx is the main ornamental floral whorl. It is the most abundant native woody plant and covers the greatest land area in Britain.

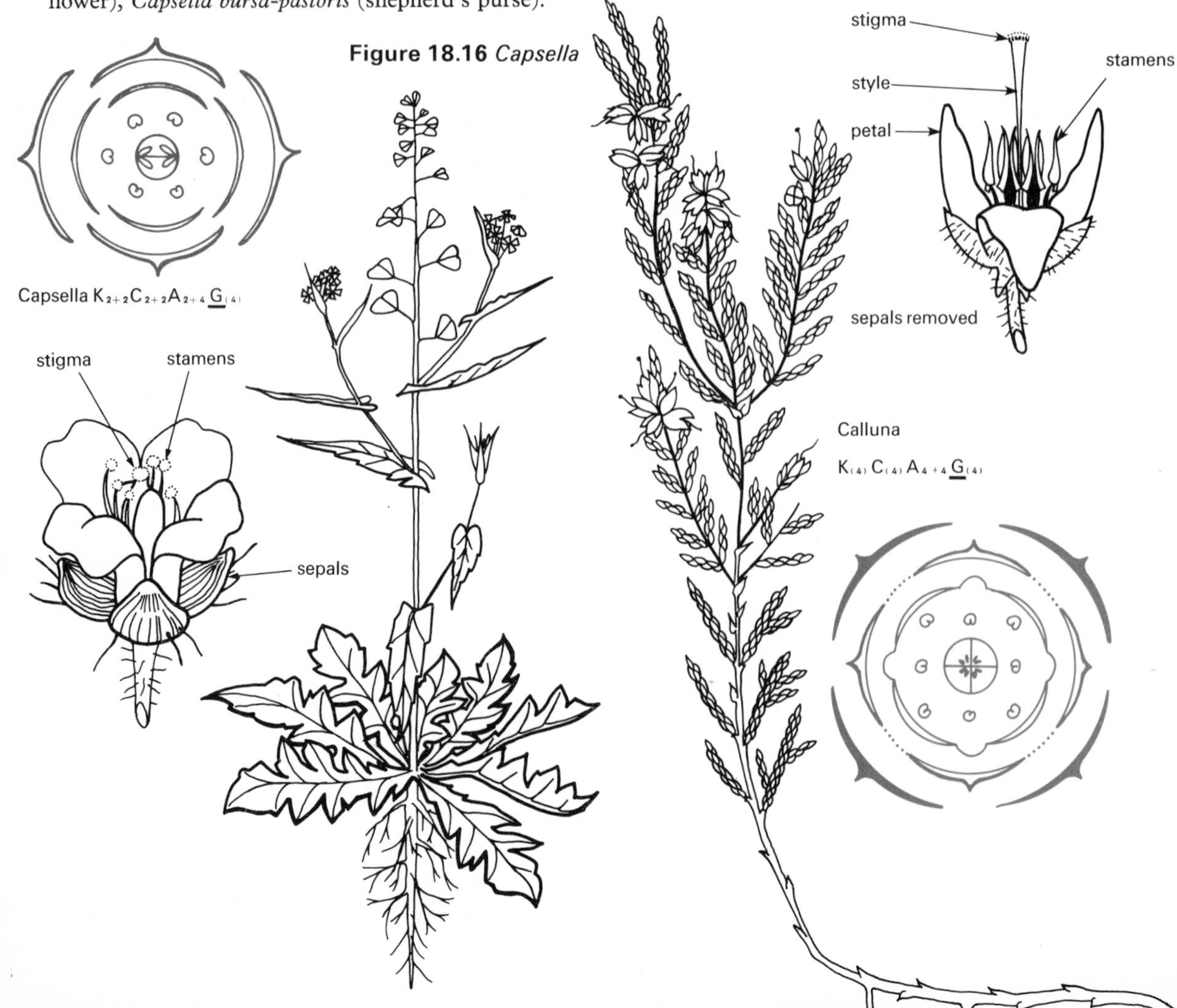

Figure 18.16 *Capsella*

Family Fagaceae

Trees, never normally shrubs in Britain; separate male and female flowers inconspicuous, female flowers generally P3 + 3 G(3), but male flowers more variable; flowers aggregated into separate male and female inflorescences known as catkins, female inflorescence has distinctive cup or cupule composed of many fused bracts containing one or three flowers; usually wind pollinated and open before or when leaves unfold; fruit a large nut; e.g. *Fagus sylvatica* (beech), *Castanea sativa* (sweet chestnut), *Quercus robur* (pedunculate oak).

Figure 18.18 *Quercus*

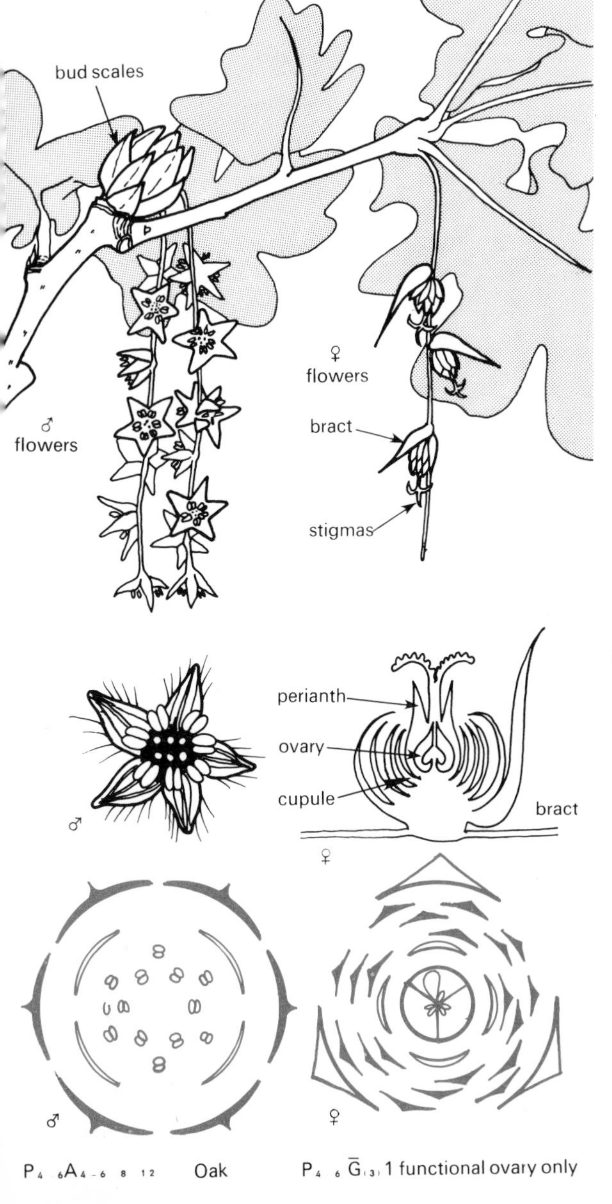

Class Monocotyledonae

The smaller of the two classes. Seeds have only one seed leaf or cotyledon which frequently acts to transfer nutrients from the endosperm to the embryo; leaves have parallel venation; vascular tissue of stem occurs as scattered vascular bundles without cambium, therefore secondary thickening lacking and are never woody perennials, except for a few palms which attain the tree habit via a diffuse mass of cambium; flower parts usually in threes or multiples of threes; root system typically fibrous, without large tap root; considered to have developed from the dicotyledons.

Family Liliaceae

One family of a very large order of monocotyledons (the Liliflorae) which contains all the common bulb-producing plants. The Liliaceae are a very large and diverse family containing some plants which are only distantly related to one another, therefore there are few general features applicable to all types. However, all are herbaceous; perianth segments are petaloid and very conspicuous; perennation by bulbs, tubers, corms and rhizomes; e.g. *Tulipa* (tulip), *Allium* (onion), *Lilium* (lilies), *Endymion non-scriptus* (bluebell).

Figure 18.19 *Endymion*

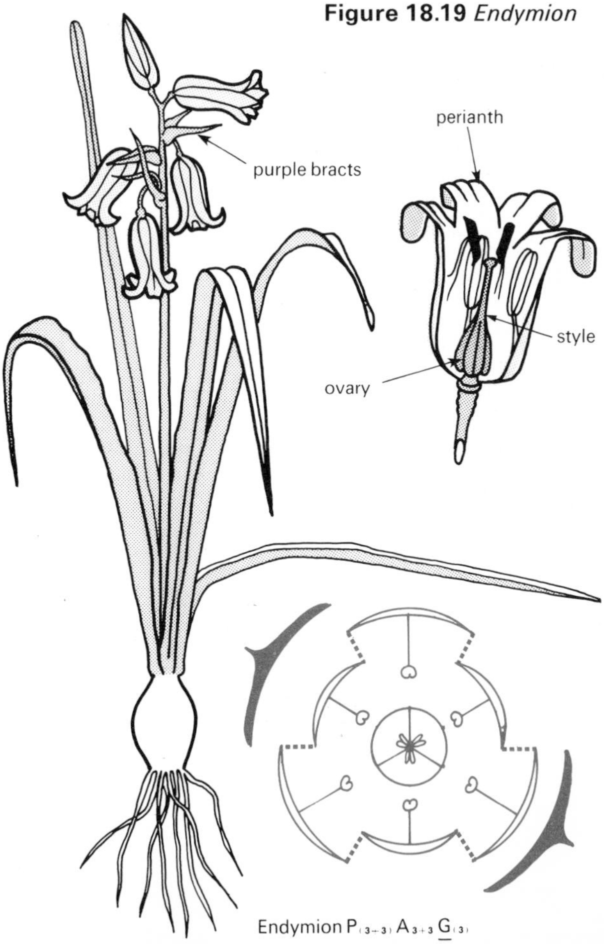

Endymion $P_{(3+3)}$ A_{3+3} $\underline{G}_{(3)}$

THE USE OF KEYS IN THE IDENTIFICATION OF LIVING ORGANISMS

A key is essentially a list of questions about observable features of living organisms that enable them to be sorted out from each other and eventually identified. At each stage, two or more alternative statements are presented, only one of which should apply to the particular specimen being examined. This will lead to the next set of alternatives, and so on until the organism is identified. The simplest type offers only two alternatives at each stage and is known as a **dichotomous key**. It is difficult to construct good keys, especially when attempting to differentiate between species, and at this level a key can only act as a guide.

Figure 18.20 *Animal key—invertebrates in turf* (After Jenkins and Wright, *School Grounds* (London: Heinemann, 1973).)

Look for and count the jointed legs					
no legs	Body is ∩ shaped in cross-section	No tentacles	→	PLATYHELMINTHES TURBELLARIA	TURBELLARIA 
		Tentacles	Shell	MOLLUSCA GASTROPODA	SNAILS 
			no shell	MOLLUSCA GASTROPODA	SLUGS 
	Body is O shaped in cross-section	Segments few up to 13	→	INSECTA DIPTERA	FLY LARVAE 
		Segments numerous	→	ANNELIDA OLIGOCHAETE	EARTHWORM 
3 pairs of true legs	Insects with larval form	No prolegs on abdomen	→	INSECTA COLEOPTERA	BEETLE LARVAE 
		Prolegs present on abdomen	→	INSECTA LEPIDOPTERA	CATERPILLAR 
	Insects with adult form	No waist between abdomen and thorax	'Spring' on last segment	INSECTA COLLEMBOLA	SPRING-TAIL 
			Pair of cornicles on abdomen	INSECTA HOMOPTERA	APHID 
			Front wings are hard and protective	INSECTA COLEOPTERA	ADULT BEETLE 
		'Wasp' waisted	Waist with 1 or 2 nodes	INSECTA HYMENOPTERA	ANT 
			No nodes	INSECTA HYMENOPTERA	WASP 
4 pairs of legs	Body not in 2 clearly distinguished parts		→	ARACHNIDA ACARINA	MITE
	Body of 2 clear parts and abdomen not segmented		→	ARACHNIDA ARANEIDA	SPIDER 
7 pairs of legs	Body flattened from above		→	CRUSTACEA ISOPODA	WOOD-LOUSE 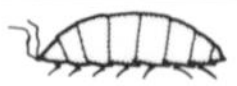
Numerous pairs of legs	Each segment has 2 pairs of legs		→	MYRIAPODA DIPLODA	MILLIPEDE 
	Each segment has 1 pair of legs		→	HYRIAPODA CHILOPODA	CENTIPEDE 

Appendix 3
Histograms, graphs, log scales, and exponents

Histogram

A histogram is essentially a frequency distribution diagram in which, for example, the number or frequency of individuals within a certain range of measurements is represented by a rectangle of a certain area. The area of each rectangle is proportional to the frequency, and if the rectangles are of equal width then their heights are also proportional to the frequency.

If results are listed in the order in which they are obtained, it is often difficult to gain much information from the figures. For example, consider the set of measurements of stem length in a population of plants. If the results are listed as given below in the order in which they were obtained, then little information can be derived from the figures.

Length of stem in mm

30, 41, 53, 42, 36, 50, 61, 56, 35, 46, 58, 61, 37, 66, 76, 28, 40, 20, 48, 60, 52, 22, 57, 70, 43, 57, 62, 63, 73, 31, 49, 52, 47, 37, 53, 63, 56, 71, 23, 63, 24, 58, 39, 44, 56, 36, 64, 74, 55, 66, 75, 29, 48, 26, 46, 61, 53, 26, 59, 72, 45, 25, 68, 51, 66, 21, 75, 30, 47, 54, 46, 31, 59, 62, 51, 34, 46, 54, 69, 37, 69, 75, 46, 53, 45, 62, 57, 68, 53, 79, 47, 53, 63, 81, 82, 79, 37, 48, 56, 43, 33, 58, 69, 52, 31, 48, 52, 61, 31, 83, 43, 56, 42, 68, 51, 63, 54, 78, 49, 51, 68, 84, 85, 76, 38, 42, 57, 41, 32, 51, 62, 53, 33, 40, 50, 78, 33, 86, 78, 41, 58, 44, 69, 52, 64, 78, 76, 45, 51, 79, 71, 87, 77, 88, 59, 69, 68, 89, 89.

However, if they are grouped into classes then it becomes easier to notice any trends or patterns.

The range of frequencies in each class interval is purely a matter of choice, for example each length could constitute a class of its own, or all the measurements could be placed in one large class. The choice of these extreme examples would obliterate any emergent patterns in the results, and therefore a happy medium must be sought so that the class intervals are large enough to reduce the effects of chance variations in frequency, but not so large as to obscure any patterns.

The results above can be classified as follows:

From		*Up to*		*Number in class*
20	–	30	=	10
30	–	40	=	20
40	–	50	=	30
50	–	60	=	40
60	–	70	=	30
70	–	80	=	20
80	–	90	=	10
Total				160

A diagrammatic representation of these groupings in the form of a histogram makes them even clearer.

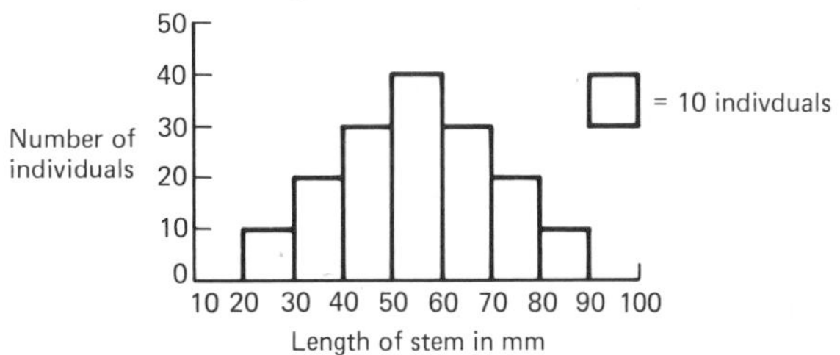

Figure 19.1 *Histogram showing distribution of plants according to stem length*

Such a distribution that builds up gradually from a low number at the two extremes to a maximum in the middle is known as a **normal distribution.**

Some distributions are asymmetrical or **skewed**, e.g.

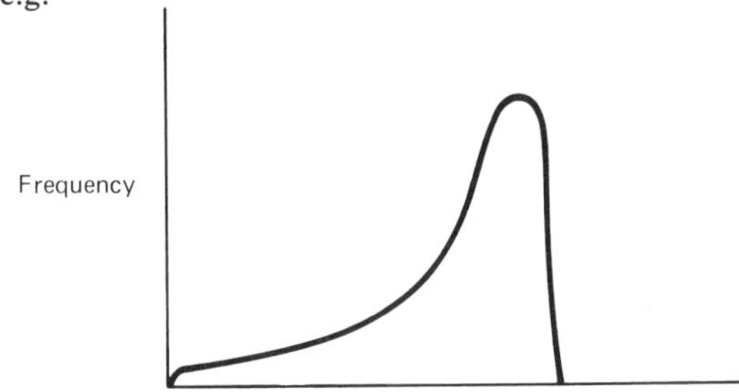

Some distributions have a double peak; for example when a population contains two distinct types or varieties which have been sampled together. This is known as the bimodal distribution.

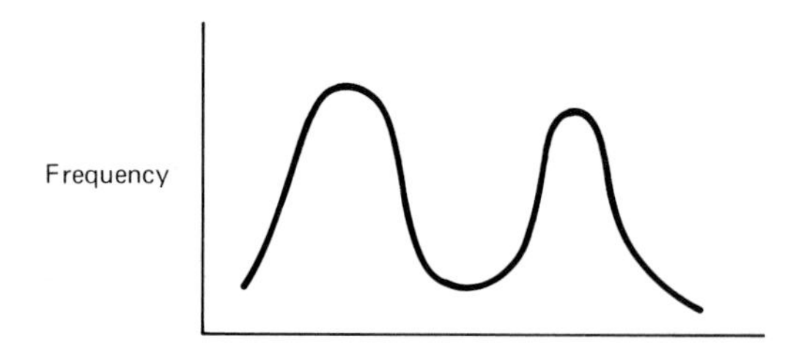

The information in such histograms can be further summarized by calculation of the **centering constant** which attempts to define the centre of distribution. Three such centering constants are commonly used, the **mean** or average, the **median**, and the **mode**.

Average or mean ($\bar{x}$)

This is the most easily calculated and generally understood centering constant, and is equally weighted by each observation. Living organisms are very variable in their activity and therefore it is usually necessary to take the average or mean of a set of readings, as it is indeed in most experiments. The average of several values is obtained by adding the values together and dividing by the total number of values. For example, if it was required to find the average or mean number of heartbeats per minute ($\bar{x}$) from a set of readings of 80, 90, 89, then:

$$\bar{x} = \frac{80 + 90 + 89}{3} = \frac{258}{3} = 86$$

The larger the number of values taken for consideration, the less 'distorting' an effect the extreme values will have on the average figure. For example,

$$\bar{x}\,(101, 79, 81) = \frac{261}{3} = 87$$

$$\bar{x}\,(101, 79, 83, 82, 78, 79) = \frac{502}{6} = 84$$

However, the average or mean gives no idea of the extent of the variation between the different values. For example, in the set of heart rate values used to illustrate the definition of an average value, the reading 101 stands out from the cluster of 79, 83, 82, 78, and 79. This atypical reading could be of great interest within the scope of the experiment; or it could be the result of factors outside the scope of the experiment, in which case it is in some ways a false reading which 'distorts' the mean.

Median

Some distributions are asymmetrical or skewed, and in these cases the median should be used as the centering constant, as it should when the data include occasional extreme values. The median is the middle value when all the values are arranged in order of magnitude, e.g.

81, 86, 88, (90), 91, 99, 100.
Median

Mode

The mode is the most frequent value or class, e.g.

81, 86, 88, 90, 91, 91, 100.
Mode

For the symmetrical curve of the normal distribution, the mean, the median, and the mode are all equal to each other.

Graphs

A graph shows the relationship between two quantities in a way that is easily understood. Every graph should be self-explanatory, and should therefore be correctly titled, should have the axes fully labelled, and the scales used fully explained. Standard graph paper is divided by thick lines into centimetre squares or two centimetre squares, and by finer lines into millimetre squares. On such graph paper equal increments on the scale must represent equal numerical values.

Generally, values such as frequency or rate, in other words values which 'fluctuate' during an experiment, are represented on the vertical scale (y-axis); and the other values, usually those that are changed in a regular manner, such as time, are represented on the horizontal scale (x-axis).

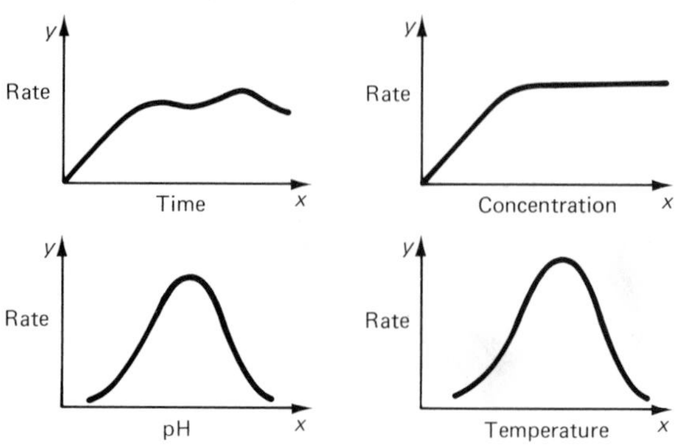

The scale representing frequency should always start with zero. If the range of figures is very wide then the scale should still start at zero, but a break may be made across the range of figures that are not used.

The points of the graph should be plotted as accurately as possible, using either a cross (×) or a circled dot (⊙).

There are no hard and fast rules about drawing the line of a graph. Generally if there is a theoretical expected shape to the curve then the line may be drawn to this shape, even if it does not actually pass through all the points. For example, the line may be either straightened up by drawing the 'best straight line' between the points, or rounded off to a shape.

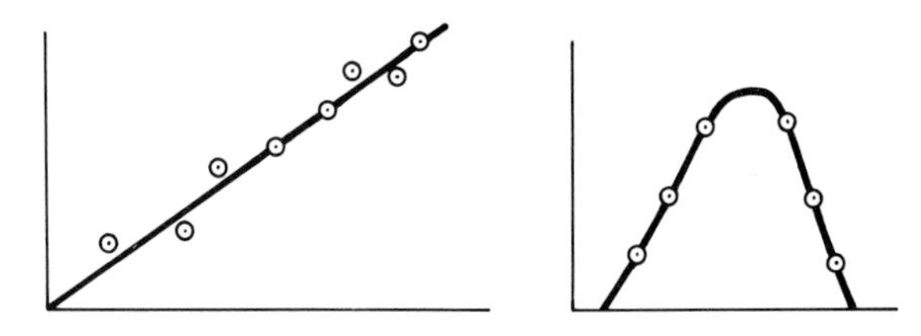

Log scales

On a log scale the distance between two points is equivalent to the difference between the logarithms of the coordinate of the point. For example, the distance between 10 and 20 is $1.3010 - 1.000 \simeq .3$, and is the same as the distance between 1 and 2. The use of log scales shortens the scale required for a certain set of values, and also confers other advantages. Log scales are mainly used in biology in relation to population growth curves. Under certain conditions populations can grow exponentially, thus giving typical exponential growth curves when plotted on normal graph paper.

For example, consider a population which grows with a doubling every day:

Time in days	0	1	2	3	4	5	6	7	8	9	10
Number in population	1	2	4	8	16	32	64	128	256	512	1024

When plotted on normal graph paper the following exponential curve is obtained.

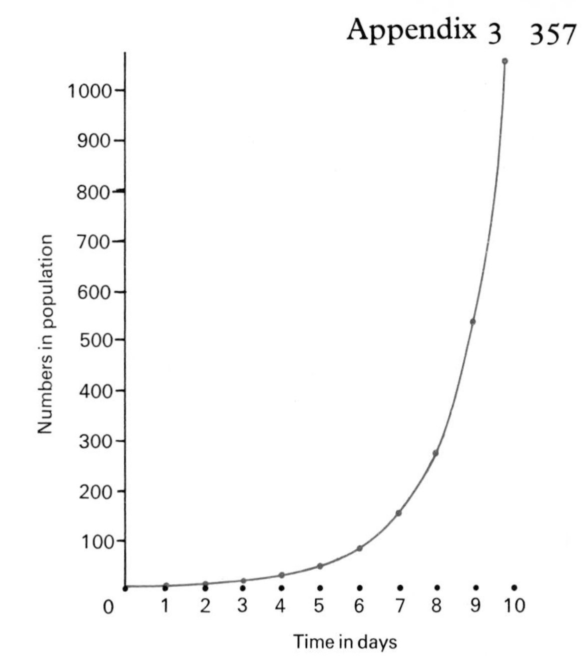

Figure 19.2 *Exponential population growth curve*

However, when these figures are plotted on log/linear graph paper, a straight line is obtained.

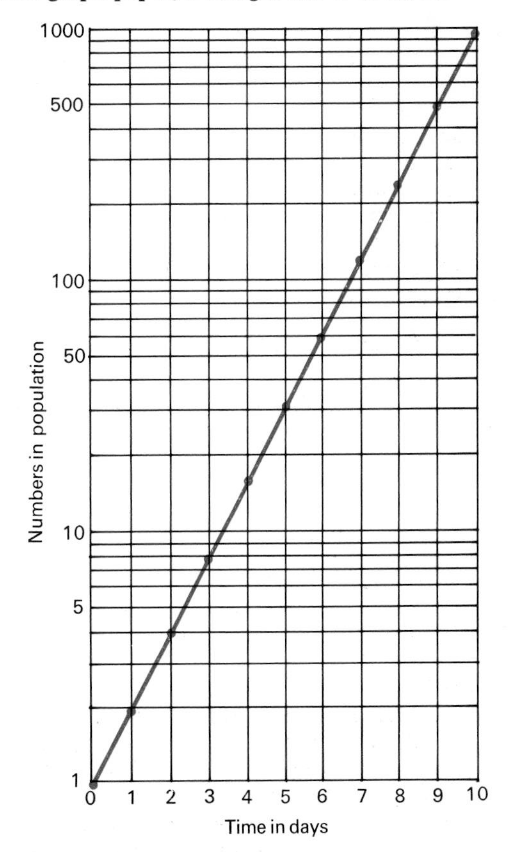

Figure 19.3 *Log/linear population growth curve*

If it is known that a population is growing exponentially, then only two readings are required to give the straight line on the log/linear graph paper, and any other values can be read off. In most cases involving resource usage, the use of resources also shows an exponential increase. In these cases the log of the annual production, etc., is plotted against time and this then provides a straight line which may be extended into the curue with greater simplicity and accuracy.

The main difficulty with log scales is that a slight deviation from the linearity of a log plot can result in a very large difference in projection ('prediction'), and the greater the time involved in the extrapolation the more inaccurate does the 'prediction' become.

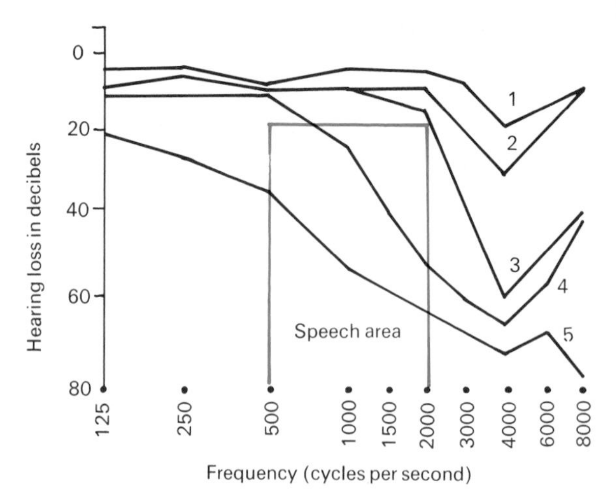

Figure 19.5 *Noise levels*

Log scale response of sense organs

The eye, ear, and other sense organs operate on a log scale of response, in which equal fractional changes in stimulus give equal increments of response. For example, an increase of 26 per cent in the level of stimulation is necessary for any difference to be detected by the ear; and this increase always sounds the same, regardless whether it is 26 per cent of a low level or 26 per cent of a high level.

This log response of the sense organs is necessary to cover the tremendous range of stimulation to which they can respond. For example, if the range that the ear has to deal with was plotted on a linear scale with 1 cm representing the range between 0 and the threshold of hearing, then the upper end of the scale, at the threshold of pain, would be well beyond the Sun.

threshold of hearing — well beyond the Sun — threshold of pain

0 1 cm

2×10^{-5} N m^{-2} — 2×10^{2} N m^{-2}

This log-response of the ear is reflected in the use of the **decibel scale**, which is a log scale, to express sound intensities. The 26 per cent increase in sound pressure (which always sounds the same) is designated as being 1 decibel. The decibel scale is a log scale on which an increase of 26 per cent is always represented by 1 decibel, even though it may be 26 per cent of a large or small sound pressure.

Exponents

An exponent is the number which indicates the power of a quantity, for example the exponent of x in x^6 is 6; x^6 means x multiplied by itself 6 times. Exponents are used because they are concise, and once understood have an immediate meaning. For example, 10^6 is 10 multiplied by itself 6 times, which equals 1 000 000. In this example the exponent of 10, namely 6, is the same as the number of noughts. Thus 10^2 is 1 with 2 noughts (100), 10^3 is 1 with 3 noughts (1000), etc. This is only true with powers of 10, which, however, are the most commonly used in biology. If a number is given as 3.9×10^9, then the exponent tells us to move the decimal point 9 places to the right.

$$3.9 \times 10^9 = 3.9 \times 1\,000\,000\,000 = 3\,900\,000\,000$$

or three thousand, nine hundred million, which was the estimated world population in 1979.

A negative exponent, such as in 10^{-1}, tells us the number of places to move the decimal point to the left, for example:

$$1.0 \times 10^{-1} = 1.0 \times 0.1 = 0.1 = \frac{1}{10}$$

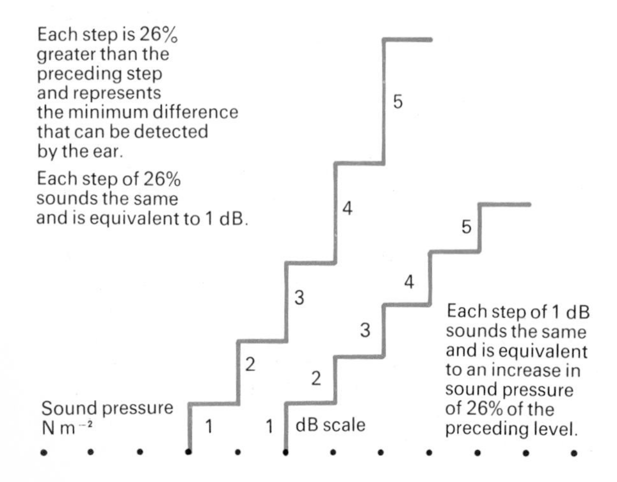

Figure 19.4 *Log response of the ear*

Appendix 4

SI Units

This is the revised and extended form of the metric system, finally agreed and adopted by the Conference Générale des Poids et Mesures (CGPM). It is a coherent system based on seven units. A simplified and amended version of the fully comprehensive SI units system of base and derived units will be described here.

The SI contains three classes of units: the base units; the derived units; and the supplementary units (these will not be described here).

Base units

Physical quantity	*SI unit*	*Symbol*
Length	metre	m
Mass	kilogram	kg
Time	second	s
Electric current	ampere	A
Temperature*	kelvin	K
Luminous intensity	candella	cd
Amount of substance	mole	mol

*The customary international unit is the degree Celsius or Centigrade. Although 0 °C corresponds to 273.16 K, the degree centigrade has the same interval as the kelvin or degree absolute, e.g. 0 °C = 273.16 K, 100 °C = 373.16 K.

Derived units

These are formed by combining two or more base units, for example velocity is produced by combining length with time. Similarly, the unit of force can be produced by combining three mechanical base units: length, mass, and time. Frequently such derived units are given their own special name, often to honour a famous scientist; for example in the case of force the unit is called the newton.

Definitions of the fundamental mechanical units

Length: the metre (m) This is now defined in terms of a particular wavelength of light, although it was originally intended to be one ten-millionth part of the distance from the North Pole to the Equator at sea level through Paris. It was later defined as the distance between two marks on a special alloy bar kept in Paris.

Mass: the kilogram (kg) This is equal to the mass of a platinum-uridium bar kept at the Bureau International des Poids et Mesures at Sèvres. It is the only SI base unit which is still defined in terms of an original prototype rather than in terms of reproducible measurements of physical phenomena (1000 kg = 1 tonne).

Time: the second (s) This was originally defined as a specified fraction of a particular time period, but since 1968 it has been defined in terms of an atomic caesium 'clock'.

Derived units in SI

Area: the metre squared (m^2) 10 000 m^2 = 1 hectare

Volume: the metre cubed (m^3) For most practical laboratory purposes this unit of volume is far too large. For example:

$$m^3 = \text{one cubic metre}$$
$$1\ m^3 = (100\ cm)^3 = 1\ 000\ 000\ cm^3$$

1 m^3 of water weighs 1 tonne (1000 kg), which is greater than the weight of a mini car.

The common units of volume are the centimetre cubed (cm^3), formerly called the millilitre (ml); and the decimetre cubed (dm^3), formerly called the litre (l); however, the litre is still widely used.

$$dm^3 = (\text{decimetre})^3 = \left(\frac{100\ cm}{10}\right)^3 = (10\ cm)^3 = 1000\ cm^3 = 1\ \text{litre}$$

Force: the newton (N) The newton is the force required to accelerate a mass of 1 kg through an acceleration of 1 $m\ s^{-2}$.

Pressure: in newtons per square metre Pressure is a force per unit area, and is expressed in newtons per square metre ($N\ m^{-2}$) or pascals (Pa), where 1 $N\ m^{-2}$ = 1 Pa. Thus, strictly speaking, standard atmospheric pressure should now be expressed as 101 325 $N\ m^{-2}$, or as 101.325 kPa; instead of as 760 mmHg, which is the height of a mercury column that atmospheric pressure will support.

Energy: the joule (J) This was formally defined as the ability to do work, that is the force × the distance moved in the direction of the force. The SI unit of energy is the joule (J) which can be defined in a variety of equivalent ways. The simplest defines it as that work done when a force of 1 newton moves its point of application through 1 metre. The joule is rapidly replacing the former unit, the calorie, but the calorie is still a useful unit of heat energy. The calorie was defined as the amount of heat required to raise 1 g of water through 1 °C, strictly from 14.5 to 15.5 °C (1 calorie is equal to 4.18 J).

SI prefixes

Number	10	100	1000	1 000 000
Factor	10^1	10^2	10^3	10^6
Prefix	deca	hecto	kilo	mega
Symbol	da	h	k	M

Number	$\frac{1}{10}$	$\frac{1}{100}$	$\frac{1}{1000}$	$\frac{1}{1\,000\,000}$	$\frac{1}{1\,000\,000\,000}$
Factor	10^{-1}	10^{-2}	10^{-3}	10^{-6}	10^{-9}
Prefix	deci	centi	milli	micro	nano
Symbol	d	c	m	μ	n

Some conversions

Weight

1 g = 0.035 oz
1 kg = 2.205 lb
1 tonne = 0.9842 ton

1 oz = 28.35 g
1 lb = 454 g
1 ton = 1.0160 tonne

Area

1 hectare = 2.471 acres (≈1$\frac{1}{5}$ football pitch)
1 acre = 0.4047 hectare (≈$\frac{1}{2}$ football pitch)
1 square mile = 259 hectares
1 acre = 4840 square yards
1 square mile = 640 acres

Volume

1 litre = 1.760 pints
1 pint = 0.868 litre

Glossary of chemical and physical terms

This glossary explains chemical and physical terms used in the text. It is organised so that related terms appear together.

Element An element is a substance that cannot be broken down chemically to simpler substances. An element consists entirely of atoms of the same atomic number. Amongst the most commonly found in living organisms are: carbon ($_6C$), hydrogen ($_1H$), oxygen ($_8O$), and nitrogen ($_7N$).

Compound These are substances consisting of two or more elements combined together chemically in definite proportions by weight. The combining elements in compounds no longer show their characteristic physical properties. For example, the compound glucose is a white crystalline substance ($C_6H_{12}O_6$), but its constituent elements are carbon, which is normally a black powder; hydrogen, which is the lightest gas; and oxygen, another gas.

Mixtures These are simply physical mixtures of different substances involving no chemical reactions, e.g. sand and iron filings.

Atoms An atom is the smallest particle of an element that can enter into chemical combination (or can exhibit the chemical properties of that element). It consists of a positively charged nucleus, surrounded by one or more negatively charged electrons which can be imagined as moving in orbits around the nucleus. The nucleus consists of two different types of stable particles of almost equal mass: the positively charged protons and the electrically neutral neutrons. The number of electrons determines the chemical behaviour of the atom, the number of protons equals the number of electrons. This number is known as the **atomic number**, and is unique and constant for a particular element. The number of protons and neutrons together is the **mass number**.

Number of protons and neutrons = mass number → $^{12}_{6}C$ $^{1}_{1}H$ $^{16}_{8}O$ $^{14}_{7}N$

Number of protons = atomic number

Atoms of an element which have the same number of protons but different numbers of neutrons are known as **isotopes** of that element:

6 protons, 6 neutrons $^{12}_{6}C$ 6 protons, 8 neutrons $^{14}_{6}C$

Both ^{12}C and ^{14}C are isotopes of carbon but ^{12}C is the most commonly occurring. Some nuclear configurations of protons and neutrons are less stable than others and emit various radiations in order to achieve stability, these are known as **radioactive isotopes**. Radioactive isotopes behave the same chemically as the non-radioactive isotopes of that element, and their radioactive emissions enable them to be detected and traced through complex metabolic pathways, thus they are known in this role as **radioactive tracers**. Such radioactive isotopes gradually decrease in their activity, that is they decay over a period of time which is characteristic for each one. The term **half-life** is used to describe that period of time over which the activity of a particular isotope is halved. Some radioactive isotopes are capable of undergoing splitting, or **fission**, whilst others are not.

Radioactive non-fissionable isotope
$^{238}_{92}$Uranium
92 protons, 146 neutrons
Emits α-particles
Half-life 4.5×10^9 years

Radioactive fissionable isotope
$^{235}_{92}$Uranium
92 protons, 142 neutrons
Occurs as 0.7 per cent of natural ores
Emits α-particles and γ-rays
Only naturally-occurring isotope that is spontaneously fissionable

When the fissionable isotope ^{235}U is struck by a free-moving slow or thermal neutron the neutron is absorbed by the nucleus and the nucleus undergoes fission.

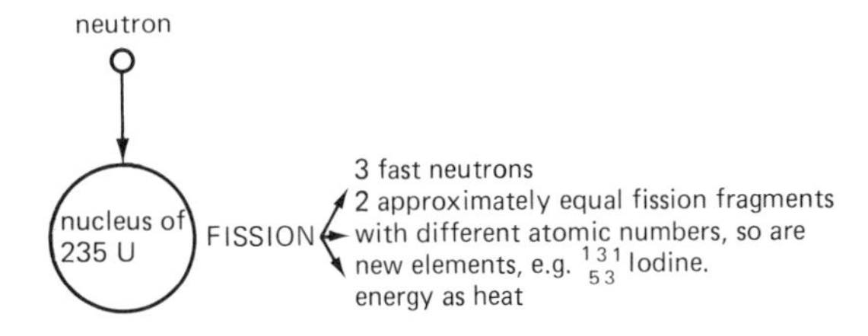

The fission fragments move rapidly, but they quickly come to a stop due to collision with surrounding atoms. This causes atomic displacement and movement which creates more heat. The neutrons, being smaller, travel further, but some lose energy by collision and become slow neutrons which are then capable of causing the fission of other ^{235}U nuclei, thus initiating a **chain reaction**. For this chain reaction to occur, the amount of ^{235}U must exceed a certain minimum or critical size. If sub-critical, the neutrons escape. If super-critical, that is too much above the critical, then the chain reaction will be explosive. This occurs when two sub-critical masses are brought suddenly together to produce an atomic explosion. The critical mass of an unconfined sphere of ^{235}U metal is about 10.5 cm in diameter and about 48 kg in weight. There is an upper limit to the amount of ^{235}U that can be brought together instantaneously which limits the size of fission explosions. The fission process is responsible for the large quantities of nuclear energy released in an atomic reactor or weapon. The fission products can contain about 200 different radioactive isotopes of about 35 elements. Nuclear fission and radioactive isotopes emit ionizing radiations which are of great biological significance. Such ionizing radiations include α-and β-particles, and γ- and X-rays.

Nuclear fusion In nuclear fission temperatures of at least a million degrees are produced. At these temperatures the nuclei are stripped of most of their surrounding cloud of electrons and they move at high speeds and undergo many collisions. Under these conditions the nuclei of the isotopes of hydrogen known as deuterium (^2_1H) and tritium (^3_1H) have sufficient energy of motion to overcome the repulsive forces of their single positive charges, and they fuse. As a result a nucleus of a helium atom is formed (^4_2He), and a high speed neutron and much energy is released from each pair of reacting nuclei.

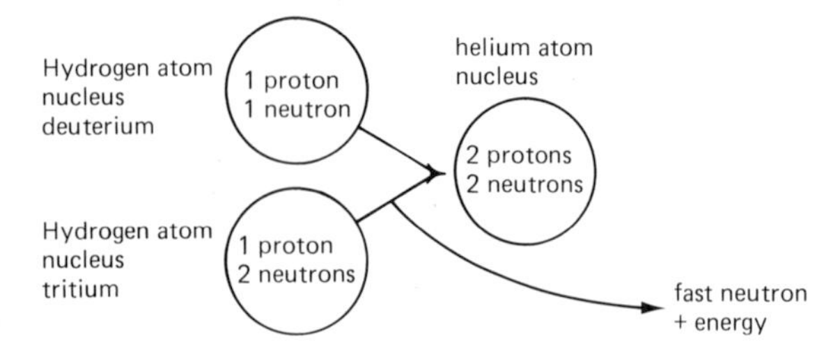

This process of nuclear fusion is the basis of the hydrogen bomb or thermo-nuclear weapon, to which, unlike the fission weapon, there is no theoretical upper limit in size. Nuclear fusion could also be the basis of a vast unlimited supply of energy for man, should it ever be possible to harness its energy in a controlled way. As the main product of the reaction is helium, which is not radioactive, fusion is much cleaner than fission. However, the fast neutrons that are released induce radioactivity in materials with which they collide. Fusion has many advantages over fission as a potential source of energy; the raw materials are relatively abundant, and although the problem of induced radioactivity remains, there are no high level radioactive wastes, the disposal of which represents the main hazard to man and the living world in the use of the fission process.

Molecules These are the smallest particles of an element or a compound that can have a stable existence. Some substances, for example helium (He), have monatomic molecules, that is the atom is the smallest particle of helium that can have a stable existence. More usually two or more atoms combine to make a molecule, for example H_2 (hydrogen), O_2 (oxygen), N_2 (nitrogen).

Chemical formulae When two or more atoms combine to form a molecule, the chemical formula tells one which atoms, and how many, are involved. For example the formula for glucose is $C_6H_{12}O_6$, this means that there are 6 carbon, 12 hydrogen, and 6 oxygen atoms combined together.

Chemical equations These describe chemical reactions in a short-hand way, for example

$$\underset{\text{glucose}}{C_6H_{12}O_6} + \underset{\text{oxygen}}{6O_2} \longrightarrow \underset{\text{carbon dioxide}}{6CO_2} + \underset{\text{water}}{6H_2O}$$

The numbers in front of some of the molecules are necessary to balance the numbers of atoms on either side of the equation.

Valency Chemical combination between atoms involves the transfer or sharing of outer electrons. The valency of an atom is a measure of its combining power. For example, the valency of oxygen in water is two and the valency of hydrogen is one, and these combine to form water which is H_2O. **Covalent bonds** are formed when electrons are shared, and are typical of organic compounds. **Electrovalent bonds** are formed by the transfer of electrons from one atom to another. The atom that loses the electron(s) becomes a **positive ion**, and the one that gains the electron(s) becomes a **negative ion**.

Electrolytes There are two types of electrolytes: electrovalent and covalent. **Electrovalent** types have charged particles or ions already existing in their solid (crystalline) state. Salts such as NaCl (sodium chloride) are electrovalent compounds whose crystals are composed of electrically-balanced numbers of anions and cations held in a lattice structure. When they dissolve in water the ions become hydrated, the electrostatic attraction between ions of opposite charge is weakened, and they are dispersed. **Covalent** ones only yield ions when they undergo chemical reactions with water molecules or with molecules of some other suitable solvent. The extent to which the molecules of an electrolyte are ionized in an aqueous solution is

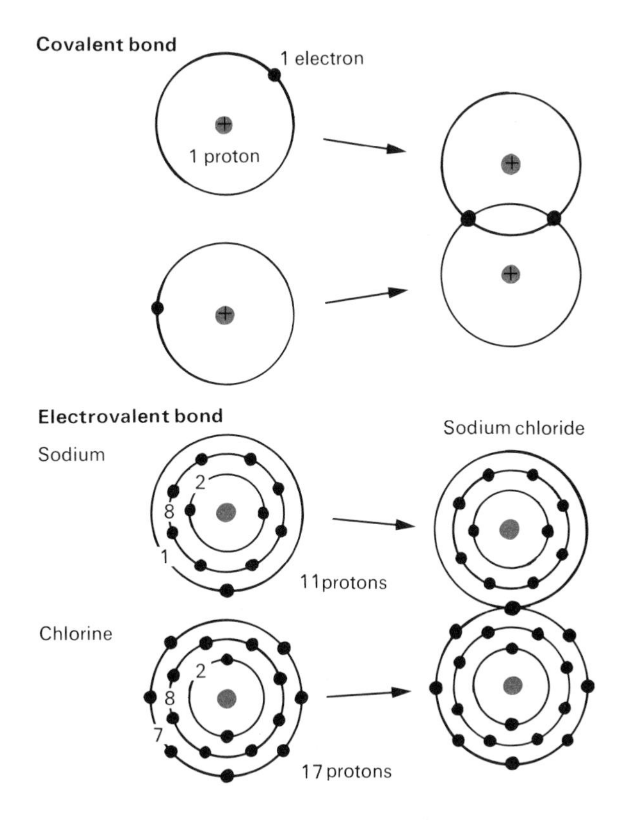

Covalent and electrovalent chemical bonds

known as the **degree of dissociation** or the **degree of ionization**.

Oxidation and reduction Oxidation is the combination of oxygen with a substance or the removal of hydrogen from it; and reduction is the opposite, that is the removal of oxygen or the addition of hydrogen. For example, in cellular respiration

$$\underset{\text{(reduced cytochrome)}}{\text{cytochrome } H_2} + \tfrac{1}{2}O_2 \xrightarrow{\text{oxidase enzyme}} \underset{\text{(oxidized cytochrome)}}{\text{cytochrome}} + H_2O$$

$$AH_2 + \text{cytochrome} \xrightarrow{\text{dehydrogenase enzyme}} \underset{\text{(oxidised A)}}{A} + \underset{\text{(reduced cytochrome)}}{\text{cytochrome } H_2}$$

In fact, all oxidations in biological systems are by the removal of hydrogen.

Acids, bases and pH One definition of **acids** is that they are substances with a tendency to furnish a proton (hydrogen ion) in chemical reactions. They turn litmus paper red. Conversely, one definition of **bases** is that they are substances with a tendency to accept a proton in chemical reaction. They turn litmus paper blue.

$$\underset{\text{acid}}{HCl} + \underset{\text{base}}{NaOH} \longrightarrow \underset{\text{salt}}{NaCl} + \underset{\text{water}}{H_2O}$$

$$(\underset{\text{proton}}{H^+} + Cl^-)(Na^+ + OH^-) \longrightarrow (Na^+Cl^-) + (H_2O)$$

In this way acids and bases neutralize each other. The strength of acids and bases depends on the degree of their dissociation in water; strong acids and bases dissociate completely in water, and weaker ones, usually organic compounds, only dissociate slightly in water. **pH** is a measure of acidity given by the concentration of hydrogen ions present. The pH scale is the negative of the exponent of the hydrogen ion concentration. Thus water, which is normally slightly dissociated to the extent of 10^{-7}M, has a pH of 7 which is taken as neutral. Acids have a pH of less than 7 and alkalis a pH greater than 7, up to a maximum of 14. Enzymes are very sensitive to pH, only working maximally within a narrow band of pH values referred to as the optimum pH range.

Very acid pH 0	pH 1	*Neutral* pH 7	*Very basic* pH 14
1 M acids if completely ionized	stomach acid (HCl) concentration	optimum range for enzyme action	

Buffers A buffer resists a change of pH when either acids or alkalis are added to it, in other words it has the capacity to 'buffer' or 'soak up' changes of acidity or alkalinity, so that the pH remains relatively constant. Biologically, buffers are very important in the blood of vertebrates, particularly mammals.

The metabolism of cells is extremely sensitive to pH, with the optimum range for most enzymes being between 6.8 and 7.4. However, in mammals (and others) metabolism tends to produce an excess acidity, which is eventually excreted over the lungs (as carbon dioxide, an acid gas) and through the kidneys. This acidity must be buffered in the blood while it is transported from all parts of the body to the lungs and the kidneys. Buffers in the blood include the plasma proteins and the respiratory pigment haemoglobin.

Metabolic processes Metabolism is the sum total of all the chemical reactions occurring in the body. There is generally a balance between breakdown reactions or **catabolism**, and synthetic reactions, or **anabolism**. One of the commonest type of catabolic reaction in living systems is hydrolysis.

Hydrolysis This is the 'splitting by water' of compounds. In biological systems it is particularly seen in the digestion and breakdown of large complex organic compounds. During the process the water itself is also broken down, e.g.

$$\underset{\text{peptide bond combining amino acids in proteins}}{-CO-NH-} + H_2O \xrightarrow{\text{enzymes}} -COOH + HNH-$$

Condensation reactions Condensation reactions are in effect the 'reverse' of hydrolytic reactions. In condensation reactions molecules are combined with the elimination of water, e.g.

$$-\text{COOH} + \text{HNH}- \xrightarrow{\text{enzymes}} \underset{\text{peptide bond}}{-\text{CO}-\text{NH}-} + \text{H}_2\text{O}$$

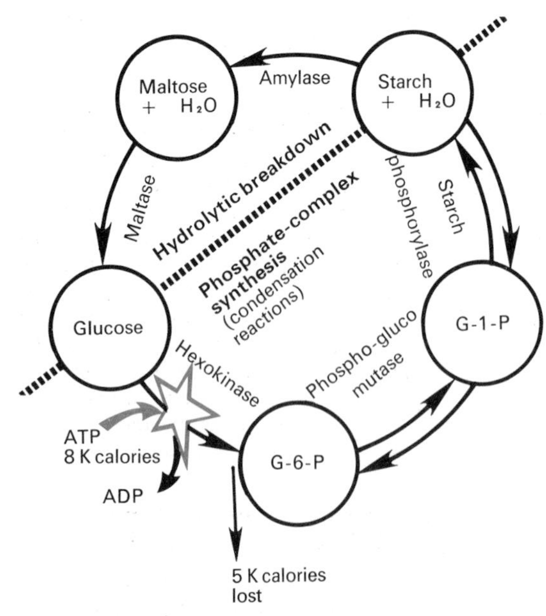

Catabolic and anabolic pathways
Although all reactions are theoretically reversible, the general principle in many biological reactions is for hydrolytic breakdown in catabolism, and phosphate complex condensation reactions in the synthetic pathways of anabolism using different enzyme systems

Water Water is essential to life, or rather, water is a part of life. It is a constituent of protoplasm, being closely attached to molecules and ions in a way that contributes to their physical and chemical behaviour. It is a 'universal' solvent in which most biologically active materials dissolve, and in which the metabolic reactions occur. It participates directly in many chemical reactions of the cell, including particularly hydrolysis and condensation reactions. It is also the source of hydrogen for the reduction of carbon dioxide to organic compounds in photosynthesis; and it is a product of respiration.

Water is important as a transport medium; it forms a continuous fluid system throughout plant and animal bodies in which materials can move, it aids diffusion, or undergoes mass flow movements by which materials are physically carried around the living body. It is also important in the support of cells as it is responsible for their turgidity, particularly in plant cells with their strong but elastic cellulose walls. This turgidity is also important in the extension phase of growth in plant cells.

Water has a number of special properties that particularly suit it to its many roles in living systems. When compared to other substances that are chemically similar, water shows some unexpected characteristics, due mainly to hydrogen bonding increasing the cohesive forces among the molecules.

Water molecules are asymmetric so that one end is more positive and the other more negative. This formation is known as a dipole, and it is this dipole arrangement that causes the hydrogen bonding.

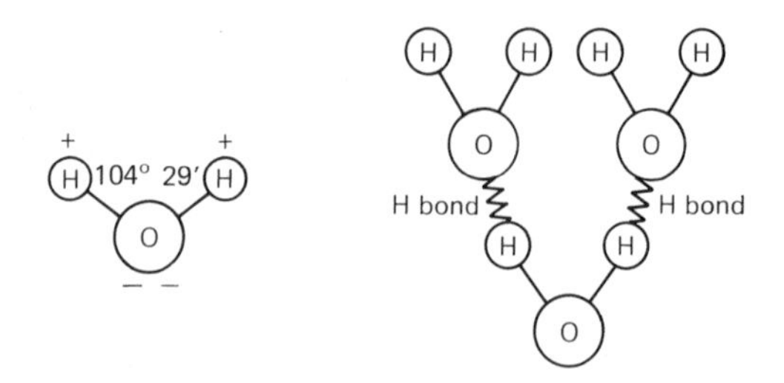

These ring-like configurations produced by the hydrogen bonding actually exist in the structure of ice and persist in the liquid state.

As a result of these molecular associations, extra heat is required to separate the molecules to the point when they vaporize, thus the boiling point is greatly raised above that expected from its low molecular weight. As it does not evaporate so readily it has a high **latent heat of evaporation**, which means that it can exert a pronounced cooling effect on evaporation.

The **specific heat** of water is also relatively high, which means that it can resist sudden fluctuations in temperature. This is important both intra-cellularly (within the cells), inter-cellularly (around the cells), and when it acts as the environmental medium.

Water has a high surface tension and viscosity, due to its high **cohesion** between molecules, and this is particularly important in the transpiration stream in plants and in binding sheet and fibrils of complex organic compounds together. It also has a 'high dielectric constant', which reduces the electric field between pairs of oppositely charged ions, and makes it a good solvent for the dissociation or ionization of electrolytes.

It has a maximum density at 4 °C, which means that ice floats, an important factor in sustaining aquatic life in cold climates. Because the ice floats, the deeper-lying water takes longer to freeze as the surface ice insulates it from the lower air temperature.

Carbohydrates These are composed of carbon, hydrogen, and oxygen joined together by covalent bonds to form a complex covalent molecule, as, for example, in the simplest sugars or **monosaccharides**, e.g. $C_6H_{12}O_6$.

Two such monosaccharides or simple sugars combine to form a **disaccharide**, such as sucrose (table sugar), with the elimination of a water molecule, in what is known as a **condensation reaction**.

$$C_6H_{12}O_6 + C_6H_{12}O_6 \longrightarrow \underset{\text{disaccharide}}{C_{12}H_{22}O_{11}} + H_2O$$

If more monosaccharides join together to form **polysaccharides**, this process of the elimination of water continues with each bond made. Such long chain repetitions of a basic unit molecule or monomer are known as **polymers**, and the process of combination is called polymerization.

The complex carbohydrates can be broken down to monosaccharides by hydrolysis, that is by the 're-addition' of water. In the laboratory this requires acid conditions and high temperatures, but in the living cell it is catalysed by enzymes under neutral conditions at relatively low temperatures.

Monosaccharides These are the simple sugars containing either 3, 4, 5, or 6 carbon atoms, known as trioses, tetroses, pentoses, and hexoses, respectively. Only the hexoses occur in appreciable amounts in the diet of most animals.

Hexoses These are crystalline, sweet, and readily soluble in water, for example glucose and fructose. They give a green to red precipitate when boiled with Fehling's solutions A and B, or Benedict's solution. This reaction occurs because they are reducing agents and for this reason they may be referred to as **reducing sugars**.

Both glucose and fructose have the formula $C_6H_{12}O_6$, but the shape of their molecules differ, glucose forming a 6-membered ring, and fructose a 5-membered ring.

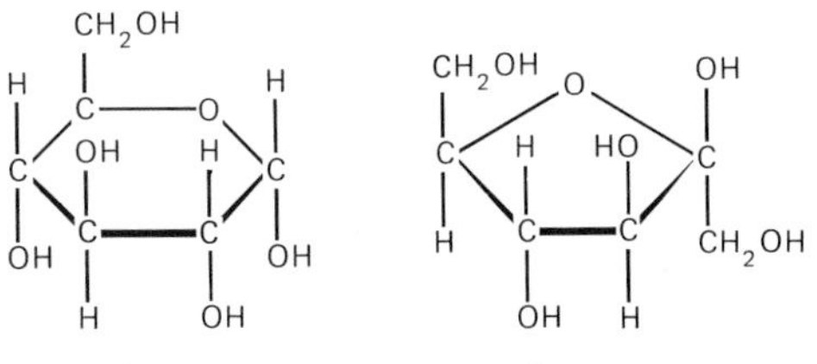

glucose

fructose

The different shape of the molecules results in solutions of glucose and of fructose affecting the passage of a special type of light, known as polarized light, through them. Glucose rotates the plane of the light to the right and can therefore also be known as dextrose (meaning to the right); fructose rotates it to the left and can therefore also be known as laevulose (meaning to the left). Another difference is that they have different 'reducing groups' in their molecules, thus glucose has an aldehyde group and is therefore known as an aldose, and fructose has a ketone group and is therefore known as a ketose. Fructose, which occurs naturally in honey and many fruits, is sweeter than glucose.

Thus we see that it is not only the chemical composition of the molecules of a substance that is important, but also their 3-dimensional structure or **stereochemistry**. Two substances with the same chemical composition but different shaped molecules (stereoisomers) can have very different properties which can be of great biological importance. Other examples of hexoses include galactose and mannose.

The **pentoses**, ribose and deoxyribose, are components of the genetic material, and another pentose, arabinose, is found in a wide range of fruits and root vegetables.

Disaccharides These consist of two monosaccharides linked together, with the elimination of water (a condensation reaction). They are crystalline, sweet, and readily soluble in water, but they do not give a positive test with Fehling's A and B or Benedict's solutions (except maltose). This is because in the non-reducing disaccharides, the reducing group of the two joining monosaccharides are involved in the glycoside bond that joins them together. The different shapes of the monosaccharide hexose molecules, that is the stereoisomers, in turn influence the shape of the disaccharide produced, which again alters their chemical activities and biological significance. For example, lactose is unique to mammals, and sucrose commonly occurs in plants.

glucose + fructose ⟶ sucrose + water
$C_6H_{12}O_6 \quad C_6H_{12}O_6 \quad C_{12}H_{22}O_{11} \quad H_2O$

glucose + glucose ⟶ maltose + water
$C_6H_{12}O_6 \quad C_6H_{12}O_6 \quad C_{12}H_{22}O_{11} \quad H_2O$

glucose + galactose ⟶ lactose + water
(milk sugar unique to mammals)
$C_6H_{12}O_6 \quad C_6H_{12}O_6 \quad C_{12}H_{22}O_{11} \quad H_2O$

Polysaccharides These consist of variably large numbers of glucose units linked together by glycoside bonds. More than three but fewer than ten monosaccharides joined together are known as oligosaccharides, more than ten joined together are known as polysaccharides, although conventions differ. Any differences in chain length above a certain maximum do not affect the chemical properties of a polysaccharide, and substances such as starch have no fixed maximum number of monosaccharides in their molecules. Apart from their size they share few common chemical characteristics.

Starch This is a product of photosynthesis and is stored as starch grains in many parts of plants, especially in food storage organs. It is amorphous, that is it lacks any definite crystal structure, and is relatively insoluble. The starch grain consists of two polysaccharides derived from glucose: the straight-chained amylose and the branched amylopectin. Amylose constitutes about 20 per cent of the total starch and is the part that gives the blue-black reaction with iodine. Its hydrolysis is catalysed by the enzyme amylase.

Amylopectin constitutes the major part of starch and is not affected by amylase.

Inulin This is a polymer of fructose found in some plants. Humans do not possess an enzyme capable of digesting this, and therefore it has no nutritional value.

Agar This is a polymer of galactose found in seaweeds, and is widely used as the basis for a culture medium for micro-organisms.

Cellulose Like amylose, this is also composed of long unbranched chains of glucose molecules. However, amylose is made up of chains of one form of glucose known as α-glucose, whereas cellulose is made up of chains of another form or stereoisomer known as β-glucose.

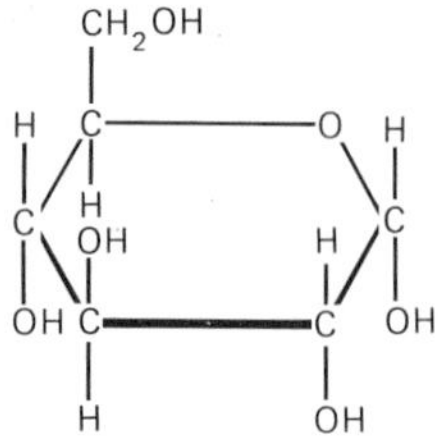

α-glucose

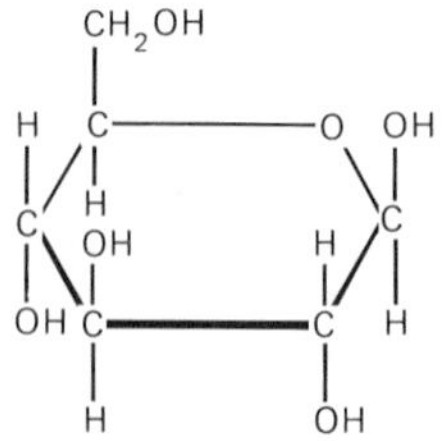

β-glucose

Cellulose fibres are the basis of all plant cell walls. Cellulose fibres have a strength somewhat greater than a thread of high grade steel of the same diameter. Vertebrates lack a cellulose enzyme and must harbour invertebrate symbionts that secrete cellulases in the gut.

Hemicelluloses These are a mixed group of polysaccharides closely associated with cellulose in plant tissues.

Pectin This is an amorphous polymer of galacturonic acid important in the middle lamella of plant cell walls.

Glycogen This is found in animals as a polysaccharide food store; some refer to it as 'animal starch'. However, unlike starch, it is soluble in water and readily broken down by enzyme action. It is made up of branched chains of glucose sub-units.

Chitin This is a complex polysaccharide that is used by some animals as a structural material, particularly in the exoskeleton of the Arthropoda. Other complex animal polysaccharides include mucus and other lubricating fluids.

Proteins These are compounds of carbon, hydrogen, oxygen, and nitrogen; most proteins also contain sulphur. They consist of long chains of amino acids joined together by characteristic peptide bonds to form polypeptide chains. A polypeptide chain of more than about fifty amino acids constitutes a protein. Amino acids have the general formula of:

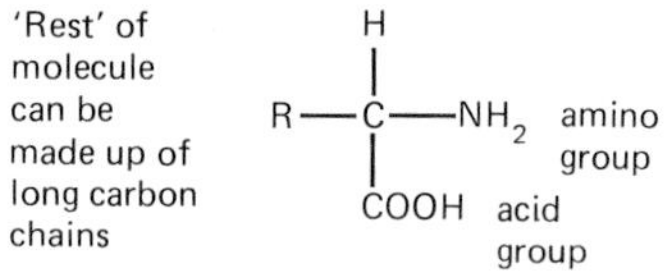

'Rest' of molecule can be made up of long carbon chains

amino group

acid group

All naturally-occurring amino acids are α-amino acids, i.e. the NH_2 group is always attached to the C next to the COOH group. If this were not so, they would not form chains They combine with the elimination of water.

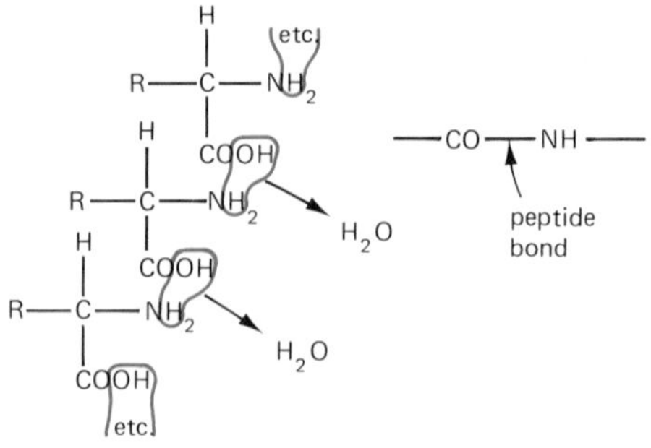

peptide bond

There are about twenty different amino acids which combine together in varying proportions to produce all the proteins of the living world, under the guidance of the genetic message encoded in the nucleic acids of all living things. All proteins contain all the 20, but in a different order and proportion. Protein structure can be considered at four different levels: primary, secondary, tertiary, and quaternary.

The primary structures of a protein refers to the sequence of amino acids linked together by peptide bonds to form a polypeptide chain.

The secondary structure refers to the folding or coiling of the polypeptide chain which occurs as a result of various attractions between different parts of the molecule.

The tertiary structure refers to the further folding and coiling of the chain, to give an overall 3-dimensional configuration of the protein molecule. (This could be compared to the tying of a knot in a coiled spring.)

The quaternary structure refers to the spatial relationship between two or more such protein molecules which sometimes occurs to give a multimolecular complex.

Broadly speaking there are two main groupings of biological proteins, insoluble **fibrous** proteins which are cytoskeletal and contractile in function, e.g. keratin in hair; collagen in bone and tendons; elastin in ligaments; and actin and myosin in muscles; and soluble **globular** proteins which have a more intricate tertiary structure and which are biologically active, for example all enzymes, antibodies, and some hormones. Conjugated proteins consist of a globular protein associated with a non-protein group, as for example in haemoglobin.

Proteins undergo irreversible changes in their 3-dimensional structure under certain conditions of pH, temperature, and in the presence of heavy metals such as lead and cadmium. The globular proteins, which tend to have a more intricate tertiary structure than the fibrous proteins, are more liable to disruption of their structure. Such disruption changes the configuration of the reactive site or active centre of enzymes and renders them inactive or denatured. Extreme disruption of protein structure results in their coagulation.

Lipids These are a mixed group of substances, but they are all insoluble in water and soluble in alcohol, acetone, chloroform, and ether. There are three main groups of lipids: neutral fats, phospholipids, and steroids.

Neutral fats These are combinations of glycerol and fatty acids, and are also known as triglycerides, as three fatty acids are involved.

Glycerol < fatty acid, fatty acid, fatty acid

They have but one biochemical role, and that is as an energy reserve; they yield more energy per unit weight than carbohydrates and are used as the major energy reserves in animals. Plants generally utilize carbohydrates as their energy reserve materials.

Oils are fats with low melting points, so that they are liquid at room temperature. The melting points of fats is directly related to the number of double bonds there are in the fatty acids, the more double bonds the lower the melting point, and the greater their chemical activity. Those with no double bonds are known as **saturated fats** and are chemically inert and stable. Those with a double bond are known as **unsaturated fats**, and those with more than two double bonds are known as **polyunsaturated**.

Again, these chemical structural differences have a profound effect on the biological role of these compounds. Thus saturated fats accumulate in the fat stores of the body, while unsaturated fats are very reactive in metabolism.

Waxes are more complicated than the fats and oils, and serve in water conservation as in the cuticle of insects and the cuticle of leaves of plants.

Phospholipids These are neutral fats in which one fatty acid is replaced by a phosphorus-containing compound. They are an important constituent of cell membranes.

Steroids These are very comples, and include such biologically important substances as sex hormones, adrenal cortex hormones, vitamin C, and cholesterol.

Exothermic and endothermic reactions All chemical reactions are accompanied by energy changes. In exothermic reactions energy is released to the surroundings, but endothermic reactions require a supply of energy in order to proceed. For example, cellular respiration is exothermic:

$$C_6H_{12}O_6 + 6O_2 \xrightarrow{\text{enzymes}} 6CO_2 + 6H_2O + \textbf{energy}$$

Photosynthesis is endothermic:

$$6CO_2 + 6H_2O + \textbf{energy} \xrightarrow{\text{enzymes}} C_6H_{12}O_6 + 6O_2$$

Adenosine tri-phosphate (ATP)
ATP occurs in all living cells, where it is of central importance to the complex of reactions known as metabolism. Conventionally, it is considered to be an energy carrier, capable of 'trapping' energy in energy-rich phosphate bonds and subsequently releasing it, as required, to energy-dependent reactions by the hydrolysis or breaking of these bonds.

$$\text{Adenosine} + \underset{(H_3PO_4)}{\text{phosphate}} \xrightarrow{\text{free energy}} \text{adenosine Monophosphate or AMP or } A \sim P \ (12\text{ kJ})$$

$$\text{AMP} + \text{P} \xrightarrow{\text{free energy}} \text{adenosine diphosphate or ADP or } A - P \sim P \ (12\text{ kJ}, 34\text{ kJ})$$

$$\text{ADP} + \text{P} \xrightarrow{\text{free energy}} \text{adenosine triphosphate or ATP or } A - P \sim P \sim P \ (12\text{ kJ}, 34\text{ kJ}, 34\text{ kJ})$$

Only the final phosphate bond is considered as being hydrolysed in living systems, therefore the available energy in ATP is taken as being 34 kJ mol^{-1}.

$$\text{ATP} + H_2O \xrightarrow[\text{ATPase enzyme}]{} \text{ADP} + \text{P} + 34\text{ kJ } \textbf{energy}$$

It is considered that ATP can be used as a 'pure' energy carrier, whereby the energy is utilized in a synthetic reaction directly, without the phosphorylation of the substrate; for example

$$\text{fatty acid} + \text{coenzyme A} \xrightarrow{\text{ATP} \rightarrow \text{AMP} + \text{P} + \text{P} + \textbf{energy}} \text{fatty acid CoA}$$

Also, it is considered that ATP can be used as a phosphate carrier to produce a phosphorylated and therefore 'energy rich' substance, e.g.

$$\text{glucose} \xrightarrow{\text{ATP} \rightarrow \text{ADP},\ \text{P}} \text{glucose} \sim \text{P}$$

The product is a phosphorylated molecule, it may or may not have a high energy phosphate bond, depending on its structure, but it has a total energy content exceeding that of the non-phosphorylated molecule by 3–12 kcal mol^{-1}.

This concept of ATP and its function is still widely used in biology, even though it is now considered as being very misleading. Indeed, formation of chemical bonds is always accompanied by a release of energy and the breaking of bonds always requires energy. Moreover, high energy bonds are stable and unreactive whereas the so-called 'high energy bonds' of ATP are considered as being very reactive and unstable. In addition there is no evidence that ATP ever undergoes simple hydrolytic breakdown when participating in metabolic processes.

It is clearer not to consider the phosphate bond of ATP as being a high energy bond, but to consider the energy as an attribute of the molecule as a whole and of its reaction with other substances. A substance in which bonds are broken is only a source of energy when it is involved in chemical reactions by which stronger bonds are formed. If this does not happen then no energy can be released.

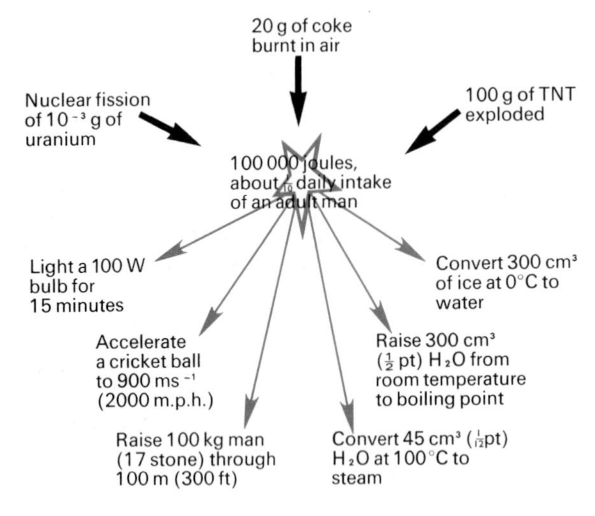

Energy equivalents
(From Raffan *et al.*, *Chemistry*, Hodder and Stoughton.)

Heat Heat is a form of energy measured in joules or calories and, although a calorie is the older, non-SI unit, it is still widely used in biology, particularly in relation to the energy values of foods. A calorie is equivalent to 4.18 J, and is that amount of heat required to raise the temperature of 1 g of water by 1°C. A kilocalorie is a 1000 such calories, and a kilojoule is a 1000 joules.

Temperature This is a measure of the relative hotness or coldness of a substance measured in degrees centigrade. Heat flows from regions of high temperature to regions of low temperature by means of conduction, convection, and radiation.

Specific heat capacity This is the quantity of heat required to raise the temperature of a unit mass of a substance by one degree. It is expressed in calories per gram per °C, or joules per kilogram per kelvin. Put more simply, it is a measure of the capacity of a substance to absorb heat. Water has a high specific heat ($4.2\ J\ g^{-1}\ °C^{-1}$), which means that it does not undergo rapid fluctuations in temperature, and therefore provides a medium of fairly constant temperature for living cells.

Effect of temperature on the rate of chemical reactions The rate of chemical reactions increases with increased temperature. The Q_{10} is a measure of this increase with every 10°C rise, e.g.

$$Q_{10} = \frac{\text{rate at } (T + 10)}{\text{rate at } T}$$

In those chemical reactions found in living systems the rate generally doubles for every 10°C rise in temperature, therefore in these cases $Q_{10} = 2$. Also, under conditions found in living systems, temperature has little or no effect on photochemical reactions, so in these cases $Q_{10} = 1$; an example in nature being the light reaction of photosynthesis.

Chemical change In any chemical change the original substance is changed into different substances as a result of a chemical reaction, e.g. Hydrolysis, oxidation, and reduction.

Physical change A physical change produces no new substances, only alters the original state of the elements or compounds, e.g. evaporation, freezing, and melting.

States of matter There are three physical states of matter: solid, liquid, and gas. A solid has a definite shape and volume; a liquid has a definite volume but no definite shape; and a gas has neither definite shape nor definite volume. Changes in pressure (force per unit area) or temperature are needed for a substance to change its state.

The actual change in state of water requires a relatively large amount of energy. For example, 80 cal are required to melt 1 g of ice without any change in temperature; conversely, 80 cal have to be removed to freeze 1 g of water at 0°C. This quantity of heat, required to change the state with no change in temperature, is known as the **latent heat of fusion of ice**. Similarly, 540 cal are required to change 1 g of water to steam with no change in temperature, and this is known as the **latent heat of evaporation of water**. This is very important in the regulation of the body temperature by means of the evaporation of sweat from the skin. Conversely, 540 cal need to be removed for 1 g of steam to be condensed to water.

Diffusion This is the net movement of a substance from a region of high concentration to a region of lower concentration down a diffusion gradient until the two concentrations become equal. The term 'net movement' is used rather than just 'movement', as some of the substance does move from a region of low concentration to a region of high concentration, but more moves from a region of high concentration to a region of low concentration. Thus there is an overall or 'net' movement from high to low.

Diffusion occurs as a result of the inherent motion or kinetic energy of atoms or molecules. The smaller the particles and the higher the temperature, the faster is the rate of diffusion.

Diffusion is of great importance in the transport of substances, both within living organisms and between living organisms and the environment. In living organisms, diffusion is mainly a surface phenomenon, occurring between different regions separated by living membranes. In these cases the larger the surface area and the thinner the membrane, the more rapidly will larger amounts of substance diffuse across it, if there is a difference in concentration between the two sides.

Evaporation When a liquid is in contact with a gas, some molecules of the liquid escape into the gaseous phase, as a result of their kinetic energy, to become vapour. When the gas can accommodate no more vapour it is said to be saturated, as there is a dynamic equilibrium between the number of molecules evaporating and the number of molecules of vapour condensing back into liquid, so that there is no net evaporation of liquid.

Evaporation increases with increase in temperature, and liquids with a lower latent heat of evaporation evaporate more readily than those with a higher latent heat of evaporation.

The loss of water by evaporation is one of the major hazards for plants and animals in the terrestrial environment. However, the evaporation of water is of great importance in the lowering of the temperature of living organisms, and in the transpiration stream of higher plants.

Pressure and volume The volume of a gas is affected by temperature and pressure. If the temperature increases the gas expands (i.e. its volume increases), and vice versa. If the temperature is kept constant, then, if the pressure goes up, the volume goes down proportionately; for example if the pressure is doubled the volume is halved, and vice versa. This relationship is expressed by **Boyle's Law**, which states that for a given mass of gas: **pressure × volume = a constant**. For example

$$10 \times 20 = 200$$
$$20 \times 10 = 200$$
$$1 \times 200 = 200$$

During breathing in mammals, for example, when the thoracic cavity is increased in volume the thoracic pressure drops below the external atmospheric pressure and air is forced into the lungs. When the thoracic cavity is decreased in volume the thoracic pressure increases above the external atmospheric pressure and air is forced out of the lungs.

Solutions A solution is an even or homogenous mixture of the molecules of two or more different substances. The term is usually applied to either solids or gases dissolved in liquids. The dissolved substance is termed the **solute**, and the dissolving liquid is termed the **solvent**. The solubility of solids increases with increase in temperature, but the solubility of gases decreases with increase in temperature. The decreased solubility of oxygen in water with increasing temperature has important ecological effects, as warmer waters hold less oxygen. It is because of this that thermal pollution has its major impact on ecosystems.

Mole (gram-molecule) This is the quantity of a compound, the mass in grams, which is equal to its molecular weight. For example, the molecular weight of water is 18, therefore one mole of water weighs 18 g. 1 mole of every compound contains 6.023×10^{23} particles (the **Avogadro number**).

Molar solution 1 mole per litre litre of solution.

Molal solution 1 mole plus 1000 g H_2O.

Avogadro's number This is the number of atoms or molecules in a **mole** of a substance, and equals 6.023×10^{23}. For example, 1 mole water = 18 g, therefore 18 g water contains 6.02×10^{23} water molecules.

To give some idea of the size of this number, various calculations have been made. For example, it has been estimated that it would take 6.023×10^{23} water 'drops' 126000 years to pass over Niagara Falls at a flow rate of 120 million gallons/minute. Another striking analogy is that 6.023×10^{23} cubic centimetres of water is equal to about twice the volume of the Atlantic Ocean.

Colligative properties These are those properties of a solution which depend on, and which are directly proportional to, the number of solute 'particles' introduced per unit volume of solvent, i.e. the concentration. Such properties depend only on the number of particles and are independent of the special nature of these particles, e.g. size, shape, chemical composition, and electrical charge. They include the lowering of the vapour pressure, the raising of the boiling point, and the lowering of the freezing point, of the solvent.

Examples of colligative properties which are of biological interest include the lowering of the freezing point of a solvent, and osmosis. The lowering of the freezing point of a solvent such as water by the presence of solute particles prevents the body fluids of living organisms from freezing under certain circumstances. Osmosis is the way in which water is moved across living membranes.

The measured value of the colligative properties of a solution of any electrolyte will be greater than the value predicted on the basis of the molar concentration of the solute, as dissociation of a mole of an electrolyte gives more particles than a mole of non-electrolyte, e.g.

$$\underset{\text{1 particle}}{NaCl} \longrightarrow \underset{\text{2 particles}}{Na^+ + Cl^-}$$

$$\underset{\text{1 particle}}{C_6H_{12}O_6} \text{ does not dissociate}$$

Osmosis This is the passage of water from a more dilute or **hypotonic** solution to a more concentrated or **hypertonic** solution through a semi-permeable membrane. If uninterrupted this will continue until the concentration on both sides of the membrane are the same, or **isotonic**. A semi-permeable membrane is one that is more permeable to solvent than solute particles.

At a given temperature the rate of osmosis is directly proportional to the difference in concentration of the solutions, and at a given concentration difference it is directly proportional to the temperature.

There is still no perfectly satisfactory explanation of osmosis, and the nature of the semi-permeability of the membranes is not understood. One explanation of osmosis is that water molecules have a 'chemical potential' and that they will move from regions of high chemical potential to one of lower chemical potential. In other words, water flows down its chemical concentration gradient, in the same way that other substances flow down their concentration gradients in the process of diffusion. The presence of solutes lowers the chemical potential of water, or 'water potential' so water flows in. Water potential is numerically equal to the older terms of suction pressure (SP) and diffusion pressure deficit (DPD), used in discussions of plant–water relations, but is of opposite sign. All water in living organisms moves by osmosis; there is no evidence for the active transport of water by living cells.

Colloidal state A colloid consists of a suspension of particles or disperse phase in a liquid known as its continuous phase. The suspended or dispersed particles are larger than those of substances in true solu- usually being in the range 10^{-4}–10^{-6} mm in diameter. The cytoplasm of cells acts as a colloidal system in which there is a tremendously large internal surface area between the disperse phase and the continuous phase. This is very important in living cells where there are very many surface-limited reactions that require this surface area.

Control experiment In order to study the behaviour of living systems it is nearly always essential to have a control experiment, the use of which allows more reliable conclusions to be reached. For example, in attempting to establish the enzymatic nature of the breakdown of amylose by saliva, the following tubes would need to be set up:

(a) starch and saliva,
(b) starch and boiled saliva (control),
(c) starch and water (control);

and the following tests would have to be made both at the start and finish of the experiment:

(i) iodine test for the presence of starch,
(ii) Benedict's test for the presence of reducing sugars.

A negative test for starch and a positive test for reducing sugars in tube (a) alone at the end of the experiment would not *prove* that saliva contained the enzyme that catalysed the hydrolysis of starch to reducing sugars. Alternative explanations could be that the starch was hydrolysed by an inorganic catalyst in the saliva, or by a spontaneous reaction with water. The control tube (b) allows the catalyst to be identified as an enzyme, since enzymes are typically denatured by boiling; the control tube (c) demonstrates that starch will not react spontaneously with water.

The tests for reducing sugars at the beginning of the experiment demonstrated that there were no reducing sugars present in the starch or saliva, and that therefore those that appeared in tube (a) did so as the result of some reaction occurring during the experiment.

Surface area to volume ratio This is of the utmost importance in biological systems. The smaller an organism, organ, cell, or organelle is, the larger is its surface area to volume ratio, and vice versa. With increasing size the surface area of a certain volume increases by the square of its linear dimensions, whereas the volume increases by the cube of its linear dimensions. For example, consider two cubes, one with a of length 1 cm and the other of 2 cm.

Length of side (L) cm	1	2
Surface area ($6 \times L^2$) cm^2	6	24
Volume (L^3) cm^3	1	8
$\dfrac{\text{Surface area}}{\text{Volume}}$	6	3

For a given volume, a spherical shape provides the least possible surface area, therefore organisms, organs, cells, and organelles tend to 'avoid' this shape and are often elongated, flattened, or have many extensions or invaginations. These shapes increase the surface area of a given volume and increase the surface area to volume ratio. This is of importance to living systems, where vital exchanges occur over surfaces; materials have to be transported to all parts of the volume from these surfaces; and where sequential reactions are carried out at surfaces by enzymes physically attached to membranes.

Mass A measure of the quantity of matter.

Weight A force equal to mass × acceleration. Normally the acceleration is due to gravity, and gravity can vary; for example it is less at the poles than at the equator. Therefore a given mass of a substance will weigh less at the poles than at the equator; and in the absence of gravity, although its mass would stay the same, it would weigh nothing.

Sound Sound is a subjective auditory experience of a pressure wave motion through a material such as air. This means that the waves themselves are not sound but they are detected as such by the ear. Sound waves, as opposed to light waves, are a mechanical phenomenon and require a material substance for their propagation. Thus they travel through gases (air), liquids, and solids, but not across a vacuum.

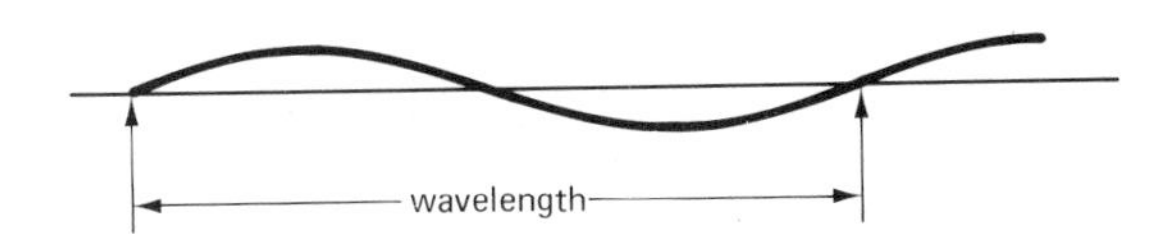

Wavelength and frequency

$$\text{Frequency} = \frac{\text{velocity}}{\text{wavelength}} \text{ (in cycles per second or hertz)}$$

(Frequency is defined as the number of waves that pass a given point in one second.) At any point along the path of a wave motion a periodic displacement or vibration about a mean position takes place. For example, a cork floating on water disturbed by waves bobs up and down, but does not travel in the direction of the wave.

Electromagnetic radiation These are radiations which, unlike sound waves for example, do not need the movement of any particles to enable them to travel, and which can therefore travel across a vacuum. All the different wavelengths of electromagnetic radiations, including those of the visible spectrum (light), travel at the same speed, i.e. 2.9979×10^8 m s^{-1}.

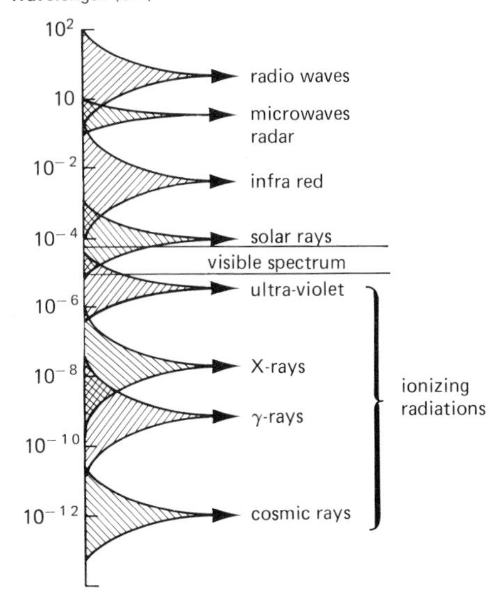

Ultra-violet light This is electromagnetic radiation with a wavelength range between visible light and X-rays, that is in the range of about 4×10^{-7} m to 5×10^{-9} m (400 nm – 5 nm). The longest u.v. waves are just shorter than the shortest light waves visible to most human eyes. Lens-less eyes can detect u.v. light, as can the compound eyes of insects such as bees.

U.v. acts on a substance (7-dehydro-cholesterol or ergosterol) in the skin to produce vitamin D. Very few foods contain vitamin D, and under natural conditions man obtains almost all his supply from the skin. In fact a dietary supply is not essential. U.v. light is an ionizing radiation, causing ionizations in substances through which it passes. When it passes through living cells ionization can occur in the genetic material DNA and in this way mutations may be induced.

Certain wavelengths of ultra-violet light (297 nm) absorbed by the skin have a physiological action which is similar to that produced by physical training. The resting respiratory ventilation rate slows, the blood pressure is lowered, the red blood corpuscles increase in number, the measured muscular strength increases more under training in ultra-violet light than without it, and the adrenal glands are stimulated.

Microscopy The power to define detail, that is the ability to produce separate images of structures very close to each other, is referred to as the **resolving power**. In order to make microscopic detail visible it must be **magnified** sufficiently for the eye of the observer to resolve it, and up to a point the greater the magnification the greater the resolving power, although past a certain optimum point further magnification is of no value. The wavelength of the 'illuminating' radiations is also involved in determining the degree of resolution possible. Shorter wavelengths allow greater resolution, thus u.v. is better than light in the visible spectrum, and electrons (used in the electron microscope) are even better.

The microscope magnifies length but leaves time unaffected, it therefore has the effect of apparently increasing the velocity of moving organisms, as velocity is the rate of motion in a given direction measured as length or distance travelled in unit time. Thus microscopic organisms such as the ciliated protozoa *Paramecium* appear to move very rapidly across the field of view, whereas the actual speed of motion is about twelve times its own length per second; about 10 metres in 1 hour.

Further reading

Baldwen E., *The Nature of Biochemistry* (Cambridge: Cambridge University Press, 1962).

Brown E. G., *An Introduction to Biochemistry*, Royal Institute of Chemistry Monographs for Teachers No. 17 (London, 1971).

Edelman J. and Chapman J., *Basic Biochemistry* (London: Heinemann, 1978).

Jones, Netterville, Johnston, and Wood, *Chemistry, Man, and Society* (London: Saunders, 1976).

Rose. S., *The Chemistry of Life* (Harmondsworth: Penguin Books, 1966).

Index

Figures in italics refer to diagrams.